T0269239

Lecture Notes in Physics

Volume 921

The Lecture Notes in Physics

The series Lecture Notes in Physics (LNP), founded in 1969, reports new developments in physics research and teaching-quickly and informally, but with a high quality and the explicit aim to summarize and communicate current knowledge in an accessible way. Books published in this series are conceived as bridging material between advanced graduate textbooks and the forefront of research and to serve three purposes:

- to be a compact and modern up-to-date source of reference on a well-defined topic
- to serve as an accessible introduction to the field to postgraduate students and nonspecialist researchers from related areas
- to be a source of advanced teaching material for specialized seminars, courses and schools

Both monographs and multi-author volumes will be considered for publication. Edited volumes should, however, consist of a very limited number of contributions only. Proceedings will not be considered for LNP.

Volumes published in LNP are disseminated both in print and in electronic formats, the electronic archive being available at springerlink.com. The series content is indexed, abstracted and referenced by many abstracting and information services, bibliographic networks, subscription agencies, library networks, and consortia.

Proposals should be sent to a member of the Editorial Board, or directly to the managing editor at Springer:

Christian Caron
Springer Heidelberg
Physics Editorial Department I
Tiergartenstrasse 17
69121 Heidelberg/Germany
christian.caron@springer.com

More information about this series at http://www.springer.com/series/5304

Stefano Lepri
Editor

Thermal Transport in Low Dimensions

From Statistical Physics to Nanoscale Heat Transfer

Editor
Stefano Lepri
Istituto Sistemi Complessi
Consiglio Nazionale delle Ricerche
Sesto Fiorentino (FI), Italy

ISSN 0075-8450 ISSN 1616-6361 (electronic)
Lecture Notes in Physics
ISBN 978-3-319-29259-5 ISBN 978-3-319-29261-8 (eBook)
DOI 10.1007/978-3-319-29261-8

Library of Congress Control Number: 2016933766

Printed on acid-free paper

This Springer imprint is published by Springer Nature
The registered company is Springer International Publishing AG Switzerland

Preface

The behavior of classical and quantum many-particle systems out of equilibrium is among the most prominent problems in modern statistical-mechanics. Besides its own interest for the theoretical foundations of irreversible thermodynamics, this topic is also relevant to develop innovative ideas for nanoscale thermal management with possible future applications to nanotechnologies and effective energetic resources.

On the other hand, the physics of many-body systems constrained in reduced spatial dimensions (1 and 2D) display many unusual properties. According to familiar statistical arguments, this should be traced back to the predominant role of fluctuations that, for instance, inhibits long-range order at equilibrium. In the context of nonequilibrium processes, the presence of anomalies is signaled by the appearance of long-ranged dynamical correlations often leading to the breakdown of standard hydrodynamics.

Within this general context, one of the issues that attracted a certain interest in the last decade is the problem of anomalous heat conduction in low-dimensional lattice models. Historically, this originates from the attempt to construct a minimal, non-perturbative theory of nonequilibrium stationary states and the quest for a rigorous microscopic foundation of phenomenological relations (the Fourier's law in this case). Moreover, many of the peculiarities of low-dimensional models turned out to be of interest by themselves, as examples of highly correlated, and thus complex, behavior. This leads to various forms of violation of the ordinary laws of diffusive heat conduction.

Theoretical studies of low-dimensional transport were recently paralleled by those on phonon thermal transfer in nano- and microscale systems (nanotubes, nanowires, and even graphene). Thus, the fundamental aspects of this research are currently finding a great interest in the domain of thermal conduction on small scales.

The aim of this volume is to provide an updated, basic reference for this research field. Although the content is advanced, the presentation is as self-contained as possible and addresses to a broad audience interested in approaching the topic from both the statistical-mechanics/mathematical physics and condensed-matter/engineering sides. In this spirit, the contents range from more fundamental issues up to more realistic applications.

The first part (Chaps. 1–6) discusses and summarizes the basic models, the phenomenology, and the various theoretical approaches. The reference systems are arrays of oscillators coupled on a lattice in interaction with two or more external reservoirs. The methods will employ equilibrium and nonequilibrium molecular dynamic simulations, hydrodynamic and kinetic approaches, and the solution of specific stochastic models.

In the second part (Chaps. 7–9), emphasis will be on application to nano- and microscale heat transfer, as, for instance, phononic heat conduction in carbon-based nanomaterials, including the prominent cases of nanotubes and graphene. A review of realistic simulations and experiments will be presented throughout. Finally (Chap. 10), some possible future developments on heat flow control and new ideas on thermoelectric energy conversion that originates from the studies of heat transport in low-dimensional materials will be reviewed.

Sesto Fiorentino, Italy Stefano Lepri
November 2015

Contents

Chapter 1
Heat Transport in Low Dimensions: Introduction and Phenomenology

Stefano Lepri, Roberto Livi, and Antonio Politi

Abstract In this chapter we introduce some of the basic models and concepts that will be discussed throughout the volume. In particular we describe systems of nonlinear oscillators arranged on low-dimensional lattices and summarize the phenomenology of their transport properties.

1.1 Introduction

In this first chapter we review the main properties of low-dimensional lattices of coupled classical oscillators. We will describe how reduced dimensionality and conservation laws conspire in giving rise to unusual relaxation and transport properties. The aim is to provide both a general introduction to the general phenomenology and to guide the reader in the volume reading (where appropriate we indeed point to the more detailed analyses developed in the subsequent chapters).

For the sake of concreteness, one may think of quasi-1D objects, like long molecular chains or nanowires, suspended between two contacts which play the role of thermal reservoirs. These experimental setups, repeatedly discussed throughout the volume, are schematically depicted in Fig. 1.1.

We start Sect. 1.2 by introducing the main models without technicalities and providing the relevant definitions. Section 1.3 contains a summary of the different properties that is worth testing to characterize heat transport in a physical system.

S. Lepri (✉)
Consiglio Nazionale delle Ricerche, Istituto dei Sistemi Complessi, via Madonna del Piano 10, I-50019 Sesto Fiorentino, Italy
e-mail: stefano.lepri@isc.cnr.it

R. Livi
Dipartimento di Fisica e Astronomia and CSDC, Università di Firenze, via G. Sansone 1, I-50019 Sesto Fiorentino, Italy
e-mail: roberto.livi@unifi.it

A. Politi
Institute for Complex Systems and Mathematical Biology & SUPA University of Aberdeen, Aberdeen AB24 3UE, UK
e-mail: a.politi@abdn.ac.uk

S. Lepri (ed.), *Thermal Transport in Low Dimensions*, Lecture Notes in Physics 921, DOI 10.1007/978-3-319-29261-8_1

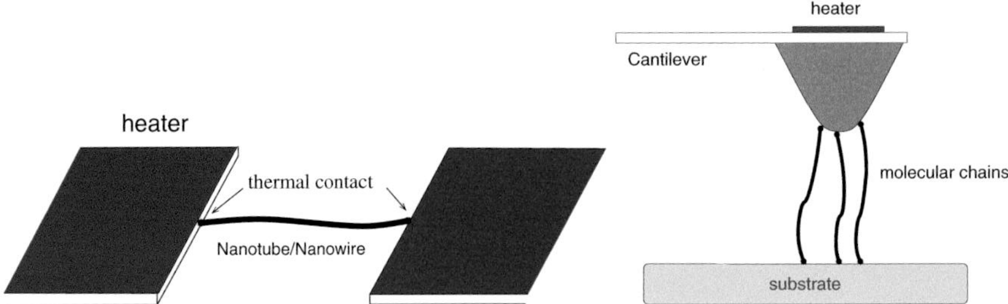

Fig. 1.1 Sketch of two experimental setups illustrating the physical setting. *Left*: a nanotube or nanowire suspended between two contacts acting as heaters and probes, see [14] and Chap. 8. *Right*: a scanning thermal microscopy setup whereby an assembly of molecular chains with one end attached on a substrate is heated through a cantilever tip [74]

The natural starting point is the effective conductivity in finite systems, which diverges with the system-size in the case of anomalous transport. The existence of long-time tails in the equilibrium correlation functions is another way of probing the system dynamics, together with the diffusion of localized perturbations and the relaxation of spontaneous fluctuations. Another, not much explored property, is the shape of the temperature profile that is strictly nonlinear even in the limit of small temperature differences, when heat transport is anomalous.

In Sect. 1.4, we present the overall scenario, making reference to the universality classes unveiled by the various theoretical approaches. More specifically, we emphasize the relationship with the evolution of rough interfaces and thereby the Kardar-Parisi-Zhang equation. Coupled rotors represent an important subclass of 1D systems where heat conduction is normal in spite of momentum conservation: their behavior is reviewed in Sect. 1.5.

The expected scenario in two-dimensions (namely the logarithmic divergence of heat conductivity) is discussed in Sect. 1.6, while the peculiar behavior of integrable systems is briefly reviewed in Sect. 1.7. In Sect. 1.8, we discuss the more general physical setup, where another quantity is being transported besides energy. This is the problem of coupled transport, where the interaction between the two processes may give rise to unexpected phenomena even when the transport is altogether normal. In particular, we consider a chain of coupled rotors in the presence of an additional torque, where the second quantity is angular momentum and the discrete nonlinear Schrödinger equation, where the second quantity is the norm (or mass). Finally, the still open problems are recalled in Sect. 1.9.

1.2 Models

The simplest microscopic dynamical model for the characterization of heat conduction consists of a chain of N classical point-like particles with mass m_n and position q_n, described by the Hamiltonian

$$H = \sum_{n=1}^{N} \left[\frac{p_n^2}{2m_n} + U(q_n) + V(q_{n+1} - q_n) \right] \quad . \tag{1.1}$$

The potential $V(x)$ accounts for the nearest-neighbour interactions between consecutive particles, while the on-site potential $U(q_n)$ takes into account the possible interaction with an external environment (either a substrate, or some three-dimensional matrix). The corresponding evolution equations are

$$m_n \ddot{q}_n = -U'(q_n) - F(r_n) + F(r_{n-1}) \quad , \quad n = 1, \ldots, N \,, \tag{1.2}$$

where $r_n = q_{n+1} - q_n$, $F(x) = -V'(x)$, and the prime denotes a derivative with respect to the argument. Usually q_n denotes the longitudinal position along the chain, so that

$$L = \sum_{n=1}^{N} r_n \,, \tag{1.3}$$

represents the total length of the chain (which, in the case of fixed b.c., is a constant of motion). Different kinds of boundary conditions may and will be indeed used in the various cases. For instance, if the particles are confined in a simulation "box" of length L with periodic boundary conditions,

$$q_{n+N} = q_n + L \quad . \tag{1.4}$$

Alternatively one can adopt a lattice interpretation, in which case, the (discrete) position is $z_n = an$ (where a is lattice spacing), while q_n is a transversal displacement. Thus, the chain length is obviously equal to Na.

The Hamiltonian (1.1) is generally a constant of motion. In the absence of an on-site potential ($U = 0$), the total momentum is conserved, as well,

$$P = \sum_{n=1}^{N} p_n \equiv \sum_{n=1}^{N} m_n \dot{q}_n \,. \tag{1.5}$$

Since we are interested in heat transport, one can set $P = 0$ (i.e., we assume to work in the center-of-mass reference frame) without loss of generality. As a result, the relevant state variables of microcanonical equilibrium are the specific energy (i.e., the energy per particle) $e = H/N$ and the elongation $\ell = L/N$ (i.e., the inverse of the particle density). On a microscopic level, one can introduce three local densities, namely r_n, p_n and

$$e_n = \frac{p_n^2}{2m_n} + \frac{1}{2}\Big[V(r_n) + V(r_{n-1})\Big] \quad , \tag{1.6}$$

which, in turn, define a set of currents through three (discrete) continuity equations. For instance, the energy current is defined as

$$\dot{e}_n = j_{n-1} - j_n \tag{1.7}$$

$$j_n = \frac{1}{2}a(\dot{q}_{n+1} + \dot{q}_n)\, F(r_n) \quad . \tag{1.8}$$

The definition (1.8) is related to the general expression, originally derived by Irving and Kirkwood that is valid for every state of matter (see e.g. [53]) that, in one dimension, reads

$$j_n = \frac{1}{2}(q_{n+1} - q_n)(\dot{q}_{n+1} + \dot{q}_n)\, F(r_n) + \dot{q}_n e_n \, . \tag{1.9}$$

In the case of lattice systems, where we assume the limit of small oscillations (compared to the lattice spacing) or in the lattice field interpretation, one can recover formula (1.8) setting $q_{n+1} - q_n = a$ in the first term and neglecting the second one [64]. The expression (1.9) is useful in the opposite limit of freely colliding particles, where the only relevant interaction is the repulsive part of the potential, that is responsible for elastic collisions. There, the only contribution to the flux arises from the kinetic term of e_n, i.e.

$$j_n \approx \frac{1}{2} m_n \dot{q}_n^3 \quad . \tag{1.10}$$

Having set the basic definitions, let us now introduce some specific models. A first relevant example is the harmonic chain, where the potential V is quadratic (and $U = 0$). From the point of view of transport properties, we expect this system to behave like a ballistic conductor. The heat flux decomposes into the sum of independent contributions associated to the various eigenmodes. This notwithstanding, this model proves useful, as it allows addressing general questions about the nature of stationary nonequilibrium states. This includes the role of disorder (either in the masses or the spring constants), of boundary conditions, and quantum effects. Since the linear case (classical and quantum) will be treated in detail in Chap. 2, here we focus on the anharmonic problem. In this context, the most paradigmatic example is the Fermi–Pasta–Ulam (FPU) model [50, 76, 79]

$$V(r_n) = \frac{k_2}{2}(r_n - a)^2 + \frac{k_3}{3}(r_n - a)^3 + \frac{k_4}{4}(r_n - a)^4 \quad . \tag{1.11}$$

Following the notation of the original work [32], the couplings k_3 and k_4 are denoted by α and β respectively: historically this model is sometimes referred to as the "FPU-$\alpha\beta$" model. Also, the quadratic plus quartic ($k_3 = 0$) potential is termed the "FPU-β" model. Notice that upon introducing the displacement $u_n = q_n - na$ from the equilibrium position, r_n can be rewritten as $u_{n+1} - u_n + a$, so that the lattice spacing a disappears from the equations.

Another interesting model is the Hard Point Gas (HPG), where the interaction potential is [9, 40, 41]

$$V(y) = \begin{cases} \infty & y = 0 \\ 0 & \text{otherwise} \end{cases} .$$

The dynamics consist of successive collisions between neighbouring particles,

$$v_n' = \frac{m_n - m_{n+1}}{m_n + m_{n+1}} v_n + \frac{2m_{n+1}}{m_n + m_{n+1}} v_{n+1} \quad , \quad v_{n+1}' = \frac{2m_n}{m_n + m_{n+1}} v_n - \frac{m_n - m_{n+1}}{m_n + m_{n+1}} v_{n+1} \quad , \tag{1.12}$$

where m_n is the mass of the nth particle, $v_n = \dot{q}_n$ and the primed variables denote the values after the collision. For equal masses the model is completely integrable, as the set of initial velocities is conserved during the evolution. In order to avoid this peculiar situation, it is customary to choose alternating values, such as $m_n = m$ (rm) for even (odd) n. This type of dynamical systems are particularly appropriate for numerical computation as they do not require the numerical integration of nonlinear differential equations. In fact, it is sufficient to determine the successive collision times and update the velocities according to Eq. (1.12). The only errors are those due to machine round-off. Moreover, the simulation can be made very efficient by resorting to fast updating algorithms. In fact, since the collision times depend only on the position and velocities of neighbouring particles, they can be arranged in a heap structure and thereby simulate the dynamics with an event driven algorithm [40].

Another much studied model involves the Lennard–Jones potential, that in our units reads [66, 71]

$$V(y) = \frac{1}{12}\left(\frac{1}{y^{12}} - \frac{2}{y^6} + 1\right) . \tag{1.13}$$

For computational purposes, the coupling parameters have been fixed in such a way as to yield the simplest form for the force. With this choice, V has a minimum in $y = 1$ and the resulting dissociation energy is $V_0 = 1/12$. For the sake of convenience, the zero of the potential energy is set in $y = 1$. In one-dimension, the repulsive term ensures that the ordering is preserved (the particles do not cross each other).

In the presence of a substrate potential U, the invariance $q_l \to q_l + const.$ is broken and the total momentum P is no longer a constant of motion. Accordingly, all branches of the dispersion relation have a gap at zero wavenumber. We therefore refer to them as *optical* modes. An important subclass is the one in which V is quadratic, which can be regarded as a discretization of the Klein-Gordon field: relevant examples are the Frenkel-Kontorova [39, 44] and "ϕ^4" models [1] which, in suitable units, correspond to $U(y) = 1 - \cos(y)$ and $U(y) = y^2/2 + y^4/4$, respectively. Another toy model that has been studied in some detail is the ding-a-ling system [11], where U is quadratic and the nearest-neighbor interactions are replaced by elastic collisions.

We will always deal with genuine nonintegrable dynamics. For the FPU model this means working with high enough energies/temperatures to avoid all the difficulties induced by quasi-integrability and the associated slow relaxation to equilibrium. For the diatomic HPG this requires fixing a mass-ratio r not too close to unity.

1.3 Signatures of Anomalous Transport

The results emerged from a long series of works can be summarized as follows. Models of the form (1.2) with $U(q) = 0$ typically display *anomalous* transport and relaxation features, this meaning that (at least) one of the following phenomena has been reported:

- The finite-size heat conductivity $\kappa(L)$ diverges in the limit of a large system size $L \to \infty$ [62] as[1]

$$\kappa(L) \propto L^\alpha$$

  This means that this transport coefficient is ill-defined in the thermodynamic limit;
- The equilibrium correlator of the energy current displays a nonintegrable power-law decay,

$$\langle J(t)J(0)\rangle \propto t^{-(1-\delta)} \tag{1.14}$$

  with $0 \leq \delta < 1$, for long times $t \to \infty$ [63]. Accordingly, the Green-Kubo formula yields an infinite value of the conductivity;
- Energy perturbations propagate superdiffusively [15, 24]: a local perturbation of the energy broadens and its variance σ^2 grows in time as

$$\sigma^2(t) \propto t^\beta \tag{1.15}$$

  with $\beta > 1$;

[1] For historical reasons two of the scaling exponents introduced in this section are conventionally denoted by the same Greek letters, α and β, adopted for the FPU models described in Sect. 1.2.

- Relaxation of spontaneous fluctuations is fast (i.e. superexponential) [66]: at variance with standard hydrodynamics, the typical decay rate in time of fluctuations at wavenumber k, $\tau(k)$, is found to scale as

$$\tau(k) \sim |k|^{-z}$$

(with $z < 2$).
- Temperature profiles in the nonequilibrium steady states are nonlinear, even for vanishing applied temperature gradients.

Altogether, these features can be summarized by saying that the usual Fourier's law *does not hold*: the kinetics of energy carriers is so correlated that they are able to propagate *faster* than in the the standard (diffusive) case.

Numerical studies [64] indicate that anomalies occur generically in 1 and 2D, whenever the conservation of energy, momentum and length holds. This is related to the existence of long-wavelength (Goldstone) modes (an acoustic phonon branch in the linear spectrum of (1.2) with $U = 0$) that are very weakly damped. Indeed, it is sufficient to add external (e.g. substrate) forces, to make the anomalies disappear.

Let us now discuss these features in more detail.

1.3.1 Diverging Finite-Size Conductivity

A natural way to simulate a heat conduction experiment consists in putting the system in contact with two heat reservoirs operating at different temperatures T_+ and T_- (see Fig. 1.2). This requires a suitable modeling of interaction with the environment. Several methods, based on both deterministic and stochastic algorithms, have been proposed. A more detailed presentation can be found in [26, 64]. A simple and widely used choice consists in adding Langevin-type forces on some chain subsets. If this is done on the first and the last site of a finite chain ($n = 1, \ldots, N$), it is obtained

$$\ddot{q}_n = -F_n + F_{n-1} + \delta_{n1}(\xi_+ - \lambda \dot{q}_1) + \delta_{nN}(\xi_- - \lambda \dot{q}_N) \quad , \qquad (1.16)$$

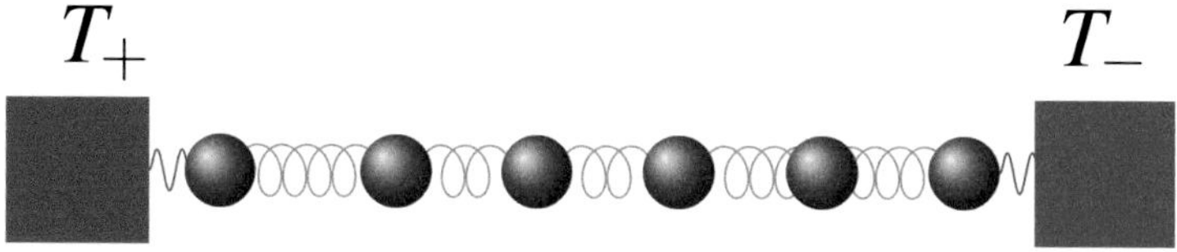

Fig. 1.2 A one-dimensional chain of coupled oscillators interacting with two thermal reservoirs ad different temperatures T_+ and T_-

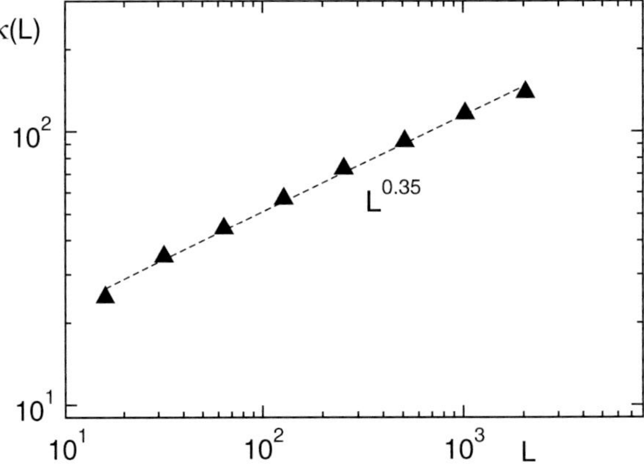

Fig. 1.3 Scaling of the finite-size conductivity for the FPU-$\alpha\beta$ model: with energy $e = 1$ and cubic coupling constant $\alpha = 0.1$

where we assume unitary-mass particles, while $\xi_{\pm}$'s are two independent Gaussian processes with zero mean and variance $2\lambda k_B T_{\pm}$ (k_B is the Boltzmann constant). The coefficient λ is the coupling strength with the heat baths.

After a long enough transient, an off-equilibrium stationary state sets in, with a net heat current flowing through the lattice.[2] The thermal conductivity κ of the chain is then estimated as the ratio between the time-averaged flux $\bar{j}$ and the overall temperature gradient $(T_+ - T_-)/L$, where L is the chain length. Notice that, by this latter choice, κ amounts to an effective transport coefficient, including both boundary and bulk scattering mechanisms. The average $\bar{j}$ can be estimated in several equivalent ways, depending on the employed thermostatting scheme. One possibility is to directly measure the energy exchanges with the two heat reservoirs [26, 64]. A more general (thermostat-independent) definition consists in averaging the heat flux as defined by (1.9).

As a result of many independent simulations performed with the above-described methods, it is now established that $\kappa \propto L^\alpha$ for L large enough. Figure 1.3 illustrates the typical outcome of simulations for the FPU chain.

1.3.2 *Long-Time Tails*

In the spirit of linear-response theory, transport coefficients can be computed from equilibrium fluctuations of the associated currents. More precisely, by introducing

---

[2]From the mathematical point of view, the existence of a unique stationary measure is a relevant question and has been proven in some specific cases models of this class, see the review [8, 28, 29].

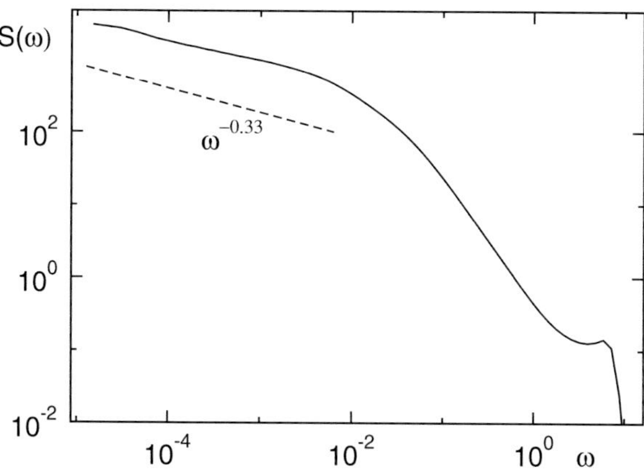

Fig. 1.4 Spectrum of energy current for the FPU-$\alpha\beta$ model, same parameters as in previous figure

the total heat flux

$$J = \sum_n j_n \quad , \tag{1.17}$$

the Green-Kubo formalism tells us that heat conductivity is given by the expression

$$\kappa = \frac{1}{k_B T^2} \lim_{t\to\infty} \lim_{N\to\infty} \frac{1}{N} \int_0^t dt' \, \langle J(t')J(0)\rangle \quad , \tag{1.18}$$

where the average is performed in a suitable equilibrium ensemble, e.g. microcanonical with zero total momentum ($P = 0$).

A condition for the formula (1.18) to give a well-defined heat conductivity is that the time integral is convergent. This is clearly not the case when the current correlator vanishes as in (1.14) with $0 \le \delta < 1$. Here, the integral diverges as t^δ and we may thus define a finite-size conductivity $\kappa(L)$ by truncating the time integral in the above equation to $t \approx L/c$, where c is the sound velocity. Consistency with the definition of the power-law divergence of $\kappa(L)$ implies $\alpha = \delta$. The available data agrees with this expectation, thus providing an independent method for estimating the exponent α.

For later purposes, we mention that, by means of the Wiener–Khintchine theorem, one can equivalently extract δ from the low-frequency behavior of the spectrum of current fluctuations

$$S(\omega) \equiv \int d\omega \langle J(t)J(0)\rangle \mathrm{e}^{i\omega t} \tag{1.19}$$

that displays a low-frequency singularity of the form $S(\omega) \propto \omega^{-\delta}$ (see Fig. 1.4). From the practical point of view, this turns out to be the most accurate numerical strategy, as divergencies are better estimated than convergences to zero.

1.3.3 Diffusion of Perturbations

Consider an infinite system at equilibrium with a specific energy e_0 per particle and a total momentum $P = 0$. Let us perturb it by increasing the energy of a subset of adjacent particles by some preassigned amount Δe and denote with $e(x,t)$ the energy profile evolving from such a perturbed initial condition (for simplicity, we identify x with the average particle location $n\ell$). We then ask how the perturbation

$$\delta e(x,t) = \langle e(x,t) - e_0 \rangle \tag{1.20}$$

behaves in time and space [42], where the angular brackets denote an ensemble average over independent trajectories. Because of energy conservation, $\sum_n \delta e(n\ell, t) = \Delta e$ remains constant at any time: $\delta e(x,t)$ can be interpreted as a probability density (provided it is also positive-defined and normalized).

For sufficiently long time t and large x, one expects $\delta e(x,t)$ to scale as

$$\delta e(x,t) = t^{-\gamma} \mathcal{G}(x/t^{\gamma}) \tag{1.21}$$

for some probability distribution $\mathcal{G}$ and a scaling parameter $0 \leq \gamma \leq 1$. The case $\gamma = 1/2$ corresponds to a normal diffusion and to a normal conductivity. On the other hand, $\gamma = 1$ corresponds to a ballistic motion and to a linear divergence of the conductivity. Consequently, a γ-value larger than $1/2$ implies a superdiffusive behavior of the macroscopic evolution of the energy perturbation [24]. In Fig. 1.5, the evolution of infinitesimal energy perturbations is reported in the case of the HPG [15]: a very good data-collapse is reported for $\gamma = 3/5$.

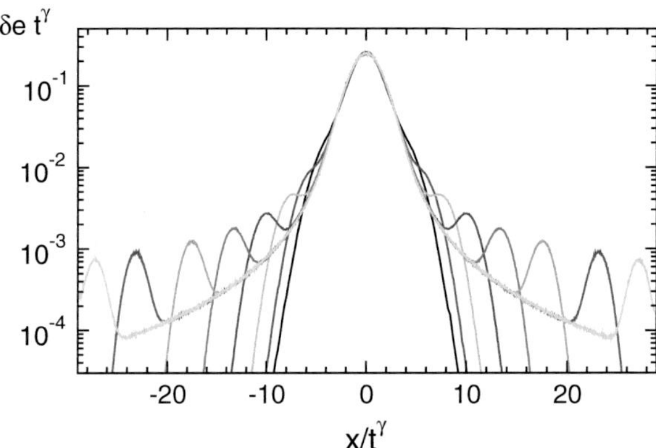

Fig. 1.5 Spreading of infinitesimal perturbations in the HPG model: rescaled perturbation profiles at different times $t = 40, 80, 160, 320, 640, 1280, 2560, 3840$ (the width increases with time), with $\gamma = 3/5$

Remarkably, the above results can be rationalized in terms of a very simple random dynamics: the *Lévy walk model* [7, 95]. Consider a point particle that moves ballistically in between successive "collisions", whose time separation is distributed according to a power law, $\psi(t) \propto t^{-\mu-1}$, $\mu > 0$, while its velocity is chosen from a symmetric distribution $\Psi(v)$. By assuming a δ-like distribution, $\Psi(v) = (\delta(v-\tilde{v}) + \delta(v+\tilde{v}))/2$, the propagator $P(x,t)$ (the probability distribution function to find in x at time t, a particle initially localized at $x = 0$) can be written as $P(x,t) = P_L(x,t) + t^{1-\mu}[\delta(x-\tilde{v}t) + \delta(x-\tilde{v}t)]$ where [7]

$$P_L(x,t) \propto \begin{cases} t^{-1/\mu} \exp\left[-(\eta x/t^{1/\mu})^2\right] & |x| < t^{1/\mu} \\ t x^{-\mu-1} & t^{1/\mu}, |x| < \tilde{v}t \\ 0 & |x| > \tilde{v}t \end{cases}, \quad (1.22)$$

where η is a generalized diffusion coefficient. From the evolution of the perturbation profile, it is possible to infer the exponent α of the thermal conductivity. In fact, in [24] it has been argued that the exponents α, β (the growth rate of the mean square displacement, $\sigma^2(t) = \sum_n n^2 \delta e(x = n\ell, t) \propto t^{\beta}$) and $\gamma = 1/\mu$ are linked by the following relationships,

$$\alpha = \beta - 1 = 2 - \frac{1}{\gamma}. \quad (1.23)$$

In particular, we see that the case $\gamma = 1/2$ corresponds to normal diffusion ($\beta = 1$) and to a normal conductivity ($\alpha = 0$). On the other hand, $\gamma = 1$ corresponds to a ballistic motion ($\beta = 2$) and to a linear divergence of the conductivity ($\alpha = 1$). The numerically observed value $\gamma = 3/5$ corresponds to an anomalous divergence with $\alpha = 1/3$.

The spreading of the wings can be accounted by means of a model which allows for velocity fluctuations [25, 94], which originates from wave dispersion. Assigning smoother velocity distributions $\Psi(v)$ leads to broadening of δ side-peaks, but does not affect the shape and the scaling behavior of the bulk contribution $P_L(x,t)$, which scales, as predicted in Eq. (1.21), with the exponent $\gamma = 1/\mu$.

An alternative way to study finite amplitude perturbations is by looking directly at the behavior of the nonequilibrium correlation function of the energy density [96],

$$C_e(x,t) = \langle \delta e(y,\tau) \delta e(x+y, t+\tau) \rangle \quad , \quad (1.24)$$

where the angular brackets denote a spatial as well as a temporal average over the variables y and τ, respectively. At $t = 0$, $C_e(x,0)$ is a δ function in space. Moreover, in the microcanonical ensemble, energy conservation implies that the area $\int dx C_e(x,t)$ is constant at any time. By assuming that $C_e(x,t)$ is normalized to a unit area, its behavior is formally equivalent to that of a diffusing probability distribution. This allows one to determine the scaling behavior of the heat conductivity from the growth rate of the variance of $C_e(x,t)$ [96]. As the determination of the variance is troubled by the fluctuating tails, it is preferable to proceed by looking at the decay of the maximum $C_e(0,t)$, that is statistically more reliable. An interesting relation between correlation function and anomalous heat transport has been pointed out recently [69] and is reviewed in Chap. 6.

1.3.4 Relaxation of Spontaneous Fluctuations

The above discussion suggests that scaling concepts can be of great importance in dealing with thermal fluctuations of conserved quantities. The evolution of a fluctuation of wavenumber k excited at $t = 0$ is described by its correlation function. For 1D models like (1.1) one of such functions is defined by considering the relative displacements $u_n = q_n - n\ell$ and defining the collective coordinates through the discrete transform

$$U(k,t) = \frac{1}{N}\sum_{n=1}^{N} u_n \exp(-ikn) \quad . \tag{1.25}$$

By virtue of the periodic boundaries, the allowed values of the wavenumbers k are integer multiples of $2\pi/N$. We then define the dynamical structure factor, namely the square modulus of the temporal Fourier transform of the particle displacements as

$$S(k,\omega) = \left\langle \left| U(k,\omega) \right|^2 \right\rangle \quad . \tag{1.26}$$

The angular brackets denote an average over an equilibrium ensemble.

For sufficiently small wavenumbers k, the dynamical structure factor $S(k,\omega)$ usually displays sharp peaks at finite frequency, whose position is proportional to the wavenumber $\omega_{max} = c|k|$; c is naturally interpreted as the phonon sound speed. The data in Fig. 1.6 show that long-wavelength correlations, $k \to 0$, obey *dynamical*

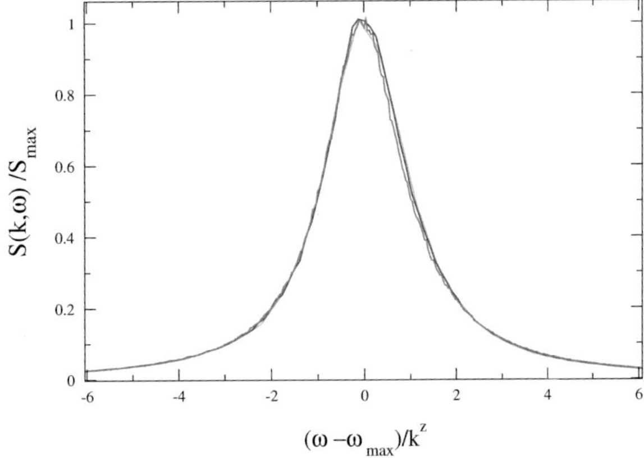

Fig. 1.6 FPU$\alpha\beta$ model: check of dynamical scaling for the dynamical structure factors $\alpha = 0.1$, $N = 4096$, $e = 0.5$ and four different wavenumbers $k = 2, 4, 8, 16$ (in units of $2\pi/N$). The best estimate of the dynamical exponent is $z = 1.5$

scaling, i.e. there exist a function f such that

$$S(k,\omega) \sim f\left(\frac{\omega - \omega_{max}}{k^z}\right). \tag{1.27}$$

for ω close enough to ω_{max}. The associated linewidths are a measure of the fluctuation's inverse lifetime. Simulations indicate that these lifetimes scale as k^{-z} with $z \approx 1.5$. Thus the behavior is different from the diffusive one where one would expect $z = 2$. As explained above, one may think of this as a further signature of an underlying superdiffusive process, intermediate between standard Brownian motion and ballistic propagation.

Other correlation functions can be defined similarly and obey some form of dynamical scaling. For instance, one could consider the structure factor $S_e(k,\omega)$ associated with the local energy density e_n, defined in (1.6). It has a large central component (as a result of the heat modes) and a ballistic one (following from the sound modes). If we assume that the low-frequency part is dominated by the heat-mode scaling, we should have for $\omega \to 0$

$$S_e(k,\omega) \sim g(\omega/q^{5/3}) \,, \tag{1.28}$$

with g being a suitable scaling function.

The origin of the nontrivial dynamical exponents is to be traced back to the nonlinear interaction of long-wavelength fluctuations. For a chain of coupled anharmonic oscillators with three conserved fields (H, L, and P), a linear theory would yield two propagating sound modes and one diffusing heat mode, all of the three diffusively broadened. In contrast, the nonlinear theory predicts that, at long times, the sound mode correlations satisfy Kardar-Parisi-Zhang scaling, while the heat mode correlations follow a Lévy-walk scaling. Various spatiotemporal correlation functions of Fermi-Pasta-Ulam chains and a comparison with the theoretical predictions can be found in [17].

1.3.5 Temperature Profiles

Anomalous transport manifests itself also in the shape of the steady-state temperature profiles. For chains in contact with two baths like in Eq. (1.16), one typically observes that the kinetic temperature profile $T_n = \langle p_n^2 \rangle$ is distinctly nonlinear also for small temperature differences ΔT. For fixed ΔT, the profile typically satisfies a "macroscopic" scaling, $T_n = T(n/L)$ for $L \to \infty$ with $T(0) = T_+$ and $T(1) = T_-$.[3]

[3]Temperature discontinuities may appear at the chain boundaries. This is a manifestation of the well-known Kapitza resistance, the temperature discontinuity arising when a heat flux is maintained across an interface among two substances. This discontinuity is the result of a boundary

In view of the above correspondence with Lèvy processes it may be argued that this feature too could be described in terms of anomalous diffusing particles in a finite domain and subject to external sources that steadily inject particles through its boundaries. The idea is to interpret the local temperature $T(x)$ as the density $P(x)$ of suitable random walkers. A general stochastic model can be defined as follows [61]. Let n denote the position of a discrete-time random walker on a finite one-dimensional lattice ($1 \le n \le N$). In between consecutive scattering events, the particle either jumps instantaneously (Lévy flight—LF) or moves with unit velocity (Lévy walk—LW) over a distance of m sites, that is randomly selected according to the step-length distribution

$$\lambda_m = \frac{q}{|m|^{1+\mu}}, \qquad \lambda_0 = 0 , \tag{1.29}$$

which is the discrete analogous of the ψ distribution defined above, with μ ($1 \le \mu \le 2$) being the Lévy exponent and q a normalization constant. The process can be formulated by introducing the vector $\mathbf{W} \equiv \{W_n(t)\}$, where W_n is the probability for the walker to undergo a scattering event at site n and time t. It satisfies a master equation, which, for LFs, writes

$$\mathbf{W}(t+1) = \mathbf{Q}\mathbf{W}(t) + \mathbf{S}, \tag{1.30}$$

where $\mathbf{S}$ accounts for the particles steadily injected from external reservoirs; $\mathbf{Q}$ is a matrix describing the probability of paths connecting pair of sites. In the simple case of absorbing BC, it is readily seen that Q_{ji} is equal to the probability λ_{j-i} of a direct flight, as from Eq. (1.29). In the LW case, the $\mathbf{W}$ components in the r.h.s. must be estimated at different times (depending on the length of the path followed from j to i) [52]. Since, the stationary solution is the same in both cases, this difference is immaterial, and is easier to refer to LFs, since Eq. (1.30) can be solved iteratively. Note that in the LF case, W_i is equal to the density P_i of particles at site i, while for the LW, P_i includes those particles that are transiting at the ith site during a ballistic step.

The source term is fixed by assuming that the reservoir is a semi-infinite lattice, homogeneously filled by Lévy walkers of the same type as those residing in the domain. This amounts to defining $S_m = s m^{-\mu}$, where s measures the density of particles and m the distance from the reservoir. It is easy to verify that in the presence of two identical reservoirs at the lattice ends, the density is constant (for any N), showing that our definition satisfies a kind of "zeroth principle", as it should.

In the nonequilibrium case, it is not necessary to deal with two reservoirs. The linearity of the problem teaches us that it is sufficient to study the case of a single reservoir, that we assume to be in $n = 0$: the effect of, say, a second one on the

---

resistance, that is explained as a "phonon mismatch" between the two media: see [2] for a discussion of the class of models at hand.

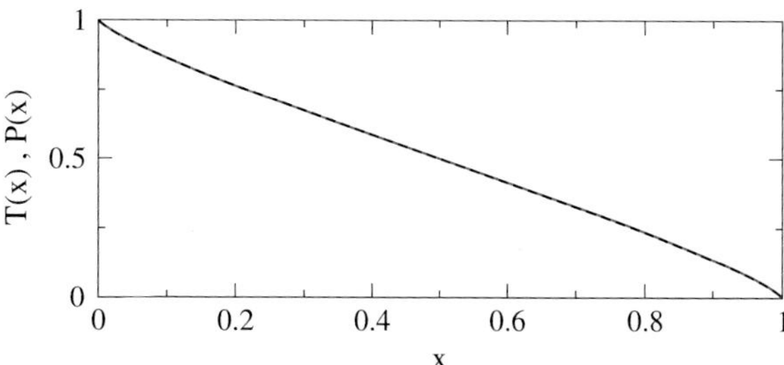

Fig. 1.7 Temperature profile $T(x)$ of the oscillator chain with conservative noise with free boundary condition and $\lambda = \gamma = 1$ (*solid line*) and density profile $P(x)$ for the master equation with reflection coefficient $r = -0.1$ (*dashed line*)

opposite side can be accounted for by a suitable linear combination. For large-enough N values, the steady-state density depends on n and N through the combined variable $x = n/N$, i.e. $P(x) = P_n$. As seen in Fig. 1.7, P vanishes for $x \to 1$ because on that side the absorbing boundary is not accompanied by an incoming flux of particles.

Altogether, upon identifying the particle density with the temperature, the profile can be viewed as a stationary solution of the stationary Fractional Diffusion Equation (FDE)

$$D_x^{\mu} P = -\sigma(x) \tag{1.31}$$

on the interval $0 \le x \le 1$ (see e.g. [98] and references therein for the definition of the integral operator D_x^{μ}). The source term $\sigma(x)$ must be chosen so as to describe the effect of the external reservoirs. A condition to be fulfilled is that two identical reservoirs yield a homogeneous state $T(x) = const$. Using the integral definition of D_x^{μ} [98], it can be shown that this happens for $\sigma(x) = \sigma_{eq}(x) \equiv x^{-\mu} + (1-x)^{-\mu}$ (we, henceforth, ignore irrelevant proportionality constants). It is thus natural to associate $\sigma(x) = x^{-\mu}$ to the nonequilibrium case with a single source in $x = 0$. The numerical solution of the FDE agrees perfectly with the stationary solution of the discrete model, thus showing that long-ranged sources are needed to reproduce the profiles in the continuum limit.

A distinctive feature of the profile is that it is not analytic at the boundaries. Indeed, the data for $x \to 0$ are well fitted by

$$P(x) \;=\; P(0) + Cx^{\mu_m} \tag{1.32}$$

(the same behavior occurs for $x \to 1$, as the profiles are symmetric). In view of the similarity with the shape of the liquid surface close to a wall, we metaphorically term μ_m as the *meniscus exponent*. Such nonanalytic behavior is peculiar of anomalous kinetics, as opposed to the familiar linear shape in standard diffusion. For the above discussed case of absorbing BC, we find that $\mu_m \approx \mu/2$. This value is consistent with the singular behavior of the eigenfunctions of D_x^μ [98]. In the general case, by assuming a linear dependence of μ_m on both r, and μ, it has been conjectured that [61]

$$\mu_m = \frac{\mu}{2} + r\left(\frac{\mu}{2} - 1\right) . \qquad (1.33)$$

This expression is consistent with the $\mu_m = \alpha/2$ value found above for $r = 0$. Moreover, for $\alpha = 2$ (normal diffusion) it yields $\mu_m = 1$, as it should.

Let us now compare this probability distribution of the above process with the temperature profiles in one-dimensional systems displaying anomalous energy transport. It is convenient to refer to a chain of harmonic oscillators coupled with two Langevin heat baths (with a damping constant λ), and with random collisions that exchange the velocities of neighboring particles with a rate γ [23]. On the one hand, this model has the advantage of allowing for an exact solution of the associated Fokker-Planck equation [67]; on the other hand it is closely related to a model that has been proved to display a Lévy-type dynamics [4].

In Fig. 1.7 we compare the temperature profile $T(x)$ (suitably shifted and rescaled) of the heat-conduction model [67] with free BC and the solution of our discrete Lévy model with a reflection coefficient $r = -0.1$. Since they are essentially indistinguishable, we can conclude that the Lévy interpretation does not only allow explaining the anomalous scaling of heat conductivity [15], but also the peculiar shape of $T(x)$. The weird (negative) value of r can be justified a posteriori by introducing two families of walkers and interpreting the reflection as a change of family. The relevant quantity to look at is the difference between the densities of the two different families. The reason why it is necessary to invoke the presence of such two families and their physical meaning in the context of heat conductivity is an open problem.

In the case of a chain with fixed BC, the temperature profile $T(x)$ can be computed analytically [67] and it is thereby found that $\mu_m = 1/2$. By inserting this value in Eq. (1.33) and recalling that $\mu = 3/2$, we find that $r = 1$, i.e. the fixed-BC $T(x)$ corresponds to the case of perfectly reflecting barriers. Unfortunately, this (physically reasonable) result could not be tested quantitatively. Indeed, it turns out that finite-size corrections become increasingly important upon increasing r, and for r close to 1, it is practically impossible to achieve convergence to the steady-state.

The description of the steady state in terms of Lévy walk has been further investigated in [27]. The authors calculate exactly the average heat current, the large deviation function of its fluctuations, and the temperature profile in the steady state. The current is nonlocally connected to the temperature gradient. Also, all the cumulants of the current fluctuations have the same system-size dependence in the open geometry as those of deterministic models like the HPG. The authors investigated also the case of a ring geometry and argued that a size-dependent cutoff time is necessary for the Lévy-walk model to behave like in the deterministic case. This modification does not affect the results on transport in the open geometry for large enough system sizes.

1.4 Universality and Theoretical Approaches

In view of their common physical origin, it is expected that the exponents describing the different processes will be related to each other by some "hyperscaling relations". Their value should be ultimately dictated by the dynamical scaling of the underlying dynamics. Moreover, one can hope that they are largely independent of the microscopic details, thus allowing for a classification of anomalous behavior in terms of "universality classes". This crucial question is connected to the predictive power of simplified models and to the possibility of applying theoretical results to real low-dimensional materials.

1.4.1 Methods

Various theoretical approaches to account for the observed phenomenology have been developed and implemented. In the rest of the volume they will be exposed in detail; here we limit ourselves to a brief description. The methods discussed are

1. *Fluctuating hydrodynamics* approach: here the models are described in terms of the random fields of deviations of the conserved quantities with respect to their stationary values. The role of fluctuations is taken into account by renormalization group or some kind of self-consistent theory.
2. *Mode-coupling* theory: this is closely related to the above, as it amounts to solving (self-consistently) some approximate equations for the correlation functions of the fluctuating random fields.
3. *Kinetic theory*: it is based on the familiar approach to phonon transport by means of the Boltzmann equation.
4. *Exact solution* of specific models: typically in this case the original microscopic Hamiltonian dynamics is replaced by some suitable stochastic one which can be treated by probabilistic methods.

A sound theoretical basis for the idea that the above described anomalies are generic and universal for all momentum-conserving system was put forward in [77]. The authors treated the case of a fluctuating d-dimensional fluid and applied renormalization group techniques to evaluate the contribution of noisy terms to transport coefficients. The calculation predicts that the thermal conductivity exponent is $\alpha = (2-d)/(2+d)$. From the arguments exposed above, it follows that in 1D the exponents are

$$\alpha = \delta = \frac{1}{3}, \qquad \beta = \frac{4}{3}, \qquad z = \frac{3}{2} \quad . \tag{1.34}$$

According to this approach, any possible additional term in the noisy Navier-Stokes equation yields irrelevant corrections in the renormalization procedure, meaning that the above exponents are model independent, provided the basic conservation laws are respected.

Next we give a flavour of one of the other approaches: the Mode-Coupling Theory (MCT). This type of theories has been traditionally invoked to estimate long-time tails of fluids [82] and to describe the glass transition [87]. In the simplest version, it involves the normalized correlator of the particle displacement [see Eq.(1.25)], where the discrete wavenumber k has been turned to the continuous variable q)

$$G(q,t) = \frac{\langle U^*(q,t)U(q,0)\rangle}{\langle |U(q)|^2\rangle} \ .$$

$G(q,t)$ is akin to the density–density correlator, an observable routinely used in condensed-matter physics. The main idea is to write a set of approximate equations for $G(q,t)$ that must be solved self-consistently. For the problem at hand, the simplest version of the theory amounts to consider the equations [60, 86]

$$\ddot{G}(q,t) + \varepsilon \int_0^t \Gamma(q,t-s)\dot{G}(q,s)\, ds + \omega^2(q)G(q,t) = 0 \quad , \tag{1.35}$$

where the memory kernel $\Gamma(q,t)$ is proportional to $\langle \mathscr{F}(q,t)\mathscr{F}(q,0)\rangle$, with $\mathscr{F}(q)$ being the nonlinear part of the fluctuating force between particles. Equation (1.35) is derived within the well-known Mori–Zwanzig projection approach [53]. It must be solved with the initial conditions $G(q,0) = 1$ and $\dot{G}(q,0) = 0$.

The mode-coupling approach basically amounts to replacing the exact memory function Γ with an approximate one, where higher-orders correlators are written in terms of $G(q,t)$. In the generic case, in which k_3 is different from zero [see Eq. (1.11)], the lowest-order mode coupling approximation of the memory kernel turns out to be [60, 86]

$$\Gamma(q,t) = \omega^2(q)\frac{2\pi}{N} \sum_{p+p'-q=0,\pm\pi} G(p,t)G(p',t) \quad . \tag{1.36}$$

Here p and p' range over the whole Brillouin zone (from $-\pi$ to π in our units) . This yields a closed system of nonlinear integro-differential equations. Both the coupling constant ε and the frequency $\omega(q)$ are temperature-dependent input parameters, which should be computed independently by numerical simulations or approximate analytical estimates. For the present purposes it is sufficient to restrict ourselves to considering their bare values, obtained in the harmonic approximation. In the adopted dimensionless units they read $\varepsilon = 3k_3^2 k_B T/2\pi$ and $\omega(q) = 2|\sin\frac{q}{2}|$. Of course, the actual renormalized values are needed for a quantitative comparison with specific models. The long-time behavior of G can be determined by looking for a solution of the form

$$G(q,t) \;=\; C(q,t)e^{i\omega(q)t} + c.c. \tag{1.37}$$

with $\dot{G} \ll \omega G$. It can thus be shown [19, 21] that, for small q-values and long times $C(q,t) = g(\sqrt{\varepsilon}tq^{3/2})$ i.e. $z = 3/2$ in agreement with the above mentioned numerics. Furthermore, in the limit $\sqrt{\varepsilon}tq^{3/2} \to 0$ one can explicitly evaluate the functional form of g, obtaining

$$C(q,t) = \frac{1}{2}\exp\left(-Dq^2|t|^{\frac{4}{3}}\right) \quad , \tag{1.38}$$

where D is a suitable constant of order unity. The correlation displays a "compressed exponential" behavior in this time range. This also means that the lineshapes of the structure factors $S(q,\omega)$ are non-Lorentzian but rather exhibit an unusual faster power-law decay $(\omega - \omega_{max})^{-7/3}$ around their maximum. Upon inserting this scaling result into the definition of the heat flux, one eventually concludes that the conductivity exponent is $\alpha = 1/3$, in agreement with (1.34).

A more refined theory requires considering the mutual interaction among *all* the hydrodynamic modes associated with the conservation laws of the system at hand. The resulting calculations are considerably more complicated but they can be worked out [88, 89]. As a result, the same values of the scaling exponents are found, but also a more comprehensive understanding is achieved (see Chap. 3 for a detailed account).

1.4.2 Connection with the Interface Problem

Relevant theoretical insight comes from the link with one of the most important equations in nonequilibrium statistical physics, the Kardar-Parisi-Zhang (KPZ) equation. This is a nonlinear stochastic Langevin equation which was originally introduced in the (seemingly unrelated) context of surface growth [3]. Let us first consider the fluctuating Burgers equation for the random field $\rho(x,t)$

$$\frac{\partial\rho}{\partial t} = \frac{\lambda}{2}\frac{\partial\rho^2}{\partial x} + D\frac{\partial^2\rho}{\partial x^2} + \frac{\partial\eta}{\partial x}, \tag{1.39}$$

where $\eta(x,t)$ represents a Gaussian white noise with $\langle\eta(x,t)\eta(x',t')\rangle = 2D\delta(x-x')\delta(t-t')$. As it is well-known, Eq. (1.39) can be transformed into the KPZ equation by introducing the "height function" h such that $\rho = \frac{\partial h}{\partial x}$,

$$\frac{\partial h}{\partial t} = \frac{\lambda}{2}\left(\frac{\partial h}{\partial x}\right)^2 + D\frac{\partial^2 h}{\partial x^2} + \eta. \tag{1.40}$$

It has been shown [89] that the mode-coupling approximation for the correlator of ρ obeying (1.39) is basically identical to the equation for C described in the previous paragraph. Thus one may argue that the dynamical properties are those of the KPZ equation in one dimension. Loosely speaking, we can represent the displacement

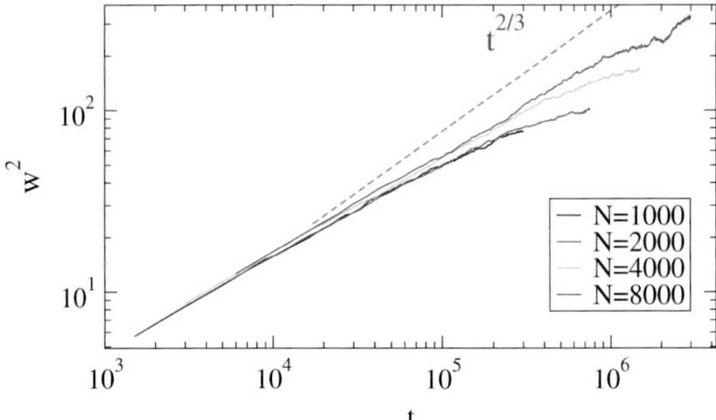

Fig. 1.8 Evolution of the variance (1.41) for the FPU $\alpha\beta$ chain with $e = 0.5$ $\alpha = 0.1$ and for increasing chain lengths N (bottom to top solid lines). The *dashed line* is the expected KPZ growth rate

field as the superposition of counterpropagating plane waves modulated by an envelope that is ruled, at large scales, by Eq. (1.40).

In order to illustrate this, we have performed a typical "KPZ numerical experiment" [3] for the for the FPU $\alpha\beta$ chain. In practice, we monitored

$$w^2(t,N) = \left\langle \frac{1}{N}\sum_n h_n^2 - (\frac{1}{N}\sum_n h_n)^2 \right\rangle \tag{1.41}$$

where $h_n(t) = q_n(t) - q_n(0)$, $q_n(0)$ is an equilibrium configuration and the angular brackets denote an average over an ensemble of different trajectories. The results are reported in Fig. 1.8. The only difference with respect to the usual setup is that here the square-width is plotted only at times t multiples of L/c, where c is the effective sound speed. These are the only moments, when the effect of counterpropagating sound waves cancel out, offering the chance to identify a KPZ-like behavior. In fact, one can see that the growth in Fig. 1.8 is compatible with the expected KPZ exponent 2/3 (actually, a bit smaller) followed by a saturation due to the finite size of the chain. A more rigorous discussion of the above topics can be found in Chap. 3.

1.4.3 Other Universality Classes

In the previous section we argued that the scaling properties of anomalous transport are independent of the microscopic details and correspond to those of the KPZ universality class. One might wonder whether other classes exist and under which conditions they can be observed. A reasonable argument, that can be invoked to delimit the KPZ universality class, is the *symmetry* of the interaction potential with respect to the equilibrium position. With reference to the MCT, one realizes that the symmetry of the fluctuations implies that the quadratic kernel in (1.36) should be replaced by a cubic one,[4] thus yielding different values of the exponents [21]. In the language of KPZ interfaces, whenever the coefficient of the nonlinear term vanishes, the evolution equation reduces to the Edwards-Wilkinson equation that is indeed characterized by different scaling exponents. The argument can be made more precise in the framework of the full hydrodynamic theory [88, 89]. There, different dynamical exponents can arise if the coupling between some modes vanishes (we refer again the reader to Chap. 3 for a detailed discussion). A thermodynamic interpretation of this difference is given in [57, 58].

The FPU model is a natural instance to test this working hypothesis. In fact, systematically larger values of the scaling exponent α have been reported for the FPU-β case where the cubic term of the potential is absent [65]. The existence of two universality classes for thermal transport in one-dimensional oscillator systems has been also demonstrated in [55], where it was further proposed that the criterion for being out of the KPZ class is the condition $\gamma = c_P/c_V = 1$, where c_P and c_V are the specific heat capacities at constant pressure and volume, respectively.

The scenario can be further illustrated by considering a modification of the HPG model, the so-called Hard-Point Chain (HPC) [20], characterized by a square-well potential in the relative distances

$$V(y) = \begin{cases} 0 & 0 < y < a \\ \infty & \text{otherwise} \end{cases} . \tag{1.42}$$

The infinite barriers at $y = a$ imply an elastic "rebounding" of particles as if they were linked by an inextensible and massless string of fixed length a. The string has no effect on the motion, unless it reaches its maximal length, when it exerts a restoring force that tends to rebound the particles one against the other. The potential (1.42) introduces the physical distance a as a parameter of the model.

As it is well known, the thermodynamics of models like the HPC can be solved exactly and the equation of state is found to be

$$L = N\left[\frac{1}{\beta P} - \frac{a}{\exp(\beta P a) - 1}\right]$$

where P is the pressure of the HPC. Note that, for large values of a, the equation of state is the same of an HPC i.e. the one of an ideal gas in 1D. The important point here is that we can choose the parameter a such as $P = 0$. In this particular point the interaction is symmetric ($L/N = a/2$).

A peculiarity of the HPC model is that energy transfer occurs also at rebounding "collisions" at distance a, this means that besides the contribution defined by Eq. (1.10) one should include a term j_i' as from Eq. (1.9). However, one cannot

[4] In fact, the quadratic kernel corresponds to a quadratic force originating from the leading cubic nonlinearity of any asymmetric interaction potential, while a quartic leading nonlinearity of a symmetric interaction potential yields a cubic kernel (force).

proceed directly, since the force is singular, there. By defining the force between two particles as the momentum difference induced by a collision, j'_i can be written as the kinetic energy variation times the actual distance a, i.e. $j'_i = am_i(u_i'^2 - u_i^2)/2$, divided by a suitable time-interval Δt. In order to get rid of the microscopic fluctuations, it is necessary to consider a sufficiently long Δt, so as to include a large number of collisions. Since the number of collisions is proportional to the system size, it is only in long systems that fluctuations can be removed without spoiling the slow dynamics of the heat flux. Equilibrium simulations show that for $L/N = a/2$ the leading contribution to the heat flux is given by the term j'_i which exhibits a low-frequency divergence with an exponent $\delta = 0.45$, that is not only definitely larger than 1/3 (the value predicted for the KPZ class), but is also fairly close to the results found for the FPU-β model [65].

In out-of-equilibrium simulations, a compatible exponent $\alpha = 0.4$ has been measured [81]. Those values should be compared with $\alpha = 1/2$, the prediction of mode-coupling theory, thus supporting the conjecture that the case $P = 0$ belongs to a universality class different from KPZ.

To conclude this section, let us mention that further support to this scenario comes from a stochastic model of a chain of harmonic oscillators, subject to momentum and energy-conserving noise [4]. Indeed, one can prove that the dynamical exponents are different from the KPZ class, e.g. $\delta = 1/2$ [4] and $\alpha = 1/2$ [67]. Details about this class of models can be found in Chap. 5. The qualitative explanation is that, as the stochastic collisions occur independently of the actual positions, the effective interaction among particles is symmetric and thus equivalent to the $P = 0$ case. Notably, this remains true even if the harmonic potential is replaced by an anharmonic one, like the FPU-$\alpha\beta$ [5]. Finally, the application of kinetic theories to the β-FPU model [70, 78, 80] yields a non-KPZ behavior, $\alpha = 2/5$. We refer the reader to Chap. 4 for a detailed account.

1.4.4 Comparison with Simulations and Experiments

The theoretical predictions have been intensively investigated in the recent literature. A direct validation by numerical simulations is, to some extent, challenging and has been debated through the years [22]. Generally speaking, the available numerical estimates of α and δ may range between 0.25 and 0.44 [64]. As a matter of fact, even in the most favorable cases of computationally efficient models as the HPG, finite-size corrections to scaling are sizeable. In this case, α-values as diverse as 0.33 [40] and 0.25 [10] for comparable parameter choices have been reported. On the other hand, a numerically convincing confirmation of the $\alpha = 1/3$ prediction comes from the diffusion of perturbations [15]. We refer to Chap. 6 for some detailed numerics.

The ultimate goal would be of course the validation of the universality hypothesis in more realistic systems, possibly characterized by more than one degree of freedom per lattice site. The first remarkable attempt was the application to the vibrational dynamics of individual single-walled carbon nanotubes, which can be in many respect considered as one-dimensional objects. Signature of anomalous thermal transport was first reported in molecular dynamics simulations in [72]. Note that this type of simulations involve complicated three-body interactions among carbon atoms, thus supporting the claim that toy models like ours can indeed capture some general features. We refer the reader to Chap. 7 for a critical discussion of molecular dynamics results on carbon-based material. Chapter 8 will report some experimental data on nanotubes and nanowires and discuss the current state of the art.

1.5 The Coupled Rotors Model

As discussed in the previous sections, one-dimensional anharmonic chains generically display anomalous transport properties. A prominent exception is the coupled rotors chain described by the equation of motion

$$\dot{q}_n = p_n\,, \quad \dot{p}_n = \sin(q_{n+1} - q_n) - \sin(q_n - q_{n-1})\,. \tag{1.43}$$

The model is sometime referred to as the Hamiltonian version of the XY spin chain. The energy flux is $j_n^e = \langle p_n \sin(q_{n+1} - q_n) \rangle$. As the interaction depends only on the angle differences, angular momentum is conserved and one may expect anomalous transport to occur. Nevertheless, molecular dynamics simulations have convincingly demonstrated normal diffusion [36, 38, 92].

There are two complementary views to account for this difference. In the general perspective of nonlinear fluctuating hydrodynamics, the chain "length" L defined as $L = \sum_n (q_{n+1} - q_n)$ is not even well defined, because of the phase slips of $\pm 2\pi$, so the corresponding evolution equation breaks down and normal transport is eventually expected. From a dynamical point of view, one can invoke that normal transport sets in due to the spontaneous formation of local excitations, the so-called *rotobreathers*, that behave like scattering centers [33]. Phase slips (jumps over the energy barrier), on their side, may effectively act as localized random kicks, that contribute to scatter the low-frequency modes, thus leading to a finite conductivity. In order to test the validity of this conjecture, one can study the temperature dependence of κ for low temperatures T, when jumps across barriers become increasingly rare. Numerics indicates that the thermal conductivity behaves as $\kappa \approx \exp(\eta/T)$ with $\eta \approx 1.2$. The same kind of dependence on T (although with $\eta \approx 2$) is found for the average escape time τ across the potential barrier: this can be explained by assuming that the phase slips are the results of activation processes.

An important extension is the 2D case, i.e. rotors coupled to their neighbors on a square lattice, akin to the celebrated XY-model. As it is well known, the latter is characterized by the presence of the so called Kosterlitz-Thouless-Berezinskii phase transition at a temperature T_{KTB}, between a disordered high-temperature phase and a low-temperature one, where vortices condensate. It is likely that transport properties are qualitatively different in the two phases. Numerical simulations [18] performed on a finite lattice indeed show that they are drastically different in the high-temperature and in the low-temperature phases. In particular, thermal conductivity is finite in the former case, while in the latter it does not converge up to lattice sizes of order 10^4. In the region where vorticity is negligible ($T < 0.5$) the available data suggest a logarithmic divergence with the system size, analogous to the one observed for coupled oscillators (see next section). Close to T_{KTB}, where a sizable density of bounded vortex pairs are thermally excited, numerical data still suggests a divergence, but the precise law has not be reliably estimated.

1.6 Two-Dimensional Lattices

Heat conduction in $2D$ models of anharmonic oscillators coupled through momentum-conserving interactions is expected to exhibit different properties from those of $1D$ systems. In fact, extension of the arguments discussed in the previous sections predicts a logarithmic divergence of κ with the system size N at variance with the power-law predicted for the $1D$ case. Consideration of this case is not only for completeness of the theoretical framework, but is also of great interest for almost-2D materials, like graphene, that will be treated in the Chaps. 7 and 9.

Although the theory in this case if far less developed, there are several numerical evidences in favor of such logarithmic divergence. In [68], a square lattice of oscillators interacting through the FPU-β (see Eq. (1.11), with $k_3 = 0$) or the Lennard-Jones [see Eq. (1.13)] potentials, was investigated by means of both equilibrium and nonequilibrium simulations. The models are formulated in terms of two-dimensional vector displacements u_{ij} and velocities and $\dot{u}_{ij}$, defined on a square lattice containing $N_x \times N_y$ atoms of equal masses m and nearest-neighbor interactions. Periodic and fixed boundary conditions have been adopted in the direction perpendicular (y) and parallel (x) to the thermal gradient, respectively. Simulations for different lattice sizes have been performed by keeping the ratio N_y/N_x constant and not too small to observe genuine $2D$ features (in [68] $N_y/N_x = 1/2$ was chosen).

The simulations reveal several hallmarks of anomalous behavior: temperature profiles display deviations from the linear shape predicted by Fourier law and the size dependence of the thermal conductivity is well-fitted by a logarithmic law

$$\kappa = A + B \log N_x \, , \tag{1.44}$$

with A and B being two unknown constants. A consistent indication comes from the evaluation of the Green-Kubo integrand in the microcanonical ensemble. Indeed, the energy-current autocorrelation is compatible with a decay $1/t$ at large times.

Despite these first indications, the numerics turns out to be very difficult, which is not surprising in view of the very weak form of the anomaly, peculiar of the 2D case.

As a matter of fact very robust finite-size effects are observed in the calculations for other lattices, which well exemplify the difficulties in observing the true asymptotic behavior with affordable computational resources [90].

Another interesting issue concerns dimensional crossover, namely how the divergence law of the thermal conductivity will change from the 2D class to 1D class as N_y/N_x decreases. This issue has been studied for the two-dimensional FPU lattice in [93]. We refer to Chap. 6 for a further detailed discussion.

1.7 Integrable Nonlinear Systems

The harmonic crystal behaves as an ideal conductor, because its dynamics can be decomposed into the superposition of independent "channels". This peculiarity can be generalized to the broader context of integrable nonlinear systems. They are mostly one-dimensional models characterized by the presence of "mathematical solitons", whose stability is determined by the interplay of dispersion and nonlinearity. This interplay is expressed by the existence of a macroscopic number of *conservation laws*, constraining the dynamical evolution. Intuitively, the existence of freely travelling solitons is expected to yield ballistic transport, i.e. an infinite conductivity. From the point of view of the Green-Kubo formula, this ideal conducting behavior is reflected by the existence of a nonzero flux autocorrelation at arbitrarily large times. This, in turn, implies that the finite-size conductivity *diverges linearly with the system size*.

Although integrable models are, in principle, exactly solvable, the actual computation of dynamic correlations is technically involved. A more straightforward approach is nevertheless available to evaluate the asymptotic value of the current autocorrelation. This is accomplished by means of an inequality due to Mazur [73] that, for a generic observable $\mathscr{A}$, (with $\langle \mathscr{A} \rangle = 0$, where $\langle \ldots \rangle$ denotes the equilibrium thermodynamic average) reads

$$\lim_{\tau \to \infty} \frac{1}{\tau} \int_0^\tau \langle \mathscr{A}(t)\mathscr{A}(0) \rangle \, dt \geq \sum_n \frac{\langle \mathscr{A} \mathscr{Q}_n \rangle^2}{\langle \mathscr{Q}_n^2 \rangle} \quad , \tag{1.45}$$

where $\mathscr{Q}_n$ denote a set of conserved and mutually orthogonal quantities, $(\langle \mathscr{Q}_n \mathscr{Q}_m \rangle = \langle \mathscr{Q}_n^2 \rangle \delta_{n,m})$.

In the present context the most relevant example is the equal-masses Toda chain with periodic boundary conditions, defined, in reduced units, by the Hamiltonian

$$H = \sum_{n=1}^{N} \left[\frac{p_n^2}{2} + \exp(-r_n) \right] , \tag{1.46}$$

where $r_n = q_{n+1} - q_n$ is the relative position of neighboring particles. The model is completely integrable, since it admits N independent constants of the motion

[34, 43]. Lower bounds on the long time value of $\langle J(t)J(0)\rangle$ can be calculated through the inequality (1.45) [99]. The resulting lower bound to the conductivity is found to increase monotonously with the temperature. At low T, the growth is linear with a slope comparable to the density of solitons $N_s/N = (\ln 2/\pi^2)T$. This trend is interpreted as an evidence for the increasing contribution of thermally excited nonlinear modes to ballistic transport.

To conclude, let us also mention that Mazur-type of inequalities have been recently used as a theoretical basis for the study of thermoelectric coefficients. This is discussed in Chap. 10 of the present volume.

1.8 Coupled Transport

Up to this point we have restricted the discussion to models where just one quantity, the energy, is exchanged with external reservoirs and transported across the system. In general, however, the dynamics can be characterized by more than one conserved quantity. In such cases, it is natural to expect the emergence of coupled transport phenomena, in the sense of ordinary linear irreversible thermodynamics. Works on this problem are relatively scarce [6, 39, 54, 75]. Interest in this field has been revived by recent works on thermoelectric phenomena [12, 13, 84] in the hope of identifying dynamical mechanisms that could enhance the efficiency of thermoelectric energy conversion. This will be treated in detail in Chap. 10.

Here, we briefly discuss two models: a chain of coupled rotors and the discrete nonlinear Schrödinger equation, where the second conserved quantity is the momentum and the norm (number of particles), respectively.

1.8.1 Coupled Rotors

The evolution equation defined in (1.43) must be augmented to include the exchange of momentum with the external reservoirs,

$$\begin{aligned} \dot{p}_n &= \sin(q_{n+1} - q_n) - \sin(q_n - q_{n-1}) \\ &+ \delta_{1n}\left(\gamma(F_+ - p_1) + \sqrt{2\gamma T_+}\,\eta_+\right) + \delta_{1N}\left(\gamma(F_- - p_N) + \sqrt{2\gamma T_-}\,\eta_-\right) \end{aligned} \tag{1.47}$$

where $F_\pm$ and $T_\pm$ denote the torque applied to the chain boundary and the corresponding temperature, respectively; γ is the coupling strength with the external baths and $\eta_\pm$ is a Gaussian white noise with unit variance. The effect of external forces on the Hamiltonian XY model has been preliminary addressed in [31, 46, 49].

As discussed in Sect. 1.5, (angular) momentum is conserved and one can, in fact, define the corresponding flux as $j_n^p = \sin(q_{n+1} - q_n)$. A chain of rotors is perhaps the simplest model where one can exert a gradient of forces that couples to heat transport, giving rise to nontrivial phenomena, even though the transport itself is normal. For $F_+ = F_-$, all the oscillators rotate with the same frequency $\omega = F$, no matter which force is applied: no momentum flux is generated. In fact, what matters is the difference between the forces applied at the two extrema of the chain. Therefore, from now on we consider the case of zero-average force, i.e. $F_+ = -F_-$. In the presence of such a gradient of forces, the oscillators may rotate with different frequencies and, as a result, a coupling between angular momentum and energy transport may set in. In principle, one could discuss the same setup for general chains of kinetic oscillators, as (linear) momentum is conserved in that context too. However, nothing interesting is expected to arise. For a binding potential, like in the FPU model, the presence of an external force is akin to the introduction of a homogeneous, either positive or negative, pressure all along the chain. In fact, the pressure P is, by definition, equal to the equilibrium average of the momentum flux, $P = \langle j^p \rangle$ (at equilibrium, the r.h.s. is independent of n). On the other hand, if the potential is not binding [e.g., the Lennard-Jones chain (1.13)] and the applied force is equivalent to a negative pressure, the system would break apart.

In the presence of two fluxes, the linear response theory implies that they must satisfy the equations [84] (angular brackets denote an ensemble, or equivalently, a time average, assuming ergodicity)

$$\begin{aligned} \langle j^p \rangle &= -L_{pp}\frac{d(\beta\mu)}{dy} + L_{pe}\frac{d\beta}{dy} \\ \langle j^e \rangle &= -L_{ep}\frac{d(\beta\mu)}{dy} + L_{ee}\frac{d\beta}{dy} \,, \end{aligned} \tag{1.48}$$

where $y = n/N$, β is the inverse temperature $1/T$ (in units of the Boltzmann constant) and μ is the chemical potential, which, in the case of the coupled rotors, coincides with the average angular frequency $\omega_n = \langle p_n \rangle$. Finally, $\mathbf{L}$ is the symmetric, positive definite, 2×2 Onsager matrix. If $L_{ep} = 0$, the two transport processes are uncoupled.

In the case of the rotor chain, it is important to realize that a correct definition of the kinetic temperature requires subtracting the coherent contribution due to the nonzero angular velocity, i.e.

$$T_n = \langle (p_n - \omega_n)^2 \rangle \,.$$

The effect of coupling between energy and momentum transport is better understood by considering a setup where the two thermal baths operate at the same temperature T. Because of the flux of momentum, the temperature profile deviates from the value imposed at the boundaries. In Fig. 1.9 we show the results for $T = 0.5$ and $F = 1.5$ and two different system sizes. Notably, the temperature profile displays a peak in the central region [46], where it reaches a value around 1.2; the average frequency varies nonuniformly across the sample with a steep region in correspondence of the central hot spot. At the same time, the energy flux j^e is zero, so that the anomalous

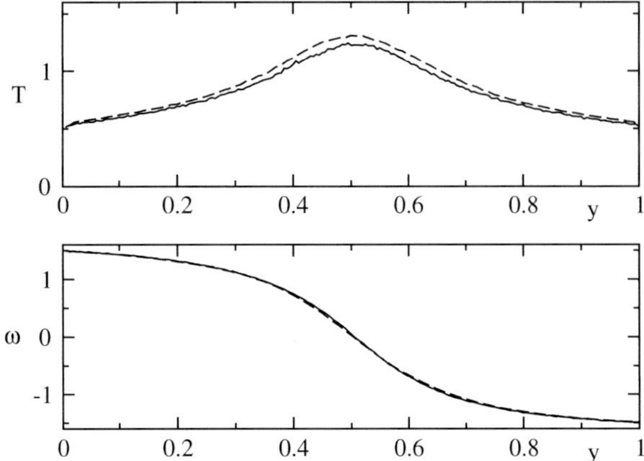

Fig. 1.9 Stationary profile of the temperature (*upper panel*) and of the average frequency (*lower panel*) for $T(0) = T(1) = 0.5$, $F = 1.5$, and $\gamma = 1$; $y = n/N$. The *dashed* and *solid curves* correspond to $N = 100$, and 200, respectively

behavior of the temperature profile is entirely due to the coupling with the nonzero momentum flux.

This behavior can be traced back to the existence of a (zero-temperature) boundary-induced transition. In fact, for $T = 0$, there exists a critical torsion $F_c = 1/\gamma$ [49] such that for $F < F_c$ the ground state is a twisted fully-synchronized state, whereby each element is at rest and is characterized by a constant phase gradient. Here, $T_n = 0$ throughout the whole lattice. For $F > F_c$ the fully synchronized state turns into a chaotic asynchronous dynamics with $\omega_1 = F = -\omega_N$. Remarkably, even though both heat baths operate at zero temperature and the equations are deterministic and dissipative, the temperature in the middle raises to a finite value (see Fig. 1.10) even in the thermodynamic limit.

The phenomenon can be interpreted as the onset of an interface (the hot region) separating two different phases: the oscillators rotating with a frequency F (on the left) from those rotating with a frequency $-F$ (on the right). The phenomenon is all the way more interesting in view of the anomalous scaling of the interface width with the system size (it grows as $N^{1/2}$, see Fig. 1.10) and its robustness (it is independent of the value of the torsion F, provided it is larger than the critical value F_c [49]).

Accordingly, the interface is neither characterized by a finite width nor it is extensive. A more careful inspection reveals that the $N^{1/2}$ width is due to a spatial Brownian-like behavior of an instantaneously much thinner interface. Nevertheless, even the instantaneous interface extends over a diverging number of sites, of order $N^{1/5}$, thus leaving the anomaly fully in place. Such a state can neither be predicted within a linear-response type of theory, nor traced back to some underlying equilibrium transition. Even more remarkably, it constitutes an example of a highly inhomogeneous, unusual chaotic regime. Indeed, while the fractal dimension is

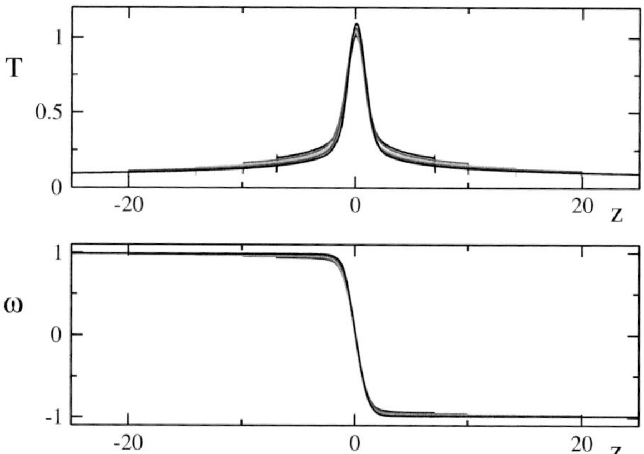

Fig. 1.10 Stationary profile of the temperature (*upper panel*) and of the average frequency (*lower panel*) for $T(0) = T(1) = 0.$, $F = 1.05$, and $\gamma = 1$; $z = (n - N/2)/N^{1/2}$. The various curves correspond to $N = 200$, 400, 800, 1600, and 3200

extensive (i.e. proportional to the number of oscillators) the Kolmogorov-Sinai (KS) entropy is not: it increases only as $N^{1/2}$. The KS entropy measures the diversity of the "ground state" non-equilibrium configurations that are compatible with the given thermal baths. Its lower-than-linear increase with N implies that we are not in the presence of a macroscopic degeneracy, as in spin glasses.

The anomaly of the regime is finally reinforced by the scaling behavior of the momentum flux, which scales as $N^{-1/5}$. A theoretical explanation of this behavior is still missing. All of these anomalies disappear as soon as the temperature at the boundaries is selected to be strictly larger than zero. In particular, the width of the hot spot suddenly becomes extensive and the scaling of the momentum is normal ($j^p \simeq 1/N$). The nonmonotonous behavior of the temperature is nevertheless a nontrivial consequence of the coupling between heat and momentum transport.

1.8.2 The Discrete Nonlinear Schrödinger Equation

The above discussed non-equilibrium transition is not a peculiarity of the rotor model. A similar scenario can be observed also in the Discrete Nonlinear Schrödinger (DNLS) equation [30, 51], a model with important applications in many domains of physics. In one dimension, the DNLS Hamiltonian is

$$H = \frac{1}{4}\sum_{n=1}^{N}\left(p_n^2 + q_n^2\right)^2 + \sum_{n=1}^{N-1}\left(p_n p_{n+1} + q_n q_{n+1}\right) \quad , \tag{1.49}$$

where the sum runs over the N sites of the chain. The sign of the quartic term is positive, while the sign of the hopping term is irrelevant, due to the symmetry associated with the canonical (gauge) transformation $z_n \to z_n e^{i\pi n}$ (where $z_n \equiv (p_n + \imath q_n)/\sqrt{2}$ denotes the amplitude of the wave function). The equations of motion are

$$i\dot{z}_n = -z_{n+1} - z_{n-1} - 2|z_n|^2 z_n \tag{1.50}$$

with $n = 1, \cdots, N$, and fixed boundary conditions ($z_0 = z_{N+1} = 0$). The model has two conserved quantities, the energy and the total norm (or total number of particles)

$$A = \sum_{n=1}^{N} (p_n^2 + q_n^2) = \sum_{n=1}^{N} |z_n|^2 \quad , \tag{1.51}$$

so that it is a natural candidate for the study of coupled transport.

Since the Hamiltonian is not the sum of a kinetic and potential energy, the thermal baths cannot be described by standard Langevin equations. An effective strategy has been proposed in [48]. Here below we report the evolution equation for the first oscillator, in contact with a thermal bath at temperature T_+ and with a chemical potential μ_+ (a similar equation holds for the last particle at site N)

$$\begin{aligned} \dot{p}_1 &= -(p_1^2 + q_1^2)q_1 - q_2 - \gamma\left[(p_1^2 + q_1^2)p_1 + p_2 - \mu_+ p_1\right] + \sqrt{2\gamma T_+}\xi_1' \\ \dot{q}_1 &= (p_1^2 + q_1^2)p_1 + p_2 - \gamma\left[(p_1^2 + q_1^2)q_1 + q_2 - \mu_+ q_1\right] + \sqrt{2\gamma T_+}\xi_1'' \quad , \end{aligned} \tag{1.52}$$

where γ measures the coupling strength with the thermal bath, while ξ_1' and ξ_1'' define two independent white noises with unit variance. It can be easily seen that the deterministic components of the thermostat, are gradient terms. As a result, in the absence of thermal noise, they would drive the system towards a state characterized by a minimal $(H - \mu A)$. Notice the nonlinear structure of the dissipation terms in Eq. (1.52).

An additional problem of the DNLS model is the determination of the temperature, as one cannot rely on the usual kinetic definition (this is again a consequence of the nonseparable Hamiltonian). An operative definition can be, however, given by adopting the microcanonical approach [83], i.e. by invoking the thermodynamic relationships,

$$T^{-1} = \frac{\partial \mathcal{S}}{\partial H} \quad , \quad \frac{\mu}{T} = -\frac{\partial \mathcal{S}}{\partial A} \, ,$$

where $\mathcal{S}$ is the entropy. As shown in [35, 47], the partial derivative $\partial\mathcal{S}/\partial C_i$ ($i = 1, 2$, with $C_1 = H$ and $C_2 = A$) can be computed by exploiting the fact that C_i is a conserved quantity,

$$\frac{\partial \mathcal{S}}{\partial C_i} = \left\langle \frac{W\|\boldsymbol{\xi}\|}{\nabla C_i \cdot \boldsymbol{\xi}} \nabla \cdot \left(\frac{\boldsymbol{\xi}}{\|\boldsymbol{\xi}\| W} \right) \right\rangle \tag{1.53}$$

where $\langle\ \rangle$ stands for the microcanonical average,

$$\xi = \frac{\nabla C_1}{\|\nabla C_1\|} - \frac{(\nabla C_1 \cdot \nabla C_2)\nabla C_2}{\|\nabla C_1\|\|\nabla C_2\|^2} \tag{1.54}$$

$$W^2 = \sum_{\substack{m,n=1\\ m<n}}^{2N} \left[\frac{\partial C_1}{\partial x_m}\frac{\partial C_2}{\partial x_m} - \frac{\partial C_1}{\partial x_n}\frac{\partial C_2}{\partial x_m} \right]^2 ,$$

and $x_{2n} = q_n, x_{2n+1} = p_n$. The resulting definitions of T and μ have the unpleasant property of being nonlocal: numerical simulations, however, show that they give meaningful results even when they are implemented for relatively short subchains.

As for the fluxes, they are naturally defined from the continuity equations for energy and norm

$$j_n^e = \dot{q}_n q_{n-1} + \dot{p}_n q_{n-1} \qquad j_n^p = q_n p_{n-1} - p_n q_{n-1} , \tag{1.55}$$

Notice that for the sake of simplicity we still use the same notations as in the previous setup although here j_n^p denotes the flux of norm/mass rather than momentum.

If one sets $T_+ = T_- = 0$, as in the XY model, the control parameter, i.e. the driving force, is given by $\delta\mu = |\mu_- - \mu_+|/2$ [48]. When $\delta\mu$ is larger than a critical value (that here depends on A), a bumpy temperature profile spontaneously emerges. As shown in Fig. 1.11, the left-right symmetry of the profile found in the XY model is lost, but the width of the peak still scales as $N^{1/2}$. A second crucial difference is the scaling behavior of the norm-flux, which decreases as $N^{-2/5}$ instead of $N^{-1/5}$. This suggests that more than one universality class is presumably present: the symmetry of the profile might play a crucial role.

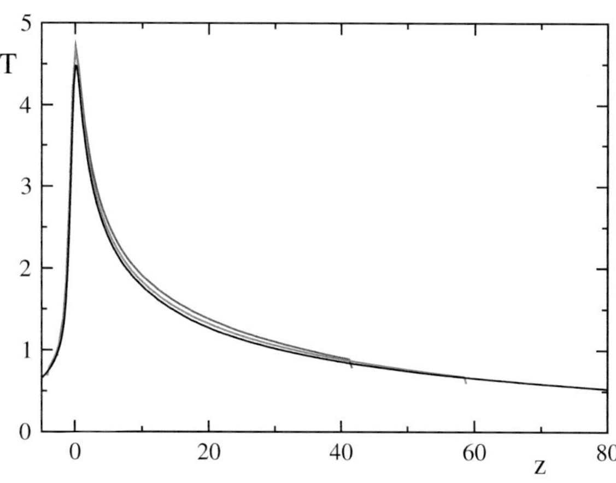

Fig. 1.11 Temperature profiles of the DNLS equation for 2000, 4000, 8000 and $T = 0$, $\mu_+ = 2$ and $\mu_- = 5$; $z = (n - \hat{n})/\sqrt{N}$, where $\hat{n}$ is the site with the highest temperature

In coupled transport, each conservation law implies the presence of a corresponding thermodynamic variable. In the case of the DNLS equation, there are two of them: the temperature T (or, equivalently β) and the chemical potential μ. If the extrema of a given system are "attached" to two different points in the (μ, T) space, a new question arises with respect to the transport of just one variable: the selection of the path in the phase plane. This problem can be solved with the help of the linear transport equations (1.48), which can be rewritten as

$$\frac{d\beta}{d\mu} = \frac{\langle j^e \rangle \beta L_{pp} - \langle j^p \rangle \beta L_{ep}}{\langle j^e \rangle \left(L_{pe} - \mu L_{pp}\right) - \langle j^p \rangle \left(L_{ee} - \mu L_{ep}\right)} \,. \tag{1.56}$$

The above first order differential equation can be solved once the Onsager matrix is known across the thermodynamics phase-diagram and the ratio of the two fluxes is given. This determines unambiguously the resulting temperature and chemical-potential profiles.

It is worth recalling that in the absence of a mutual coupling between the two transport processes (zero off-diagonal elements of the Onsager matrix) such curves would be vertical and horizontal lines in the latter representation. It is remarkable that the solid lines, which correspond to $j^e = 0$, are almost vertical for large μ: this means that in spite of a large temperature difference, the energy flux is very small. This is an indirect but strong evidence that the nondiagonal terms are far from negligible.

The condition of a vanishing particle flux $j^p = 0$ defines the Seebeck coefficient which is $S = -d\mu/dT$. Accordingly, the points in Fig. 1.12, where the dashed

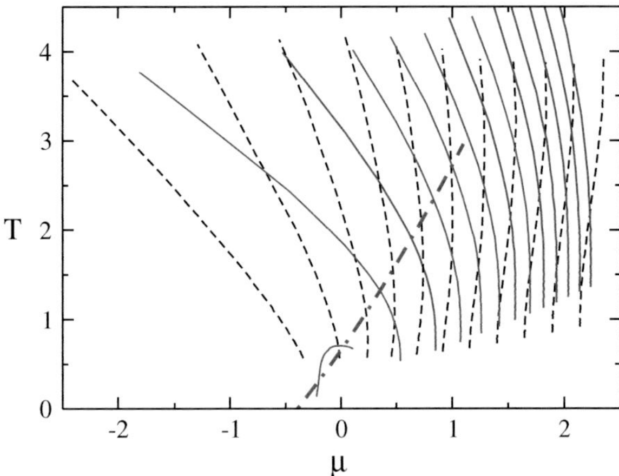

Fig. 1.12 Zero-flux curves in the (μ, T) planes. *Black dashed lines* correspond to $j^p = 0$ and are obtained with norm-conserving thermostats upon fixing the total norm density a_{tot}, T_L and T_R. *Blue solid lines* are for $j^e = 0$ using energy-conserving thermostats with fixed total energy density h_{tot}, μ_L/T_L and μ_R/T_R. Simulations are for a chain of length $N = 500$. The *thick dot-dashed lines* identify the locus where S changes sign (see text) (Color figure online)

curves are vertical identify the locus where S changes sign. The $j^e = 0$ curves have no direct interpretation in terms of standard transport coefficients.

1.9 Conclusions and Open Problems

In the previous sections we have seen that various theoretical approaches predict the existence of two universality classes for the divergence of heat conductivity in systems characterized by momentum conservation. Although this scenario is generally confirmed by numerical simulations, some exceptions have been found as well. The most notable counterexample is the normal conduction which emerges in chains of coupled rotors. As we have already discussed in Sect. 1.5, it is quite clear that the peculiarity of this model is to be traced back to the 2π-slips of the angles q_n.

Further, less-understood, anomalies have been found in models where q_n is a genuine displacement variable. One example is a momentum conserving modification of the famous "ding-a-ling" model. The system is composed of two kinds of alternating point particles (A and B): the A particles mutually interact via nearest-neighbour harmonic forces; the B particles are free to move and collide elastically with the A particles. Equilibrium and non-equilibrium numerical simulations indicate that the thermal conductivity κ is finite [59].

Normal heat transport in accordance to Fourier law has been claimed also in simulations of the FPU-$\alpha\beta$model (and of other asymmetric potentials), at low-enough energies/temperatures [97]. More detailed numerical simulations, however, indicate that the unexpected results for asymmetric potentials do not represent the asymptotic behavior [16, 91], but rather follow from an insufficient chain length. This if further strengthened in [56] where mode-coupling arguments have been used to determine the frequency below which finite-size effects are negligible. It turns out that, in some cases, the asymptotic behavior may only be seen at exceedingly low frequencies (and thereby exceedingly large system-sizes).

More recent studies report a finite thermal conductivity in the thermodynamic limit for potentials that allow for bond dissociation (like e.g Lennard-Jones, Morse, and Coulomb potentials) [37, 85]. This is explained by invoking phonon scattering on the locally strongly-stretched loose interatomic bonds at low temperature and by the many-particle scattering at high temperature. Nevertheless, the hard-point gas, a model where "dissociation" arises automatically, without the need to overcome an energy barrier, is found to exhibit a clean divergence of the conductivity. On the other hand, the universality of scaling in this model has been recently challenged by numerical studies of the hard-point gas with alternate masses and thermal baths at different temperatures acting at the boundaries. When the mass ratio is varied, the anomalous exponent is found to depart significantly from the value 1/3 predicted by the nonlinear fluctuating hydrodynamics [45].

Irrespective whether the above discrepancies are a manifestation of strong finite-size corrections, or of the existence of other universality classes, where the standard hydrodynamic theories do not apply, they have to be explained.

Acknowledgements We wish to thank L. Delfini and S. Iubini for their effective contribution to the achievement of several results summarized in this chapter.

References

1. Aoki, K., Kusnezov, D.: Bulk properties of anharmonic chains in strong thermal gradients: non-equilibrium ϕ^4 theory. Phys. Lett. A **265**(4), 250 (2000)
2. Aoki, K., Kusnezov, D.: Fermi-Pasta-Ulam β model: boundary jumps, Fourier's law, and scaling. Phys. Rev. Lett. **86**(18), 4029–4032 (2001)
3. Barabási, A.L., Stanley, H.E.: Fractal Concepts in Surface Growth. Cambridge University Press, Cambridge (1995)
4. Basile, G., Bernardin, C., Olla, S.: Momentum conserving model with anomalous thermal conductivity in low dimensional systems. Phys. Rev. Lett. **96**, 204303 (2006)
5. Basile, G., Delfini, L., Lepri, S., Livi, R., Olla, S., Politi, A.: Anomalous transport and relaxation in classical one-dimensional models. Eur. Phys. J.: Spec. Top. **151**, 85–93 (2007)
6. Basko, D.: Weak chaos in the disordered nonlinear Schrödinger chain: destruction of Anderson localization by Arnold diffusion. Ann. Phys. **326**(7), 1577–1655 (2011)
7. Blumen, A., Zumofen, G., Klafter, J.: Transport aspects in anomalous diffusion: Lévy walks. Phys. Rev. A **40**(7), 3964–3973 (1989)
8. Bonetto, F., Lebowitz, J.L., Rey-Bellet, L.: Fourier's law: a challenge to theorists. In: Fokas, A., Grigoryan, A., Kibble, T., Zegarlinsky, B. (eds.) Mathematical Physics 2000, p. 128. Imperial College, London (2000)
9. Casati, G.: Energy transport and the Fourier heat law in classical systems. Found. Phys. **16**(1), 51–61 (1986)
10. Casati, G., Prosen, T.: Anomalous heat conduction in a one-dimensional ideal gas. Phys. Rev. E **67**(1), 015203 (2003)
11. Casati, G., Ford, J., Vivaldi, F., Visscher, W.M.: One-dimensional classical many-body system having a normal thermal conductivity. Phys. Rev. Lett. **52**(21), 1861–1864 (1984)
12. Casati, G., Mejía-Monasterio, C., Prosen, T.: Increasing thermoelectric efficiency: a dynamical systems approach. Phys. Rev. Lett. **101**(1), 016601 (2008). doi:10.1103/PhysRevLett.101.016601
13. Casati, G., Wang, L., Prosen, T.: A one-dimensional hard-point gas and thermoelectric efficiency. J. Stat. Mech.: Theory Exp. **2009**(03), L03004 (2009)
14. Chang, C.W., Okawa, D., Garcia, H., Majumdar, A., Zettl, A.: Breakdown of Fourier's law in nanotube thermal conductors. Phys. Rev. Lett. **101**(7), 075903 (2008). doi:10.1103/PhysRevLett.101.075903
15. Cipriani, P., Denisov, S., Politi, A.: From anomalous energy diffusion to Lévy walks and heat conductivity in one-dimensional systems. Phys. Rev. Lett. **94**(24), 244301 (2005)
16. Das, S., Dhar, A., Narayan, O.: Heat conduction in the α-β Fermi-Pasta-Ulam chain. J. Stat. Phys. **154**(1–2), 204–213 (2014)
17. Das, S.G., Dhar, A., Saito, K., Mendl, C.B., Spohn, H.: Numerical test of hydrodynamic fluctuation theory in the Fermi-Pasta-Ulam chain. Phys. Rev. E **90**(1), 012124 (2014)
18. Delfini, L., Lepri, S., Livi, R.: A simulation study of energy transport in the Hamiltonian XY model. J. Stat. Mech.: Theory Exp. **2005**, P05006 (2005)
19. Delfini, L., Lepri, S., Livi, R., Politi, A.: Self-consistent mode-coupling approach to one-dimensional heat transport. Phys. Rev. E **73**(6), 060201(R) (2006)

20. Delfini, L., Denisov, S., Lepri, S., Livi, R., Mohanty, P.K., Politi, A.: Energy diffusion in hard-point systems. Eur. Phys. J.: Spec. Top. **146**, 21–35 (2007)
21. Delfini, L., Lepri, S., Livi, R., Politi, A.: Anomalous kinetics and transport from 1D self-consistent mode-coupling theory. J. Stat. Mech.: Theory Exp. **2007**, P02007 (2007)
22. Delfini, L., Lepri, S., Livi, R., Politi, A.: Comment on "equilibration and universal heat conduction in Fermi-Pasta-Ulam chains". Phys. Rev. Lett. **100**(19), 199401 (2008)
23. Delfini, L., Lepri, S., Livi, R., Politi, A.: Nonequilibrium invariant measure under heat flow. Phys. Rev. Lett. **101**(12), 120604 (2008)
24. Denisov, S., Klafter, J., Urbakh, M.: Dynamical heat channels. Phys. Rev. Lett. **91**(19), 194301 (2003)
25. Denisov, S., Zaburdaev, V., Hänggi, P.: Lévy walks with velocity fluctuations. Phys. Rev. E **85**(3), 031148 (2012)
26. Dhar, A.: Heat transport in low-dimensional systems. Adv. Phys. **57**, 457–537 (2008)
27. Dhar, A., Saito, K., Derrida, B.: Exact solution of a Lévy walk model for anomalous heat transport. Phys. Rev. E **87**, 010103 (2013)
28. Eckmann, J.P., Hairer, M.: Non-equilibrium statistical mechanics of strongly anharmonic chains of oscillators. Commun. Math. Phys. **212**(1), 105–164 (2000)
29. Eckmann, J.P., Pillet, C.A., Rey-Bellet, L.: Non-equilibrium statistical mechanics of anharmonic chains coupled to two heat baths at different temperatures. Commun. Math. Phys. **201**, 657 (1999)
30. Eilbeck, J.C., Lomdahl, P.S., Scott, A.C.: The discrete self-trapping equation. Physica D **16**, 318–338 (1985)
31. Eleftheriou, M., Lepri, S., Livi, R., Piazza, F.: Stretched-exponential relaxation in arrays of coupled rotators. Physica D **204**(3), 230–239 (2005)
32. Fermi, E., Pasta, J., Ulam, S.: Studies of nonlinear problems. Los Alamos Report LA-1940, p. 978 (1955)
33. Flach, S., Miroshnichenko, A., Fistul, M.: Wave scattering by discrete breathers. Chaos **13**(2), 596–609 (2003)
34. Flaschka, H.: The toda lattice. II. Existence of integrals. Phys. Rev. B **9**(4), 1924 (1974)
35. Franzosi, R.: Microcanonical entropy and dynamical measure of temperature for systems with two first integrals. J. Stat. Phys. **143**, 824–830 (2011)
36. Gendelman, O.V., Savin, A.V.: Normal heat conductivity of the one-dimensional lattice with periodic potential of nearest-neighbor interaction. Phys. Rev. Lett. **84**(11), 2381–2384 (2000)
37. Gendelman, O., Savin, A.: Normal heat conductivity in chains capable of dissociation. Europhys. Lett. **106**(3), 34004 (2014)
38. Giardiná, C., Livi, R., Politi, A., Vassalli, M.: Finite thermal conductivity in 1D lattices. Phys. Rev. Lett. **84**(10), 2144–2147 (2000)
39. Gillan, M., Holloway, R.: Transport in the Frenkel-Kontorova model 3: thermal-conductivity. J. Phys. C **18**(30), 5705–5720 (1985)
40. Grassberger, P., Nadler, W., Yang, L.: Heat conduction and entropy production in a one-dimensional hard-particle gas. Phys. Rev. Lett. **89**(18), 180601 (2002)
41. Hatano, T.: Heat conduction in the diatomic toda lattice revisited. Phys. Rev. E **59**, R1–R4 (1999)
42. Helfand, E.: Transport coefficients from dissipation in a canonical ensemble. Phys. Rev. **119**(1), 1–9 (1960). doi:10.1103/PhysRev.119.1
43. Hénon, M.: Integrals of the Toda lattice. Phys. Rev. B **9**(4), 1921 (1974)
44. Hu, B., Li, B., Zhao, H.: Heat conduction in one-dimensional chains. Phys. Rev. E **57**(3), 2992 (1998)
45. Hurtado, P.I., Garrido, P.L.: Violation of universality in anomalous Fourier's law (2015). arXiv preprint arXiv:1506.03234
46. Iacobucci, A., Legoll, F., Olla, S., Stoltz, G.: Negative thermal conductivity of chains of rotors with mechanical forcing. Phys. Rev. E **84**(6), 061108 (2011)
47. Iubini, S., Lepri, S., Politi, A.: Nonequilibrium discrete nonlinear Schrödinger equation. Phys. Rev. E **86**(1), 011108 (2012)

48. Iubini, S., Lepri, S., Livi, R., Politi, A.: Off-equilibrium Langevin dynamics of the discrete nonlinear Schrödinger chain. J. Stat. Mech.: Theory Exp. **2013**(08), P08017 (2013)
49. Iubini, S., Lepri, S., Livi, R., Politi, A.: Boundary-induced instabilities in coupled oscillators. Phys. Rev. Lett. **112**, 134101 (2014)
50. Kaburaki, H., Machida, M.: Thermal-conductivity in one-dimensional lattices of Fermi-Pasta-Ulam type. Phys. Lett. A **181**(1), 85–90 (1993)
51. Kevrekidis, P.G.: The Discrete Nonlinear Schrödinger Equation. Springer, Berlin (2009)
52. Klafter, J., Blumen, A., Shlesinger, M.F.: Stochastic pathway to anomalous diffusion. Phys. Rev. A **35**(7), 3081–3085 (1987). doi:10.1103/PhysRevA.35.3081
53. Kubo, R., Toda, M., Hashitsume, N.: Statistical Physics II. Springer Series in Solid State Sciences, vol. 31. Springer, Berlin (1991)
54. Larralde, H., Leyvraz, F., Mejía-Monasterio, C.: Transport properties of a modified Lorentz gas. J. Stat. Phys. **113**, 197–231 (2003)
55. Lee-Dadswell, G.: Universality classes for thermal transport in one-dimensional oscillator systems. Phys. Rev. E **91**(3), 032102 (2015)
56. Lee-Dadswell, G.R.: Predicting and identifying finite-size effects in current spectra of one-dimensional oscillator chains. Phys. Rev. E **91**, 012138 (2015)
57. Lee-Dadswell, G.R., Nickel, B.G., Gray, C.G.: Thermal conductivity and bulk viscosity in quartic oscillator chains. Phys. Rev. E **72**(3), 031202 (2005)
58. Lee-Dadswell, G.R., Nickel, B.G., Gray, C.G.: Detailed examination of transport coefficients in cubic-plus-quartic oscillator chains. J. Stat. Phys. **132**(1), 1–33 (2008)
59. Lee-Dadswell, G., Turner, E., Ettinger, J., Moy, M.: Momentum conserving one-dimensional system with a finite thermal conductivity. Phys. Rev. E **82**(6), 061118 (2010)
60. Lepri, S.: Relaxation of classical many-body Hamiltonians in one dimension. Phys. Rev. E **58**(6), 7165–7171 (1998)
61. Lepri, S., Politi, A.: Density profiles in open superdiffusive systems. Phys. Rev. E **83**(3), 030107 (2011)
62. Lepri, S., Livi, R., Politi, A.: Heat conduction in chains of nonlinear oscillators. Phys. Rev. Lett. **78**(10), 1896–1899 (1997)
63. Lepri, S., Livi, R., Politi, A.: On the anomalous thermal conductivity of one-dimensional lattices. Europhys. Lett. **43**(3), 271–276 (1998)
64. Lepri, S., Livi, R., Politi, A.: Thermal conduction in classical low-dimensional lattices. Phys. Rep. **377**, 1 (2003)
65. Lepri, S., Livi, R., Politi, A.: Universality of anomalous one-dimensional heat conductivity. Phys. Rev. E **68**(6, Pt 2), 067102 (2003). doi:10.1103/PhysRevE.68.067102
66. Lepri, S., Sandri, P., Politi, A.: The one-dimensional Lennard-Jones system: collective fluctuations and breakdown of hydrodynamics. Eur. Phys. J. B **47**(4), 549–555 (2005)
67. Lepri, S., Mejía-Monasterio, C., Politi, A.: Stochastic model of anomalous heat transport. J. Phys. A: Math. Theor. **42**, 025001 (2009)
68. Lippi, A., Livi, R.: Heat conduction in two-dimensional nonlinear lattices. J. Stat. Phys. **100**(5–6), 1147–1172 (2000)
69. Liu, S., Hänggi, P., Li, N., Ren, J., Li, B.: Anomalous heat diffusion. Phys. Rev. Lett. **112**(4), 040601 (2014)
70. Lukkarinen, J., Spohn, H.: Anomalous energy transport in the FPU-β chain. Commun. Pure Appl. Math. **61**(12), 1753–1786 (2008). doi:http://dx.doi.org/10.1002/cpa.20243
71. Mareschal, M., Amellal, A.: Thermal-conductivity in a one-dimensional Lennard-Jones chain by molecular-dynamics. Phys. Rev. A **37**(6), 2189–2196 (1988)
72. Maruyama, S.: A molecular dynamics simulation of heat conduction in finite length SWNTs. Physica B: Condens. Matter **323**(1), 193–195 (2002)
73. Mazur, P.: Non-ergodicity of phase functions in certain systems. Physica **43**(4), 533–545 (1969)
74. Meier, T., Menges, F., Nirmalraj, P., Hölscher, H., Riel, H., Gotsmann, B.: Length-dependent thermal transport along molecular chains. Phys. Rev. Lett. **113**(6), 060801 (2014)

75. Mejía-Monasterio, C., Larralde, H., Leyvraz, F.: Coupled normal heat and matter transport in a simple model system. Phys. Rev. Lett. **86**(24), 5417–5420 (2001)
76. Nakazawa, H.: On the lattice thermal conduction. Prog. Theor. Phys. Suppl. **45**, 231–262 (1970)
77. Narayan, O., Ramaswamy, S.: Anomalous heat conduction in one-dimensional momentum-conserving systems. Phys. Rev. Lett. **89**(20), 200601 (2002)
78. Nickel, B.: The solution to the 4-phonon Boltzmann equation for a 1D chain in a thermal gradient. J. Phys. A-Math. Gen. **40**(6), 1219–1238 (2007). doi:10.1088/1751-8113/40/6/003
79. Payton, D., Rich, M., Visscher, W.: Lattice thermal conductivity in disordered harmonic and anharmonic crystal models. Phys. Rev. **160**(3), 706 (1967)
80. Pereverzev, A.: Fermi-Pasta-Ulam β lattice: Peierls equation and anomalous heat conductivity. Phys. Rev. E **68**(5), 056124 (2003). doi:10.1103/PhysRevE.68.056124
81. Politi, A.: Heat conduction of the hard point chain at zero pressure. J. Stat. Mech.: Theory Exp. **2011**, P03028 (2011)
82. Pomeau, Y., Résibois, P.: Time dependent correlation functions and mode-mode coupling theories. Phys. Rep. **19**(2), 63–139 (1975)
83. Rugh, H.H.: Dynamical approach to temperature. Phys. Rev. Lett. **78**(5), 772 (1997)
84. Saito, K., Benenti, G., Casati, G.: A microscopic mechanism for increasing thermoelectric efficiency. Chem. Phys. **375**, 508–513 (2010)
85. Savin, A.V., Kosevich, Y.A.: Thermal conductivity of molecular chains with asymmetric potentials of pair interactions. Phys. Rev. E **89**(3), 032102 (2014)
86. Scheipers, J., Schirmacher, W.: Mode-coupling theory for the lattice dynamics of anharmonic crystals: self-consistent damping and the 1D Lennard-Jones chain. Z. für Phys. B Condens. Matter **103**(3), 547–553 (1997)
87. Schilling, R.: Theories of the structural glass transition. In: Collective Dynamics of Nonlinear and Disordered Systems, pp. 171–202. Springer, Berlin (2005)
88. Spohn, H.: Nonlinear fluctuating hydrodynamics for anharmonic chains. J. Stat. Phys. **154**(5), 1191–1227 (2014)
89. van Beijeren, H.: Exact results for anomalous transport in one-dimensional hamiltonian systems. Phys. Rev. Lett. **108**, 180601 (2012)
90. Wang, L., Hu, B., Li, B.: Logarithmic divergent thermal conductivity in two-dimensional nonlinear lattices. Phys. Rev. E **86**, 040101 (2012)
91. Wang, L., Hu, B., Li, B.: Validity of fourier's law in one-dimensional momentum-conserving lattices with asymmetric interparticle interactions. Phys. Rev. E **88**, 052112 (2013)
92. Yang, L., Hu, B.: Comment on "normal heat conductivity of the one-dimensional lattice with periodic potential of nearest-neighbor interaction". Phys. Rev. Lett. **94**, 219404 (2005)
93. Yang, L., Grassberger, P., Hu, B.: Dimensional crossover of heat conduction in low dimensions. Phys. Rev. E **74**(6), 062101 (2006)
94. Zaburdaev, V., Denisov, S., Hänggi, P.: Perturbation spreading in many-particle systems: a random walk approach. Phys. Rev. Lett. **106**(18), 180601 (2011)
95. Zaburdaev, V., Denisov, S., Klafter, J.: Lévy walks. Rev. Mod. Phys. **87**(2), 483 (2015)
96. Zhao, H.: Identifying diffusion processes in one-dimensional lattices in thermal equilibrium. Phys. Rev. Lett. **96**(14), 140602 (2006)
97. Zhong, Y., Zhang, Y., Wang, J., Zhao, H.: Normal heat conduction in one-dimensional momentum conserving lattices with asymmetric interactions. Phys. Rev. E **85**, 060102 (2012)
98. Zoia, A., Rosso, A., Kardar, M.: Fractional Laplacian in bounded domains. Phys. Rev. E **76**(2), 21116 (2007)
99. Zotos, X.: Ballistic transport in classical and quantum integrable systems. J. Low Temp. Phys. **126**(3–4), 1185–1194 (2002)

Chapter 2
Heat Transport in Harmonic Systems

Abhishek Dhar and Keiji Saito

Abstract In this chapter we study heat transport in the simplest model of a solid, namely the harmonic crystal. We consider an open system, consisting of a harmonic crystal coupled to heat reservoirs at different temperatures, and examine properties of the nonequilibrium steady state. Both the average heat current and the current fluctuations are studied. We show that the formalisms, of quantum Langevin equations and non-equilibrium Green's function techniques, allows one to obtain many exact formal results for the current, as well as fluctuations, in terms of the phonon transmission matrix, in systems in arbitrary dimensions. We then show how these formal results can be used to obtain explicit results in many cases of interest, especially the case of heat transport in mass-disordered harmonic crystals. Apart from throwing light on the question of validity of Fourier's law, it is explained how the study of the harmonic crystal also provides insight on recent theories of non-equilibrium current fluctuatons.

2.1 Introduction

Energy transport in dielectric crystals at low temperatures is via phonons and the two main sources of scattering of the energy carriers are isotopic mass disorder and anharmonicities. For a crystal, say in the form of a rectangular slab, that is coupled to thermal heat baths at two opposite faces, the total energy transported may in addition depend on how the baths are coupled to the boundaries of the bulk. In this chapter we ignore the effect of anharmonicity on heat transport and focus on the two other effects in depth. Thus, in this chapter, we discuss various aspects of heat transport in a purely harmonic crystal.

A. Dhar (✉)
International Centre for Theoretical Sciences (TIFR), Survey No. 151, Shivakote, Hesaraghatta, Bengaluru 560089, India
e-mail: abhishek.dhar@icts.res.in

K. Saito
Department of Physics, Keio University, Yokohama 223-8522, Japan
e-mail: saitoh@rk.phys.keio.ac.jp

S. Lepri (ed.), *Thermal Transport in Low Dimensions*, Lecture Notes in Physics 921, DOI 10.1007/978-3-319-29261-8_2

The first aspect that we address is related to the question of the validity of Fourier's law of heat conduction. Fourier's law of heat conduction states that the heat current density $\mathbf{J}(\mathbf{x}, t)$ in a solid is proportional to the local temperature gradient, i.e., $\mathbf{J} = -\kappa \nabla T$, where the proportionality constant κ is the thermal conductivity of the material. It follows from Fourier's law that, for a slab in contact with heat reservoirs kept at a small temperature difference ΔT, the average energy current density, J (in the direction of transport), should be equal to $\kappa \Delta T/N$, with N being the slab length. There have been many attempts to derive Fourier's law from microscopic dynamics. In a crystal, heat is carried by lattice vibrations or phonons. Using very reasonable physical assumptions of local equilibrium and the notion of the mean free path traveled by phonons between collisions, leads to a heuristic derivation of Fourier's law. There is however no fully convincing derivation [1] and in fact, the results from a number of studies over the last 50 years, both numerical and analytical, suggest that Fourier's law may not be valid in low-dimensional systems [2, 3]. For example, for one-dimensional momentum conserving anharmonic crystals, one of the key findings is that the heat current density scales with system size as $J \sim 1/N^{1-\alpha}$ with $\alpha > 1$. This means that the effective thermal conductivity $\kappa_N = JN/\Delta T \sim N^{\alpha}$ diverges with the size of the system, a very surprising and unexpected result.

What do we expect for a purely harmonic crystal? As mentioned in the beginning paragraph, ignoring anharmonicity, there are two sources of scattering of phonons in the harmonic crystal, one due to some form of disorder in the system and the other due to scattering at the boundaries (between system and reservoirs). In the absence of disorder, the phonons pass ballistically through the bulk and we expect their mean free path to be of order N, hence the exponent $\alpha = 1$. With the introduction of disorder, do we get diffusive heat transport ($\alpha = 0$)? This question will be addressed in detail, and as we will see that this is not the case, even perhaps in dimensions three.

Two standard frameworks for studying heat transport are through:

(a) study of the non-equilibrium steady state. The system is connected to heat baths at two different temperatures and one measures properties in the non-equilibrium steady state, in particular the temperature profile and the energy current. The size-dependence of current indicates whether or not Fourier's law is valid.
(b) The second approach is to study equilibrium correlations of conserved quantities such as energy and see if they exhibit diffusive behavior. In particular, using the Green-Kubo formula, one expects to extract the thermal conductivity of the system from the energy current autocorrelation function.

For anharmonic systems, approach (a) is usually implemented via numerical simulations and analytic progress is difficult. However, for the harmonic case, the situation is very different, and there is a general framework whereby one can express all steady state properties in terms of non-equilibrium phonon Green's functions. This approach is closely related to the Landauer treatment for transport of non-interacting electrons. It is possible for the harmonic system basically because here we are looking at non-interacting phonons. Here we will focus almost entirely on this approach, and only briefly discuss method (b), and comment on its relation to the approach (a).

In this chapter we study not only the average heat current but also properties of current fluctuations in the non-equilibrium steady state. Recently much progress has been made on current fluctuations in the steady state, with significant developments such as non-equilibrium fluctuation relations, macroscopic fluctuation theory, and so on. The harmonic crystal is a tractable and realistic physical model for understanding and developing these theories. Analytic treatment of current fluctuations can be performed up to some level and finally one can incorporate numerical calculations for detailed investigations with high accuracy. We note that most studies on current fluctuation are based on stochastic models. Harmonic lattices are however Hamiltonian-based model and hence provide us with insights on the nature of current fluctuations in realistic systems. In this chapter, we stress that the cumulant generating function (CGF) plays a crucial role in the investigation of current fluctuations. The CGF systematically generates arbitrary order of cumulants of current and has full information on current fluctuations. We will outline the derivation of a compact formula of the CGF. With the use of this formula, we can illustrate universal properties, which are in accordance with recent predictions on non-equilibrium fluctuations.

Early work on the non-equilibrium steady state of harmonic chains connected to heat baths include the work of Rieder, Lebowitz and Lieb (RLL) [4], Casher and Lebowitz [5], Rubin and Greer [6] and O'Connor and Lebowitz [7]. RLL considered an ordered harmonic chain connected to white noise Langevin type heat baths, noted that the full steady state distribution is a Gaussian for which they were able to obtain the exact correlation matrix. From this, one could obtain the average energy current and temperature profile. The papers of Casher and Lebowitz [5] and Rubin and Greer [6] looked at disordered harmonic chains with two different models of baths. While Casher and Lebowitz [5] considered white noise Langevin baths, Rubin and Greer [6] had baths which were themselves semi-infinite harmonic chains prepared in equilibrium at different temperatures. All these studies were on classical systems. For the quantum description, the available methods that have been used include phenomenological Landauer type approach [8–10], Langevin equations approach [11, 12], non-equilibrium Green's function approach using the Keldysh formalism [13, 14], equation of motion approach [15, 16], master equation approach [17] and finally path-integral approach [18].

In this chapter, for the study of steady state correlations, we will focus on the approach, developed in [11], involving generalized Langevin equations. The Langevin equations can be obtained by starting from a purely Hamiltonian model for the bath, and the formalism is valid for both classical and quantum systems. The earlier models in [4–6] can be seen as special cases. In the quantum case, this approach gives the same result as the non-equilibrium Green's function methods [13, 14]. For the study of current fluctuations, the Langevin equations approach continues to be useful only for the case of classical systems where the baths generate un-correlated noise—for the quantum case, we will see that it becomes necessary to use the non-equilibrium Green's function method.

The plan of this chapter is as follows. In Sect. 2.2 we describe the generalized Langevin equations approach and use it to obtain formal expressions for the energy current and other correlations in the non-equilibrium steady state. In Sect. 2.3 we explain the importance of studying current fluctuations and derive formal expressions for the corresponding cumulant generating function in general harmonic systems. In Sect. 2.4, we discuss explicit results for various special cases of physical interest, that can be obtained from the formal expressions for current and fluctuations. We summarize our results in Sect. 2.5.

2.2 Steady State Current from Langevin Equations and Green's Function Formalism

In this section we discuss the open system set-up for studying transport, where the basic idea is to consider a finite system connected to two reservoirs with infinite degrees of freedom. The complete model of system and reservoirs is described by a Hamiltonian. Both the system and baths and their couplings are assumed to be described by terms that are quadratic in the positional and momentum degrees of freedom. Initially the system and reservoirs are decoupled and the left and right reservoirs are prepared in thermal equilibrium *at different temperatures*. The coupling between system and reservoirs is then switched on and the system evolves through Hamiltonian dynamics (unitary evolution in the quantum case). If we just look at the system, then at long times, it reaches a non-equilibrium steady state. This is possible only because one has infinite reservoirs which can serve as heat sources and sinks. Our aim is to understand the properties of this steady state. As we will see, for the harmonic system, one can essentially find the full reduced density matrix of the system corresponding to the non-equilibrium steady state. This naturally contains information about steady state current and temperature profiles.

In this section we illustrate the Langevin equation and Greens function (LEGF) method and show how it can lead to the full set of steady state correlations. In the following section, Sect. 2.3, we will see, how with the same open system set-up, we can also study fluctuations.

2.2.1 Derivation of the Langevin Equation and It's Steady-State Solution

We consider here the case of a system which is coupled to two heat reservoirs, labeled as L (for left) and R (right), which are at two different temperatures. Let us assume that the system has N Cartesian positional degrees of freedom $\{x_l\}$, $l = 1, 2 \ldots, N$ with corresponding momenta $\{p_l\}$. These satisfy the usual commutation relations $[x_l, p_m] = i\hbar\delta_{l,m}$ and $[x_l, x_m] = [p_l, p_m] = 0$. Similarly, the left reservoir degrees of freedom are denoted by $\{x^L_\alpha, p^L_\alpha\}$, $\alpha = 1, \ldots, N_L$ and the right reservoirs by $\{x^R_{\alpha'}, p^R_{\alpha'}\}$, $\alpha' = 1, \ldots, N_R$. Eventually we will consider the limits $N_L \to \infty, N_R \to \infty$. It is convenient to use the vector notation

$X^T = (x_1, x_2, \ldots, x_N)$, $P^T = (p_1, p_2, \ldots, p_N)$ and similarly X^L, X^R, P^L, P^R. The general Hamiltonian for the system coupled to harmonic reservoirs is then given by:

$$H = H_S + H_L + H_R + H_{SL} + H_{SR} , \tag{2.1}$$

$$\text{where } H_S = \frac{1}{2} P^T \boldsymbol{M}^{-1} P + \frac{1}{2} X^T \boldsymbol{K} X , \tag{2.2}$$

$$H_L = \frac{1}{2} {P_L}^T \boldsymbol{M}_L^{-1} P_L + \frac{1}{2} {X_L}^T \boldsymbol{K}_L X_L , \tag{2.3}$$

$$H_R = \frac{1}{2} {P_R}^T \boldsymbol{M}_R^{-1} P_R + \frac{1}{2} {X_R}^T \boldsymbol{K}_R X_R , \tag{2.4}$$

$$H_{SL} = X^T \boldsymbol{K}_{SL} X_L , \;\; H_{SR} = X^T \boldsymbol{K}_{SR} X_R , \tag{2.5}$$

where $\boldsymbol{M}$, $\boldsymbol{M}_L$, $\boldsymbol{M}_R$ and $\boldsymbol{K}$, $\boldsymbol{K}_L$, $\boldsymbol{K}_R$ denote respectively the mass matrix and the force-constant matrix of the system, left reservoir and right reservoir, while $\boldsymbol{K}_{SL}$ and $\boldsymbol{K}_{SR}$ denote the linear coupling coefficients between the two reservoirs and the system.

We assume that at time $t = t_0$, the system and reservoirs are decoupled and the reservoirs are in thermal equilibrium at temperatures T_L and T_R respectively. This means that at time t_0, the two reservoirs are described by thermal density matrices $\rho_L = e^{-\beta_L H_L}/Tr[e^{-\beta_L H_L}]$ and $\rho_R = e^{-\beta_R H_R}/Tr[e^{-\beta_R H_R}]$ respectively. Once the reservoir system coupling is switched on, it is expected that the system will in the long time limit reach a steady state, provided the reservoirs are infinite. Our main aim will be to compute the heat current and other properties of this non-equilibrium steady state. In our approach we will consider the limits $N_L, N_R \to \infty$ followed by $t_0 \to -\infty$ to obtain the steady state.

We now proceed to obtain the Langevin equations of motion for the system degrees of motion by eliminating the reservoir degrees. For this we consider a Heisenberg representation where each operator A is replaced by its Heisenberg form $e^{iHt/\hbar} A e^{-iHt/\hbar}$. The Heisenberg equations of motion for the system variables (for $t > t_0$) are then

$$\boldsymbol{M} \ddot{X} = -\boldsymbol{K} X - \boldsymbol{K}_{SL} X_L - \boldsymbol{K}_{SR} X_R , \tag{2.6}$$

and the equations of motion for the two reservoirs are

$$\boldsymbol{M}_L \ddot{X}_L = -\boldsymbol{K}_L X_L - \boldsymbol{K}_{SL}^T X , \tag{2.7}$$

$$\boldsymbol{M}_R \ddot{X}_R = -\boldsymbol{K}_R X_R - \boldsymbol{K}_{SR}^T X . \tag{2.8}$$

The strategy is now to solve the reservoir equations by considering them as linear inhomogeneous equations. Consider the left reservoir and let us perform a normal-mode transformation $\boldsymbol{U}_L$ which simultaneously diagonalizes both $\boldsymbol{M}_L$ and $\boldsymbol{K}_L$. Thus they satisfy the equations

$$\boldsymbol{U}_L^T \boldsymbol{K}_L \boldsymbol{U}_L = \boldsymbol{\Omega}_L^2 , \;\; \boldsymbol{U}_L^T \boldsymbol{M}_L \boldsymbol{U}_L = \boldsymbol{I} ,$$

where $\boldsymbol{\Omega}_L$ is a diagonal matrix whose elements $[\Omega_L]_q$ are the normal mode frequencies of the left reservoir. Note that transforming to normal mode coordinates $X_L = \boldsymbol{U}_L\tilde{X}_L$ and $P_L = \boldsymbol{M}_L U_L \tilde{P}_L$ transforms the Hamiltonian in Eq. (2.3) to the form

$$H_L = \frac{\tilde{P}_L^T \tilde{P}_L}{2} + \frac{\tilde{X}_L^T \boldsymbol{\Omega}_L^2 \tilde{X}_L}{2} , \tag{2.9}$$

which now implies a collection of uncoupled oscillators. The left-reservoir equation of motion Eq. (2.7), in normal-coordinates looks like

$$\ddot{\tilde{X}}_L = -\boldsymbol{\Omega}_L^2 \tilde{X}_L - \boldsymbol{U}_L^T \boldsymbol{K}_{SL}^T X ,$$

which has the general solution

$$\begin{aligned} \tilde{X}_L(t) = \cos[\Omega_L(t-t_0)]\tilde{X}_L(t_0) + \frac{\sin[\Omega(t-t_0)]}{\Omega_L(t-t_0)}\tilde{P}_L(t_0) \\ - \int_{t_0}^{t} dt' \frac{\sin[\Omega_L(t-t')]}{\Omega_L(t-t')} \boldsymbol{U}_L^T \boldsymbol{K}_{SL}^T X(t') . \end{aligned} \tag{2.10}$$

Transforming back to the original coordinates we get

$$\begin{aligned} X_L(t) = \boldsymbol{U}_L \cos[\Omega_L(t-t_0)]\tilde{X}_L(t_0) + \boldsymbol{U}_L \frac{\sin[\Omega_L(t-t_0)]}{\Omega_L(t-t_0)}\tilde{P}_L(t_0) \\ - \int_{t_0}^{\infty} dt' \boldsymbol{g}_L^+(t-t') \boldsymbol{K}_{SL}^T X(t') , \end{aligned} \tag{2.11}$$

$$\text{with } \boldsymbol{g}_L^+(t) = \boldsymbol{U}_L \frac{\sin(\boldsymbol{\Omega}_L t)}{\boldsymbol{\Omega}_L} \boldsymbol{U}_L^T \, \theta(t) , \tag{2.12}$$

where $\theta(t)$ is the Heaviside function. A similar solution is obtained for the right reservoir. Plugging these solutions back into the equation of motion for the system, Eq. (2.6), we get the following effective equations of motion for the system:

$$\begin{aligned} \boldsymbol{M}\ddot{X} = -\boldsymbol{K}X + \eta_L + \int_{t_0}^{\infty} dt' \boldsymbol{\Sigma}_L(t-t')X(t') \\ + \eta_R + \int_{t_0}^{\infty} dt' \boldsymbol{\Sigma}_R(t-t')X(t'), \end{aligned} \tag{2.13}$$

$$\text{where } \boldsymbol{\Sigma}_L(t) = \boldsymbol{K}_{SL}\, \boldsymbol{g}_L^+(t)\, \boldsymbol{K}_{SL}^T, \quad \boldsymbol{\Sigma}_R(t) = \boldsymbol{K}_{SR}\, \boldsymbol{g}_R^+(t)\, \boldsymbol{K}_{SR}^T,$$

$$\text{and } \eta_L = \boldsymbol{K}_{SL} \left[\boldsymbol{U}_L \cos[\boldsymbol{\Omega}_L(t-t_0)]\tilde{X}_L(t_0) + \boldsymbol{U}_L \frac{\sin[\boldsymbol{\Omega}_L(t-t_0)]}{\boldsymbol{\Omega}_L(t-t_0)}\tilde{P}_L(t_0) \right] \tag{2.14}$$

$$\eta_R = \boldsymbol{K}_{SR} \left[\boldsymbol{U}_R \cos[\boldsymbol{\Omega}_R(t-t_0)]\tilde{X}_R(t_0) + \boldsymbol{U}_R \frac{\sin[\boldsymbol{\Omega}_R(t-t_0)]}{\boldsymbol{\Omega}_R(t-t_0)}\tilde{P}_L(t_0) \right] . \tag{2.15}$$

This equation has the form of a generalized quantum Langevin equation. The properties of the noise terms η_L and η_R will be determined from the condition that, at time t_0, the two isolated reservoirs are described by equilibrium phonon distribution functions. At time t_0, the left reservoir is in equilibrium at temperature T_L and the population of the normal modes (of the isolated left reservoir) is given by the distribution function $f(\omega, T_L) = 1/[e^{\hbar\omega/k_B T_L} - 1]$. This means in particular that

$$\begin{aligned}\langle \tilde{X}_{Lq}(t_0)\tilde{X}_{Lq'}(t_0)\rangle &= \delta_{qq'}\,[f(\Omega_{Lq}, T_L) + 1/2]\,\hbar/\Omega_{Lq}\,,\\ \langle \tilde{P}_{Lq}(t_0)\tilde{P}_{Lq'}(t_0)\rangle &= \delta_{qq'}\,[f(\Omega_{Lq}, T_L) + 1/2]\,\hbar\Omega_{Lq}\\ \text{and } \langle \tilde{X}_{Lq}\tilde{P}_{Lq'}\rangle &= -\langle \tilde{P}_{Lq}\tilde{X}_{Lq'}\rangle = \delta_{qq'}\, i\hbar/2 \end{aligned} \tag{2.16}$$

From these, we immediately get the noise correlations of the left reservoir noise:

$$\begin{aligned}\langle \eta_L(t)\eta_L^T(t')\rangle = \boldsymbol{K}_{SL}\boldsymbol{U}_L &\left[\cos \boldsymbol{\Omega}_L(t-t')\frac{\hbar}{2\boldsymbol{\Omega}_L}\coth\left(\frac{\hbar\boldsymbol{\Omega}_L}{2k_BT_L}\right)\right.\\ &\left. -i\sin \boldsymbol{\Omega}_L(t-t')\frac{\hbar}{2\boldsymbol{\Omega}_L}\right]\boldsymbol{U}_L^T\boldsymbol{K}_{SL}^T\,,\end{aligned} \tag{2.17}$$

and a similar expression for the right reservoir. The limits of infinite reservoir sizes $(N_L, N_R \to \infty)$ and $t_0 \to -\infty$ are now taken. One can reduce Eq. (2.13) to the formal classical form of the Langevin equation. Defining $\boldsymbol{\gamma}$ such that $\boldsymbol{\Sigma}(t) = -d\boldsymbol{\gamma}(t)/dt$, performing integration by parts and assuming that $\boldsymbol{\gamma}(t \to \infty) = 0$, we get

$$\boldsymbol{M}\ddot{X} = -\boldsymbol{K}_S X + \sum_{\alpha=L,R}\left[-\int_{-\infty}^{t} dt'\boldsymbol{\gamma}^{(\alpha)}(t-t')\dot{X}(t') + \eta_\alpha\right], \tag{2.18}$$

where $\boldsymbol{K}_S = \boldsymbol{K} - \boldsymbol{K}_{SL}\boldsymbol{K}_L^{-1}\boldsymbol{K}_{SL}^T - \boldsymbol{K}_{SR}\boldsymbol{K}_R^{-1}\boldsymbol{K}_{SR}^T$ and $\boldsymbol{\gamma}^{(\alpha)}(t) = \boldsymbol{K}_{S\alpha}\boldsymbol{U}_\alpha\frac{\cos[\Omega_\alpha t]}{\Omega_\alpha^2}\boldsymbol{U}_\alpha^T\boldsymbol{K}_{S\alpha}^T$. One can think of $\boldsymbol{K}_S$ as the effective coupling constant matrix for the system.

One can then solve Eq. (2.13) by taking Fourier transforms. Let us define the Fourier transforms $\tilde{X}(\omega) = (1/2\pi)\int_{-\infty}^{\infty} dt\; X(t)e^{i\omega t}$, $\tilde{\eta}_{L,R}(\omega) = (1/2\pi)\int_{-\infty}^{\infty} dt\;\eta_{L,R}(t)e^{i\omega t}$, $\boldsymbol{g}_{L,R}^+(\omega) = \int_{-\infty}^{\infty} dt\;\boldsymbol{g}_{L,R}^+(t)e^{i\omega t}$. One then gets the following stationary solution to the equations of motion Eq. (2.13):

$$\begin{aligned} X(t) &= \int_{-\infty}^{\infty} d\omega\tilde{X}(\omega)e^{-i\omega t}\,,\\ \text{with}\quad \tilde{X}(\omega) &= \boldsymbol{G}^+(\omega)\,[\tilde{\eta}_L(\omega) + \tilde{\eta}_R(\omega)]\,,\\ \text{where}\quad \boldsymbol{G}^+(\omega) &= \frac{1}{[-\omega^2\boldsymbol{M} + \boldsymbol{K} - \boldsymbol{\Sigma}_L^+(\omega) - \boldsymbol{\Sigma}_R^+(\omega)]}\,,\\ \text{and}\quad \boldsymbol{\Sigma}_L^+(\omega) &= \boldsymbol{K}_{SL}\boldsymbol{g}_L^+(\omega)\boldsymbol{K}_{SL}^T,\quad \boldsymbol{\Sigma}_R^+(\omega) = \boldsymbol{K}_{SR}\boldsymbol{g}_R^+(\omega)\boldsymbol{K}_{SR}^T\,.\end{aligned} \tag{2.19}$$

Taking a Fourier transform of Eq. (2.12) we get:

$$\mathbf{g}_L^+(\omega) = \frac{1}{-\boldsymbol{M}_L(\omega + i\epsilon)^2 + \boldsymbol{K}_L}, \tag{2.20}$$

which is the usual definition of the Phonon Green's function. The noise correlations, in the frequency domain, can be obtained from Eq. (2.17) and we get (for the left reservoir):

$$\langle \tilde{\eta}_L(\omega)\tilde{\eta}_L^T(\omega')\rangle = \delta(\omega + \omega')\, \boldsymbol{\Gamma}_L(\omega)\, \frac{\hbar}{\pi}[1 + f(\omega, T_L)] \tag{2.21}$$

$$\text{where} \quad \boldsymbol{\Gamma}_L(\omega) = Im[\boldsymbol{\Sigma}_L^+(\omega)]$$

which is a fluctuation-dissipation relation. This also leads to the more commonly used correlation:

$$\frac{1}{2}\langle\, \tilde{\eta}_L(\omega)\tilde{\eta}_L^T(\omega') + \tilde{\eta}_L(\omega')\tilde{\eta}_L^T(\omega)\,\rangle = \delta(\omega + \omega')\, \boldsymbol{\Gamma}_L(\omega)\, \frac{\hbar}{2\pi} \coth(\frac{\hbar\omega}{2k_B T_L}).$$

Similar relations hold for the noise from the right reservoir. The identification of $\boldsymbol{G}^+(\omega)$ as a phonon Green function, with $\boldsymbol{\Sigma}_{L,R}^+(\omega)$ as self energy contributions coming from the baths, is one of the main steps that enables a comparison of results derived by the Langevin equations and Green's function approach (LEGF), with those obtained from the non-equilibrium Green's function method. More details of this demonstration can be found in [11].

Some Examples of Baths We mention here two well-known bath models. We only discuss local baths which are applied independently to every boundary particle. They are specified by the following bath functions:

(a) Ohmic (also referred to as white noise or delta-correlated noise): $\Sigma^+(\omega) = i\gamma\omega$.
(b) Rubin model: $\Sigma^+(\omega) = k\{1 - m\omega^2/2k + i\omega(m/k)^{1/2}[1 - m\omega^2/(4k)]^{1/2}\}$. In this model the bath is an ordered harmonic chain with inter-particle spring constant k and equal masses m and coupling to the system is also through a spring with stiffness k. Note that by defining $m\omega^2 = 2k(1 - \cos q)$, we can write $\Sigma^+ = ke^{iq}$.

2.2.2 *Steady State Properties*

Current The steady state current will be the same if we obtain it on any cross-section of the system. It is simplest to evaluate the current expression at the system-reservoir interface. Let us consider the interface with the left reservoir. The steady state current can be shown [3] to be just the expectation value of the rate at which the Langevin forces (which come from the bath) do work on the particles of the system

$$I = \left\langle \sum_{l=1}^{N} \dot{x}_l(t) \left[[\eta_L]_l(t) + \int_{-\infty}^{\infty} dt' \, [\boldsymbol{\Sigma}_L^+]_{l,m}(t-t') \, x_m(t') \right] \right\rangle .$$

It is most convenient to carry out the computations by writing the above equation in matrix form as

$$\begin{aligned} I &= \langle \dot{X}^T(t) \, \eta_L(t) \rangle + \int_{-\infty}^{\infty} dt' \langle \dot{X}^T(t) \, \boldsymbol{\Sigma}_L^+(t-t') \, X(t') \rangle \\ &= -i \int_{-\infty}^{\infty} d\omega \int_{-\infty}^{\infty} d\omega' \, e^{-i(\omega+\omega')t} \, \omega \, \langle \, [\tilde{X}^T(\omega) \, \eta_L(\omega') + \tilde{X}^T(\omega) \, \boldsymbol{\Sigma}_L^+(\omega')\tilde{X}(\omega') \,] \, \rangle \\ &= -i \int_{-\infty}^{\infty} d\omega \int_{-\infty}^{\infty} d\omega' \, e^{-i(\omega+\omega')t} \, \omega \, \langle \, Tr[\, \eta_L(\omega') \, \tilde{X}^T(\omega) \,] \\ &\quad + Tr[\, \boldsymbol{\Sigma}_L^+(\omega') \, \tilde{X}(\omega') \, \tilde{X}^T(\omega) \,] \, \rangle . \end{aligned}$$

Using the solution in Eq. (2.19) we then get

$$\begin{aligned} I = -i \int_{-\infty}^{\infty} d\omega \int_{-\infty}^{\infty} d\omega' \, e^{-i(\omega+\omega')t} \, \omega \, \Big(Tr\Big[\, \langle \, \eta_L(\omega') \, \{ \, \eta_L^T(\omega) + \eta_R^T(\omega) \, \} \, \rangle \, \boldsymbol{G}^+(\omega) \Big] \\ + \, Tr\Big[\, \boldsymbol{\Sigma}_L^+(\omega') \, \boldsymbol{G}^+(\omega') \, \langle \{ \, \eta_L(\omega') + \eta_R(\omega') \, \} \{ \, \eta_L^T(\omega) + \eta_R^T(\omega) \, \} \rangle \, \boldsymbol{G}^+(\omega) \Big] \Big) , \end{aligned}$$

where we have used the fact that $\boldsymbol{G}^+$ is a symmetric matrix. If we expand the above, and use the properties of the noise correlations (including the fact that the noise from the left and right baths are uncorrelated) , we notice that all terms depend either on the left, or on the right bath temperature, and no term depends on both temperatures. Let us collect those terms in I which depend only on the temperature T_R of the right reservoir, and not on that of the left. These give

$$\begin{aligned} I_R = -i \int_{-\infty}^{\infty} d\omega \int_{-\infty}^{\infty} d\omega' \, e^{-i(\omega+\omega')t} \, \omega \, Tr \\ [\, \boldsymbol{\Sigma}_L^+(\omega') \, \boldsymbol{G}^+(\omega') \, \langle \, \eta_R(\omega') \, \eta_R^T(\omega) \, \rangle \, \boldsymbol{G}^+(\omega) \,] \, . \end{aligned}$$

Using the noise correlations in Eq. (2.21) we then get

$$I_R = -\frac{i}{\pi} \int_{-\infty}^{\infty} d\omega \, \hbar\omega \, Tr[\, \boldsymbol{\Sigma}_L^+(-\omega) \, \boldsymbol{G}^+(-\omega) \, \boldsymbol{\Gamma}_R(\omega) \, \boldsymbol{G}^+(\omega) \,] f(\omega, T_R) \, ,$$

where we used the properties $1+f(-\omega) = -f(\omega)$ and $\Gamma_R(-\omega) = -\Gamma_R(\omega)$. Taking the complex conjugate of the above expression, using $Tr[A]^* = Tr[A^\dagger]$, $G^\dagger(\omega) = G(-\omega)$, and the cyclic property of trace, we get

$$I_R^* = \frac{i}{\pi}\int_{-\infty}^{\infty} d\omega\, \hbar\omega\, Tr[\,\boldsymbol{\Sigma}_L^-(-\omega)\,\boldsymbol{G}^+(-\omega)\,\boldsymbol{\Gamma}_R(\omega)\,\boldsymbol{G}^+(\omega)\,]\, f(\omega, T_R)\,.$$

The physical current is real, so we take $I_R = (I_R + I_R^*)/2$ and, noting that $Im[\boldsymbol{\Sigma}_L^+(-\omega)] = \boldsymbol{\Gamma}_L(-\omega) = -\boldsymbol{\Gamma}_L(\omega)$, we get

$$I_R = -\frac{1}{\pi}\int_{-\infty}^{\infty} d\omega\, \hbar\omega\, Tr[\boldsymbol{G}^+(\omega)\,\boldsymbol{\Gamma}_L(\omega)\,\boldsymbol{G}^-(\omega)\,\boldsymbol{\Gamma}_R(\omega)\,]\, f(\omega, T_R)\,.$$

From symmetry, and the requirement that $I = 0$ when $T_L = T_R$, it is clear that the contribution of the left reservoir to the current must have a similar form (this can be verified directly also) and hence we can now write the net current as

$$I = \frac{1}{4\pi}\int_{-\infty}^{\infty} d\omega\, \hbar\omega\, \mathcal{T}(\omega)\, [f(\omega, T_L) - f(\omega, T_R)]\,, \tag{2.22}$$

$$\text{where } \mathcal{T}(\omega) = 4\, Tr[\boldsymbol{G}^+(\omega)\,\boldsymbol{\Gamma}_L(\omega)\,\boldsymbol{G}^-(\omega)\,\boldsymbol{\Gamma}_R(\omega)\,]\,. \tag{2.23}$$

We shall refer to this as the Landauer formula for the current and this is an important and very useful expression. The formula above can also be written as an integral over only positive frequencies using the fact that the integrand is an even function of ω. The factor $\mathcal{T}(\omega)$ is the transmission coefficient of phonons at frequency ω through the system, from the left to right reservoir and can be related to the transmittance of plane waves of frequency ω across the system [19]. Note that the transmission depends not only on the system parameters, such as the mass matrix and the force constant matrix, but also the on bath properties and system-reservoir couplings.

The general expression for current is of identical form as that obtained by the non-equilibrium Green's function approach for electron current (see for example [20–22]) and has also been derived for phonons using the non-equilibrium Green's function approach in [13, 14]. In fact this expression was first proposed in [8] and derived in [9, 10] for a $1D$ channel using the Landauer approach. The usual Landauer result for a $1D$ channel precisely corresponds to Rubin's model of bath, to be discussed in Sect. 2.4.1.2. For the case of classical one-dimensional systems the above result can be shown to lead to expressions for current and temperature profiles obtained in earlier studies of heat conduction in disordered harmonic chains [5–7, 23].

Two special limiting cases of Eq. (2.22) are:

(1) Linear response: for small temperature differences $\Delta T = T_L - T_R << T$, where $T = (T_L + T_R)/2$, the above expression reduces to:

$$I = \frac{\Delta T}{2\pi}\int_0^{\infty} d\omega\, \mathcal{T}(\omega)\hbar\omega\, \frac{\partial f(\omega, T)}{\partial T}\,. \tag{2.24}$$

(2) Classical limit: this is obtained by taking the high temperature limit $\hbar\omega/k_BT \to 0$. This gives:

$$I = \frac{k_B\, \Delta T}{2\pi} \int_0^\infty d\omega\, \mathcal{T}(\omega)\, . \tag{2.25}$$

Other Steady State Correlations One can similarly evaluate various other quantities such as velocity-velocity correlations and position-velocity correlations. These are given by Dhar and Roy [11], Yamamoto and Watanabe [13] and Wang et al. [14]

$$\begin{aligned} \langle XX^T \rangle &= \int_{-\infty}^{\infty} d\omega \frac{\hbar}{2\pi} \sum_{a=L,R} \boldsymbol{G}^+ \boldsymbol{\Gamma}_a \boldsymbol{G}^-\, g(\omega, T_a)\, , \\ \langle PP^T \rangle &= \int_{-\infty}^{\infty} d\omega \frac{\hbar\omega^2}{2\pi} \sum_{a=L,R} \boldsymbol{M}\boldsymbol{G}^+ \boldsymbol{\Gamma}_a \boldsymbol{G}^- \boldsymbol{M}\, g(\omega, T_a)\, , \\ \frac{1}{2}\langle XP^T + [PX^T]^T \rangle &= \int_{-\infty}^{\infty} d\omega \frac{i\hbar\omega}{\pi} \sum_{a=L,R} \boldsymbol{G}^+ \boldsymbol{\Gamma}_a \boldsymbol{G}^- \boldsymbol{M}\, g(\omega, T_a)\, , \end{aligned} \tag{2.26}$$

where $g(\omega, T) = \coth(\hbar\omega/2k_BT)$. The second set of correlation functions above can be used to define the local kinetic energy density which can in turn be used to define the temperature profile in the non-equilibrium steady state of the wire. Note that the correlations between X and P contain the local energy current on various bonds and one can recover from this the steady state current expression in Eq. (2.22). Also, since this is essentially a Gaussian problem, the steady state reduced density matrix is known exactly once we know the full set of correlations. The procedure for obtaining the system density matrix is described in [24].

In our derivation of the LEGF results we have *implicitly assumed that a unique steady state will be reached*. One of the necessary conditions for this is that no modes outside the bath spectrum are generated for the combined model of system and baths. These modes, when they exist, are localized near the system and any initial excitation of the mode is unable to decay. This has been demonstrated and discussed in detail in the electronic context [22].

We note that unlike other approaches such as the Green-Kubo formalism and Boltzmann equation approach, the Langevin equation approach explicitly includes the reservoirs. The Langevin equation is physically appealing since it gives a nice picture of the reservoirs as sources of noise and dissipation. Also, just as the Landauer formalism and non-equilibrium Green's function have been extremely useful in understanding electron transport in mesoscopic systems, it is likely that a similar description will be useful for the case of heat transport in (electrically) insulating nanotubes, nanowires, etc.

2.3 Steady State Current Fluctuations

2.3.1 *General Theory*

2.3.1.1 Large Deviations and Cumulant Generating Functions

In general, heat is a stochastic variable and hence it shows fluctuations in time. In this section, we consider the fluctuation of heat transfer by introducing several variables. In the steady state we measure heat transfer at the contact between the system and one heat reservoir during a finite measurement time τ. The length of measurement time is in general arbitrary, but usually the asymptotic behavior at large τ is most interesting. Let Q be the amount of heat transferred during time τ. Let us consider the probability of heat transfer $P(Q)$. Moments of heat transfer of arbitrary orders can be calculated from this probability

$$\langle Q^n \rangle = \int_{-\infty}^{\infty} dQ\, Q^n P(Q) \,. \tag{2.27}$$

If $\hat{j}$ is the heat transfer between the reservoir (say the left one) and the system, then $Q = \int_0^\tau dt \hat{j}(t)$, and we then expect the asymptotic distribution of Q to satisfy the large deviation form

$$P(Q) \sim e^{-\tau F(q)} \tag{2.28}$$

$$q = Q/\tau \,. \tag{2.29}$$

The function $F(q)$ is called the *large deviation function* (LDF) which is analogous to the free energy in equilibrium statistical mechanics. We define the *cumulant generating function* (CGF) as

$$\mu(\lambda) = \lim_{\tau\to\infty} \frac{1}{\tau} \ln\langle e^{\lambda Q}\rangle \,. \tag{2.30}$$

The CGF systematically generates arbitrary orders of cumulants on taking derivative with respect to λ

$$I_n := \langle Q^n \rangle_c/\tau = \frac{\partial^n \mu(\lambda)}{\partial \lambda^n}\Big|_{\lambda=0} \,. \tag{2.31}$$

For large measurement time, heat transfer Q is of order τ. A reasonable idea to characterize fluctuations of heat transfer is to consider the cumulants of Q, instead of the moments, since they are also of order τ.

The LDF and the CGF are the central functions in the investigation of fluctuations [25]. These quantities are in general connected to each other by the Legendre-Fenchel transform [26].

$$F(q) = \max_{\lambda} [\lambda q - \mu(\lambda)] \,, \tag{2.32}$$

$$\mu(\lambda) = \max_{q} [\lambda q - F(q)] \,. \tag{2.33}$$

This relation is referred to as the Gärtner-Ellis theorem [27, 28], which states that, when the scaled cumulant generating function $\mu(\lambda)$ is differentiable, then Q obeys a large deviation principle with a LDF given by the Legendre-Fenchel transform.

2.3.1.2 Fluctuation-Response Relation and Fluctuation Relation

The thermodynamic stochastic variables such as work and heat are not completely random but satisfy several universal relations. For instance, current fluctuations in equilibrium is connected to the response to external thermodynamic forces [29, 30]. This general relationships is referred to as a fluctuation-response relation. Consider heat transfer from the left reservoir to the system for the case of $T_L \geq T_R$. In the case of heat conduction, the fluctuation-response relation is mathematically formulated as a relationship between second order cumulant of heat transfer in equilibrium and thermal conductance defined in the linear response regime

$$I_2^{\mathrm{eq}} = 2k_B T^2 G(T) \,, \tag{2.34}$$

where I_2^{eq} is the second order cumulant of heat transfer in equilibrium at temperature T and $G(T) = I_1/\Delta T$, with $\Delta T = T_L - T_R$, the thermal conductance. This relation is similar to the Green-Kubo formula [29, 30] relating thermal conductivity κ to the equilibrium auto-correlation function of current. The difference of Eq. (2.34) with the usual Green-Kubo formula is that here the fluctuations are for a system coupled to equal temperature heat baths while, in the Green-Kubo case, one considers an isolated system [31].

The steady state fluctuation relation (SSFR) is expected to hold even in the far-from-equilibrium regime [32–38] and is described by the relation

$$\lim_{\tau \to \infty} \frac{1}{\tau} \ln [P(Q)/P(-Q)] = \mathcal{A} q \,, \tag{2.35}$$

where $\mathcal{A} = \frac{1}{k_B T_R} - \frac{1}{k_B T_L}$ is a thermodynamic force or affinity, which plays the role of a force driving the system into a non-equilibrium state. Since this is a relation for asymptotic behavior of large τ, we can use the LDF expression to get the expression in terms of the LDF, namely, $-F(q) = -F(-q) + \mathcal{A} q$. In addition, the relation (2.33) leads to [25, 33]

$$\mu(\lambda) = \mu(-\lambda - \mathcal{A}) \,. \tag{2.36}$$

This is another expression of the SSFR and is referred to as fluctuation relation symmetry. However, we should note that the relation (2.36) as well as the expression with the LDF for the SSFR does not always hold for the whole regime and the validity is limited for finite ranges $\lambda_- \leq \lambda \leq \lambda_+$ [39, 40]. This effect arises when the function $\langle e^{\lambda Q} \rangle = g(\lambda) e^{\tau \mu(\lambda)}$ has singularities in $g(\lambda)$.

The fluctuation relation symmetry is not only mathematically beautiful but yields physically important relations [41–43]. To see this, we expand the nth order of cumulant of heat transfer with respect to the affinity $\mathcal{A}$

$$I_n = \sum_{k=0}^{\infty} L_{n,k} \frac{\mathcal{A}^k}{k!} . \tag{2.37}$$

The coefficient $L_{n,k}$ is a nonlinear response coefficient. For instance $L_{1,1} = T^2 \kappa(T)$ and $L_{2,0} = I_2^{\rm eq}$. Obviously the nonlinear response coefficient $L_{n,k}$ is calculated by the relation $L_{n,k} = \partial^k I_n / \partial \mathcal{A}^k |_{\mathcal{A}=0}$. Then using the CGF and the fluctuation relation symmetry (2.36) lead to

$$\begin{aligned} L_{n,k} &= \frac{\partial^k}{\partial \mathcal{A}^k} \frac{\partial^n}{\partial \lambda^n} \mu(\lambda) \Big|_{\mathcal{A}=\lambda=0} \\ &= \frac{\partial^k}{\partial \mathcal{A}^k} \frac{\partial^n}{\partial \lambda^n} \mu(-\lambda - \mathcal{A}) \Big|_{\mathcal{A}=\lambda=0} \\ &= (-1)^n \frac{\partial^n}{\partial \xi^n} \left[\frac{\partial}{\partial \mathcal{A}} - \frac{\partial}{\partial \xi} \right]^k \mu(\xi) \Big|_{\mathcal{A}=\xi=0} \qquad (\xi = -\lambda - \mathcal{A}) \\ &= \sum_{m=0}^{k} (-1)^{n+m} {}_kC_m \, L_{n+m,k-m} . \end{aligned} \tag{2.38}$$

This gives us highly nontrivial relations between nonlinear response coefficients and provides a lot of information on the non-equilibrium process. By setting $n = 2\ell + 1$ and $k = 0$, one finds

$$L_{2\ell+1,0} = 0. \tag{2.39}$$

This relation implies that odd orders of cumulants at equilibrium must vanish. Intriguingly the relation (2.38) reproduces the fluctuation-response relation (2.34). This can be checked by setting $n = k = 1$ yielding

$$L_{1,1} = \frac{1}{2} L_{2,0} , \tag{2.40}$$

which is equivalent to (2.34). In addition, $n = 1$ and $k = 2$ yields the simple relation

$$L_{1,2} = L_{2,1} . \tag{2.41}$$

In this manner, the fluctuation relation symmetry produces nontrivial relations. These kind of relations can be extended to the case of electrical conduction in the presence of magnetic fields [43] and experiments have been conducted to demonstrate them [44].

2.3.1.3 Macroscopic Fluctuation Theory and Additivity Principle

For the case of dissipative transport where Fourier's law is satisfied, macroscopic fluctuation theory (MFT) was proposed by Bertini et al. [45]. The MFT describes fluctuation of temperature and current fields for large system size. Let $q(x,t)$ and $T(x,t)$ be the heat current density and the local temperature respectively, at the position x and time t, and $\kappa(T)$ be the thermal conductivity at the temperature T. Then, MFT predicts that the probability of finding $\{q(x,t), T(x,t)\}$ is

$$P(\{T(x,t), q(x,t)\}) \sim \exp\Big[-N \int_0^{\tau} dt \int_0^1 dx \, \frac{[q(x,t) + \kappa(T(x,t))\partial T(x,t)/\partial x]^2}{2\kappa(T(x,t))\, T^2(x,t)}\Big]. \tag{2.42}$$

Here N is the system size and x is a position scaled by N. This expression has been used in many systems [46]. To find the LDF of heat transfer for open boundary condition with $T(x=0) = T_L$ and $T(x=1,t) = T_R$, we minimize (2.42) with respect to the temperature and current fields, with the constraint of the conservation law of energy

$$F(\{T(x)\}) = \lim_{\tau\to\infty} \min_{\substack{\{T(x,t), q(x,t)\} \\ T(0)=T_L, T(1)=T_R}} \frac{N}{\tau} \int_0^{\tau/N^2} dt \int_0^1 dx \, \frac{[q(x,t) + \kappa(T(x,t))\partial T(x,t)/\partial x]^2}{2\kappa(T(x,t))\, T^2(x,t)}. \tag{2.43}$$

Variational problem is in general difficult to solve, but Bertini et al. succeeded in deriving the expression in terms of Hamilton-Jacobi type of equation for fluctuating variables [45, 46].

If current and temperature fields in the optimal profile do not depend on time, then the MFT is reduced to the additivity principle (AP) proposed by Bodineau and Derrida [47]. The LDF here is given by

$$F_{AP}(q) = \frac{1}{N} \min_{T(x)} \Big[\int_0^1 dx \, \frac{[qN + \kappa(T(x))\partial T(x)/\partial x]^2}{2\kappa(T(x))\, T^2(x)}\Big]. \tag{2.44}$$

From the relation (2.33), the CGF is derived

$$\mu_{AP}(\lambda, T_L, T_R) = -\frac{K}{N}\Big[\int_{T_R}^{T_L} dT \frac{\kappa(T)}{\sqrt{1+4K\kappa(T)T^2}}\Big] \tag{2.45}$$

$$\lambda = \int_{T_R}^{T_L} dT \frac{1}{2T^2}\Big[\frac{1}{\sqrt{1+4K\kappa(T)T^2}} - 1\Big]. \tag{2.46}$$

The additivity principle enables one to compute all higher orders of current cumulants given just the temperature dependent thermal conductivity $\kappa(T)$ of a system.

However the condition to get the AP from the MFT is highly nontrivial. For the ring geometry in equilibrium situation, Bertini and co-workers derived the following sufficient condition to get AP from the viewpoint of the MFT [48, 49]

$$\kappa(T)[\kappa(T)T^2]'' \leq [\kappa(T)]'[\kappa(T)T^2]' . \tag{2.47}$$

However, the necessary and sufficient condition for the validity of the AP in general case has not yet been clarified. It was confirmed that the AP is consistent with the exact expressions of several orders of current cumulants in the symmetric simple exclusion process [47]. The AP was numerically verified in another stochastic system, namely for heat transport in the Kipnis-Marchioro-Presutti model [50], by measuring rare events with a sophisticated algorithm. Surprisingly, it was found that the 3D harmonic crystal also exhibits the AP and gives further indications on the necessary conditions required for its validity [51]. This is further discussed in Sect. 2.4.3.

2.3.2 *Cumulant Generating Function for Harmonic Lattices*

We here derive the expression of the cumulant generating function (CGF) for harmonic lattices. Since there are several issues on which one needs to be careful in the quantum regime, we divide this section into two parts, classical [52, 53] and quantum [43, 54]. In the quantum case, even the definition of heat transfer in a given time τ needs careful interpretation, and as we will see, the analytic treatment requires use of different techniques. In the classical case, the LEGF approach suffices, while in the quantum case, the non-equilibrium Green's function method (NEGF) has to be used.

2.3.2.1 Classical Regime

The classical case was first considered in [52] for a one-dimensional chain connected to white-noise baths, and later generalized to arbitrary dimensions in [53]. Here we outline the treatment of Saito and Dhar [53].

We consider the harmonic lattices attached to the left and right heat reservoirs via surfaces of the left and right edges. The number of sites attached to the left and right reservoirs are respectively N_L and N_R. The harmonic network structure in the bulk system is arbitrary. The dynamics is given by Eq. (2.18) with memory-less friction and classical noises and one can write the equations of motion in the following form

$$\begin{aligned} \dot{X} &= V \\ \boldsymbol{M}\dot{V} &= -\boldsymbol{K}X - \left[\boldsymbol{\gamma}^{(L)} + \boldsymbol{\gamma}^{(R)}\right] V + \eta_L + \eta_R \,, \end{aligned} \tag{2.48}$$

where $\boldsymbol{\gamma}^{(\alpha)}$ ($\alpha = L, R$) is a diagonal friction matrix. The friction matrix $\boldsymbol{\gamma}^{(\alpha)}$ is a constant matrix, where only the element representing the sites attached to reservoirs have finite value. The classical noise η_α satisfies the fluctuation dissipation relation

$$\langle \eta_\alpha(t)\eta_{\alpha'}^T(t')\rangle = 2\delta_{\alpha,\alpha'}k_B T_\alpha \boldsymbol{\gamma}^{(\alpha)}\delta(t-t') \qquad \alpha,\alpha' = L, R \tag{2.49}$$

We measure heat transfer from the left reservoir to the system during the measurement time τ which is represented as

$$Q = \int_0^\tau dt\, V^T(-\boldsymbol{\gamma}^{(L)}V + \eta_L) \,. \tag{2.50}$$

Then we derive the CGF for this heat transfer. We take the following discrete Fourier transforms

$$\{X(t), V(t), \eta_\alpha(t)\} = \sum_{n=-\infty}^{\infty} \{\tilde{X}(\omega_n), \tilde{V}(\omega_n), \tilde{\eta}_\alpha(\omega_n)\} e^{-i\omega_n t} \,, \tag{2.51}$$

where $\omega_n = 2\pi n/\tau$. We multiply Eq. (2.48) by $e^{i\omega_n t}$, and integrate by parts. Solving the resulting equations for $\tilde{X}(\omega_n)$, $\tilde{V}(\omega_n)$, one gets

$$\begin{aligned} \tilde{V}(\omega_n) &= -i\omega_n \boldsymbol{G}^+(\omega_n)\left[\tilde{\eta}_L(\omega_n) + \tilde{\eta}_R(\omega_n)\right] \\ &\quad + \frac{1}{\tau}\boldsymbol{G}^+(\omega_n)\ \left[\boldsymbol{K}\Delta X + i\omega_n \boldsymbol{M}\Delta V\right] \,, \end{aligned} \tag{2.52}$$

$$\boldsymbol{G}^+(\omega_n) = \left[-\boldsymbol{M}\omega_n^2 + \boldsymbol{K} - \boldsymbol{\Sigma}_L(\omega_n) - \boldsymbol{\Sigma}_R(\omega_n)\right]^{-1} \,, \tag{2.53}$$

where the self-energy matrices are given by $\boldsymbol{\Sigma}_{L,R}(\omega) = i\omega\boldsymbol{\gamma}^{(L,R)}$, $\Delta X = X(\tau) - X(0)$, and $\Delta V = V(\tau) - V(0)$. The matrix $\boldsymbol{G}^{+}$ is the Green's function connecting bulk variables with reservoir properties, which is the classical version of the green function in Eq. (2.19). The noise correlations in the Fourier space are given by

$$\langle \tilde{\eta}_{\alpha}(\omega_n)\tilde{\eta}^{T}_{\alpha'}(\omega_{n'})\rangle = 2\delta_{\alpha,\alpha'}\delta_{n,-n'}\boldsymbol{\gamma}^{(\alpha)}k_B T_{\alpha}/\tau\,. \tag{2.54}$$

Since the noise strength $\tilde{\eta}_{\alpha}(\omega_n) \sim O(1/\tau^{1/2})$ and $\Delta X, \Delta V \sim O(1)$ we see that the second term in Eq. (2.52) is $\sim 1/\tau^{1/2}$ order smaller than the first and so can be dropped [52]. We note that $\tilde{V}^{*}(\omega_n) = \tilde{V}(-\omega_n)$ and $\tilde{\eta}^{*}(\omega_n) = \tilde{\eta}(-\omega_n)$. Then the heat transferred Q can be expressed in terms of the Fourier-modes with $n \geq 0$ as

$$Q = \sum_{n=0}^{\infty}[-2\tilde{V}^{T}(\omega_n)\boldsymbol{\gamma}^{(L)}\tilde{V}^{*}(\omega_n) + \tilde{\eta}^{T}_{L}(\omega_n)\tilde{V}^{*}(\omega_n) + \tilde{V}^{T}(\omega_n)\tilde{\eta}^{*}_{L}(\omega_n)]. \tag{2.55}$$

We define $\boldsymbol{G}^{-}(\omega) = \boldsymbol{G}^{+}(-\omega) = [\boldsymbol{G}^{+}(\omega)]^{*}$ and $\boldsymbol{\Gamma}_{L,R}(\omega) = \mathrm{Im}\{\boldsymbol{\Sigma}_{L,R}(\omega)\} = \omega\boldsymbol{\gamma}^{(L,R)}$. Noting the relation $V(\omega_n) = -i\omega_n\boldsymbol{G}^{+}(\omega_n)\,[\tilde{\eta}_L(\omega_n) + \tilde{\eta}_R(\omega_n)]$, we can express Q as

$$Q = \tau\sum_{n=0}^{\infty}(\tilde{\eta}^{T}_{L}(\omega_n), \tilde{\eta}^{T}_{R}(\omega_n))\,\boldsymbol{A}\begin{pmatrix}\tilde{\eta}^{*}_{L}(\omega_n)\\ \tilde{\eta}^{*}_{R}(\omega_n)\end{pmatrix}, \tag{2.56}$$

where the column and row vectors composed of $\tilde{\eta}_L$ and $\tilde{\eta}_R$ noises are now defined only with nonzero elements, and hence these vectors have $(N_L + N_R)$ elements. The matrix $\boldsymbol{A}$ is $(N_L + N_R) \times (N_L + N_R)$ square Hermitian matrix which is given by

$$\boldsymbol{A} = \begin{pmatrix} 2\omega_n[\boldsymbol{G}^{+}\boldsymbol{\Gamma}_R\boldsymbol{G}^{-}]_{LL} & i\omega_n[\boldsymbol{G}^{-}]_{LR} - 2\omega_n[\boldsymbol{G}^{+}\boldsymbol{\Gamma}_L\boldsymbol{G}^{-}]_{LR} \\ -i\omega_n[\boldsymbol{G}^{+}]_{RL} - 2\omega_n[\boldsymbol{G}^{+}\boldsymbol{\Gamma}_L\boldsymbol{G}^{-}]_{RL} & -2\omega_n[\boldsymbol{G}^{+}\boldsymbol{\Gamma}_L\boldsymbol{G}^{-}]_{RR} \end{pmatrix}. \tag{2.57}$$

The ω_n dependence of $\boldsymbol{G}$ and $\boldsymbol{\Gamma}$ have been suppressed. In what follows in this section, the ω_n dependence in variables is omitted unless it is necessary. The subscript L and R in the matrices respectively represent the space of the sites attached to the left and right reservoirs, respectively. In obtaining the (L, L) element of the matrix $\boldsymbol{A}$, we have used the following Green's function identity

$$\boldsymbol{G}^{+} - \boldsymbol{G}^{-} = 2i\boldsymbol{G}^{-}(\boldsymbol{\Gamma}_L + \boldsymbol{\Gamma}_R)\boldsymbol{G}^{+} = 2i\boldsymbol{G}^{+}(\boldsymbol{\Gamma}_L + \boldsymbol{\Gamma}_R)\boldsymbol{G}^{-}. \tag{2.58}$$

Using the expression (2.56), we calculate the CGF. We formally express the CGF with integration with respect to noise variables

$$\mu(\lambda) = \frac{1}{\tau}\ln\langle e^{\lambda Q}\rangle = \frac{1}{\tau}\ln\Big\{\mathcal{N}\prod_{n\geq 0}\int d[\tilde{\eta}_L^T, \tilde{\eta}_R^T, (\tilde{\eta}_L^*)^T, (\tilde{\eta}_R^*)^T]$$
$$\times \exp\Big[\ \tau\,(\tilde{\eta}_L, \tilde{\eta}_R)\Big[\lambda \boldsymbol{A} - \begin{pmatrix} \frac{1}{2k_BT_L}[\bar{\boldsymbol{\gamma}}^{(L)}]^{-1} & 0 \\ 0 & \frac{1}{2k_BT_R}[\bar{\boldsymbol{\gamma}}^{(R)}]^{-1}\end{pmatrix}\Big]\begin{pmatrix}\tilde{\eta}_L^* \\ \tilde{\eta}_R^*\end{pmatrix}\Big]\Big\}, \tag{2.59}$$

where $\mathcal{N}$ denotes the normalization factor of the noise distribution and $\bar{\boldsymbol{\gamma}}^{(L,R)}$ are diagonal matrices which are composed only with the nonzero matrix elements of $\boldsymbol{\gamma}^{(L,R)}$. In the above expression, the Gaussian measure of Langevin noises are involved. By performing the Gaussian integral, one obtains the formal expression of the CGF.

$$\mu(\lambda) = -\frac{1}{\tau}\sum_{n\geq 0}\ln\det \boldsymbol{B}\Big|_{\tau\to\infty}, \tag{2.60}$$

$$\boldsymbol{B} = \boldsymbol{1} - \lambda\begin{pmatrix}\mathcal{T}_L & \mathcal{T}_{LR} \\ \mathcal{T}_{RL} & -\mathcal{T}_R\end{pmatrix}\begin{pmatrix} k_BT_L\boldsymbol{1} & \boldsymbol{0} \\ \boldsymbol{0} & k_BT_R\boldsymbol{1}\end{pmatrix}, \tag{2.61}$$

where

$$\begin{aligned}
\mathcal{T}_L &= 4[\boldsymbol{G}^+\boldsymbol{\Gamma}_R\boldsymbol{G}^-\boldsymbol{\Gamma}_L]_{LL}, \\
\mathcal{T}_R &= 4[\boldsymbol{G}^+\boldsymbol{\Gamma}_L\boldsymbol{G}^-\boldsymbol{\Gamma}_R]_{RR}, \\
\mathcal{T}_{LR} &= -4[\boldsymbol{G}^+\boldsymbol{\Gamma}_L\boldsymbol{G}^-\boldsymbol{\Gamma}_R]_{LR} + 2i[\boldsymbol{G}^-\boldsymbol{\Gamma}_R]_{LR}, \\
\mathcal{T}_{RL} &= -4[\boldsymbol{G}^+\boldsymbol{\Gamma}_L\boldsymbol{G}^-\boldsymbol{\Gamma}_L]_{RL} - 2i[\boldsymbol{G}^+\boldsymbol{\Gamma}_L]_{RL}.
\end{aligned} \tag{2.62}$$

The matrices $\mathcal{T}_L$ and $\mathcal{T}_R$ are respectively $N_L \times N_L$ and $N_R \times N_R$ square matrices, and $\mathcal{T}_{LR}$ and $\mathcal{T}_{RL}$ are respectively $N_L \times N_R$ and $N_R \times N_L$ rectangular matrices. The matrices $\mathcal{T}_L$ and $\mathcal{T}_R$ can be regarded as transmission amplitude of energy with the mode ω_n from one reservoir to the other. These matrices appear in the Landauer formula for the steady state current (2.23). Although the physical meaning of $\mathcal{T}_{LR}$ and $\mathcal{T}_{RL}$ is not very clear, these are closely related to $\mathcal{T}_L$ and $\mathcal{T}_R$. The relations can be revealed by using the relation (2.58) iteratively. Through tedious but straightforward calculations, one finds the following nontrivial relations:

$$\mathcal{T}_{LR}\mathcal{T}_R = \mathcal{T}_L\mathcal{T}_{LR}, \tag{2.63}$$

$$\mathcal{T}_{RL}\mathcal{T}_L = \mathcal{T}_R\mathcal{T}_{RL}, \tag{2.64}$$

$$\mathcal{T}_{RL}\mathcal{T}_{LR} = \mathcal{T}_R(\boldsymbol{1} - \mathcal{T}_R). \tag{2.65}$$

$$\mathcal{T}_{LR}\mathcal{T}_{RL} = \mathcal{T}_L(\boldsymbol{1} - \mathcal{T}_L). \tag{2.66}$$

In order to get simple form of the CGF, we need to simplify the determinant of $\boldsymbol{B}$ in Eq. (2.60). The relations (2.63)–(2.66) play a central role in simplifying the determinant of $\boldsymbol{B}$ and in deriving the final expression of the CGF. We heuristically introduce the matrix $\boldsymbol{C}$:

$$\boldsymbol{C} = \begin{pmatrix} \mathbf{1} & \boldsymbol{C}_{LR} \\ \mathbf{0} & \boldsymbol{C}_{RR} \end{pmatrix}$$

$$\boldsymbol{C}_{LR} = \lambda k_B T_R \mathcal{T}_{LR} + \frac{T_R}{T_L} \mathcal{T}_L \mathcal{T}_{LR} \tag{2.67}$$

$$\boldsymbol{C}_{RR} = \mathbf{1} + \left(\frac{1}{\lambda k_B T_L} - \lambda k_B T_L - 1 \right) \mathcal{T}_R + \mathcal{T}_{RL} \mathcal{T}_{LR} = \left(\mathbf{1} + \frac{\mathcal{T}_R}{\lambda k_B T_L} \right) (\mathbf{1} - \lambda k_B T_L \mathcal{T}_R) . \tag{2.68}$$

The advantage of introducing the matrix $\boldsymbol{C}$ is that the product $\boldsymbol{BC}$ has a simple form, and this is useful to simplify $\det \boldsymbol{B}$ given by $\det \boldsymbol{BC} / \det \boldsymbol{C} = \det \boldsymbol{BC} / \det \boldsymbol{C}_{RR}$. With the relations (2.63)–(2.65), one finds the following form for the product

$$\boldsymbol{BC} = \begin{pmatrix} \mathbf{1} - \lambda k_B T_L \mathcal{T}_L & \mathbf{0} \\ -\lambda k_B T_L \mathcal{T}_{RL} & \left(\mathbf{1} + \frac{\mathcal{T}_R}{\lambda k_B T_L} \right) \left(\mathbf{1} - \mathcal{T}_R k_B^2 T_L T_R \lambda (\lambda + 1/(k_B T_R) - 1/(k_B T_L)) \right) \end{pmatrix} , \tag{2.69}$$

and hence

$$\det \boldsymbol{B} = \det \left(\mathbf{1} - \mathcal{T}_R k_B^2 T_L T_R \lambda (\lambda + 1/(k_B T_R) - 1/(k_B T_L)) \right) \frac{\det(\mathbf{1} - \lambda k_B T_L \mathcal{T}_L)}{\det(\mathbf{1} - \lambda k_B T_L \mathcal{T}_R)} . \tag{2.70}$$

Now by taking the singular value decomposition of the matrix $[(\boldsymbol{\Gamma}_L)^{1/2} \boldsymbol{G}^+ (\boldsymbol{\Gamma}_R)^{1/2}]_{LR}$ it can be shown that $\mathcal{T}_L$ and $\mathcal{T}_R$ have the same set of non-zero eigenvalues. Hence $\det(\mathbf{1} - \lambda k_B T_L \mathcal{T}_L) = \det(\mathbf{1} - \lambda k_B T_L \mathcal{T}_R)$ and on using this in Eq. (2.70) we get the CGF in the large τ limit

$$\mu(\lambda) = -\frac{1}{2\pi} \int_0^\infty d\omega \, \mathrm{Tr} \ln \left[\mathbf{1} - \mathcal{T}(\omega) k_B^2 T_L T_R \lambda \, (\lambda + 1/(k_B T_R) - 1/(k_B T_L)) \right] . \tag{2.71}$$

One can use either $\mathcal{T}_L$ and $\mathcal{T}_R$ for the transmission matrix $\mathcal{T}(\omega)$, both of which generate the same values of current cumulants. Note that we so far do not specify the numbers of sites attached to the left and right reservoirs. Interestingly, the formula (2.71) is valid even when different numbers of particles are attached to left and right reservoirs.

First and second derivatives with respect to λ yield the average current described by the Landauer formula (2.22) and the second cumulants

$$I_1 = \frac{k_B(T_L - T_R)}{2\pi} \int_0^\infty d\omega \, \mathrm{Tr}[\mathcal{T}(\omega)] , \tag{2.72}$$

$$I_2 = \frac{k_B^2}{2\pi} \int_0^\infty d\omega \, \mathrm{Tr}[\mathcal{T}^2(\omega)(T_R - T_L)^2 + 2\mathcal{T}(\omega)T_L T_R]. \tag{2.73}$$

Higher order cumulants are also systematically given. One can easily check that the fluctuation relation symmetry (2.36) is satisfied. As already mentioned, this symmetry of the CGF implies linear response results (2.34) and further nontrivial relations such as (2.41) and more generally Eq. (2.38). A Fokker-Planck equation approach for computing heat current fluctuations in small systems is given in [40, 55] where the validity of fluctuation relations has been discussed. The fluctuation relation for heat has been experimentally demonstrated [56] and theoretical analysis of this experiment was made in [57].

2.3.2.2 Quantum Regime

Standard Protocol for the CGF In quantum case, we start with the total Hamiltonian (2.1)

$$H = H_S + H_L + H_R + H_{SL} + H_{SR} . \tag{2.74}$$

We here consider only one-dimensional harmonic system composed of N particles, where the first and the Nth particles respectively are attached to the left and right reservoir. We use reservoir's Hamiltonian in the normal mode representation. The left and right reservoir are described as

$$H_L = \sum_\ell p_\ell^2/2m_\ell + m_\ell \omega_\ell^2 x_\ell^2/2 ,$$

$$H_R = \sum_r p_r^2/2m_r + m_r \omega_r^2 x_r^2/2 . \tag{2.75}$$

The coupling Hamiltonian is given by

$$H_{SL} = -\sum_\ell \lambda_\ell x_\ell x_1 ,$$

$$H_{SR} = -\sum_r \lambda_r x_r x_N . \tag{2.76}$$

To calculate fluctuation of heat transfer, one needs to set measurement protocol for observing heat transfer. Choice of the observation process is not unique and hence fluctuation of quantum heat transfer is somewhat unclear. This aspect is very different from classical case. In general, higher order of cumulants of heat transfer and distribution are affected by quantum observation procedure. We here show the standard protocol to get the CGF, which satisfies fluctuation relation symmetry. First we measure subsystems composed of the system, left and right reservoirs. Let us assume that this measurement makes the wave-function collapse onto the product state $|i\rangle = \phi_i^S \otimes \phi_i^L \otimes \phi_i^R$. Here ϕ_i^α is the eigenstate of the Hamiltonian $\mathcal{H}_\alpha$ for the eigenvalue $\epsilon_{\alpha\, i}$. After this measurement, the total system freely evolves in time during τ. During this time-evolution, heat is transferred from one reservoir to another. After this free time-evolution, we again observe three parts of systems. Let us assume that the collapsed wave function is $|f\rangle = \phi_f^S \otimes \phi_f^L \otimes \phi_f^R$. For this process, the probability that heat transferred from the left reservoir to the system is Q, is given by

$$P_{i\to f}(Q) = \delta(Q - (\epsilon_{L\,i} - \epsilon_{L\,f}))\, |\langle f|U_\tau|i\rangle|^2\,, \tag{2.77}$$

where U_τ is the time evolution operator given by the total Hamiltonian. Iterating this process for the same initial density matrix, the distribution of Q is defined as

$$P(Q) = \sum_{i,f} \langle i|\rho_{\rm ini}|i\rangle P_{i\to f}(Q)\,. \tag{2.78}$$

$\rho_{\rm ini}$ is the initial density matrix which is assumed to be decoupled form $\rho_{\rm ini} = \rho_L \otimes \rho_S \otimes \rho_R$ where $\rho_\alpha = e^{-H_\alpha/(k_B T_\alpha)}/Z_\alpha$ $(\alpha = S, L, R)$. Then the distribution of heat transfer is written in simpler form

$$P(Q) = \frac{1}{2\pi}\int_{-\infty}^{\infty} d\xi \left\langle e^{\frac{i}{\hbar}H_-(\xi)\tau} e^{-\frac{i}{\hbar}H_+(\xi)\tau}\right\rangle e^{-i\xi Q}\,, \tag{2.79}$$

where $H_\pm(\xi) = e^{\mp i\xi H_L/2} H e^{\pm i\xi H_L/2}$. The bracket $\langle \ldots \rangle$ implies taking average over the initial density matrix $\rho_{\rm ini}$ From this we can identify the formal expression of the CGF

$$\mu(i\xi) = \frac{1}{\tau}\ln\!\left\langle e^{\frac{i}{\hbar}H_-(\xi)\tau} e^{-\frac{i}{\hbar}H_+(\xi)\tau}\right\rangle. \tag{2.80}$$

Derivation of the CGF We now derive a more detailed expression for the CGF $\mu(i\xi)$. Replacing $i\xi$ by λ in μ leads to the CGF, $\mu(\lambda)$, the classical version of which was already given in the previous subsection. For deriving the quantum version of the CGF using the expression (2.80), we employ the non-equilibrium Green's function technique where the fictitious time along the Keldysh contour depicted in Fig. 2.1 has crucial roles. Several elemental properties on the Green's function are

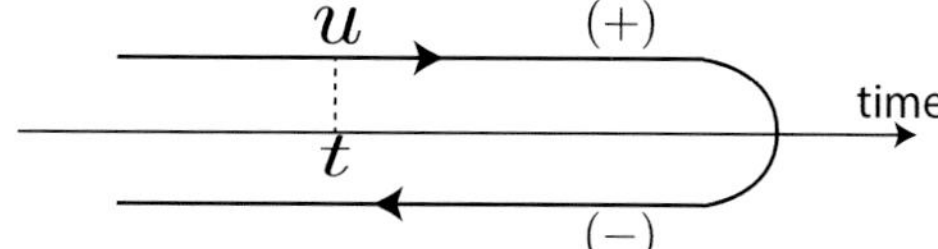

Fig. 2.1 Schematic picture of Keldysh contour. A fictitious time u is defined on (+) and (−) branch. The real time for u is t

explained in the Appendix. We introduce the modified Hamiltonian H_φ

$$\begin{aligned}H_\varphi(u) &= e^{-i\varphi(u)H_L} H e^{i\varphi(u)H_L} ,\\ &= H_S + H_L - \sum_\ell \lambda_\ell \sqrt{\frac{\hbar}{2m_\ell\omega_\ell}} (b_\ell^\dagger e^{-i\hbar\omega_\ell\varphi(u)} + b_\ell e^{i\hbar\omega_\ell\varphi(u)})x_1 + H_{SR} ,\end{aligned} \tag{2.81}$$

where b and $b^\dagger$ are respectively the annihilation and creation bosonic operator for the harmonic oscillators. The function $\varphi(u)$ is defined as

$$\varphi(u) = \begin{cases} \xi/2 & \cdots (+) \text{ branch} \\ -\xi/2 & \cdots (-) \text{ branch} \end{cases} , \tag{2.82}$$

We start with the path-integral expression for the CGF.

$$\begin{aligned}\mu(i\xi) &= \frac{1}{\tau} \ln\langle e^{-\frac{i}{\hbar}\int_c du\, H_\varphi(u)}\rangle \\ &= \frac{1}{\tau} \ln \int \mathcal{D}[\{x\}] \int \mathcal{D}\left[\{b\},\{\bar{b}\}\right] \exp\Big[\frac{i}{\hbar} S_{\mathrm{tot}}\left(\{x,b,\bar{b}\}\right)\Big],\end{aligned} \tag{2.83}$$

where $\int_c du$ implies contour-ordered integral along the Keldysh contour and the action S_{tot} is given by

$$\begin{aligned}S_{\mathrm{tot}}\left(\{x,b,\bar{b}\}\right) = &\int_c du\, X^T(u)\Big[-\frac{\boldsymbol{M}}{2}\frac{\partial^2}{\partial u^2} - \boldsymbol{K}\Big]X(u) \\ &+ \sum_\ell \bar{b}_\ell(u)\,(i\hbar\partial/\partial u - \hbar\omega_\ell)\,b_\ell(u) + \sum_r \bar{b}_r(u)\,(i\hbar\partial/\partial u - \hbar\omega_r)\,b_r(u) \\ &+ \sum_\ell \lambda_\ell \sqrt{\frac{\hbar}{2m_\ell\omega_\ell}} (\bar{b}_\ell(u)x_1(u)e^{-i\hbar\omega_\ell\varphi(u)} + x_1(u)b_\ell(u)e^{i\hbar\omega_\ell\varphi(u)}) \\ &+ \sum_r \lambda_r \sqrt{\frac{\hbar}{2m_r\omega_r}} (\bar{b}_r(u)x_N(u) + x_N(u)b_r(u)) .\end{aligned} \tag{2.84}$$

By integrating out with respect to reservoir's degrees of freedom, one gets a reduced action which contains only system's degrees of freedom:

$$\mu(i\xi) \propto \frac{1}{\tau} \ln \int \mathcal{D}[\{x\}]\, e^{\frac{i}{\hbar}S(\{x\})}, \tag{2.85}$$

$$S(\{x\}) = \int_c du_1 \int_c du_2\, X^T(u_1) \left[\mathbf{g}^{-1}(u_1,u_2) - \boldsymbol{\Sigma}(u_1,u_2)\right] X(u_2), \tag{2.86}$$

where[1]

$$\Sigma_{i,j}(u_1,u_2) = \delta_{i,1}\delta_{j,1}\, \Sigma_{L\varphi}(u_1,u_2) + \delta_{i,N}\delta_{j,N}\, \Sigma_R(u_1,u_2), \tag{2.87}$$

$$\Sigma_{L\varphi}(u_1,u_2) = \sum_\ell \frac{\lambda_\ell^2 \hbar}{2m_\ell \omega_\ell} g_\ell(u_1,u_2)\, e^{i\hbar\omega_\ell(\varphi(u_1)-\varphi(u_2))}, \tag{2.88}$$

$$\Sigma_R(u_1,u_2) = \sum_r \frac{\lambda_r^2 \hbar}{2m_r \omega_r} g_r(u_1,u_2). \tag{2.89}$$

Here we used the contour ordered Green's function where the time evolution is driven with each part of Hamiltonian

$$\mathbf{g}(u_1,u_2) = (-i/\hbar)\, \langle \hat{T}_c X(u_1)\, X^T(u_2)\rangle, \tag{2.90}$$

$$g_\ell(u_1,u_2) = (-i/\hbar)\, \langle \hat{T}_c\, b_\ell(u_1)\, b_\ell^\dagger(u_2)\rangle, \tag{2.91}$$

$$g_r(u_1,u_2) = (-i/\hbar)\, \langle \hat{T}_c\, b_r(u_1)\, b_r^\dagger(u_2)\rangle, \tag{2.92}$$

where $X(u) = e^{iH_S t/\hbar} X e^{iH_S t/\hbar}$, $b_\ell(u) = e^{iH_L t/\hbar} b_\ell e^{-iH_L t/\hbar}$ and $b_r(u) = e^{iH_R t/\hbar} b_r e^{-iH_R t/\hbar}$ and $\hat{T}_c$ is the contour-ordering operator that orders the operators according to the position on the Keldysh contour (See Appendix).[2] In real time representation, the reduced action is then written as follows

$$S = \int_{-\tau/2}^{\tau/2} dt_1 \int_{-\tau/2}^{\tau/2} dt_2 \sum_{\sigma,\sigma'=\pm} [X^\sigma]^T(t_1)\sigma \left[[\mathbf{g}^{-1}]^{\sigma\sigma'}(t_1,t_2) - \boldsymbol{\Sigma}^{\sigma\sigma'}(t_1,t_2)\right] \sigma' X^{\sigma'}(t_2) \tag{2.93}$$

We introduce discrete Fourier expansion

$$X^\sigma(t) = \frac{1}{\tau} \sum_{n=-\infty}^{\infty} X^\sigma(\omega_n) e^{i\omega_n t}, \tag{2.94}$$

[1] The inverse here implies that it satisfies $\int_c du' \mathbf{g}^{-1}(u,u')\mathbf{g}(u',u'') = \mathbf{1}\,\delta(u-u'')$, where $\mathbf{1}$ is $N\times N$ identity matrix.

[2] For system's Green's function $\mathbf{g}(u_1,u_2)$, one can neglect the $(+,-)$ and $(-,+)$ components which are related to the initial state, since system's initial state does not affect on the steady state.

where $\omega_n = 2\pi n/\tau$. Similarly we use the Fourier representation for the other variables. We define the column vector $Y = (x_1^+, x_1^-, x_2^+, x_2^-, \cdots, x_N^+, x_N^-)^T$. Noting $Y^T(-\omega_n) = Y^\dagger(\omega_n)$ we can express (2.93) using positive Fourier modes

$$S = 2\sum_{n\geq 0} Y^\dagger(\omega_n)\boldsymbol{A}_\xi(\omega_n)Y(\omega_n)\,. \tag{2.95}$$

The matrix $\boldsymbol{A}_\xi$ is defined as follows

$$\boldsymbol{A}_\xi(\omega) = \begin{pmatrix} \left[\boldsymbol{G}_\xi^{-1}\right]_{1,1}, \left[\boldsymbol{G}_\xi^{-1}\right]_{1,2}, \cdots \left[\boldsymbol{G}_\xi^{-1}\right]_{1,N-1}, \left[\boldsymbol{G}_\xi^{-1}\right]_{1,N} \\ \vdots \qquad\qquad\qquad\qquad \vdots \\ \left[\boldsymbol{G}_\xi^{-1}\right]_{N,1}, \left[\boldsymbol{G}_\xi^{-1}\right]_{N,2}, \cdots \left[\boldsymbol{G}_\xi^{-1}\right]_{N,N-1}, \left[\boldsymbol{G}_\xi^{-1}\right]_{N,N} \end{pmatrix}, \tag{2.96}$$

where the Green function $[\boldsymbol{G}_\xi^{-1}]_{k,\ell}$ is 2×2 matrix for (k,ℓ) element $(k,\ell = 1,\cdots,N)$[3]

$$\left[\boldsymbol{G}_\xi^{-1}\right]_{k,\ell} = [\boldsymbol{g}^{-1}]_{k,\ell}(\omega_n) - \boldsymbol{\sigma}_z[\boldsymbol{\Sigma}_\xi]_{k,\ell}(\omega_n)\boldsymbol{\sigma}_z\,, \tag{2.97}$$

$$\left[\boldsymbol{\Sigma}_\xi\right]_{k,\ell}(\omega_n) = \delta_{k,1}\delta_{\ell,1}e^{i\xi\hbar\omega_n\boldsymbol{\sigma}_z/2}\boldsymbol{\Sigma}_L(\omega_n)e^{-i\xi\hbar\omega_n\boldsymbol{\sigma}_z/2} + \delta_{k,N}\delta_{\ell,N}\boldsymbol{\Sigma}_R(\omega_n)\,, \tag{2.98}$$

where $\boldsymbol{\sigma}_z$ is the z-component of Pauli matrix, i.e., $\boldsymbol{\sigma}_z = \mathrm{diag}(1,-1)$.

$$[\Sigma_L]^{\sigma\sigma'}(\omega_n) = \frac{1}{2}\sum_\ell \frac{\lambda_\ell^2\hbar}{2m_\ell\omega_\ell}\left[g_\ell^{\sigma\sigma'}(\omega_n) + g_\ell^{\sigma'\sigma}(-\omega_n)\right], \tag{2.99}$$

$$[\Sigma_R]^{\sigma\sigma'}(\omega_n) = \frac{1}{2}\sum_r \frac{\lambda_r^2\hbar}{2m_r\omega_r}\left[g_r^{\sigma\sigma'}(\omega_n) + g_r^{\sigma'\sigma}(-\omega_n)\right]. \tag{2.100}$$

In the limit of $\tau \to \infty$, we can use continuous Fourier mode. We perform the Gaussian integration and by taking account of normalization $\mu(\xi = 0) = 0$, one can express the CGF with the following formal expression

$$\mu(i\xi) = \frac{1}{2\pi}\int_0^\infty d\omega \ln\left[\frac{\det\boldsymbol{A}_{\xi=0}(\omega)}{\det\boldsymbol{A}_\xi(\omega)}\right] = \frac{-1}{2\pi}\int_0^\infty d\omega \ln\det\left[\boldsymbol{A}_\xi(\omega)\boldsymbol{A}_{\xi=0}^{-1}(\omega)\right]. \tag{2.101}$$

[3]The inverse implies that it satisfies $\sum_{\ell=1}^N \left[\boldsymbol{G}_\xi^{-1}\right]_{k,\ell}[\boldsymbol{G}_\xi]_{\ell,k'} = \delta_{k,k'}\mathbf{1}$, where $\mathbf{1}$ is the 2×2 identity matrix. This is equivalent to $\boldsymbol{A}_\xi\boldsymbol{A}_\xi^{-1} = \mathbf{1}$, where $\mathbf{1}$ is the $2N\times 2N$ identity matrix.

We now simplify the formal expression (2.101). To this end, we note the following expression derived from Eqs. (2.97) and (2.98).

$$\sum_{\ell=1}^{N} \left[\boldsymbol{G}_{\xi}^{-1} \right]_{k,\ell} [\boldsymbol{G}_{\xi=0}]_{\ell,k'} = \boldsymbol{1}\delta_{k,k'} - \delta_{k,1}\, \boldsymbol{\Sigma}'_{\xi}\, [\boldsymbol{G}_{\xi=0}]_{1,k'}\,, \tag{2.102}$$

$$\boldsymbol{\Sigma}'_{\xi} = \begin{pmatrix} 0, & (1-e^{i\xi\hbar\omega})\Sigma_{\rm L}^{+-}(\omega) \\ (1-e^{-i\xi\hbar\omega})\Sigma_{\rm L}^{-+}(\omega), & 0 \end{pmatrix}. \tag{2.103}$$

Using this relation, $\boldsymbol{A}_{\xi}(\omega)\boldsymbol{A}_{\xi=0}^{-1}(\omega)$ is written as

$$\boldsymbol{A}_{\xi}(\omega)\left[\boldsymbol{A}_{\xi=0}^{-1}\right](\omega) = \begin{pmatrix} \boldsymbol{1}-\boldsymbol{\Sigma}'_{\xi}[\boldsymbol{G}_{\xi=0}]_{1,1}, & -\boldsymbol{\Sigma}'_{\xi}[\boldsymbol{G}_{\xi=0}]_{1,2}, & -\boldsymbol{\Sigma}'_{\xi}[\boldsymbol{G}_{\xi=0}]_{1,3}, & \cdots, & -\boldsymbol{\Sigma}'_{\xi}[\boldsymbol{G}_{\xi=0}]_{1,N} \\ \boldsymbol{0}, & \boldsymbol{1}, & \boldsymbol{0}, & \cdots, & \boldsymbol{0} \\ \vdots & & & & \vdots \\ \boldsymbol{0}, & \boldsymbol{0}, & \boldsymbol{0}, & \cdots, & \boldsymbol{1} \end{pmatrix}. \tag{2.104}$$

Hence the determinant is simplified as

$$\begin{aligned} \det\left(\boldsymbol{A}_{\xi}(\omega)\left[\boldsymbol{A}_{\xi=0}^{-1}\right](\omega)\right) &= \det\left[\boldsymbol{1}-\boldsymbol{\Sigma}'_{\xi}[\boldsymbol{G}_{\xi=0}]_{1,1}\right] \\ &= 1 - G_{1N}^{R}\Sigma_{\rm R}^{<}G_{1N}^{A}\Sigma_{\rm L}^{>}(1-e^{-i\xi\hbar\omega}) \\ &\quad - G_{1N}^{R}\Sigma_{\rm R}^{>}G_{1N}^{A}\Sigma_{\rm L}^{<}(1-e^{i\xi\hbar\omega}), \end{aligned} \tag{2.105}$$

where $R, A, <, >$ are the retarded, advanced, lesser, and greater parts in the Green's functions (See Appendix).[4] In deriving Eq. (2.105), we used the relations

$$\text{Det}[\boldsymbol{G}_{\xi=0}]_{1,1} = -G_{11}^{R}G_{11}^{A}\,, \tag{2.106}$$

$$\left[G_{\xi=0}\right]_{11}^{<,>} = G_{11}^{R}\Sigma_{\rm L}^{<,>}G_{11}^{A} + G_{1N}^{R}\Sigma_{\rm R}^{<,>}G_{N1}^{A}\,. \tag{2.107}$$

[4]In LEGF method, we use the notation G^{+} and G^{-} for the Green's function. This notation $G^{+}(G^{-})$ is equivalent to the advanced (retarded) Green's function in non-equilibrium Green's function approach. See Appendix.

Thus we arrive at the general expression for the generating function,

$$\mu(i\xi) = -\int_0^\infty \frac{d\omega}{2\pi} \ln\Big[1 + \mathcal{T}(\omega)\Big[f(\omega, T_R)f(-\omega, T_L)(e^{-i\xi\hbar\omega} - 1)$$
$$+f(-\omega, T_R)f(\omega, T_L)(e^{i\xi\hbar\omega} - 1)\Big]\Big]. \tag{2.108}$$

Here we used the relations

$$\mathcal{T}(\omega) = 4|G^R_{1N}(\omega)|^2 \mathrm{Im}[\Sigma^R_L(\omega)]\, \mathrm{Im}[\Sigma^R_R(\omega)]\,, \tag{2.109}$$

$$\Sigma^{<,>}_\alpha(\omega) = \pm 2\Sigma^R_\alpha(\omega) f(\pm\omega, T_\alpha)\,, \tag{2.110}$$

$$\Sigma^{R,A}_\alpha(\omega) = \mp\frac{i}{2\pi}\mathcal{J}_\alpha(\omega) \quad \alpha = L, R\,. \tag{2.111}$$

Here $\mathcal{J}_\alpha(\omega)$ is the spectral density of the αth reservoir, $\mathcal{J}_\alpha(\omega) = (\pi/2) \sum_j \lambda_j^2/(2m_j\omega_j)\delta(\omega-\omega_j)$ where $j = \ell(r)$ for $\alpha = L(R)$.

Comments on the CGF (2.108) Several comments are listed in order. Phonons convey energy in units $\hbar\omega$ and this appears in the exponential form with the factor $i\xi$ in Eq. (2.108). The expression is similar to the CGF for non-interacting electric conduction [58] It is easily verified that Eq. (2.108) reproduces the correct first cumulant, given by the Landauer formula (2.22).

$$I_1 = \frac{1}{2\pi}\int_0^\infty d\omega\hbar\omega\mathcal{T}(\omega)\left[f(\omega, T_L) - f(\omega, T_R)\right]\,. \tag{2.112}$$

We also note that the fluctuation relation symmetry (2.36) is satisfied. The second cumulant is given by

$$I_2 = \int_0^\infty d\omega \frac{\hbar^2\omega^2}{2\pi}\Big\{\mathcal{T}^2(\omega)\left[f(\omega, T_L) - f(\omega, T_R)\right]^2$$
$$-\mathcal{T}(\omega)[f(\omega, T_L)f(-\omega, T_R) + f(-\omega, T_L)f(\omega, T_R)]\Big\}. \tag{2.113}$$

This bosonic fluctuation is also similar to the optical one [59]. Significant difference from the classical case is temperature dependence at low temperatures. Transient properties in quantum current fluctuation was discussed in [60].

2.4 Explicit Results for Steady State Properties in Ordered and Disordered Harmonic Crystal

So far we have discussed how one can obtain formal expressions for two-point correlations and current fluctuations in the non-equilibrium heat-conducting steady state. These formal expressions typically involve an integral over frequencies, with the integrand written in terms of the transmittance function $\mathcal{T}(\omega)$. The transmittance function itself is written as a trace involving the Green's function of the system and the dissipation matrices of the bath. In this section we analyze these formal expressions, for the case of ordered and mass-disordered lattices. It will be shown how one can obtain explicit results on important and interesting questions such as the system-size and boundary condition dependence of the heat current and it's dependence on dimensionality, on validity of fluctuation theorems and the additivity principle. In Sect. 2.4.1, we will discuss properties of the steady state current, we briefly discuss temperature profiles in Sect. 2.4.2 and finally explicit results on fluctuations will be discussed in Sect. 2.4.3.

2.4.1 Current

Heat conduction in the ordered harmonic chain was first studied by Rieder, Lebowitz and Lieb (RLL) [4] and its higher dimensional generalization was obtained by Nakazawa [61]. They considered classical systems connected to heat baths described by Gaussian white noise Langevin equations. One can then write Fokker-Planck equations for the time-evolution of the full phase-space distribution function of the system. Since the process is linear and the noise Gaussian, it follows that the steady state solution of the Fokker-Planck equation is a Gaussian. The approach followed in both the RLL and Nakazawa papers was to show that the correlation matrix, corresponding to the Gaussian measure, satisfies a linear matrix equation which can be solved exactly. Thus, a complete solution for the correlation matrix was obtained and from this one could find both the steady state temperature profile and the heat current.

In contrast to the RLL approach, the LEGF method, discussed in Sect. 2.2, involved solution of the linear Langevin equations by Fourier transforms. It was shown that one can again obtain exact formal integral expressions for the steady state current and the complete set of two point correlation functions for arbitrary systems of harmonic networks connected to two heat reservoirs. The general expressions involve non-equilibrium Green's functions. The advantages of the LEGF method are: (a) it leads to analytic treatment even for the case of disordered systems and (b) as we have shown, it is easily extendable to the quantum case. In this section we will see how the results of RLL and Nakazawa for the ordered case are recovered from the LEGF method and we will also discuss explicit results for the disordered and quantum cases.

Even before carrying out a detailed analysis, it is clear that heat transport in a perfectly ordered harmonic crystal should be "ballistic". We expect that the heat carriers, the phonons should pass through the system without any scattering, except at the boundaries. Hence the current is expected to be independent of system size and the conductivity should scale linearly with the linear size of the system. This case will be discussed in Sect. 2.4.1.2. Even though the conductivity, in such cases, diverges with system size, the conductance remains finite and is the quantity that is measured in experiments. The value of the conductance is fixed by scattering of phonons at the boundary between heat bath and the system.

It is possible to introduce phonon scattering, while still remaining within the harmonic approximation. This can be done through introduction of impurities or disorder into the system and a very interesting question is whether or not one then obtains diffusive heat transport in this case. In Sect. 2.4.1.3 we discuss the nature of heat transport in regular lattices, with disorder introduced through random masses.

In Sect. 2.4.1.1 we will first define the precise model for which explicit results have been obtained, and also an outline of the methods used. For simplicity we will consider here only the case where transverse vibration modes are ignored and the displacement at each site is described by a scalar variable. Also we restrict to the case of d-dimensional hypercubic lattices with nearest neighbor interactions. For our special choice of the harmonic crystal, a lot of simplification is achieved in the evaluation of the Landauer current formula. We first give the precise definition of the model and the resulting simplifications that occur for this choice of model.

2.4.1.1 Definition of Model and Transfer Matrix Representation

Let us denote the points on the d-dimensional hyper-cubic lattice by the vector $\mathbf{n} = \{n_1, n_2, \ldots, n_d\}$ with $n_\nu = 1, 2, \ldots, N$. The scalar displacement of a particle at the lattice site $\mathbf{n}$ is given by $x_\mathbf{n}$. It is assumed that only the particles on the boundary layers at $n_1 = 1$ and $n_1 = N$ are coupled to heat reservoirs, at temperatures T_L and T_R respectively. Periodic boundary conditions is assumed in the other $(d-1)$ directions. We label by $\{1, \mathbf{n}'\}$ and $\{N, \mathbf{n}'\}$, respectively, the set of sites in the first and last layers. Consider the following harmonic Hamiltonian for the system:

$$H_S = \sum_{\mathbf{n}} \frac{1}{2} m_\mathbf{n} \dot{x}_\mathbf{n}^2 + \sum_{\mathbf{n},\hat{\mathbf{e}}} \frac{k}{2} (x_\mathbf{n} - x_{\mathbf{n}+\hat{\mathbf{e}}})^2 + \sum_{\mathbf{n}} \frac{k_\mathbf{n}}{2} x_\mathbf{n}^2 , \qquad (2.114)$$

where $\hat{\mathbf{e}}$ refers to the $2d$ nearest neighbors of any site. The term with k corresponds to nearest neighbor particles connected by springs, $k_\mathbf{n}$ refers to external harmonic pinning potential at all sites. For the ordered system, the masses of all particles is taken to be the same, i.e., $m_\mathbf{n} = m$ for all $\mathbf{n}$. In Sect. 2.4.1.3 we consider a mass disordered crystal where half of the randomly chosen particles are assigned mass $m + \Delta$ and the remaining half have mass $m - \Delta$.

The full system consists of $\mathcal{N} = N^d$ sites while each of the two boundary layers ($n_1 = 1$ and $n_1 = N$) consist of $\mathcal{N}' = N^{d-1}$ particles, and only these are connected to heat baths. We further assume that each particle is connected to an independent bath. Following the results obtained in Sect. 2.2, the equations of motion for the system is thus given by

$$m_{\mathbf{n}}\ddot{x}_{\mathbf{n}} = -\sum_{\hat{\mathbf{e}}} k(x_{\mathbf{n}} - x_{\mathbf{n}+\hat{\mathbf{e}}}) - k_{\mathbf{n}}x_{\mathbf{n}} + \delta_{n_1,1}\left[\int_{-\infty}^{\infty} dt'\sigma^L_{\mathbf{n}'}(t-t')x_{\mathbf{n}'}(t') + \eta^L_{\mathbf{n}'}\right]$$
$$+ \delta_{n_1,N}\left[\int_{-\infty}^{\infty} dt'\sigma^R_{\mathbf{n}'}(t-t')x_{\mathbf{n}'}(t') + \eta^R_{\mathbf{n}'}\right], \qquad (2.115)$$

where the dissipative and noise terms are related by the fluctuation dissipation relation

$$\frac{1}{2}\langle\, \eta^a_{\mathbf{n}'}(\omega)\eta^a_{\mathbf{l}'}(\omega') + \eta^a_{\mathbf{l}'}(\omega')\eta^a_{\mathbf{n}'}(\omega)\,\rangle$$
$$= \delta_{\mathbf{n}',\mathbf{l}'}\ \delta(\omega+\omega')\ \nu^a(\omega)\ \frac{\hbar}{2\pi}\ \coth(\frac{\hbar\omega}{2k_BT_a})\ , \quad a = L, R\ ,$$

where $\nu^a(\omega) = Im[\sigma^a(\omega)]$. We define $\boldsymbol{\sigma}^{L,R}$ to be $N^{d-1} \times N^{d-1}$ diagonal matrices with $\sigma_{\mathbf{n}'}$ as the diagonal elements.

Choice of Heat Baths As discussed in Sect. 2.2.1, two possible choices of heat bath are the Ohmic bath and the Rubin model. Here we will mostly discuss the Ohmic bath example and the Rubin model will be discussed only in the context of the one-dimensional chain.

For Ohmic baths the following cases will be studied:

(1) Unpinned lattice with fixed boundary condition (BC)—this corresponds to setting $k_{n_1,\mathbf{n}'} = 0$, for $1 < n_1 < N$, and $k_{n_1=1,\mathbf{n}'} = k_{n_1=N,\mathbf{n}'} = k' \neq 0$.
(2) Unpinned lattice with fixed boundary condition (BC)—this corresponds to setting $k_{\mathbf{n}} = 0$ for $1 \leq n_1 \leq N$.
(3) Pinned lattice—this corresponds to setting $k_{\mathbf{n}} = k_o \neq 0$, for $1 \leq n_1 \leq N$.

As we will see, the presence of either bulk or boundary pinning can drastically affect transport properties.

The heat current is given by the general formula

$$I = \frac{1}{2\pi}\int_0^{\infty} d\omega\ \hbar\omega\ \mathcal{T}(\omega)\ [f(\omega,T_L) - f(\omega,T_R)]\ ,$$

where the transmission coefficient is $\mathcal{T}(\omega) = 4\ Tr[\boldsymbol{G}^+(\omega)\ \boldsymbol{\Gamma}_L(\omega)\ \boldsymbol{G}^-(\omega)\ \boldsymbol{\Gamma}_R(\omega)\ ]$. The required Green's function is given by $\boldsymbol{G}^+(\omega) = [-\omega^2\boldsymbol{M} + \boldsymbol{K} - \boldsymbol{\Sigma}^+_L - \boldsymbol{\Sigma}^+_R]^{-1}$, $\boldsymbol{\Gamma}_{L,R} = Im[\boldsymbol{\Sigma}^+_{L,R}]$. For the special case described by the equations of motion in Eq. (2.115), the various matrices have a simple block structure. This structure can be used to simplify the analysis. For simplicity we henceforth set the inter-particle spring constant to the value $k = 1$. The matrices $\boldsymbol{M}, \boldsymbol{K}$ then have the following structure

$$\boldsymbol{M} = \begin{pmatrix} \boldsymbol{M}_1 & 0 & \ldots & 0 \\ 0 & \boldsymbol{M}_2 & \ldots & 0 \\ 0 & 0 & \ldots & 0 \\ 0 & 0 & .. & \boldsymbol{M}_N \end{pmatrix}, \ \boldsymbol{K} = \begin{pmatrix} \boldsymbol{K}_1 & -\boldsymbol{I} & \ldots & 0 \\ -\boldsymbol{I} & \boldsymbol{K}_2 & \ldots & 0 \\ 0 & 0 & \ldots & 0 \\ 0 & 0 & ..-\boldsymbol{I} & \boldsymbol{K}_N \end{pmatrix}, \tag{2.116}$$

where $\boldsymbol{M}_n$ denotes the diagonal mass-matrix for the $n_1 = n$ layer, $\boldsymbol{K}_n$ is a force-constant matrix whose off-diagonal terms correspond to coupling to sites within the $N_1 = n$ layer. The matrix $\mathbf{I}$ is a $\mathcal{N}' \times \mathcal{N}'$ unit matrix, and 0 is a $\mathcal{N}' \times \mathcal{N}'$ matrix with all elements equal to zero. The matrices $\boldsymbol{\Sigma}^+_{L,R}$ take the form

$$\boldsymbol{\Sigma}^+_L = \begin{pmatrix} \boldsymbol{\sigma}^+_L & 0 \ldots 0 \\ 0 & 0 \ldots 0 \\ 0 & 0 \ldots 0 \end{pmatrix}, \boldsymbol{\Sigma}^+_R = \begin{pmatrix} 0 & 0 \ldots & 0 \\ 0 & 0 \ldots & 0 \\ 0 & 0 \ldots & \boldsymbol{\sigma}^+_R \end{pmatrix}, \tag{2.117}$$

Hence the matrix $[\boldsymbol{G}^+]^{-1} = [-\boldsymbol{M}\omega^2 + \boldsymbol{K} - \boldsymbol{\Sigma}^+_L - \boldsymbol{\Sigma}^+_R]$ has the following structure:

$$[\boldsymbol{G}^+]^{-1} = \begin{pmatrix} \boldsymbol{a}_1 - \boldsymbol{\sigma}^+_L & -\boldsymbol{I} & 0 & \ldots & 0 \\ -\boldsymbol{I} & \boldsymbol{a}_2 & -\boldsymbol{I}\ 0 & \ldots & 0 \\ \ldots & \ldots & \ldots & \ldots & \ldots \\ 0 & \ldots\ 0 & -\boldsymbol{I} & \boldsymbol{a}_{N-1} & -\boldsymbol{I} \\ 0 & \ldots & 0 & -\boldsymbol{I} & \boldsymbol{a}_N - \boldsymbol{\sigma}^+_R \end{pmatrix}, \tag{2.118}$$

where $\boldsymbol{a}_l = -\boldsymbol{M}_l\omega^2 + \boldsymbol{K}_l$. Now defining $\boldsymbol{v}_{L,R} = Im[\boldsymbol{\sigma}^+_{L,R}]$ and with the form of $\boldsymbol{\Sigma}^+_{L,R}$ given in Eq. (2.117), we find that the expression for the transmission coefficient reduces to the following form:

$$\mathcal{T}(\omega) = 4\, Tr[\boldsymbol{v}_L(\omega)\mathbf{g}^+_N(\omega)\boldsymbol{v}_R(\omega)\mathbf{g}^-_N(\omega)] \,, \tag{2.119}$$

where $\mathbf{g}^+_N$ is the $(1, N)$th block element of $\boldsymbol{G}^+(\omega)$ and $\mathbf{g}^-_N = [\mathbf{g}^+_N]^\dagger$. Thus, we have a huge simplification, since, instead of $N^d \times N^d$ matrices, we now have to deal with $N^{d-1} \times N^{d-1}$ matrices. The Green's function $\mathbf{g}^+_N$ can be expressed using recursive relations and transfer matrix forms. We give an outline of the steps for the one-dimensional case and state the results for higher-dimensional case.

One Dimensional Case Consider a one-dimensional chain of oscillators with N sites. From the general Hamiltonian in Eq. (2.114), we have for the one-dimensional chain

$$H = \sum_{n=1}^{N} \frac{p_n^2}{2m_n} + \sum_{n=1}^{N-1} \frac{1}{2}(x_{n+1} - x_n)^2 + \sum_{n=1}^{N} \frac{k_n}{2} x_n^2 \,. \tag{2.120}$$

The first and last sites are connected to heat baths. In this case, $\boldsymbol{\nu}_L$ and $\boldsymbol{\nu}_R$ are matrices whose only non-vanishing elements are either the $[\boldsymbol{\nu}_L]_{1,1} = \nu_L(\omega)$ and $[\boldsymbol{\nu}_R]_{N,N} = \nu_R(\omega)$ respectively. Hence, it is clear that the transmission coefficient is given by

$$\mathcal{T}(\omega) = 4\nu_L(\omega)\nu_R(\omega)|g_N^+|^2 \,, \qquad (2.121)$$

where the $\{1, N\}$th matrix element $g_N^+ = \boldsymbol{G}_{1,N}^+$ is, in the $1D$ case, just a complex number. The matrix $\mathbf{Z} = [\boldsymbol{G}^+]^{-1}$ is tri-diagonal, with the following structure

$$\mathbf{Z} = [\boldsymbol{G}^+]^{-1} = \begin{pmatrix} a_1 - \sigma_L^+ & -1 & 0 & 0 & \dots \\ -1 & a_2 & -1 & 0 & \dots \\ 0 & -1 & a_3 & -1 & \dots \\ \dots & \dots & \dots & \dots & \dots \\ \dots & 0 & -1 & a_{N-1} & -1 \\ 0 & 0 & \dots & -1 & a_N - \sigma_R^+ \end{pmatrix} , \qquad (2.122)$$

where $a_n = 2+k_n-m_n\omega^2$ for $1 < n < N$, $a_1 = 1+k_1-m_1\omega^2$, $a_N = 1+k_N-m_N\omega^2$. From the structure of the matrix, and using standard matrix properties, we then immediately get

$$g_N^+ = Z_{1,N}^{-1} = \frac{1}{\Delta_N} \quad \text{where } \Delta_N = \mathrm{Det}[Z] \,.$$

Let us now define $\Delta_{l,m}$ to be the determinant of the sub-matrix of $\mathbf{Z}$, beginning with the lth row and column, and ending with the mth row and column. Also let $\mathbf{Z}_0$ be the tri-diagonal matrix with of-diagonal elements equal to -1 and the nth diagonal element equal to a_n. We then define $D_{l,m}$ to be the determinant of the sub-matrix of $\mathbf{Z}_0$, beginning with the lth row and column, and ending with the mth row and column. Using recursion relations such as $\Delta_{1,N} = (a_1 - \sigma_L^+)\Delta_{2,N} - \Delta_{3,N}$ and $D_{1,N} = a_1 D_{2,N} - D_{3,N}$, it is easy to see that

$$\begin{aligned} \Delta_N(\omega) &= D_{1,N} - \sigma_L^+ D_{2,N} + \sigma_R^+ D_{1,N-1} + \sigma_L^+ \sigma_R^+ D_{2,N-1} \\ &= (1 \quad -\sigma_R^+) \begin{pmatrix} D_{1,N} & -D_{1,N-1} \\ D_{2,N} & -D_{2,N-1} \end{pmatrix} \begin{pmatrix} 1 \\ \sigma_L^+ \end{pmatrix} , \end{aligned} \qquad (2.123)$$

$$\text{where} \begin{pmatrix} D_{(1,N)} & -D_{(2,N)} \\ D_{(1,N-1)} & -D_{(2,N-1)} \end{pmatrix} = \mathbf{T}_N \mathbf{T}_{N-1} \dots \mathbf{T}_1 \,, \quad \text{with } \mathbf{T}_l = \begin{pmatrix} a_l & -I \\ I & 0 \end{pmatrix} . \qquad (2.124)$$

As we will see, this form is crucial in arriving at more explicit results for the ordered and harmonic chain in one dimensional harmonic chains. We now mention the generalizations of this form to the higher dimensional case.

Higher Dimensional Case As shown in [62, 63], the above forms can be generalized. One can obtain two possible transfer matrix representations for $\mathbf{g}_N^+$ (the crucial term appearing in the transmission formula Eq. (2.119)).

(i) **First Representation**: Analogous to the form in Eq. (2.123) for the one-dimensional case, one can derive the following form

$$\begin{aligned}[\mathbf{g}_N^+]^{-1} &= \mathbf{D}_{(1,N)} - \mathbf{D}_{(2,N)}\,\boldsymbol{\sigma}_L^+ - \boldsymbol{\sigma}_R^+\,\mathbf{D}_{(1,N-1)} + \boldsymbol{\sigma}_R^+\,\mathbf{D}_{(2,N-1)}\,\boldsymbol{\sigma}_L^+ \\ &= (\mathbf{I} \quad -\boldsymbol{\sigma}_R^+)\begin{pmatrix} \mathbf{D}_{(1,N)} & -\mathbf{D}_{(2,N)} \\ \mathbf{D}_{(1,N-1)} & -\mathbf{D}_{(2,N-1)} \end{pmatrix}\begin{pmatrix} \mathbf{I} \\ \boldsymbol{\sigma}_L^+ \end{pmatrix}. \end{aligned} \tag{2.125}$$

where $$\begin{pmatrix} \mathbf{D}_{(1,N)} & -\mathbf{D}_{(2,N)} \\ \mathbf{D}_{(1,N-1)} & -\mathbf{D}_{(2,N-1)} \end{pmatrix} = \mathbf{T}_N\mathbf{T}_{N-1}\ldots\mathbf{T}_1\,, \quad \text{with } \mathbf{T}_l = \begin{pmatrix} \mathbf{a}_l & -\mathbf{I} \\ \mathbf{I} & \mathbf{0} \end{pmatrix} \tag{2.126}$$

Note that in this representation, the system and reservoir contributions are separated.

(ii) **Second Representation**: This is particularly useful for numerical computations and requires evaluating inverses recursively. One has

$$\mathbf{g}_N^+ = \mathbf{r}_1^{-1}\mathbf{r}_2^{-1}\ldots\mathbf{r}_N^{-1}\,, \tag{2.127}$$

The matrices $\mathbf{r}_n$, for $n = 1, \ldots, N$, are determined recursively through

$$\mathbf{r}_n = \mathbf{c}_n - \mathbf{r}_{n-1}^{-1}\,, \tag{2.128}$$

where $\mathbf{c}_n = \mathbf{a}_n - \boldsymbol{\sigma}_L^+\delta_{n,1} - \boldsymbol{\sigma}_R^+\delta_{n,N}$ and one starts with the initial condition $\mathbf{r}_1 = \mathbf{c}_1$.

2.4.1.2 Ordered Harmonic Lattices

We now show how the results of the previous section can be used to calculate the heat current in ordered harmonic lattices. We will see that exact expressions for the asymptotic current ($N \to \infty$) can be obtained in many cases. We first discuss the one-dimensional case and then, briefly the extensions to higher dimensions.

One Dimensional Case Heat conduction in the one-dimensional case has been studied mostly for two models of heat baths which have been discussed in Sect. 2.2.1, and for different bulk and boundary pinning conditions. In the case where there is no disorder, the current formula can be reduced to a simpler integral (for large N), which can often be evaluated exactly. This simplification follows from the fact that the **T**-matrices in Eq. (2.124) are all identical except at the edges where boundary conditions induce some difference. As we now show, one can absorb this difference into the bath coefficients σ_L^+, σ_R^+. Thus, consider the case where all particles have the same mass m and the pinning terms $k_n = k_o$ for $n \neq 1, N$ and $k_1 = k_N = k_o + k'$. For simplicity we set $\sigma_L^+ = \sigma_R^+ = \sigma$. We then write

$$a_1 - \sigma = a_N - \sigma = 1 + k_o + k' - m\omega^2 - \sigma = a - \tilde{\sigma} ,$$
$$\text{where } a = 2 + k_o - m\omega^2, \quad \tilde{\sigma} = \sigma + 1 - k' .$$

Following the arguments as used in deriving Eq. (2.123) we get

$$\Delta_N(\omega) = D_{1,N} - \tilde{\sigma}(D_{2,N} + D_{1,N-1}) + \tilde{\sigma}^2 D_{2,N-1} . \tag{2.129}$$

Now the matrix $\mathbf{Z}_0$ has all diagonal elements equal to a and then one easily gets $D_{1,N} = \sin[q(N+1)]/\sin(q)$, where the wave-vector q is defined through the relation

$$m\omega^2 = k_0 + 2(1 - \cos q) . \tag{2.130}$$

We note that for $q \in (0, \pi)$, this just corresponds to the spectrum of the ordered harmonic chain. For ω outside the allowed spectrum, i.e. $k_0/m < \omega^2 < (k_0 + 4k)/m$, the variable q is imaginary. In terms of this new variable, the diagonal elements of the matrix $\mathbf{Z}_0$ are all equal to $2\cos q$. Then from Eq. (2.129) we get after some small algebra

$$\Delta_N = \frac{a(q)\sin(Nq) + b(q)\cos(Nq)}{\sin(q)} , \tag{2.131}$$
$$\text{where } a(q) = (1 - \tilde{\sigma}^2)\cos(q) - 2\tilde{\sigma} ,$$
$$b(q) = (1 + \tilde{\sigma}^2)\sin(q) . \tag{2.132}$$

Generally, the bath phonon spectrum (band-width) will be different from the system band-width, and for frequencies outside the system band-width the wave-vector q will be imaginary.

Rubin Bath There is one special case where they have the same spectrum and this is the case where the bath itself is an extension of the system. If we also set $k_o = 0$ this is then precisely the case where the bath is the Rubin model. We expect perfect transmission $\mathcal{T}(\omega) = 1$, *independent of system size N*, and let us see if we can recover this result. For the Rubin model $\sigma = e^{-iq}$, hence $\nu(\omega) = \sin(q)$ and, with $k' = 1$, we get $\tilde{\sigma} = e^{-iq}$. From Eq. (2.131), we then get $\Delta_N = -2i\sin(q)e^{iqN}$ and therefore

$$\mathcal{T} = 4\nu^2(\omega)\frac{1}{|\Delta_N|^2} = 1 . \tag{2.133}$$

Therefore the current is given by

$$I = \frac{1}{2\pi} \int_0^{\omega_m} d\omega \, \hbar\omega \, [f(\omega, T_L) - f(\omega, T_R)], \tag{2.134}$$

where $\omega_m = 2\sqrt{k/m}$. Assuming a small temperature difference $\Delta T = T_L - T_R$, and a mean temperature $k_B T << \hbar\omega_m$, one then gets for the conductance of the wire

$$G = \frac{I}{\Delta T} = \frac{k_B}{2\pi} \int_0^{\omega_m} d\omega \left(\frac{\hbar\omega}{k_B T} \right)^2 \frac{e^{\hbar\omega/(k_B T)}}{[e^{\hbar\omega/(k_B T)} - 1]^2} \approx \frac{\pi^2 k_B^2 T}{3h} . \tag{2.135}$$

This is usually referred to as the quantum of thermal conductance [9, 64, 65]. Note that in the high temperature limit, $k_B T >> \hbar\omega_m$, the conductance would saturate to the value $G = k_B \omega_m/(2\pi)$.

Ohmic Bath Now we discuss the more usual scenario where the bath and system can have different spectral properties. For simplicity we will consider the case of the Ohmic bath, for which one has $\sigma = i\gamma\omega$. For imaginary q, $\Delta_N(\omega)$ grows exponentially with N and so the transmission coefficient $\mathcal{T}(\omega)$ decays exponentially with N. Hence for large N we need only consider the range $0 < q < \pi$ and the current is given by:

$$I = \frac{2\gamma^2}{\pi} \int_0^{\pi} dq \left|\frac{d\omega}{dq}\right| \frac{\hbar\omega_q^3}{|\Delta_N|^2} [f(\omega_q, T_L) - f(\omega_q, T_R)], \tag{2.136}$$

with $m\omega_q^2 = k_o + 2[1 - \cos(q)]$. Now, one can prove the following result [5, 66]:

$$\lim_{N\to\infty} \int_0^{\pi} dq \frac{g_1(q)}{1 + g_2(q)\sin Nq} = \int_0^{\pi} dq \frac{g_1(q)}{[1 - g_2^2(q)]^{1/2}} , \tag{2.137}$$

where $g_1(q)$ and $g_2(q)$ are any two well-behaved functions. We the insert Eq. (2.131), with the Ohmic form of σ, in Eq. (2.136). For small ΔT, we then get, using Eq. (2.137) in the limit of large N

$$I = \frac{\gamma k^2 \hbar^2 (T_L - T_R)}{4\pi k_B m T^2} \int_0^{\pi} dq \frac{\sin^2 q}{\Lambda - \Omega\cos q} \, \omega_q^2 \, \mathrm{cosech}^2\left(\frac{\hbar\omega_q}{2k_B T}\right), \tag{2.138}$$

$$\text{where } \Lambda = 2(1 - k') + k'^2 + \frac{(k_o + 2)\gamma^2}{m} \text{ and } \Omega = 2(1 - k') + \frac{2\gamma^2}{m} .$$

One can see that, in the high temperature classical limit, the above expression gives a temperature-independent value for the conductance $I/\Delta T$. On the other hand, in the low temperature ($T << \hbar(k/m)^{1/2}/k_B$) regime, the following behavior is observed depending on parameter values:

$$I/\Delta T \sim T^3 \quad \text{for} \quad k' = k, k_o = 0 \tag{2.139}$$

$$\sim T \quad \text{for} \quad k' = 0, k_o = 0 \tag{2.140}$$

$$\sim \frac{e^{-\hbar\omega_o/(k_B T)}}{T^{1/2}} \quad \text{for} \quad k' = 0, k_o \neq 0\,, \tag{2.141}$$

where $\omega_o = (k_o/m)^{1/2}$. Note that this is different from the Landauer thermal conductance obtained in Eq. (2.135) for the Rubin model of bath. The difference comes from the resistance of the contacts which is least for the Rubin model.

In the classical limit $k_B T >> \hbar\omega_m$, we get

$$I = \frac{\gamma k^2 k_B (T_L - T_R)}{\pi m} \int_0^{\pi} \frac{\sin^2 q \, dq}{\Lambda - \Omega \cos q}$$
$$= \frac{\gamma k^2 k_B (T_L - T_R)}{m\Omega^2} (\Lambda - \sqrt{\Lambda^2 - \Omega^2})\,, \tag{2.142}$$

For the special case, $k' = k = 1$ and $k_o = 0$, we recover the result of Rieder et al. [4]:

$$I_C^{RLL} = \frac{k_B(T_L - T_R)}{2\gamma}\left[1 + \frac{\nu}{2} - \frac{\nu}{2}\sqrt{1 + \frac{4}{\nu}}\,\right] \text{ where } \nu = \frac{mk}{\gamma^2}\,. \tag{2.143}$$

In the other case of free ends, i.e. $k' = 0$, one gets the result, first obtained by Nakazawa [61]:

$$I^N = \frac{k\gamma k_B (T_L - T_R)}{2(mk + \gamma^2)}\left[1 + \frac{\lambda}{2} - \frac{\lambda}{2}\sqrt{1 + \frac{4}{\lambda}}\,\right] \text{ where } \lambda = \frac{k_o \gamma^2}{k(mk + \gamma^2)}\,. \tag{2.144}$$

Higher Dimensions It was shown by Nakazawa [61] that the problem of heat conduction in ordered harmonic lattices in more than one dimension can be reduced to an effectively one-dimensional problem. We briefly give the arguments here.

As before, let us write $\mathbf{n} = (n, \mathbf{n}')$, where $\mathbf{n} = (n_2, n_3 \ldots n_d)$ represent the transverse non-conducting direction, with $1 \leq n \leq N$ and for other directions we take $1 \leq n_\alpha \leq W$. We assume periodic boundary condition in all the transverse directions. For each of the transverse directions $\alpha = 2, 3, \ldots, d$, we can define a set of orthonormal basis functions $\phi_{q_\alpha}(n_\alpha)$ such that $2\phi_q(n) - \phi_q(n-1) - \phi_q(n+1) = 2[1 - \cos(q)]\phi_q(n)$. A possible choice is to set $q = 2s\pi/W$ and then take

$$\phi_q(n_\alpha) = \left(\frac{2}{W}\right)^{1/2} \sin(qn_\alpha) \quad \text{for} \quad -(W/2-1) \le s \le 1$$
$$= \left(\frac{2}{W}\right)^{1/2} \cos(qn_\alpha) \quad \text{for} \quad -0 \le s \le W/2\,,$$

where we have assumed W to be even. We then form the complete ortho-normal basis $\Phi_{\mathbf{q}}(\mathbf{n}') = \prod_{\alpha=2}^{d} \phi_{q_\alpha}(n_\alpha)$ and use these to transform to the new set of variables

$$\{X_n(\mathbf{q}),\ \sigma^a_{\mathbf{q}},\ \eta^a_{\mathbf{q}}\} = \sum_{\mathbf{n}'} \{X_{n,\mathbf{n}'},\ \sigma^a_{\mathbf{n}'},\ \eta^a_{\mathbf{n}'}\}\ \Phi_{\mathbf{q}}(\mathbf{n}')\,, \quad a = L, R\,. \tag{2.145}$$

Applying these to Eq. (2.115), one finds that for each fixed $\mathbf{q}$, $X_n(\mathbf{q})$ satisfies an effective $1D$ Langevin equation, with independent dissipation and noise terms, and with the onsite spring constant k_o replaced by the effective spring constants

$$k_o(\mathbf{q}) = k_o + 2\,k\,[\,d - 1 - \sum_{\alpha=2}^{d} \cos(q_\alpha)\,] \;= k_o + 4k \sum_{\alpha=2}^{d} \sin^2(q_\alpha/2)\,. \tag{2.146}$$

One can then repeat the analysis of the $1D$ case to get results for the higher dimensional case. In particular, it is clear that for large system sizes, we will again get a current density I/W^{d-1}, that is independent of the length of the system.

2.4.1.3 Disordered Harmonic Lattices

From the previous section we see that the heat current in an ordered harmonic lattice is independent of system size (for large systems) and hence transport is ballistic. This is of course expected since, in a pure harmonic crystal, there is no mechanism for scattering of the phonons. We now consider the effect of disorder on heat conduction in a harmonic system. Disorder can be introduced in various ways, for example by making the masses of the particles random as would be the case in a isotopically disordered solid, or by making the spring constants random. Here we discuss only the case of mass-disorder since the important qualitative features do not seem to vary much with the type of disorder.

It is interesting to look at the history of progress in the understanding of heat transport in this simple model. The thermal conductivity of disordered harmonic lattices was first investigated by Allen and Ford [67] who, using the Kubo formalism, obtained an exact expression for the thermal conductivity of a finite chain attached to infinite reservoirs. From this expression they concluded, erroneously as we now know, that the thermal conductivity remains finite in the limit of infinite system size.

Simulations of the disordered lattice were carried out by Payton et al. [68] and from the restricted small system sizes ($N \sim 400$) that they could study, a finite thermal conductivity was obtained.

The first paper to notice anomalous transport and the importance of localization physics was that by Matsuda and Ishii (MI) [69]. In their important work on the localization of normal modes in the disordered harmonic chain, MI showed that all high frequency modes were exponentially localized. They also observed that, for small ω, the localization length in an infinite sample varied as ω^{-2}, hence normal modes with frequency $\omega \lesssim \omega_d$ have localization length greater than N, and are thus extended. For a harmonic chain of length N, given the average mass $m = \langle m_l \rangle$, the variance $\sigma^2 = \langle (m_l - m)^2 \rangle$ and inter-particle spring constant k, it was shown that $\omega_d \sim [km/(N\sigma^2)]^{1/2}$. They also evaluated expressions for thermal conductivity of a finite disordered chain connected to two different bath models (see Sect. 2.2.1), namely (i) white noise baths and (ii) Rubin model. This was treated using the Green-Kubo formalism of Allen and Ford [67]. They found $\alpha = 1/2$ in both cases, which, as we will see is not completely correct. The other two important theoretical papers on heat conduction in the disordered chain were those by Rubin and Greer [6] who considered model (ii) and of Casher and Lebowitz [5] who used model (i) for baths. A lower bound $[J] \geq 1/N^{1/2}$ was obtained for the disorder averaged current $[J]$ in [6] who also gave numerical evidence for an exponent $\alpha = 1/2$. This was later proved rigorously by Verheggen [70]. On the other hand, for model (i), [5] found a rigorous bound $[J] \geq 1/N^{3/2}$ and it was suggested that $\alpha = -1/2$. This was recently proved rigorously by Ajanki and Huveneers [71]. The work in [23] gave a unified treatment of the problem of heat conduction in disordered harmonic chains connected to baths modeled by generalized Langevin equations and showed that models (i,ii) were two special cases. In [66] it was pointed out that the difference in exponents obtained for these two cases is because they effectively correspond to choice of two different boundary conditions. Model (i) used fixed BC, while model (ii) corresponds to free BC. These results show that the asymptotic properties of current in the disordered chain depends on the boundary conditions (fixed or free) and the low-frequency form of the $\nu(\omega)$ ($Im[\sigma(\omega)]$).

The extension of these studies to higher dimensions was carried out in [62, 63]. The discussions of this section are based on the framework of these papers. We start with some phenomenological considerations of the kind of physics we expect in the presence of disorder. In particular let us discuss the kinetic theory picture and the phenomenon of Anderson localization of phonons.

(i) **Kinetic Theory Picture**: For weak disorder one can imagine that the phonons (of the pure crystal) are weakly scattered, and hence, some sort of kinetic theory approach might be applicable. The low frequency phonons in particular are less effected by disorder and the normal modes can approximately be described by perturbed plane waves. The Rayleigh scattering theory of phonons gives an effective mean free path $\ell_K(\omega) \sim \omega^{-(d+1)}$, for dimensions $d > 1$. The phonon diffusion constant is given by $D(\omega) = v\ell_K(\omega)$ where v, the sound velocity, can be taken to be a constant. For a finite system of linear dimension N we have $\ell_K(\omega) > N$ for $\omega \lesssim \omega_c^K = N^{-1/(d+1)}$ and in this case, the mean free path is basically set by boundary scattering and so we take $D = vN$. Kinetic theory then predicts

$$\kappa = \int_{N^{-1}} d\omega \rho(\omega) D(\omega) ,$$
$$\sim \int_{N^{-1}}^{\omega_c^K} d\omega \omega^{d-1} vN + \int_{\omega_c^K}^{\omega_d} d\omega \omega^{d-1} v \ell_K(\omega) \sim N^{1/(d+1)} . \tag{2.147}$$

where $\rho(\omega) \sim \omega^{d-1}$ is the density of states and we are considering the classical limit where the specific heat for each phonon mode is a constant. The divergence of the phonon mean free path at low frequencies and the resulting divergence of the thermal conductivity of a disordered harmonic crystal has been discussed in the literature. For example, in the classic book by Ziman it is argued that, even at low temperatures, anharmonicity is necessary to make κ finite [72, 73].

(ii) **Phonon Localization**: The kinetic theory picture is expected to be useful in the limit of weak disorder and also for the description of long-wavelength phonons. On the other hand, for strong disorder, if we look at the normal mode solutions of the disordered system, then one expects that these are affected by the physics of Anderson localization. Anderson localization is a phenomenon that has been studied in the context of electron transport in the presence of a disordered potential. However, localization is expected whenever one has linear waves moving in a disordered media.

In fact the problem of finding the normal modes of the harmonic lattice can be directly mapped to that of finding the eigenstates of an electron in a disordered potential (in a tight-binding model, for example) and so we expect the same kind of physics as in electron localization. To see this, consider the disordered harmonic lattices in the absence of coupling to reservoirs. The d-dimensional lattice has $p = 1, 2, \ldots, N^d$ normal modes and let us denote the displacement field corresponding to the pth mode by $a_{\mathbf{n}}(p)$ and the corresponding eigenvalue by ω_p^2. The normal mode equation corresponding to the Hamiltonian in Eq. (2.114) is given by:

$$m_{\mathbf{n}} \omega_p^2 a_{\mathbf{n}} = (2dk + k_{\mathbf{n}}) a_{\mathbf{n}} - k \sum_{\hat{\mathbf{e}}} a_{\mathbf{n}+\hat{\mathbf{e}}} , \tag{2.148}$$

where the $a_{\mathbf{n}}$ satisfy appropriate boundary conditions. Introducing variables $\psi_{\mathbf{n}}(p) = m_{\mathbf{n}}^{1/2} a_{\mathbf{n}}(p)$, $\epsilon_{\mathbf{n}} = (2d + k_{\mathbf{n}})/m_{\mathbf{n}}$ and $t_{\mathbf{n},\mathbf{l}} = k/(m_{\mathbf{n}} m_{\mathbf{l}})^{1/2}$ for nearest neighbor sites $\mathbf{n}$, $\mathbf{l}$, the above equation transforms to the following form:

$$\omega_p^2 \psi_{\mathbf{n}}(p) = \epsilon_{\mathbf{n}} \psi_{\mathbf{n}}(p) - \sum_{\mathbf{l}} t_{\mathbf{n},\mathbf{l}} \psi_{\mathbf{l}}(p) . \tag{2.149}$$

This has the usual structure of an eigenvalue equation for a single electron moving in a d-dimensional lattice corresponding to a tight-binding Hamiltonian with nearest neighbor hopping $t_{\mathbf{n},\mathbf{l}}$ and on-site energies $\epsilon_{\mathbf{n}}$. Note that $t_{\mathbf{n},\mathbf{l}}$ and $\epsilon_{\mathbf{n}}$ are, in general, correlated random variables, hence the disorder-energy diagram might differ from that of the usual single band Anderson tight-binding model.

For the case of electrons, by looking at the eigenstates and eigenfunctions of the isolated system of a single electron in a disordered potential one finds that, in contrast to the spatially extended Bloch states in periodic potentials, there are now many eigenfunctions which are exponentially localized in space. It has been shown rigorously [74–76], that in one dimension ($1D$), all states are exponentially localized. In two dimensions ($2D$), there is no proof, but it is believed that again all states are localized. In three dimensions ($3D$) there is expected to be a transition from extended to localized states as the energy is moved towards the band edges [77]. The transition from extended to localized states, which occurs when the disorder is increased, changes the system from a conductor to an insulator. Analogously, one finds that the normal mode wave-function of a disordered harmonic crystal show these same features of exponential localization in space. It is clear that such states cannot participate in transmission of energy.

One difference between the electron case and the phonon case is that, for the harmonic crystal,in the absence of an external potential, the translational invariance of the problem leads to the fact that *low frequency modes are not localized and are effective in transporting energy*. Another important difference between the electron and phonon problems is that *electron transport is dominated by electrons near the Fermi level while in the case of phonons, all frequency phonons participate in transport*. These two differences lead to the fact that the disordered harmonic crystal in one and two dimensions is not a heat insulator, unlike its electronic counterpart (except when we introduce an external pinning potential).

The heat current carried by a mode which is localized on a length scale ℓ, decays with system length N as $e^{-N/\ell}$. This ℓ depends on the phonon frequency and, low frequency modes for which $\ell \sim N$, will therefore be carriers of the heat current. A renormalization group study of phonon localization in a continuum vector displacement model was carried out by John et al. [78]. They found that much of the predictions of the scaling theory of localization for electrons carry over to the phonon case. Specifically they showed that in one and two dimensions all non-zero frequency phonons are localized with the low frequency localization length diverging as $\ell \sim \omega^{-2}$ and $\sim e^{c/\omega^2}$, respectively (where $c > 0$ is some constant). This means that in $1D$, all modes with $\omega \gtrsim \omega_c^L = N^{-1/2}$ are localized, while in $2D$, all modes with $\omega \gtrsim \omega_c^L = [\log(N)]^{-1/2}$ are localized. In $3D$ the prediction is that there is an ω_c^L, independent of N, above which all modes are localized.

In summary the discussions above predict a frequency cut-off ω_c^K that kinetic theory predicts, below which we expect the mean free path to be of order system size, implying ballistic phonons. Localization theory predicts a second cut-off ω_c^L above which we expect states to be localized, implying non-conducting phonons. We shall use these inputs in the general Landauer heat current formula, to build our understanding of the asymptotic behaviour of current with system size. We start with the one-dimensional case for which a lot of exact results are known.

General Remarks

(1) We will focus here on the classical case—the size-dependence of current is governed by low frequency phonons, hence, so far as this aspect is concerned, we expect the same behaviour also in the quantum case.
(2) Disorder is chosen only in the masses of the particles and this is specified by assigning half of randomly chosen particles to have masses $m - \Delta$ and other half to have masses $m + \Delta$.
(3) We will only consider Ohmic baths, where all boundary particles are coupled to reservoirs with $\nu(\omega) = \gamma\omega$, and discuss the three situations (see Sect. 2.4.1.1 after Eq. (2.115)), corresponding to unpinned lattices with fixed and free boundary conditions, and pinned lattices.
(4) In discussions on properties of disordered systems, one encounters the phenomenon of self-averaging, whereby it is seen that a single large system is sufficient to represent the whole ensemble. We will assume that the systems we study satisfy this property and so the average property is also the typical property. We then use the following notation for the disorder averaged (denoted by the symbol [...]) transmission per unit area

$$T(\omega) = \frac{[\mathcal{T}(\omega)]}{N^{d-1}} . \tag{2.150}$$

(5) We also define the average current density as

$$J = \frac{[I]}{N^{d-1}} . \tag{2.151}$$

One-Dimensional Disordered Lattice Let us briefly review the understanding of the boundary condition dependence, of the heat current exponent α, for the one dimensional case. Our model is the one-dimensional Hamiltonian given by Eq. (2.120).

For any realization of disorder, the current is given by the general expression Eq. (2.121), with the transmission given by $\mathcal{T}(\omega) = 4\gamma^2\omega^2|g_N^+(\omega)|^2$ where $g_N^+(\omega)$ is now just a complex number. The disorder averaged transmission is given by $T(\omega) = [\mathcal{T}(\omega)]$. There are three observations that enable one to determine the asymptotic system size dependence of the current. These are:

(i) $\Delta_N = [g_N^+]^{-1}$, given by Eqs. (2.125), (2.126) is a complex number which can be expressed in terms of the product of N random 2×2 matrices. From a theorem due to Furstenberg [79], it follows that for almost all disorder realizations, the large N behaviour of Δ_N, for fixed $\omega > 0$, is $|\Delta_N| \sim e^{bN\omega^2}$, where $b > 0$ is a constant. Since $T(\omega) \sim |\Delta_N|^{-2} \sim e^{-2bN\omega^2}$, this implies that transmission is significant only for low frequencies $\omega \lesssim \omega_c(N) \sim 1/N^{1/2}$. The current is therefore dominated by the small ω behaviour of $T(\omega)$.

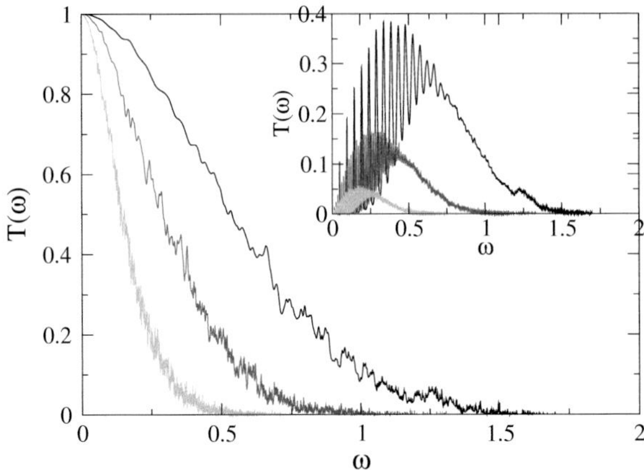

Fig. 2.2 $1D$ unpinned case with both free and fixed (*inset*) boundary conditions: plot of the disorder averaged transmission $T(\omega)$ versus ω for $\Delta = 0.4$. The various curves (*from top to bottom*) correspond to lattices of sizes $N = 64, 256, 1024$ respectively (from [62])

(ii) The second observation is that the transmission for $\omega < \omega_c(N)$ is ballistic in the sense that $T(\omega)$ is insensitive to the disorder.

(iii) The final important observation is that the form of the prefactors of $e^{-bN\omega^2}$ in $T(\omega)$ for $\omega < \omega_c(N)$ depends strongly on boundary conditions and bath properties. For the white noise Langevin baths one finds $T(\omega) \sim \omega^2 e^{-bN\omega^2}$ for fixed BC and $T(\omega) \sim \omega^0 e^{-bN\omega^2}$ for free BC. This difference arises because of the scattering of long wavelength modes by the boundary pinning potentials. With these form of $T(\omega)$, one can perform the integral over ω, to arrive at the conclusion that $J \sim N^{-3/2}$ for fixed BC and $J \sim N^{-1/2}$ for free BC. In the presence of a pinning potential the low-frequency modes are suppressed and one obtains a heat insulator with $J \sim e^{-cN}$, with c a constant.

In Fig. 2.2 we plot numerical results showing $T(\omega)$ for the $1D$ binary mass-disordered lattice with both fixed and free boundary conditions. One can clearly see the various features discussed above, namely the dependence of frequency cut-off on system size and the dependence of form of $T(\omega)$ on boundary conditions. In Fig. 2.3 we plot results for size dependence of J, obtained in [66] from a direct numerical evaluation of the Landauer formula. The computations were done for white noise baths and Rubin baths, and for two different BCs. We see the expected dependence of α on BCs, and no dependence on baths (note that this is because both baths have $\nu(\omega) \sim \omega$ for $\omega \to 0$)

Higher Dimensional Disordered Harmonic Lattice One can extend the analysis of the $1D$ case to higher dimensions, using inputs from kinetic theory and the theory of phonon localization. The main point of the arguments involves the assumption that normal modes can be classified as ballistic, diffusive or localized. Using localization theory one can estimate the frequency region where states are

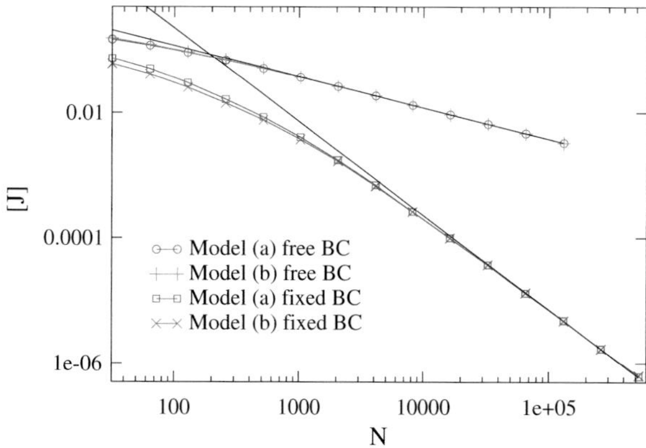

Fig. 2.3 Plot of $[J]$ versus N for free and fixed boundary conditions. Results are given for both (a) white noise baths and (b) Rubin model. The *two straight lines* correspond to the asymptotic expected forms $J \sim N^{1-\alpha}$ with $\alpha = 1/2$ (free BC) and $\alpha = -1/2$ (fixed BC). Parameters used were $m = 1,\ \Delta = 0.5,\ k = 1,\ \gamma = 1,\ T_L = 2,\ T_R = 1$ and $k_o = 1$ (from [66])

localized. The lowest frequency states with $\omega \to 0$ will be ballistic and one can use kinetic theory to determine the fraction of extended states which are ballistic. The assumption is that at sufficiently low frequencies the effective disorder is weak (even when the mass variance Δ is large) and one can use kinetic theory. Corresponding to the three observations made above for the $1D$ case, one can now make the following arguments:

(1) From localization theory one expects all fixed non-zero frequency states in a $2D$ disordered system to be localized when the size of the system goes to infinity. As discussed in Sect. 2.4.1.3 localization theory gives us a frequency cut-off $\omega_c^L = (\ln N)^{-1/2}$ in $2D$ above which states are localized. In $3D$ one obtains a finite frequency cut-off ω_c^L independent of system size above which states are localized.

(2) For the unpinned case with finite N, there will exist low frequency states below ω_c^L, in both $2D$ and $3D$, which are extended states. These states are either diffusive or ballistic. Ballistic modes are insensitive to the disorder and their transmission coefficient are almost the same as for the ordered case. To find the frequency cut-off, below which states are ballistic, one can use kinetic theory results. For the low-frequency extended states we expect kinetic theory to be reliable and this gives us a mean free path for phonons $\ell_K \sim \omega^{-(d+1)}$. This means that for low frequencies $\omega \lesssim \omega_c^K = N^{-1/(d+1)}$ we have $\ell_K(\omega) > N$ and phonons transmit ballistically. We now proceed to calculate the contribution of these ballistic modes to the total current. This can be obtained by looking at the small ω form of $T(\omega)$ for the ordered lattice.

(3) For the ordered lattice, $T(\omega)$ is typically a highly oscillatory function, with the oscillations increasing with system size. An effective transmission coefficient in

the $N \to \infty$ limit can be obtained by considering the integrated transmission. This asymptotic *effective* low-frequency form of $T(\omega)$, for the ordered lattice can be calculated using methods described in [80] and is given by:

$$\begin{aligned} T(\omega) &\sim \omega^{d+1}, \quad \text{fixed BC} \\ T(\omega) &\sim \omega^{d-1}, \quad \text{free BC}, \end{aligned} \tag{2.152}$$

the result being valid for $d = 1, 2, 3$.

Using the above arguments we then get the ballistic contribution to the total current density (for the unpinned case) as:

$$\begin{aligned} J_{\text{ball}} &\sim \int_0^{\omega_c^K} d\omega\, \omega^{d+1} \sim \frac{1}{N^{(d+2)/(d+1)}}, \quad \text{fixed BC}, \\ &\sim \int_0^{\omega_c^K} d\omega\, \omega^{d-1} \sim \frac{1}{N^{d/(d+1)}}, \quad \text{free BC}. \end{aligned} \tag{2.153}$$

We can now make predictions for the asymptotic system size dependence of total current density in two and three dimensions.

Two Dimensions From localization theory one expects that all finite frequency modes $\omega \gtrsim \omega_c^L = (\ln N)^{-1/2}$ are localized and their contribution to the total current falls exponentially with system size. Our kinetic theory arguments show that the low frequency extended states, with $\omega_c^K \lesssim \omega \lesssim \omega_c^L$ are diffusive (where $\omega_c^K = N^{-1/3}$), while the remaining modes with $\omega \lesssim \omega_c^K$ are ballistic. The diffusive contribution to total current will then scale as $J_{\text{diff}} \sim (\ln N)^{-1/2} N^{-1}$. The ballistic contribution depends on BCs and is given by Eq. (2.153). This gives $J_{\text{ball}} \sim N^{-4/3}$ for fixed BC and $J_{\text{ball}} \sim N^{-2/3}$ for free BC. Hence, adding all the different contributions, we conclude that asymptotically:

$$\begin{aligned} J &\sim \frac{1}{(\ln N)^{1/2} N}, \quad \text{fixed BC}, \; d = 2, \\ &\sim \frac{1}{N^{2/3}}, \quad \text{free BC}, \; d = 2. \end{aligned} \tag{2.154}$$

In the presence of an onsite pinning potential at all sites the low frequency modes get cut off and all the remaining states are localized, hence we expect:

$$J \sim e^{-bN}, \qquad \text{pinned}, \; d = 2, \tag{2.155}$$

where b is some positive constant.

Three-Dimensions In this case localization theory tells us that modes with $\omega \gtrsim \omega_c^L$ are localized and ω_c^L is independent of N. From kinetic theory we find that the extended states with $\omega_c^K \stackrel{<}{\sim} \omega \stackrel{<}{\sim} \omega_c^L$ are diffusive (with $\omega_c^K = N^{-1/4}$) and those with $\omega \stackrel{<}{\sim} \omega_c^K$ are ballistic. The contribution to current from diffusive modes scales as $J_{\rm diff} \sim N^{-1}$. The ballistic contribution (from states with $\omega \stackrel{<}{\sim} N^{-1/4}$) is obtained from Eq. (2.153) and gives $J_{\rm ball} \sim N^{-5/4}$ for fixed BC and $J_{\rm ball} \sim N^{-3/4}$ for free BC. Hence, adding all contributions, we conclude that asymptotically:

$$\begin{aligned} J &\sim \frac{1}{N}, \qquad \text{fixed BC}, \ d = 3, \\ &\sim \frac{1}{N^{3/4}}, \qquad \text{free BC}, \ d = 3. \end{aligned} \tag{2.156}$$

In the presence of an onsite pinning potential at all sites the low frequency modes get cut off and, since in this case the remaining states form bands of diffusive and localized states, hence we expect:

$$J \sim \frac{1}{N}, \qquad \text{pinned}, \ d = 3. \tag{2.157}$$

Thus in $3D$ both the unpinned lattice with fixed boundary conditions and the pinned lattice are expected to show Fourier type of behaviour as far as the system size dependence of the current is considered.

Note that for free BC, the prediction for the current contribution from the ballistic part $J_{\rm ball} \sim N^{-d/(d+1)}$ is identical to that from kinetic theory, Eq. (2.147). The kinetic theory argument is independent of BCs and the agreement with free BCs can be traced to the small ω form of $T(\omega) \sim \omega^{d-1}$ (see Eq. (2.152)) being identical to the form of the density of states $\rho(\omega)$ used in the kinetic theory formula. On the other hand, for fixed BCs the form of $T(\omega)$ and $\rho(\omega)$ are different. However it seems reasonable to expect that, since the transport current phonons are injected at the boundaries, in kinetic theory one needs to use the *local density of states* evaluated at the boundaries. For fixed BC this will then give rise to an extra factor of ω^2 (from the squared wave-function) and then the kinetic theory prediction matches with those given above.

Results from Numerics and Simulations In [62, 63], numerical calculations and simulations were done to check the above heuristic predictions. The authors considered heat transport across harmonic mass-disordered lattices in two and three dimensions. Slabs of dimensions $N \times W$ (in 2D) and $N \times W \times W$ (in 3D), with $W \leq N$, were studied and the heat current was evaluated from the Landauer formula using the transfer matrix approach (see Sect. 2.4.1.1). They computed the current for $2D$ and $3D$ mass-disordered lattices for different boundary conditions, and for the unpinned and pinned lattice. Various other non-equilibrium quantities such as transmission coefficients, heat current density and temperature profiles were evaluated.

In addition some properties of the normal modes of the isolated system were also examined, such as density of states and inverse participation ratios (IPR). All eigenvalues and eigenstates of Eq. (2.148) were numerically evaluated for finite cubic lattices. The degree of localization of a given mode was measured via the inverse participation ratio (IPR), defined as follows:

$$P^{-1} = \frac{\sum_{\mathbf{n}} a_{\mathbf{n}}^4}{(\sum_{\mathbf{n}} a_{\mathbf{n}}^2)^2} . \tag{2.158}$$

For a completely localized state, i.e. $a_{\mathbf{n}} = \delta_{\mathbf{n},\mathbf{n}_0}$, P^{-1} takes the value 1. On the other hand for a completely delocalized state, for which $a_{\mathbf{n}} = N^{-d/2} e^{i\mathbf{n}.\mathbf{q}}$ where $\mathbf{q}$ is a wave vector, P^{-1} takes the value N^{-d}. Another useful characterization of the disorder system is obtained from the density of states, $\rho(\omega)$, of the disordered system defined by:

$$\rho(\omega) = \sum_p \delta(\omega_p - \omega) . \tag{2.159}$$

We summarize here some of the main conclusions of Kundu et al. [62, 63].

Average Current The numerical results for the size-dependence of current, in two and three dimensional systems, are shown in Figs. 2.4 and 2.5. The results for system-size dependence for the various cases, and their comparison with the theory are summarized in Table 2.1. We see that for the pinned system, the theoretical predictions of insulating (2D) and diffusive (3D) transport seem to be validated. For the un-pinned case, the 2D system shows significant difference for different BCs but the expected size-dependence is not observed. For the 3D unpinned case both the free and fixed BCs give a size-dependence in accordance to kinetic theory prediction. It is possible that one needs to go to larger system sizes to see the true scaling limit.

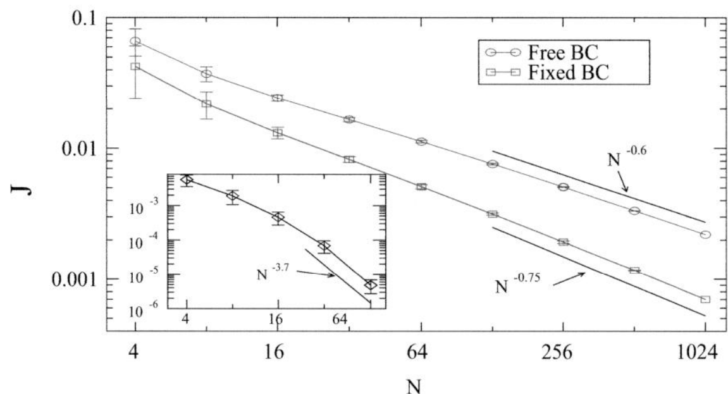

Fig. 2.4 Plot of the N-dependence of disorder averaged current J in two dimensions. The inset shows result for pinned case. For free BC $\Delta = 0.8$ and for fixed BC $\Delta = 0.95$. For pinned case $\Delta = 0.4, k_o = 10.0$. Error bars show standard deviations due to sample-to-sample fluctuations and are very small except in the $2D$ pinned case (from [63])

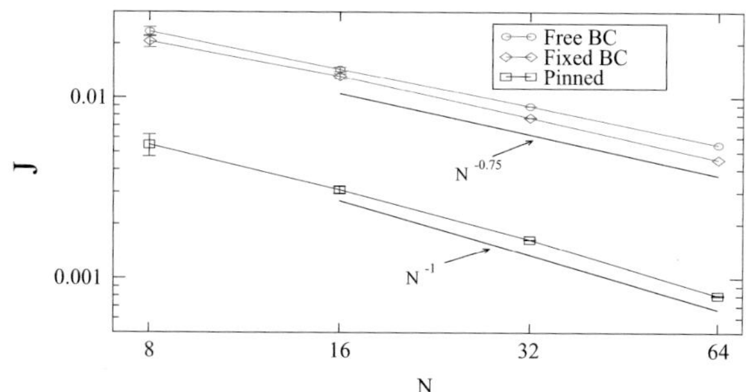

Fig. 2.5 Plot of the N-dependence of disorder averaged current J in three dimensions. For all cases $\Delta = 0.8$ and for pinned case $k_o = 10.0$. Error bars show standard deviations due to sample-to-sample fluctuations and are very small except in the 2D pinned case (from [63])

Table 2.1 Numerical results obtained in [63], for size-dependence of heat current in two and three dimensional mass-disordered harmonic lattices, are compared with the analytical predictions

	$d = 2$		$d = 3$	
	Analytical	Numerical	Analytical	Numerical
Pinned	$\exp(-bN)$	$N^{-3.7}$	N^{-1}	$N^{-1.0}$
Fixed	$N^{-1}(\ln N)^{-1/2}$	$N^{-0.75}$	N^{-1}	$N^{-0.75}$
Free	$N^{-2/3}$	$N^{-0.6}$	$N^{-3/4}$	$N^{-0.71}$

The errors bar for the numerically obtained exponent values were estimated to be of the order ± 0.02

A clearer understanding of the nature of heat transport is obtained by looking at the detailed form of the phonon transmission function $\mathcal{T}(\omega)$ and also the properties of the normal modes. We present here some of the typical results obtained in [63] for the case of the unpinned lattice with fixed boundary conditions. The reader is referred to [62, 63] for further detail of other cases.

The N-dependence of the disorder-averaged phonon transmission coefficient $T(\omega) = [\mathcal{T}_N(\omega)]/N^{d-1}$ sheds additional light on the nature of phonons in different frequency regions. In Figs. 2.6 and 2.7 we plot the frequency dependence of $T(\omega)$ for different system sizes. The size-independence of $T(\omega)$ at a given frequency implies ballistic transmission. On the other hand, diffusive transmission implies $T(\omega) \sim N^{-1}$, hence to see the diffusive frequencies, we also plot $NT(\omega)$. Also shown in Figs. 2.6 and 2.7 are the phonon density of states and the inverse participation (IPR) ratio, which are obtained from the normal modes of the isolated system (see Eqs. (2.158) and (2.159)). The inferences that we can draw from these plots can be summarized as follows:

(a) The density of states tells us the about the allowed range of phonon modes in the system. We see that both in the 2D and 3D cases the allowed modes fall into two bands. In the figure we also show the corresponding phonon bands for the case where the two masses are arranged periodically, in which case the modes

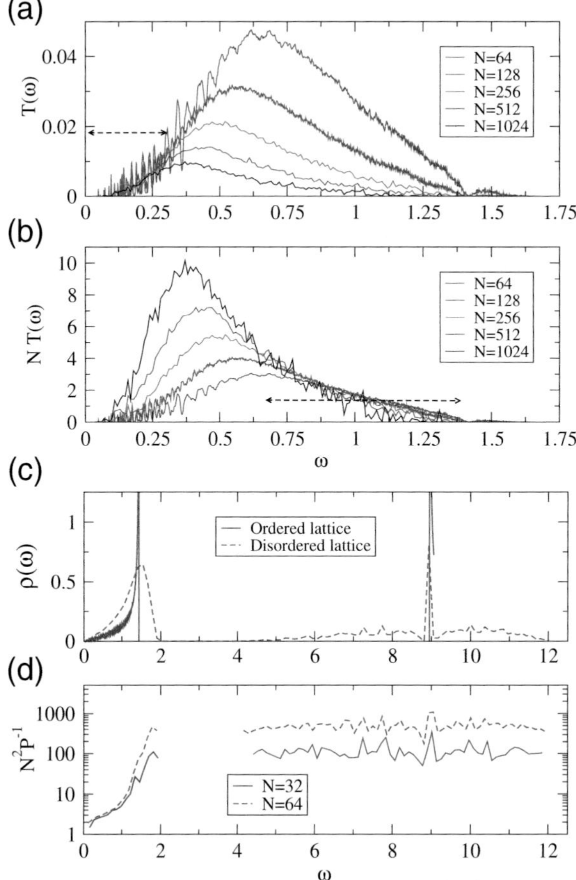

Fig. 2.6 $2D$ unpinned $N\times N$ lattice, with fixed BC, for $\Delta = 0.95$. (**a**) Plot of the disorder averaged transmission $T(\omega)$ versus ω. The *dashed line* here indicates the ballistic range where $T(\omega) \sim N^0$. (**b**) Plot of $NT(\omega)$. The range of diffusive frequencies $[T(\omega) \sim 1/N]$ is indicated by the *dashed line*. (**c**) Plot of $\rho(\omega)$ for the ordered binary mass system, and a single disordered sample. (**d**) Plot of N^2P^{-1} for single sample (smoothed data). We see that, even though the allowed normal modes occur over a large frequency band $\approx (0-12)$, transmission takes place in the small band $\approx (0-1.25)$ and is negligible elsewhere. Non-collapse of the IPR plots for different N in (**d**) indicate localized states. This confirms that the non-transmitting states correspond to localized modes. In (**a**) we see that ω_c^L is slowly decreasing with increase of N (from [63])

are basically plane-waves. In the presence of disorder the phonon bands show significant broadening, in comparison with the ordered lattice.

(b) The IPR, calculated for each normal mode, tells us whether the mode is localized or extended. The size-dependence $P^{-1} \sim 1/N^{d-1}$ implies extended states. For the $2D$ case in Fig. 2.6 we see that all modes in the high-ω band, and also a significant fraction of the lower band, are localized. This is consistent

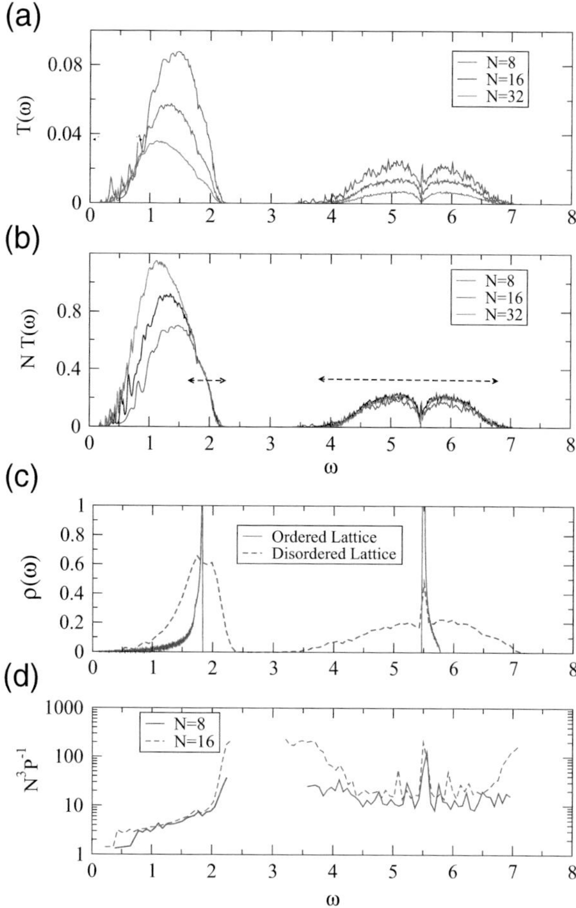

Fig. 2.7 $3D$ unpinned $N \times N \times N$ lattice, with fixed BC, for $\Delta = 0.8$. (**a**) Plot of the disorder averaged transmission $T(\omega)$ versus ω. (**b**) Plot of $NT(\omega)$. The diffusive frequencies, for which $T(\omega) \sim 1/N$, are indicated by the *dashed line*. (**c**) Plot of $\rho(\omega)$ for the ordered binary mass system, and a single disordered sample. (**d**) Plot of N^3P^{-1} for single sample (smoothed data). We see that, that in the 3D case, there is a small fraction of localized frequencies which occur in the band edges. Non-collapse of the IPR plots for different N in (**d**) indicate localized states and confirms their location at the band-edges (from [63])

with our expectation that in $2D$ (as also in $1D$), for large enough system sizes, almost all modes are localized except at the lowest frequencies. Figure 2.8 shows typical examples of extended and localized normal modes. Note that the extended state has a random structure and is very different from a plane wave. For the $3D$ case in Fig. 2.7) we see that both bands contain a large number of extended states, with a small fraction of localized states at the band edges.

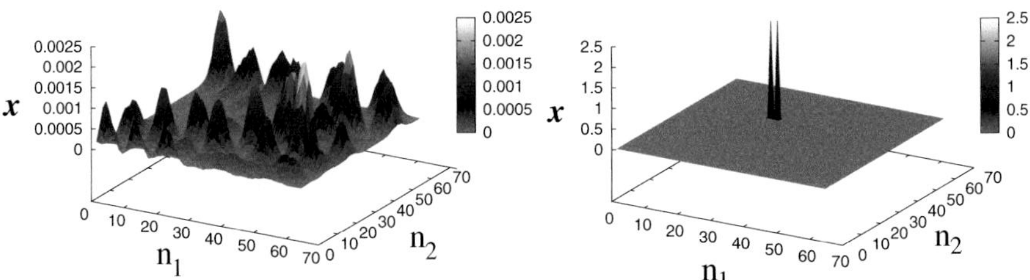

Fig. 2.8 $2D$ unpinned 64×64 lattice, with fixed BC, for $\Delta = 0.8$: Two typical normal mode amplitudes are shown, one for a small value of P^{-1} (*left*) and the other for a large value of P^{-1} (*right*) (from [62])

(c) The plot of $T(\omega)$ show a significant transmission only over the frequency ranges corresponding to extended modes, providing further confirmation on the range of frequencies where modes are localized. At small frequencies we see that transmission is independent of system size and these are the ballistic modes. We expect $T(\omega) \sim N^{-1}$ for the diffusive modes and hence, by plotting $NT(\omega)$, we can identify the set of diffusive sets. In Fig. 2.6 we see that in $2D$ the modes in the upper edge of the transmitting band is diffusive. It is difficult to verify the predictions (see discussions above) that modes in the range $\omega_c^K < \omega < \omega_c^L$ are diffusive, and the size-dependences of the two cut-off frequencies. In Fig. 2.7 for the $3D$ case, we see again that the localized states (seen from the IPR plot) show negligible transmission. The extended states in the high-ω band show diffusive property while the modes near zero-frequency show ballistic behaviour. We see that the mode properties are somewhat different from the analytic predictions, in particular, from the current system sizes that have been studied, it is difficult to see the size-dependence of ω_c^K . As expected, at small frequencies we roughly find $T(\omega) \sim \omega^{d+1}$, which is the behaviour for the ordered system.

Convergence to 3D Value for Rectangular Slabs What happens if, instead of a cubic sample, we consider a rectangular slab of dimensions $N \times W \times W$, with $W \leq N$? Results in [62] showed that there is a quick convergence of the current density, to the fully $3D$ value (for $W = N$), as we increase the ratio W/N. This point can be seen in Fig. 2.9 where results from non-equilibrium simulations show the dependence of the current densities on the ratio $W \leq N$.

2.4.2 *Temperature Profiles*

It is common (for the classical system) to define the local temperature at a site $\mathbf{n}$ through the relation $k_B T_{\mathbf{n}} = m_{\mathbf{n}} \langle \dot{\mathbf{x}}_{\mathbf{n}}^2 \rangle$. The temperature profile in the ordered classical harmonic lattices has been obtained exactly [4]. One finds that, for large system size

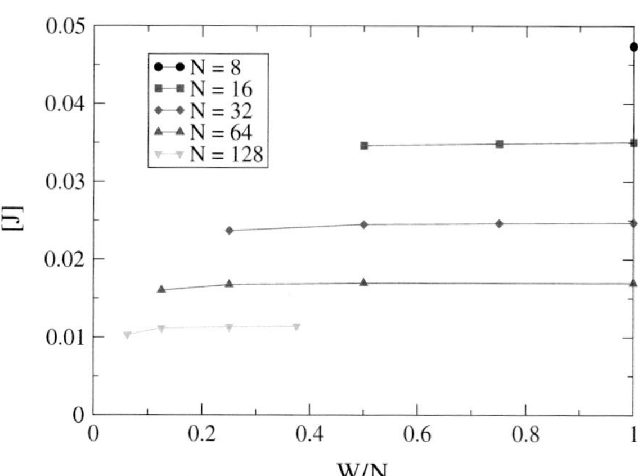

Fig. 2.9 3*D* unpinned $N \times W \times W$ lattice, with fixed BC, for $\Delta = 0.2$: Plot of disorder-averaged current density J (with the definition $J = I/W^2$) versus W/N for different fixed values of N. We see that the 3D limiting value is reached at quite small values of W/N (from [62])

and for equal system-bath coupling of system at the two ends, the temperature in the bulk of the system is given by $(T_L + T_R)/2$. Further understanding of this result is obtained if we decompose the local temperature as a sum over contributions from different normal modes. For a one-dimensional and, in the limit of weak coupling, this normal-mode analysis was carried out in [2, 69].

This analysis has also been in done the quantum mechanical case [24], again for the special case of weak coupling to reservoirs and we present this discussion. In the limit of weak system-bath coupling we can write the temperature profile in terms of the normal modes of the system. For simplicity, let us consider the one-dimensional chain in the quantum case for the case of Ohmic baths with coupling constants γ_L and γ_R at the two ends. From the results obtained in [24], we can write the following form for the local kinetic energy:

$$\left\langle \frac{p_l^2}{2m_l} \right\rangle = m_l \sum_p a_l^2(p) \left\langle \frac{P_p^2}{2} \right\rangle , \tag{2.160}$$

where $a_l(p)$ denotes the pth normal mode eigenfunction, defined in Eq. (2.148), P_p denotes normal mode momenta, and $\langle P_p^2/2 \rangle = e_p$ denotes the average kinetic energy of a normal mode in contact with two reservoirs. This is given by a weighted average of contributions from the two reservoirs:

$$e_p = \alpha_L(p) \frac{\hbar\omega_p}{4} \cot\left(\frac{\hbar\omega_p}{2k_B T_L}\right) + \alpha_R(p) \frac{\hbar\omega_p}{4} \cot\left(\frac{\hbar\omega_p}{2k_B T_R}\right) \tag{2.161}$$

$$\text{where } \alpha_L(p) = \frac{\gamma_L a_1(p)^2}{\gamma_L a_1(p)^2 + \gamma_R a_N(p)^2},$$

$$\alpha_R(p) = \frac{\gamma_R a_N(p)^2}{\gamma_L a_1(p)^2 + \gamma_R a_N(p)^2} . \tag{2.162}$$

From the local kinetic energy, one can then define an effective local temperature, based on what one gets in equilibrium. The high temperature limit of the above equation gives the effective local temperature as

$$T_l = \frac{1}{m_l} \sum_p a_l^2(p) \left[\alpha_L(p) T_L + \alpha_R(p) T_R\right] , \tag{2.163}$$

which is the result given in [2, 69] (for $\gamma_L = \gamma_R$). For the ordered chain, the wave functions at the boundaries are equal $a_1(p) = a_N(p)$, hence, using the normalization of the wave-functions, we get

$$T_l = \frac{\gamma_L T_L + \gamma_R T_R}{\gamma_L + \gamma_R} . \tag{2.164}$$

This gives the well-known result $T_l = (T_L + T_R)/2$ for the case of equal coupling.

Finally let us discuss temperature profiles for the case of disordered profiles. This is very difficult to obtain numerically in one-dimensional systems, since most of the modes are localized and it takes exponentially large (in system size) times to equilibrate them. Here we show some numerical results obtained for the case of the 3D disordered crystal discussed in the last section. This seems to show a linear temperature profile. Again one can define the local temperature at each site $\mathbf{n}$ through the relation $k_B T_{\mathbf{n}} = m_{\mathbf{n}} \langle \dot{\mathbf{x}}_{\mathbf{n}}^2 \rangle$. We expect small fluctuations in a layer transverse to the heat-conduction direction and so it is useful to study the temperature-profile in the conduction direction through $T_i = (1/N^{d-1}) \sum_{\mathbf{n}'} T_{i,\mathbf{n}'}$. In Fig. 2.10 we show the temperature profile, obtained from simulations of a single disorder realization, for lattices of different sizes and with $\Delta = 0.2$. The jumps at the boundaries again indicate that the asymptotic system size limit has not been reached even at the largest size. For strong disorder, when a significant fraction of the states are localized, the localized modes are difficult to equilibrate. Thus, steady

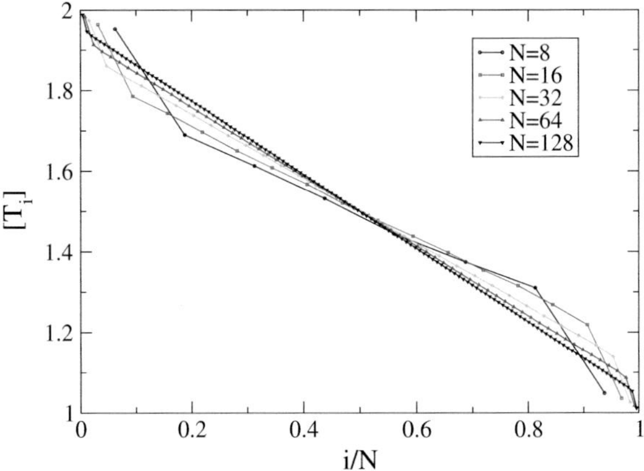

Fig. 2.10 *3D* unpinned case with fixed BC for $\Delta = 0.2$. Plot of temperature profile T_i in a single disorder realization for different system sizes. The plots are from simulations (from [62])

state temperature profiles (especially in $1D$ and $2D$ systems) become difficult to obtain from simulations or numerics and the true form of the temperature profile, in these cases, is an open problem. Temperature profiles in quantum mechanical chains have been investigated in [16, 81].

2.4.3 Fluctuation

As already discussed in the previous sections, the 3D crystal can exhibit different type of transport properties, which can be seen from studies of the average current. Here we discuss how this is manifested in current fluctuations, and study its dependence on different classes of transport [51]. We consider fluctuations of heat transfer by looking at the cumulant generating function (CGF) for the classical harmonic crystal defined in Eq. (2.114). We write again the Hamiltonian in a slightly more explicit form as

$$H = \sum_{\mathbf{n}} \frac{1}{2} m_{\mathbf{n}} \dot{x}_{\mathbf{n}}^2 + \sum_{\mathbf{n},\hat{\mathbf{e}}} \frac{k}{2} (x_{\mathbf{n}} - x_{\mathbf{n}+\hat{\mathbf{e}}})^2 + \frac{k_o}{2} \sum_{\mathbf{n}} x_{\mathbf{n}}^2 + \frac{k'}{2} \sum_{\mathbf{n}'} (x_{1,\mathbf{n}'}^2 + x_{N,\mathbf{n}'}^2) , \tag{2.165}$$

where disordered masses are taken as $m_{\mathbf{n}} = 1 - \Delta$ or $1 + \Delta$. Lattices of dimensions $N \times W^2$ are considered with white noise Langevin heat baths connected to the surface particles. As already derived, the average current $I = \langle Q \rangle / \tau$ in the harmonic crystal is given by a Landauer formula Eq. (2.22) and this gives the following expression for the size-dependent thermal conductivity defined as

$$\kappa \equiv \frac{I/W^2}{\Delta T/N} = \frac{N}{2\pi W^2} \int_0^\infty d\omega \, \mathrm{Tr}[\mathcal{T}(\omega)] , \tag{2.166}$$

where $\mathcal{T}(\omega)$ is the transmission matrix which describes transmission of phonons emitted from a site on one face attached to a reservoir to a site on another face, and is a $W^2 \times W^2$ matrix. We now characterize the properties of current fluctuations, beyond linear response, in different transport regimes characterized by the size-dependence in the thermal conductance:

$$\kappa \propto N^\alpha . \tag{2.167}$$

As seen in Sect. 2.4.1.3, in $3D$ disordered systems without pinning potentials ($k_o = 0$), low-frequency extended modes with diverging phonon mean-free-paths exist and lead to anomalous transport. However, a pinning potential removes these modes and transport is then governed by the high-frequency extended diffusive modes. Hence, a $3D$ disordered pinned crystal shows diffusive heat conduction.

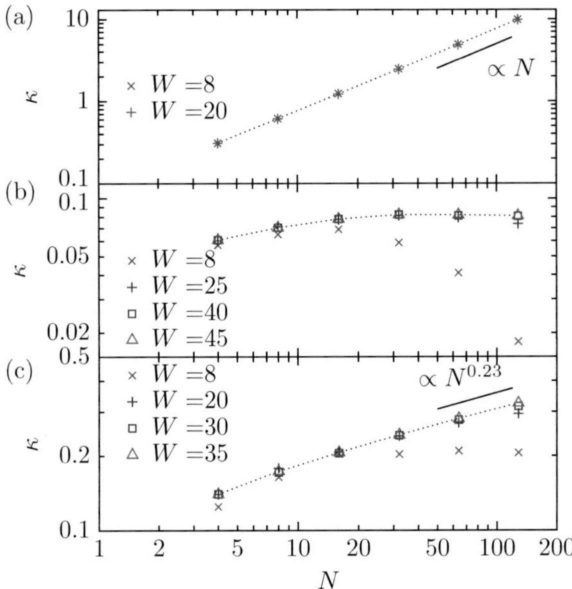

Fig. 2.11 The size-dependent thermal conductivity κ for (Δ, k_o) = (**a**) $(0, 0)$, (**b**) $(0.22, 9.0)$, and (**c**) $(0.82, 0)$. For sufficiently large W we see respectively (**a**) ballistic ($\alpha = 1$), (**b**) diffusive ($\alpha = 0$), and (**c**) anomalous transport($\alpha \sim 0.23$). Each data-point is obtained from simulations by solving the Langevin equation, while the *black dotted line* is given by Eq. (2.166) for largest W (from [51])

The following three parameter sets were chosen [51] for achieving the different regimes:

(a) ordered unpinned lattice ($\Delta = 0, k_o = 0$) for ballistic transport ($\alpha = 1$)
(b) disordered pinned lattice ($\Delta = 0.22, k_o = 9.0$) for diffusive transport ($\alpha = 0$)
(c) disordered unpinned lattice ($\Delta = 0.82, k_o = 0$) for anomalous transport ($0 < \alpha < 1$)

We first show the numerical demonstration that the different transport regimes are realized for the above choices. We show the size-dependence of thermal conductivity in Fig. 2.11. The average heat current was calculated from Eq. (2.166) by using the transfer matrix techniques, described in Sect. 2.4.1.1, to evaluate the transmission matrix $\mathcal{T}(\omega)$. Each point in Fig. 2.11 is for one disorder realization and the parameters were chosen as $\gamma = 1$, $T_L = 2.0$, and $T_R = 1.0$. From Fig. 2.11a we see that κ is independent of width and diverges linearly with N implying ballistic transport. On the other hand in Fig. 2.11b, c, we see that for small W, κ decreases for increasing N. This implies the emergence of phonon localization since the system is quasi-one dimensional. For increasing W with fixed N, the data converges to a constant value, which implies the self-averaging effect in disordered systems. Hence, one can get precise 3D behavior for sufficiently large W. Figure 2.11b shows diffusive transport ($\alpha = 0$) for sufficiently large W and N, while Fig. 2.11c shows anomalous behavior with systematic power law divergence ($\alpha \approx 0.23$). These results are similar and consistent with those discussed in the previous section.

In the following sections, we look at current fluctuation for the cases (a)–(c) and discuss the validity of the additivity principle. Recall that the expression for the full CGF of the harmonic crystal can be obtained in terms of $\mathcal{T}(\omega)$ and is given by (see Sect. 2.3)

$$\mu_{HC}(\lambda) = -\int_0^\infty \frac{d\omega}{2\pi} \operatorname{Tr}\log\Big[\mathbf{1}-\mathcal{T}(\omega)k_B^2 T_L T_R \lambda\,(\lambda + 1/(k_B T_R) - 1/(k_B T_L))\Big]. \tag{2.168}$$

For the case of the ordered crystal, a simpler expression can be obtained and we first discuss this in Sect. 2.4.3.1. In Sect. 2.4.3.2, we derive a closed expression for the CGF, assuming the validity of the additivity principle. This expression depends on the system properties only through the conductivity κ. Finally in Sect. 2.4.3.3 we numerically evaluate the CGF in all three transport regimes and compare with the AP prediction.

2.4.3.1 The CGF of the Ordered Harmonic Crystal

We here give a simplified expression of $\mu_{HC}(\lambda)$ for ordered harmonic crystal ($\Delta = 0$). The simplified expression for $\mu_{HC}(\lambda)$ is computationally much simpler to evaluate than using the recursive Green's function technique for solving the CGF. We follow the approach, discussed in Sect. 2.4.1.2 to first make a change of variables (see Eq. (2.145)), to reduce the dynamics to that of a collection of W^2 independent one-dimensional chains, each evolving with its independent Langevin dynamics. The independent chains are labeled by a mode-index $\mathbf{q} = 2\pi\mathbf{m}/W$, where $\mathbf{m} = (m2, m3)$ are integers with allowed values between $-W/2$ to $W/2$. Each chain differs by having an effective onsite spring constant, $k_o(\mathbf{q})$, given by Eq. (2.146). To use the results of Sect. 2.4.1.2, we need to set $\sigma(\omega) = i\gamma\omega$ and $k' = k = 1$.

The total heat Q is a sum of the heat carried by each chain, and hence the 3D CGF $\mu(\lambda)$ is a sum of the CGF of each of the 1D systems. For a 1D chain, the transmission matrix is a number, and for the $\boldsymbol{m}$-mode

$$\mathcal{T}^{(m)} = 4\gamma^2\omega^2 |\boldsymbol{G}_{1,N}^{+\,(m)}|^2 , \tag{2.169}$$

with the Green's function given by the inverse of a tri-diagonal matrix (see below Eq. (2.121)) The expression of $\boldsymbol{G}_{1,N}^{+\,(m)}$ is readily obtained, using

Eqs. (2.129)–(2.131). We finally get the following explicit formula of $\mu_{HC}(\lambda)$ for the ordered 3D harmonic crystal:

$$\mu_{HC}(\lambda)\Big|_{\Delta=0} = -\frac{1}{2\pi}\sum_{m}\int_0^\infty d\omega \log\Big[1-\frac{4\gamma^2\omega^2\sin^2\theta_m}{|\Lambda_m|^2}T_L T_R\lambda\left(\lambda+\frac{1}{k_BT_R}-\frac{1}{k_BT_L}\right)\Big], \tag{2.170}$$

$$\text{where}\quad \cos\theta_m = \frac{2+k_0-\omega^2}{2}+2\Big[\sin^2\left(\frac{\pi m_2}{W}\right)+\sin^2\left(\frac{\pi m_3}{W}\right)\Big],$$
$$\Lambda_m = \big[\left(1-\gamma^2\omega^2\right)\cos\theta_m - 2i\gamma\omega\big]\sin(N\theta_m) + \left(1+\gamma^2\omega^2\right)\sin\theta_m\cos(N\theta_m).$$

2.4.3.2 Explicit Expression of the Additivity Principle

We now derive an explicit expression for the CGF, using the additivity principle (AP) (see Eq. (2.45)). The AP is derived assuming time-independence of temperature and current fields in the macroscopic fluctuation theory (MFT) (see Eq. (2.43)). In the harmonic crystal, the thermal conductivity, κ, is temperature independent, and hence we can derive a simpler expression. Note also that, in general, we expect κ for the harmonic crystal to be size-dependent, and so we are here making the assumption that the AP is still applicable. Indeed, in the next section, by comparing with the exact numerical results, we will try to verify this assumption. We here use the inverse temperature $\beta_\alpha = 1/(k_BT_\alpha)$.

We think of the regime $-\beta_R \le \lambda \le \beta_L$. The AP states that the large deviation function (LDF) for the system with the size N is given by Bodineau and Derrida [47]

$$F_N(q,T_L,T_R) = \min_{T(x)}\frac{1}{N}\int_0^1 dx\frac{\left[Nq+\tilde{\kappa}(T)\frac{dT(x)}{dx}\right]^2}{4T^2\tilde{\kappa}(T)}, \tag{2.171}$$

where $\tilde{\kappa}(T) = W^2\kappa(T)$. The variational problem in Eq. (2.171) is reduced to finding the optimal profile $T(x)$ satisfying

$$\left(\frac{dT(x)}{dx}\right)^2 = \frac{(Nq)^2\left[1+4KT^2\tilde{\kappa}(T)\right]}{\tilde{\kappa}^2(T)}, \tag{2.172}$$

where the function K is determined from the boundary condition $T(0) = T_L$ and $T(1) = T_R$. From now on, we consider the case of temperature-independent thermal conductivity $\tilde{\kappa}(T) = \tilde{\kappa}$, which is the case for harmonic crystals. The CGF is given by the Legendre transformation:

$$\mu_{AP}(\lambda) = \max_q \left[q\lambda - F_N(q, T_L, T_R)\right]. \tag{2.173}$$

Suppose $T_L > T_R$ and the deviations are not too large so that the optimal profile remains monotonic. Then

$$\frac{dT(x)}{dx} = \frac{-Nq\sqrt{1 + 4KT^2\tilde{\kappa}}}{\tilde{\kappa}}. \tag{2.174}$$

In this case, from the Legendre transformation, the CGF is given by Eq. (2.45)

$$\mu_{AP}(\lambda) = -\frac{K}{N}\Big[\int_{T_R}^{T_L} dT \frac{\tilde{\kappa}}{\sqrt{1 + 4KT^2\tilde{\kappa}}}\Big]^2, \tag{2.175}$$

$$\lambda = \int_{T_R}^{T_L} dT \frac{1}{2T^2}\Big[\frac{1}{\sqrt{1 + 4KT^2\tilde{\kappa}}} - 1\Big]. \tag{2.176}$$

This expression is valid for $\lambda_- \le \lambda \le \lambda_+$ where $\lambda_\pm = (\beta_L - \beta_R \pm \sqrt{\beta_R^2 - \beta_L^2})/2$. To go beyond this regime of λ, we consider a non-monotonic optimal profile given by

$$\frac{dT(x)}{dx} = \begin{cases} \pm Nq\frac{\sqrt{1+4KT^2\tilde{\kappa}}}{\tilde{\kappa}} & 0 \le x \le x_c\,, \\ \mp Nq\frac{\sqrt{1+4KT^2\tilde{\kappa}}}{\tilde{\kappa}} & x_c \le x \le 1\,, \end{cases} \tag{2.177}$$

where x_c satisfies $dT(x_c)/dx = 0$ implying $1 + 4K\tilde{\kappa}T^2(x_c) = 0$. In these cases, the CGF is given by

$$\mu_{AP}(\lambda) = \frac{-K}{N}\Big[\int_{T_L}^{T(x_c)} dT \frac{\tilde{\kappa}}{\sqrt{1 + 4KT^2\tilde{\kappa}}} + \int_{T_R}^{T(x_c)} dT \frac{\tilde{\kappa}}{\sqrt{1 + 4KT^2\tilde{\kappa}}}\Big]^2, \tag{2.178}$$

$$\begin{aligned} \lambda = &\int_{T_L}^{T(x_c)} dT \frac{1}{2T^2}\Big[1 \pm \frac{1}{\sqrt{1 + 4KT^2\tilde{\kappa}}}\Big] \\ &+ \int_{T_R}^{T(x_c)} dT \frac{1}{2T^2}\Big[-1 \pm \frac{1}{\sqrt{1 + 4KT^2\tilde{\kappa}}}\Big]. \end{aligned} \tag{2.179}$$

These expressions are valid for the regime $\lambda_+ \le \lambda \le \beta_L$ and $-\beta_R \le \lambda \le \lambda_-$, respectively.

Simplifying Eqs. (2.175), (2.176), (2.178) and (2.179) is straightforward. By evaluating the integrations we get the following explicit expressions:

$$\mu_{AP}(\lambda) = \begin{cases} -\frac{\tilde{\kappa}}{4N}\left[\log\left(\frac{\sqrt{1+4\tilde{\kappa}KT_L^2}+\sqrt{4\tilde{\kappa}KT_L^2}}{\sqrt{1+4\tilde{\kappa}KT_R^2}+\sqrt{4\tilde{\kappa}KT_R^2}}\right)\right]^2, & \cdots \lambda_- \le \lambda \le \lambda_+, \\ \frac{\tilde{\kappa}}{4N}\left[\pi - (\theta_L + \theta_R)\right]^2 & \cdots -\beta_R \le \lambda \le \lambda_- \text{ or } \lambda_+ \le \lambda \le \beta_L, \end{cases} \tag{2.180}$$

where θ_α ($\alpha = L, R$) is given by

$$\cos\theta_\alpha = \sqrt{1 + 4\tilde{\kappa}\,K\,T_\alpha^2}\,, \quad \sin\theta_\alpha = \sqrt{4\tilde{\kappa}\,|K|\,T_\alpha^2}\,. \tag{2.181}$$

The function K is given by

$$K(\lambda, T_L, T_R) = \frac{1}{16\tilde{\kappa}}\left[(\beta_L - \beta_R - 2\lambda)^2 - 2\left(\beta_L^2 + \beta_R^2\right) + \left(\frac{\beta_L^2 - \beta_R^2}{\beta_L - \beta_R - 2\lambda}\right)^2\right]. \tag{2.182}$$

2.4.3.3 Numerical Calculation of the CGF for the Disordered Harmonic Crystals

We now show the results for current fluctuation by direct numerical evaluation of Eq. (2.168). One of the universal properties of current fluctuations which is expected to be valid irrespective of α, is the fluctuation relation. Indeed, it can be checked directly that both $\mu_{HC}(\lambda)$ in Eq. (2.168) and $\mu_{AP}(\lambda)$ exactly satisfy the fluctuation relation symmetry. Here we explore the validity of the macroscopic fluctuation theory (MFT) and especially its specific property, the additivity principle (AP), which were originally proposed for one-dimensional diffusive systems ($\alpha = 0$) (Sect. 2.3). We will now compare the additivity principle prediction from $\mu_{AP}(\lambda)$ (in last section) with the numerical result for $\mu_{HC}(\lambda)$ from Eq. (2.168).

We note that for a $3D$ disordered crystal the heat current depends on the particular realization of disorder, however for large N and W there is self-averaging, and sample-to-sample fluctuations become very small. Hence for a fixed disorder strength we get a unique current and κ from Eq. (2.166). This value of κ is then used to get $\mu_{AP}(\lambda)$. This is then compared with $\mu_{HC}(\lambda)$.

We now present results comparing $\mu_{AP}(\lambda)$ with $\mu_{HC}(\lambda)$ in the three different regimes. In these computations, the parameters were set as $\gamma = 1.0, T_L = 2.0, T_R = 0.25$. The length was fixed at $N = 128$ and the dependence of results on width W was studied. The results for ballistic, diffusive, and anomalous cases are shown in Fig. 2.12a, b, and c respectively. For ballistic case (a), we see deviations from AP curve irrespective of W as expected. On the other hand, in Fig. 2.12b, we see that the agreement between $\mu_{AP}(\lambda)$ and $\mu_{HC}(\lambda)$ improves for increasing W. For small W

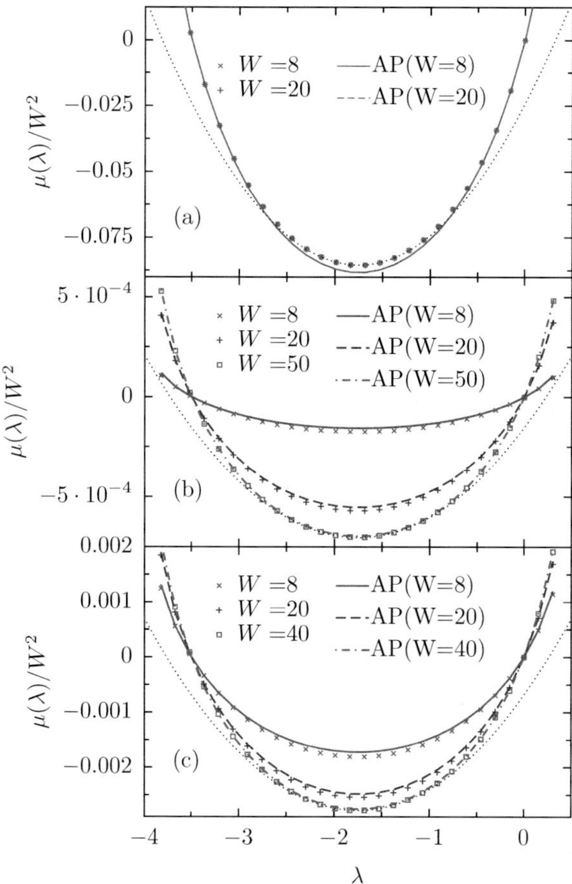

Fig. 2.12 Comparison of the numerically evaluated CGF $\mu_{HC}(\lambda)$ (points) for a $3D$ harmonic crystal and the AP predicted curve $\mu_{AP}(\lambda)$ for $N = 128$ and different widths. System parameters are the same as in Fig. 2.11 with $T_L = 2.0$ and $T_R = 0.25$. The *black dotted line* is a quadratic fit to $\mu_{HC}(\lambda)$ for the largest W. The range of λ is $(-\beta_R, \beta_L)$. Results for the cases (a)–(c) are respectively given in (**a**)–(**c**) (from [51])

where localization effect is dominant, there are clear deviations from the AP curve. For other cases with larger temperature differences, good agreement with the AP was obtained, for sufficiently large system size [51].

The degree of agreement seen in Fig. 2.12 is now quantitatively discussed. We define the following quantity

$$\delta \equiv \left| \frac{\mu_{HC}(\lambda^*) - \mu_{AP}(\lambda^*)}{\mu_{HC}(\lambda^*)} \right|, \qquad \lambda^* = -(\beta_R - \beta_L)/2, \tag{2.183}$$

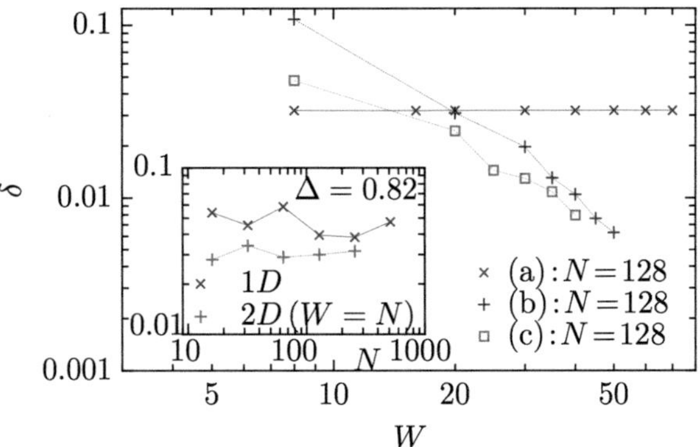

Fig. 2.13 Plot showing quantitative estimate of the degree of agreement between numerical CGF with that from AP. The quantity δ is defined in Eq. (2.183). The *inset* shows the results for low-dimensional anomalous cases. In all cases $(T_L, T_R) = (2.0, 0.25)$ (from [51])

where λ^* is the value of λ which minimizes $\mu_{AP}(\lambda)$. As seen in Fig. 2.12, the deviation becomes maximum at the minimum value of the CGF. Hence, the function δ quantitatively estimates the degree of discrepancy. In Fig. 2.13, we show δ as a function of W for the three cases in $3D$. Systematic approach to the AP is seen on increasing W for both diffusive and anomalous cases. As discussed in Sect. 2.4.1.3 harmonic lattices in $1D$ and $2D$ can show diverging thermal conductivity . Then, one may ask if low dimensional anomalous transport satisfies AP or not.

We briefly discuss some of the results for low dimensional harmonic systems obtained in [51]. In the inset of Fig. 2.13, $1D$ and $2D$ results for δ are shown. These correspond to system sizes N for $1D$ and $N \times N$ for $2D$ with open boundary condition, and hence, the x-axis is N (not W). For both $1D$ and $2D$, results for one realization of random mass are shown. Contrary to what happens in $3D$, in low dimensions we see no sign of decay of δ, and it remains almost constant. This implies that the agreement with AP for anomalous transport is true only for $3D$ systems, while in $1D$ and $2D$ we find that AP is not satisfied for the anomalous cases where $0 < \alpha < 1$. The major difference between $3D$ and lower dimensions is that in $3D$ only a small fraction of the normal modes are localized, while in lower dimensions, most of the modes are localized. Hence it is expected that there is no local equilibration in low-dimensions and could be the reason that MFT and AP are satisfied in these systems.

We should also note that the situation in 3D does not satisfy the sufficient condition for the MFT Eq. (2.47), since the thermal conductivity is independent of temperature. Moreover, both MFT and the AP conjecture assume diffusive transport. However, rather surprisingly, for the case of anomalous transport, we see from Fig. 2.12c behaviour similar to the diffusive case, with clear verification of AP at sufficiently large W. Hence, these results suggest the possibility that one can extend the conditions of applicability of both MFT and AP.

2.5 Summary

In this chapter, we have reviewed various aspects of heat conduction through a study of harmonic crystals. This includes mathematically technical details and physically important results. The general formulations, for computation of the average heat current and heat current fluctuations, in the non-equilibrium steady state, were developed and it was shown that these can be written completely in terms of the transmission matrix. We observed that there are a number of similar features between electric and heat conduction—such as ballistic transport, localization effects, and conductance quantum, and so on. However, anomalous behavior in transport is a unique property of phonon heat conduction. This is because in the electrical case, only frequency modes (or energy levels) around the Fermi frequency play an important role for transport, while in heat transport low frequency phonons determine transport properties. In this chapter, special attention was paid to heat transport in disordered crystals. We demonstrate different possible types of transport, depending on the pining potential, dimensionality, and boundary conditions. It is worth while to stress that a 3D harmonic crystal can show Fourier's law, but only in the presence of a pinning potential. Surprisingly, for an un-pinned system, we find anomalous conductivity even in 3D. These are numerical findings and hence analytical proof for these are important open problems.

We briefly mention some other relevant work on disordered harmonic systems. Properties of eigenstates, such as the IPR, have been carefully examined recently for mass-disordered harmonic crystals with scalar displacements [82]. The localization-delocalization transition in $3D$ crystals has been studied in [83]. Heat transport in disordered harmonic systems has been studied for more realistic models, in the context of understanding transport in disordered nanotubes at low temperatures [84, 85]. These studies typically use the Landauer formula for current and evaluate the nonequilibrium Green's function using transfer matrix approaches. Another set of interesting studies on more realistic models are those on model glass systems, such as packings of soft spheres and Lennard-Jones mixtures, which have been treated in the harmonic approximation [86–88]. The usual approach here has been to study energy transport in these disordered systems using the Green-Kubo formula. For harmonic systems, the Green-Kubo formula can be re-expressed, so that it appears in the form of kinetic theory, and is written in terms of normal mode wavefunctions and frequencies [89]. The question of system-size dependence of the thermal conductivity has not yet been carefully analyzed, and the validity or otherwise of Fourier's law in these systems is an open question.

Another aspect discussed in the review is the nature of current fluctuations in the different transport regimes. The main focus was to use the analytic tractability of the harmonic lattice as a test bed for the validity of predictions of fluctuation theorems, macroscopic fluctuation theory and the additivity principle in 3D disordered crystals. These will be helpful for making more precise the domain of validity of these important results in non-equilibrium physics. Right now we are in the beginning stages of investigating fluctuation properties, and hence more active research is necessary.

Modeling solids by harmonic crystals is justified in low temperatures and more so in the mesoscopic scale (where phonon mean free paths from anharmonic effects are much larger than system size). Recent experimental development for observing small system properties indicates the importance of study of the harmonic crystal. We hope that this review will be helpful not only for the insights that the harmonic crystal provides towards understanding theoretical aspects of heat conduction, but also in understanding experiment results.

Appendix

We explain some definitions and several relations between different Green's function defined in the non-equilibrium Green's function (NEGF) formalism. Let a and b be bosonic operators. Contour ordered Green's functions are defined as

$$G_{a,b}(u_1, u_2) = -(i/\hbar)\langle \hat{T}_c a(u_1) b(u_2)\rangle \,, \tag{2.184}$$

where u_1 and u_2 are fictitious times on Keldysh contour and $\hat{T}_c$ is the operator that makes the variables contour ordered. Let t_1 and t_2 be real times for u_1 and u_2, respectively. Depending on location of these fictitious times along the contour, different Green's function are defined. Let $u_1 >_c u_2$ denote that u_1 is larger than u_2 along the contour. Similarly $u_1 <_c u_2$ implies u_2 is larger than u_1 along the contour. Then one defines the lesser and greater Green's functions

$$G^{<}_{a,b}(t_1, t_2) = -(i/\hbar)\langle b(t_2) a(t_1)\rangle \qquad u_1 <_c u_2, \tag{2.185}$$

$$G^{>}_{a,b}(t_1, t_2) = -(i/\hbar)\langle a(t_1) b(t_2)\rangle \qquad u_1 >_c u_2. \tag{2.186}$$

Next we define four types of Green's functions:

$$G^{\sigma\,\sigma'}_{a,b}(t_1, t_2) = G_{a,b}(u_1, u_2) \qquad (u_1, u_2)\,, \text{ on } (\sigma, \sigma')\,\text{branch}\,, \tag{2.187}$$

where σ, σ' take $+$ or $-$. In addition, one defines the retarded, advanced, Keldysh Green's function G^R, G^A, G^K as

$$G^{R,A}_{a,b}(t_1, t_2) = \mp(i/\hbar)\theta(\pm(t_1 - t_2))\langle [a(t_1), b(t_2)]\rangle \,, \tag{2.188}$$

$$G^{K}_{a,b}(t_1, t_2) = G^{<}_{a,b}(t_1, t_2) + G^{>}_{a,b}(t_1, t_2)\,. \tag{2.189}$$

In Sect. 2.2 and other sections, we frequently use the Green's function G^{+} in the Langevin equations and Greens function approach (LEGF). This is in fact exactly the advanced Green's function. Thus, in NEGF notation, $G^+ = G^A$ and $G^- = G^R$.

The above three functions can be written in terms of $G^{\sigma,\sigma'}$ as

$$G^R_{a,b}(t_1,t_2) = G^{++}_{a,b} - G^{+-}_{a,b} = -G^{--}_{a,b} + G^{-+}_{a,b}, \tag{2.190}$$

$$G^A_{a,b}(t_1,t_2) = G^{++}_{a,b} - G^{-+}_{a,b} = -G^{--}_{a,b} + G^{+-}_{a,b}, \tag{2.191}$$

$$G^K_{a,b}(t_1,t_2) = G^{++}_{a,b} + G^{--}_{a,b} = G^{+-}_{a,b} + G^{-+}_{a,b}. \tag{2.192}$$

Using the matrix $Q = \frac{1}{\sqrt{2}}\begin{pmatrix} 1, & -1 \\ 1, & 1 \end{pmatrix}$, Green's functions with $(+-)$ basis are transformed into $G^{R,A,K}$

$$\boldsymbol{G} = \begin{pmatrix} G^{++}, & G^{+-} \\ G^{-+}, & G^{--} \end{pmatrix}, \tag{2.193}$$

$$Q\boldsymbol{G}Q^T = \begin{pmatrix} 0, & G^A \\ G^R, & G^K \end{pmatrix}. \tag{2.194}$$

From this, one finds the relation

$$\det \boldsymbol{G} = -G^R G^A. \tag{2.195}$$

Suppose we have the Dyson's equation

$$\boldsymbol{G} = \boldsymbol{g} + \boldsymbol{g}\boldsymbol{\sigma}_z \boldsymbol{\Sigma} \boldsymbol{\sigma}_z \boldsymbol{G}, \tag{2.196}$$

where either time-dependent Green's function or its Fourier transformation is considered. Utilizing Eqs. (2.190)–(2.192), which leads to $\left[G^{R,A}\right]^{-1} = \left[g^{R,A}\right]^{-1} - \Sigma^{R,A}$, and assuming the case where $g^R g^{<,>} = 0$, one readily shows

$$G^{<,>} = G^R\, \Sigma^{<,>}\, G^A. \tag{2.197}$$

References

1. Bonetto, F., Lebowitz, J.L., Rey-Bellet, L.: Fourier's law: a challenge to theorists. In: Fokas, A., et al. (eds.) Mathematical Physics 2000, pp. 128–150. Imperial College Press, London (2000)
2. Lepri, S., Livi, R., Politi A.: Thermal conduction in classical low-dimensional lattices. Phys. Rep. **377**, 1–80 (2003)
3. Dhar, A.: Heat transport in low-dimensional systems. Adv. Phys. **57**, 457–537 (2008)
4. Rieder, Z., Lebowitz, J.L., Lieb, E.: Properties of a harmonic crystal in a stationary nonequilibrium state. J. Math. Phys. **8**, 1073–1078 (1967)
5. Casher, A., Lebowitz, J.L.: Heat flow in regular and disordered harmonic chains. J. Math. Phys. **12**, 1701–1711 (1971)

6. Rubin, R.J., Greer, W.L.: Abnormal lattice thermal conductivity of a one-dimensional, harmonic, isotopically disordered crystal. J. Math. Phys. **12**, 1686–1701 (1971)
7. O'Connor, A.J., Lebowitz, J.L.: Heat conduction and sound transmission in isotopically disordered harmonic crystals. J. Math. Phys. **15**, 692–703 (1974)
8. Angelescu, D.E., Cross, M.C., Roukes, M.L.: Heat transport in mesoscopic systems. Superlattice. Microst. **23**, 673 (1998)
9. Rego, L.G.C., Kirczenow, G.: Quantized thermal conductance of dielectric quantum wires. Phys. Rev. Lett. **81**, 232–235 (1998)
10. Blencowe, M.P.: Quantum energy flow in mesoscopic dielectric structures. Phys. Rev. B **59**, 4992–4998 (1999)
11. Dhar, A., Roy, D.: Heat transport in harmonic lattices. J. Stat. Phys. **125**, 801–820 (2006)
12. Segal, D., Nitzan, A., Hänggi, P.: Thermal conductance through molecular wires. J. Chem. Phys. **119**, 6840–6855 (2003)
13. Yamamoto, T., Watanabe, K.: Nonequilibrium Green's function approach to phonon transport in defective carbon nanotubes. Phys. Rev. Lett. **96**, 255503-1–255503-4 (2006)
14. Wang, J.S., Wang, J., Zeng, N.: Nonequilibrium Green's function approach to mesoscopic thermal transport. Phys. Rev. B **74**, 033408-1–033408-4 (2006)
15. Zurcher, U., Talkner, P.: Quantum-mechanical harmonic chain attached to heat baths. II. Nonequilibrium properties. Phys. Rev. A **42**, 3278–3290 (1990)
16. Dhar, A., Shastry, B.S.: Quantum transport using the Ford-Kac-Mazur formalism. Phys. Rev. B **67**, 195405-1–195405-10 (2003)
17. Saito, K., Takesue, S., Miyashita, S.: Energy transport in the integrable system in contact with various types of phonon reservoirs. Phys. Rev. E **61**, 2397–2409 (2000)
18. Chen, Y.-C., Lebowitz, J.L., Liverani, C.: Dissipative quantum dynamics in a boson bath. Phys. Rev. B **40**, 4664–4682 (1989)
19. Das, S.G., Dhar, A.: Landauer formula for phonon heat conduction: relation between energy transmittance and transmission coefficient. Eur. Phys. J. B **85**, 1–8 (2012)
20. Caroli, C., Combescot, R., Nozieres, P., Saint-James, D.: Direct calculation of the tunneling current. J. Phys. C **4**, 916–929 (1971)
21. Meir, Y., Wingreen, N.S.: Landauer formula for the current through an interacting electron region. Phys. Rev. Lett. **68**, 2512–2515 (1992)
22. Dhar, A., Sen, D.: Nonequilibrium Green's function formalism and the problem of bound states. Phys. Rev. B **73**, 085119-1–085119-14 (2006)
23. Dhar, A.: Heat conduction in the disordered harmonic chain revisited. Phys. Rev. Lett. **86**, 5882–5885 (2001)
24. Dhar, A., Saito, K., Hänggi, P.: Nonequilibrium density matrix description of steady state quantum transport. Phys. Rev. E **85**, 011126-1–011126-11 (2012)
25. Touchette, H.: The large deviation approach to statistical mechanics. Phys. Rep. **478**, 1–69 (2009)
26. McKean, H.: Probability: The Classical Limit Theorems. Cambridge University Press, Cambridge (2014)
27. Gärtner, J.: On large deviations from the invariant measure. Theor. Probab. Appl. **22**, 24–39 (1977)
28. Ellis, R.S.: Large deviations for a general class of random vectors. Ann. Probab. **12**, 1–12 (1984)
29. Green, M.S.: Markoff random processes and the statistical mechanics of time-dependent phenomena. II. Irreversible processes in fluids. J. Chem. Phys. **22**, 398–413 (1954)
30. Kubo, R., Toda, M., Hashitsume, N.: Statistical Physics II Nonequilibrium Statistical Mechanics. Springer, Berlin (1991)
31. Kundu, A., Dhar, A., Narayan, O.: Green-Kubo formula for heat conduction in open systems. J. Stat. Mech. L03001 (2009)
32. Evans, D.J., Cohen, E.G.D., Morriss, G.P.: Probability of second law violations in shearing steady states. Phys. Rev. Lett. **71**, 2401–2404 (1993)

33. Gallavotti, G., Cohen, E.G.D.: Dynamical ensembles in nonequilibrium statistical mechanics. Phys. Rev. Lett **74**, 2694–2698 (1995)
34. Lebowitz, J.L., Spohn, H.: A GallavottiCohen-type symmetry in the large deviation functional for stochastic dynamics. J. Stat. Phys. **95**, 333–365 (1999)
35. Seifert, U.: Stochastic thermodynamics, fluctuation theorems and molecular machines. Rep. Prog. Phys. **75**, 126001-1–126001-58 (2012)
36. Campisi, M., Hänggi, P., Talkner, P.: Quantum fluctuation relations: foundations and applications. Rev. Mod. Phys. **83**, 771–791 (2011)
37. Esposito, M., Harbola, U., Mukamel, S.: Nonequilibrium fluctuations, fluctuation theorems, and counting statistics in quantum systems. Rev. Mod. Phys. **81**, 1665–1702 (2009)
38. Evans, D.J., Searles, D.J.: The fluctuation theorem. Adv. Phys. **51**, 1529–1585 (2002)
39. Zon, R.V., Cohen, E.G.D.: Extension of the fluctuation theorem. Phys. Rev. Lett. **91**, 110601-1–110601-4 (2003)
40. Visco, P.: Work fluctuations for a Brownian particle between two thermostats. J. Stat. Mech: Theory Exp. P06006 (2006)
41. Gallavotti, G.: Extension of Onsager's reciprocity to large fields and the chaotic hypothesis. Phys. Rev. Lett. **77**, 4334–4337 (1996)
42. Andrieux, D., Gaspard, P.: A fluctuation theorem for currents and non-linear response coefficients. J. Stat. Mech. P02006 (2007)
43. Saito, K., Utsumi, Y.: Symmetry in full-counting statistics, fluctuation theorem, and relations among nonlinear transport coefficients in the presence of a magnetic field. Phys. Rev. B **78**, 115429-1–115429-7 (2008)
44. Nakamura, S., Yamauchi, Y., Hashisaka, M., Chida, K., Kobayashi, K., Ono, T., Leturcq, R., Ensslin, K., Saito, K., Utsumi, Y., Gossard, A.C.: Non-equilibrium fluctuation relation in a quantum coherent conductor. Phys. Rev. Lett. **104**, 080602-1–080602-4 (2010)
45. Bertini, L., De Sole, A., Gabrieli, A., Jona-Lasinio, G., Landim, C.: Fluctuations in stationary nonequilibrium states of irreversible processes. Phys. Rev. Lett. **87**, 040601–040604 (2001)
46. Bertini, L., De Sole, A., Gabrieli, A., Jona-Lasinio, G., Landim, C.: Macroscopic fluctuation theory. Rev. Mod. Phys. **87**, 593–636 (2015)
47. Bodineau, T., Derrida, B.: Current fluctuations in nonequilibrium diffusive systems: an additivity principle. Phys. Rev. Lett. **92**, 180601–180604 (2004)
48. Bertini, L., De Sole, A., Gabrieli, A., Jona-Lasinio, G., Landim, C.: Non equilibrium current fluctuations in stochastic lattice gases. J. Stat. Phys. **123**, 237–276 (2006)
49. Bodineau, T., Derrida, B.: Cumulants and large deviations of the current through non-equilibrium steady states. C. R. Phys. **8**, 540–555 (2007)
50. Hurtado, P.I., Garrido, P.L.: Test of the additivity principle for current fluctuations in a model of heat conduction. Phys. Rev. Lett. **102**, 250601-1–250601-4 (2009)
51. Saito, K., Dhar, A.: Additivity principle in high-dimensional deterministic systems. Phys. Rev. Lett. **107**, 250601-1–250601-4 (2011)
52. Kundu, A., Sabhapandit, S., Dhar, A.: Large deviations of heat flow in harmonic chains. J. Stat. Mech. P03007 (2011)
53. Saito, K., Dhar, A.: Generating function formula of heat transfer in harmonic networks. Phys. Rev. E. **83**, 041121-1–041121-5 (2011)
54. Saito, K., Dhar, A.: Fluctuation theorem in quantum heat conduction. Phys. Rev. Lett. **99**, 180601–180604 (2007)
55. Fogedby, H.C., Imparato, A.: A bound particle coupled to two thermostats. J. Stat. Mech: Theory Exp. P05015 (2011)
56. Gomez-Solano, J.R., Bellon, L., Petrosyan, A., Ciliberto, S.: Steady-state fluctuation relations for systems driven by an external random force. Europhys. Lett. **89**, 60003-1–60003-6 (2010)
57. Sabhapandit, S.: Work fluctuations for a harmonic oscillator driven by an external random force. Europhys. Lett. **96**, 20005 (2011)
58. Levitov, L.S., Lee, H.-W., Lesovik, G.B.: Electron counting statistics and coherent states of electric current. J. Math. Phys. **37**, 4845–4866 (1996)

59. Blanter, Y.M., Büttiker, M.: Shot noise in mesoscopic conductors. Phys. Rep. **336**, 1–166 (2000)
60. Wang, J.S., Agarwalla, B.K., Li, H.: Transient behavior of full counting statistics in thermal transport. Phys. Rev. B **84**, 153412-1–153412-4 (2011)
61. Nakazawa, H.: Energy flow in harmonic linear chain. Prog. Theor. Phys. **39**, 236–238 (1968)
62. Kundu, A., Chaudhuri, A., Roy, D., Dhar, A., Lebowitz, J.L., Spohn, H.: Heat transport and phonon localization in mass-disordered harmonic crystals. Phys. Rev. B **81**, 064301-1–064301-17 (2010)
63. Kundu, A., Chaudhuri, A., Roy, D., Dhar, A., Lebowitz, J.L., Spohn, H.: Heat conduction and phonon localization in disordered harmonic crystals. Europhys. Lett. **90**, 40001-1–40001-6 (2010)
64. Chiu, H.Y., Deshpande, V.V., Postma, H.C., Lau, C.N., Miko, C., Forro, L., Bockrath, M.: Ballistic phonon thermal transport in multiwalled carbon nanotubes. Phys. Rev. lett. **95**, 226101-1–226101-4 (2005)
65. Schwab, K., Henriksen, E.A., Worlock, J.M., Roukes, M.L.: Measurement of the quantum of thermal conductance. Nature **404**, 974–977 (2000)
66. Roy, D., Dhar, A.: Role of pinning potentials in heat transport through disordered harmonic chain. Phys. Rev. E **78**, 051112-1–051112-5 (2008)
67. Allen, K.R., Ford, J.: Lattice thermal conductivity for a one-dimensional, harmonic, isotopically disordered crystal. Phys. Rev. **176**, 1046–1055 (1968)
68. Payton, D.N., Rich, M., Visscher, W.M.: Lattice thermal conductivity in disordered harmonic and anharmonic crystal models. Phys. Rev. **160**, 706–711 (1967)
69. Matsuda, H., Ishii, K.: Localization of normal modes and energy transport in the disordered harmonic chain. Prog. Theor. Phys. Suppl. **45**, 56–86 (1970)
70. Verheggen, T.: Transmission coefficient and heat conduction of a harmonic chain with random masses: asymptotic estimates on products of random matrices. Commun. Math. Phys. **68**, 69–82 (1979)
71. Ajanki, O., Huveneers, F.: Rigorous scaling law for the heat current in disordered harmonic chain. Commun. Math. Phys. **301**, 841–883 (2011)
72. Callaway, J.: Model for lattice thermal conductivity at low temperatures. Phys. Rev. **113**, 1046–1051 (1959)
73. Ziman, J.M.: Principles of the Theory of Solids. Cambridge University Press, Cambridge (1972)
74. Goldsheid, I.Y., Molchanov, S.A., Pastur, L.A.: A random homogeneous Schrdinger operator has a pure point spectrum. Funkcional. Anal. i Prilovzen. **11**, 1–10 (1977)
75. Mott, N.F., Twose, W.D.: The theory of impurity conduction. Adv. Phys. **10**, 107–163 (1961)
76. Borland, R.E.: The nature of the electronic states in disordered one-dimensional systems. Proc. R. Soc. Lond. Ser. A **274**, 529–545 (1963)
77. Lee, P.A., Ramakrishnan, T.V.: Disordered electronic systems. Rev. Mod. Phys. **57**, 287–337 (1985)
78. John, S., Sompolinsky, H., Stephen, M.J.: Localization in a disordered elastic medium near two dimensions. Phys. Rev. B **27**, 5592–5603 (1983)
79. Furstenberg, H.: Noncommuting random products. Trans. Am. Math. Soc. **108**, 377–428 (1963)
80. Roy, D., Dhar, A.: Heat transport in ordered harmonic lattices. J. Stat. Phys. **131**, 535–541 (2008)
81. Gaul, C., Büttner, H.: Quantum mechanical heat transport in disordered harmonic chains. Phys. Rev. E **76**, 011111 (2007)
82. Monthus, C., Garel, T.: Anderson localization of phonons in dimension d= 1, 2, 3: finite-size properties of the inverse participation ratios of eigenstates. Phys. Rev. B **81**, 224208 (2010)
83. Pinski, S.D., Schirmacher, W., Whall, T., Römer, R.A.: Localization-delocalization transition for disordered cubic harmonic lattices. J. Phys. Condens. Matter **24**, 405401 (2012).

84. Stoltz, G., Lazzeri M., Mauri F.: Thermal transport in isotopically disordered carbon nanotubes: a comparison between Green's functions and Boltzmann approaches. J. Phys. Condens. Matter **21**, 245302 (2009)
85. Stoltz, G., Mingo, N., Mauri, F.: Reducing the thermal conductivity of carbon nanotubes below the random isotope limit. Phys. Rev. B **80**, 113408 (2009)
86. Xu, N., et al.: Energy transport in jammed sphere packings. Phys. Rev. Lett. **102**, 038001 (2009)
87. Vitelli, V., et al.: Heat transport in model jammed solids. Phys. Rev. E **81**, 21301 (2010)
88. Wyart, M.: Scaling of phononic transport with connectivity in amorphous solids. Europhys. Lett. **89**, 64001 (2010)
89. Allen, P.B., Feldman, J.L.: Thermal conductivity of disordered harmonic solids. Phys. Rev. B **48**, 12581 (1993)

Chapter 3
Fluctuating Hydrodynamics Approach to Equilibrium Time Correlations for Anharmonic Chains

Herbert Spohn

Abstract Linear fluctuating hydrodynamics is a useful and versatile tool for describing fluids, as well as other systems with conserved fields, on a mesoscopic scale. In one spatial dimension, however, transport is anomalous, which requires to develop a nonlinear extension of fluctuating hydrodynamics. The relevant nonlinearity turns out to be the quadratic part of the Euler currents when expanding relative to a uniform background. We outline the theory and compare with recent molecular dynamics simulations.

3.1 Introduction, Long Time Tails for Simple Fluids

In the mid 1950ies, Green [1] and Kubo [2, 3] discovered that transport coefficients for simple fluids can be obtained through a time-integral over the respective total current correlation function. For tracer diffusion such a connection is more immediate and was understood much earlier. But the then novel insight was that collective transport coefficients, such as viscosity and thermal conductivity, follow the same pattern. Thus it became a central issue to determine the time decay of such current correlations. With essentially no tools available this amounted to an impossible task. The static equilibrium correlations were known to decay exponentially fast, as confirmed by a convergent series expansion. But for the dynamics one would have to deal with a huge set of coupled differential equations. At the time the only theoretical tool available was the Boltzmann equation valid at low density. Kinetic theory predicts an exponential decay for the current time correlations and it was tacitly assumed that such behavior would extend to moderate densities. Alder and Wainwright [4] tried to check the situation in a pioneering molecular dynamics simulation of 500 hard disks, resp. hard spheres, at periodized volume 2, 3, and 5 times larger than close packing. For tracer diffusion they

H. Spohn (✉)
Zentrum Mathematik and Physik Department, Technische Universität München, Boltzmannstraße 3, 85747 Garching, Germany
e-mail: spohn@ma.tum.de

S. Lepri (ed.), *Thermal Transport in Low Dimensions*, Lecture Notes in Physics 921, DOI 10.1007/978-3-319-29261-8_3

convincingly observed a power law decay as $t^{-d/2}$, dimension $d = 2, 3$, which was baptized "long time tail". They also argued that the same behavior should hold for collective transport. Theory quickly jumped in and predicted a decay as $t^{-d/2}$ for viscosity and thermal conductivity [5, 6]. A first hand historical account is provided in [7].

There are several theoretical schemes and they all arrive at the same prediction, which of course increases their confidence level. As recognized for some time, the most direct approach is linear fluctuating hydrodynamics plus small nonlinear perturbations. We refer to the recent monograph [8] for a comprehensive discussion. Here I provide only a rough sketch of the method with the purpose to explain why one dimension is so special. In the physical dimension a simple fluid has five conservation laws and correspondingly fluctuating hydrodynamics has to deal with a five component field, where the momentum components are odd, density and energy are even under time reversal. As well known [8, 9], the full structure has to be used in order to arrive at quantitative predictions. But the argument becomes even more direct for the (unphysical) case of a single conservation law.

Let us thus consider the scalar field $\rho(\mathbf{x}, t)$, which for concreteness is called density. Space is $\mathbf{x} \in \mathbb{R}^d$ and time is $t \in \mathbb{R}$. $\rho(\mathbf{x}, t)$ is a fluctuating field. On the macroscopic scale fluctuations are not visible and ρ satisfies the hyperbolic conservation law

$$\partial_t \rho + \nabla \cdot \vec{j}(\rho) = 0\,, \tag{3.1}$$

where $\vec{j}$ is the density current. To also include mesoscopic details, in particular to incorporate fluctuations, one argues that the current has, in addition to the deterministic part, also a random contribution which is essentially uncorrelated in space-time. Since fluctuations are always associated with dissipation, on a more refined scale Eq. (3.1) becomes

$$\partial_t \rho + \nabla \cdot \left(\vec{j}(\rho) - D\nabla\rho + \sigma\vec{\xi}\,\right) = 0\,. \tag{3.2}$$

Here $\vec{\xi}(\mathbf{x}, t)$ is Gaussian white noise with mean zero and covariance

$$\langle \xi_\alpha(\mathbf{x}, t)\xi_{\alpha'}(\mathbf{x}', t')\rangle = \delta_{\alpha\alpha'}\delta(\mathbf{x} - \mathbf{x}')\delta(t - t')\,, \quad \alpha, \alpha' = 1, \ldots, d\,. \tag{3.3}$$

σ is the noise strength and D is the diffusion constant. They are both treated as numbers. Physically they will depend on the density. But this would be higher order effects, which are ignored in our discussion.

The goal is to compute the density time correlations in the stationary regime, which is no easy task, since (3.2) is nonlinear. But correlations can be thought of as imposing at $t = 0$ a small density perturbation at the origin and then record how the perturbation propagates in space-time. For this purpose one might hope to get away with linearizing (3.2) at the uniform background density ρ_0 as $\rho_0 + \varrho(\mathbf{x}, t)$, which yields

$$\partial_t \varrho + \nabla \cdot \big(\vec{j}\,'(\rho_0)\varrho - D\nabla\varrho + \sigma\vec{\xi}\,\big) = 0\,, \tag{3.4}$$

where $'$ refers to differentiation w.r.t. ρ. Equation (3.4) is a linear Langevin equation, hence solved easily. Since ϱ is a deviation from the uniform background, we are interested in the space-time stationary process with zero mean. First note that (3.4) has a unique time-stationary zero mean measure, which is Gaussian white noise in the spatial variable,

$$\langle \varrho(\mathbf{x})\varrho(\mathbf{x}')\rangle = (\sigma^2/2D)\delta(\mathbf{x}-\mathbf{x}')\,. \tag{3.5}$$

For the stationary space-time covariance one obtains

$$\langle \varrho(\mathbf{x}, t)\varrho(0,0)\rangle = (\sigma^2/2D)p(\mathbf{x}-\vec{c}\,t, t)\,, \tag{3.6}$$

where p is the Gaussian transition kernel,

$$p(\mathbf{x}, t) = (4\pi D|t|)^{-d/2}\exp\big(-\mathbf{x}^2/4D|t|\big)\,, \tag{3.7}$$

and $\vec{c} = \vec{j}\,'(\rho_0)$. By (3.4) the fluctuating current is given by

$$\vec{\mathscr{J}} = \vec{c}\varrho - D\nabla\varrho + \sigma\vec{\xi}\,. \tag{3.8}$$

For the stationary correlations of the total current one arrives at

$$\int_{\mathbb{R}^d} dx \langle \vec{\mathscr{J}}(\mathbf{x}, t)\cdot\vec{\mathscr{J}}(0,0)\rangle = d\sigma^2\delta(t)\,. \tag{3.9}$$

No surprise. A density fluctuation propagates with velocity $\vec{c}$ and spreads diffusively. The currents are delta-correlated, which should translate into exponential decay for the underlying microscopic system. But before jumping at such conclusions one has to study the stability of (3.9) against including higher orders in the expansion. By power counting the next to leading term is the nonlinear current at second order, which amounts to

$$\partial_t\varrho + \nabla\cdot\big(\vec{c}\varrho + \vec{G}\varrho^2 - D\nabla\varrho + \sigma\vec{\xi}\,\big) = 0\,, \quad \vec{G} = \tfrac{1}{2}\vec{j}\,''(\rho_0)\,. \tag{3.10}$$

The task is to compute the current correlation (3.9) perturbatively in $\vec{G}$.

We first remove $\vec{c}$ by switching to a moving frame of reference. Secondly we note that, quite surprisingly, white noise is still time-stationary under the nonlinear Langevin equation (3.10). Since, as argued above, spatial white noise is time-stationary under the linear part, one only has to check its invariance under the deterministic evolution $\partial_t \varrho = -\vec{G} \cdot \nabla \varrho^2$. Formally its vector field is divergence free, since one can choose a divergence free discretization [10], compare with the discussion in Sect. 3.3 below. Hence the "volume measure" is preserved and one has to check only the time change of the logarithm of the stationary density,

$$\frac{d}{dt} \int_{\mathbb{R}^d} dx \tfrac{1}{2} \varrho(\mathbf{x})^2 = -\int_{\mathbb{R}^d} dx \varrho(\mathbf{x}) \vec{G} \cdot \nabla \varrho^2(\mathbf{x}) = \tfrac{1}{3} \int_{\mathbb{R}^d} dx \vec{G} \cdot \nabla \varrho^3(\mathbf{x}) = 0\,. \quad (3.11)$$

For an expansion in $\vec{G} \cdot \nabla \varrho^2$ it is most convenient to use the Fokker-Planck generator, denoted by $L = L_0 + L_1$, where L_0 corresponds to the Gaussian part and L_1 to the nonlinear flow. We define

$$S(\mathbf{x}, t) = \langle \varrho(\mathbf{x}, t) \varrho(0, 0) \rangle = \langle \varrho(\mathbf{x}) \mathrm{e}^{Lt} \varrho(0) \rangle\,, \quad (3.12)$$

average with respect to the time-stationary Gaussian measure, and plan to use the general second moment sum rule

$$\frac{d^2}{dt^2} \int_{\mathbb{R}^d} dx \mathbf{x}^2 S(\mathbf{x}, t) = \int_{\mathbb{R}^d} dx \langle \vec{\mathscr{J}}(\mathbf{x}, t) \cdot \vec{\mathscr{J}}(0, 0) \rangle\,, \quad (3.13)$$

which follows directly from the conservation law. By second order time-dependent perturbation theory,

$$S(\mathbf{x}, t) = \langle \varrho(\mathbf{x}) e^{L_0 t} \varrho(0) \rangle + \int_0^t dt_1 \langle \varrho(\mathbf{x}) \mathrm{e}^{L_0(t-t_1)} L_1 \mathrm{e}^{L_0 t_1} \varrho(0) \rangle$$
$$+ \int_0^t dt_2 \int_0^{t_2} dt_1 \langle \varrho(\mathbf{x}) \mathrm{e}^{L_0(t-t_2)} L_1 \mathrm{e}^{L_0(t_2-t_1)} L_1 \mathrm{e}^{L_0 t_1} \varrho(0) \rangle + \mathscr{O}(|\vec{G}|^3)\,. \quad (3.14)$$

$L_0 \varrho$ is linear, while $L_1 \varrho$ is quadratic in ϱ. Thus the second term on the right is odd in ϱ and vanishes. For any functional, F,

$$\langle \varrho(\mathbf{x}) L_1 F(\varrho) \rangle = -\langle (L_1 \varrho(\mathbf{x})) F(\varrho) \rangle\,, \quad (3.15)$$

see (3.109), and the left L_1 is swapped over to act on $\mathrm{e}^{L_0(t-t_2)} \varrho(\mathbf{x})$. By translation invariance, $\vec{G} \cdot \nabla$ can be pulled in front as acting on $\mathbf{x}$. Combining the terms one arrives at

$$S(\mathbf{x}, t) = \langle \varrho(\mathbf{x}) \mathrm{e}^{L_0 t} \varrho(0) \rangle + \int_0^t dt_2 \int_0^{t_2} dt_1 (\vec{G} \cdot \nabla)^2 \int_{\mathbb{R}^d} dx_1 \int_{\mathbb{R}^d} dx_2$$
$$\times p(\mathbf{x} - \mathbf{x}_2, t - t_2) p(\boldsymbol{x}_1, t_1) \langle \varrho(\mathbf{x}_2)^2 \mathrm{e}^{L_0(t_2-t_1)} \varrho(\boldsymbol{x}_1)^2 \rangle\,. \quad (3.16)$$

The Gaussian average is computed as

$$\langle \varrho(\mathbf{x}_2)^2 e^{L_0 t} \varrho(\boldsymbol{x}_1)^2 \rangle = 2\big((\sigma^2/2D)p(\mathbf{x}_2 - \mathbf{x}_1, t)\big)^2 + \text{ s-c}. \tag{3.17}$$

The self-contraction does not depend on $\mathbf{x}$, hence vanishes when applying $\vec{G} \cdot \nabla$. We insert in (3.13). Working out the integrals yields, including second order,

$$\int_{\mathbb{R}^d} dx \langle \vec{\mathscr{J}}(\mathbf{x}, t) \cdot \vec{\mathscr{J}}(0,0) \rangle = d\sigma^2 \delta(t) + 4(\sigma^2/2D)^2 |\vec{G}|^2 p(0, 2t), \tag{3.18}$$

which is the claimed long time tail for a scalar conserved field in d dimensions.

Equation (3.10) is singular at short distances, the worse the higher the dimension. In physical systems there is a natural cut-off at the microscopic scale which is simply ignored in (3.10). There are many possibilities to improve, but the serious constraint is to obtain a still manageable nonlinear stochastic equation. The noise should remain δ-correlated in time so to preserve the Markov property of the time evolution. One could smoothen in space, but thereby may loose the information on the time-stationary measure. To my experience the best compromise is to adopt the obvious spatial discretization. Then one has a set of coupled stochastic differential equations. For them one can check rigorously the time-stationary measure and identities as e.g. (3.15). On the perturbative level this then leads to the continuum expressions as (3.16). For a simple fluid the current correlations are bounded and will not diverge near $t = 0$. So in (3.18) only the long time prediction can be trusted.

What can be learned from the long time tails? In dimension $d \geq 3$ the decay is integrable. Thus, as assumed implicitly beforehand, the model has a finite diffusivity. Higher order terms in the expansion will modify prefactors but should not alter the exponent for the time decay. Dimension $d = 2$ is marginal. The diffusivity is weakly divergent. In principle, say, a system of hard disks has infinite viscosity and thermal conductivity. But since the divergence is only logarithmic one could convert the result into a weak system size dependence. In *one dimension* the conductivity is truly infinite. Obviously, not even the power law as based on the perturbative expansion (3.16) can be trusted and one needs to develop non-perturbative techniques.

Equation (3.10) for $d = 1$ is known as stochastic Burgers equation, which we record for later reference as

$$\partial_t \varrho - \partial_x\big(\tfrac{1}{2}\lambda\varrho^2 + \nu\partial_x\varrho + \sigma\xi\big) = 0, \quad x \in \mathbb{R}. \tag{3.19}$$

One can introduce a potential through $\varrho = \partial_x h$. Then h satisfies the one-dimensional version of the Kadar-Parisi-Zhang (KPZ) equation [11],

$$\partial_t h = \tfrac{1}{2}\lambda(\partial_x h)^2 + \nu\partial_x^2 h + \sigma\xi, \tag{3.20}$$

for the moment using the more conventional symbols for the coefficients. Over the last 15 years many properties, including rigorous results, of the KPZ equation have been obtained. While this is not the place to dive into details, I note that the solution is continuous in x, but so singular that $(\partial_x h)^2$ is ill-defined. However it is proved that the ultraviolet divergence is very mild and can be tamed by what would be an infinite energy renormalization in a $1 + 1$ dimensional quantum field theory, compare with the discussion in [12]. More precisely one chooses a regularized version of (3.20) by replacing $\xi(x,t)$ through the spatially smoothed version $\xi_\varphi(x,t) = \int \varphi(x-x')\xi(x',t)dx'$ with $\varphi \geq 0$, even, smooth, of rapid decay at infinity, and normalized to 1. It can be proved that then the solution to (3.20) is well-defined. One introduces an ultraviolet cut-off of spatial size ε by choosing the δ-function sequence $\varphi_\varepsilon(x) = \varepsilon^{-1}\varphi(\varepsilon^{-1}x)$ and substituting the noise ξ by its smoothed version $\xi_\varepsilon = \xi_{\varphi_\varepsilon}$. Let us denote the corresponding solution of the KPZ equation by $h_\varepsilon(x,t)$. Then $h_\varepsilon(x,t) - v_\varepsilon t$, i.e. $h_\varepsilon(x,t)$ viewed in the frame moving with the diverging velocity $v_\varepsilon = \varepsilon^{-1}\int \varphi(x)^2 dx$, has a non-degenerate limit as $\varepsilon \to 0$, which coincides with the Cole-Hopf solution of the KPZ equation [13, 14].

In this context Hairer recently developed a solution theory, for which he was awarded the 2014 Fields Medal. His theory works for the KPZ equation, as well as a large class of other singular stochastic partial differential equations, and for general cutoffs [15, 16]. Of course, the solution theory studies the small scale structure of solutions, and not the large scale, where the universal behaviors of interest are observed.

There is an exact formula for $S(x,t)$, which involves Fredholm determinants. In the long time limit

$$S(x,t) \simeq (\sigma^2/2\nu)(\sqrt{2}\lambda|t|)^{-2/3} f_{\mathrm{KPZ}}\big((\sqrt{2}\lambda|t|)^{-2/3}x\big). \tag{3.21}$$

The scaling function f_{KPZ} will reappear below, where more details are given. The Burgers current reads

$$\mathcal{J}(x,t) = -(\tfrac{1}{2}\lambda\varrho^2 + \nu\partial_x\varrho + \sigma\xi) \tag{3.22}$$

and for its total correlation function one obtains, using the sum rule (3.13),

$$\int_{\mathbb{R}} dx\langle \mathcal{J}(x,t)\mathcal{J}(0,0)\rangle \simeq \Big((\sigma^2/2\nu)(\sqrt{2}\lambda)^2\int_{\mathbb{R}} dx\tfrac{4}{9}x^2 f_{\mathrm{KPZ}}(x)\Big)(\sqrt{2}\lambda t)^{-2/3} \tag{3.23}$$

valid for large t. Note that the true decay turns out to be $t^{-2/3}$, to be contrasted with the perturbative result $t^{-1/2}$. In fact the power $2/3$ was argued already in [5] using a self-consistent scheme, see also [17]. As a fairly unusual feature, the non-universal coefficients are given directly in terms of the parameters of the stochastic Burgers equation. In fact, the particular form can be guessed from the scaling properties of Eq. (3.19). Only for the scaling function f_{KPZ} and the pure number $\sqrt{2}$ one has to rely on the exact solution, which is the result of an intricate analysis using lattice type models [18–21], replica computations for the KPZ equation [22, 23], and the finite time exact solution of the stationary KPZ equation itself [24].

In these notes we will explain how nonlinear fluctuating hydrodynamics, in the same spirit as already explained for a scalar field in one dimension, can be used to predict the asymptotic form of the equilibrium space-time correlations of anharmonic chains. Chains are one-dimensional objects and it may seem as if we have accomplished the task already. Well, we have not even written down a Hamiltonian. So first we have to dwell on a few general properties of anharmonic chains. In particular we will see that they have three conserved fields, generically. The corresponding Euler equations are derived, thereby identifying the macroscopic currents. But now we are forced to handle several conserved fields. While it is not so difficult to write down the multi-component generalization of (3.19), the analysis of the Langevin equation will be more complicated with no exact solution at help. A second major task will be to test the quality of these predictions by comparing with molecular dynamics simulations.

3.2 Anharmonic Chains

Conservation Laws and Equilibrium Time Correlations We consider a classical fluid consisting of particles with positions q_j and momenta $p_j, j = 1, \dots, N$, $q_j, p_j \in \mathbb{R}$, possible boundary conditions to be discussed later on. We use units such that the mass of the particles equals 1. Then the Hamiltonian is of the standard form,

$$H_N^{\mathrm{fl}} = \sum_{j=1}^{N} \tfrac{1}{2} p_j^2 + \tfrac{1}{2} \sum_{i \neq j=1}^{N} V(q_i - q_j) \,, \tag{3.24}$$

with pair potential $V(x) = V(-x)$. The potential may have a hard core and otherwise is assumed to be short ranged. The dynamics for long range potentials is of independent interest [25], but not discussed here. Furthermore the potential is assumed to be thermodynamically stable, meaning the validity of a bound as $\sum_{i \neq j=1}^{N} V(q_i - q_j) \geq A - BN$ for some constants A and $B > 0$. A substantial simplification is achieved by assuming a hard core of diameter a, i.e. $V(x) = \infty$ for $|x| < a$, and restricting the range of the smooth part of the potential to at most $2a$. Then the particles maintain their order, $q_j \leq q_{j+1}$, and in addition only nearest neighbor particles interact. Hence H_N^{fl} simplifies to

$$H_N = \sum_{j=1}^{N} \tfrac{1}{2} p_j^2 + \sum_{j=1}^{N-1} V(q_{j+1} - q_j) \,. \tag{3.25}$$

As a, at first sight very different, physical realization, we could interpret H_N as describing particles in one dimension coupled through anharmonic springs which is then usually referred to as anharmonic chain.

In the second interpretation the spring potential can be more general than anticipated so far. No ordering constraint is required and the potential does not have to be even. To have well defined thermodynamics the chain is pinned at both ends as $q_1 = 0$ and $q_{N+1} = \ell N$. It is convenient to introduce the stretch $r_j = q_{j+1} - q_j$. Then the boundary condition corresponds to the microcanonical constraint

$$\sum_{j=1}^{N} r_j = \ell N . \tag{3.26}$$

Switching to canonical equilibrium according to the standard rules, one the arrives at the obvious condition of a finite partition function

$$Z(P, \beta) = \int dx\, \mathrm{e}^{-\beta(V(x)+Px)} < \infty , \tag{3.27}$$

using the standard convention that the integral is over the entire real line. Here $\beta > 0$ is the inverse temperature and P is the thermodynamically conjugate variable to the stretch. By partial integration

$$P = -Z(P, \beta)^{-1} \int dx V'(x)\, \mathrm{e}^{-\beta(V(x)+Px)} , \tag{3.28}$$

implying that P is the average force in the spring between two adjacent particles, hence identified as thermodynamic pressure. To have a finite partition function, a natural condition on the potential is to be bounded from below and to have a one-sided linear bound as $V(x) \geq a_0 + b_0|x|$ for either $x > 0$ or $x < 0$ and $b_0 > 0$. Then there is a non-empty interval $I(\beta)$ such that $Z(P, \beta) < \infty$ for $P \in I(\beta)$. For the particular case of a hard-core fluid one imposes $P > 0$.

Note: The sign of P is chosen such that for a gas of hard-point particles one has the familiar ideal gas law $P = 1/\beta\ell$. The chain tension is $-P$.

Famous examples are the harmonic chain, $V_{\mathrm{ha}}(x) = x^2$, the Fermi-Pasta-Ulam (FPU) chain, $V_{\mathrm{FPU}}(x) = \frac{1}{2}x^2 + \frac{1}{3}\alpha x^3 + \frac{1}{4}\beta x^4$, in the historical notation [26], and the Toda chain [27], $V(x) = \mathrm{e}^{-x}$, in which case $P > 0$ is required. The harmonic chain, the Toda chain, and the hard-core potential, $V_{\mathrm{hc}}(x) = \infty$ for $|x| < a$ and $V_{\mathrm{hc}}(x) = 0$ for $|x| \geq a$, are in fact integrable systems which have a very different correlation structure and will not be discussed here. Except for the harmonic chain, one simple way to break integrability is to assume alternating masses, say $m_j = m_0$ for even j and $m_j = m_1$ for odd j.

We will mostly deal with anharmonic chains described by the Hamiltonian (3.25), including one-dimensional hard-core fluids with a sufficiently small potential range. There are several good reasons. Firstly on the level of fluctuating hydrodynamics a generic one-dimensional fluid cannot be distinguished from an anharmonic chain. Thus with the proper translation of the various terms we would also predict the large scale correlation structure of one-dimensional fluids. The second reason is that in the large body of molecular dynamics simulations there is not a single one which deals with an "honest" one-dimensional fluid. To be able to reach large system sizes all simulations are performed for anharmonic chains. In addition, from a theoretical perspective, the equilibrium measures of anharmonic chains are particularly simple in being of product form in momentum and stretch variables. Thus material parameters, as compressibility and sound speed, can be expressed in terms of one-dimensional integrals involving the Boltzmann factor $\mathrm{e}^{-\beta(V(x)+Px)}$, $V(x)$, and x.

Anharmonic chains should be thought of as a particular class of $1+1$ dimensional field theories. Thus q_j is viewed as the displacement variable at lattice site j and not necessarily the physical position of the jth particle on the real line. There is a simple translation between both pictures, but we will stick to the field theory point of view. For a fluid with unlabeled particles and, say, a bounded potential, the equivalence is lost and only the fluid picture can be used.

This being said, we follow the standard rules. We write down the dynamics of stretches and momenta and identify the conserved fields. From there we infer the microcanonical and canonical equilibrium measures. For slowly varying equilibrium parameters we deduce the Euler equations. In particular, their version linearized at uniform equilibrium will constitute the backbone in understanding the equilibrium time correlations of the conserved fields.

The dynamics of the anharmonic chain is governed by

$$\frac{d}{dt}q_j = p_j\,, \qquad \frac{d}{dt}p_j = V'(q_{j+1} - q_j) - V'(q_j - q_{j-1})\,. \tag{3.29}$$

For the initial conditions we choose a lattice cell of length N and require

$$q_{j+N} = q_j + \ell N\,, \qquad p_{j+N} = p_j \tag{3.30}$$

for all $j \in \mathbb{Z}$. This property is preserved under the dynamics and thus properly mimics a system of finite length N. The stretches are then N-periodic, $r_{j+N} = r_j$, and the single cell dynamics is given by

$$\frac{d}{dt}r_j = p_{j+1} - p_j\,, \qquad \frac{d}{dt}p_j = V'(r_j) - V'(r_{j-1})\,, \tag{3.31}$$

$j = 1, \ldots, N$, together with the periodic boundary conditions $p_{1+N} = p_1$, $r_0 = r_N$ and the constraint (3.26). Through the stretch there is a coupling to the right neighbor and through the momentum a coupling to the left neighbor. The potential is defined only up to translations, since the dynamics does not change under a simultaneous shift of $V(x)$ to $V(x-a)$ and r_j to r_j+a, in other words, the potential can be shifted by shifting the initial r-field. Note that our periodic boundary conditions are not identical to fluid particles moving on a ring, but they may become so for large system size when length fluctuations become negligible. Both equations are already of conservation type and we conclude that

$$\frac{d}{dt}\sum_{j=1}^{N} r_j = 0\,, \quad \frac{d}{dt}\sum_{j=1}^{N} p_j = 0\,. \tag{3.32}$$

We define the local energy by

$$e_j = \tfrac{1}{2}p_j^2 + V(r_j)\,. \tag{3.33}$$

Then its local conservation law reads

$$\frac{d}{dt}e_j = p_{j+1}V'(r_j) - p_jV'(r_{j-1})\,, \tag{3.34}$$

implying that

$$\frac{d}{dt}\sum_{j=1}^{N} e_j = 0\,. \tag{3.35}$$

At this point we assume that there are no further local conservation laws. Unfortunately our assumption, while reasonable, is extremely difficult to check. It certainly rules out the integrable chains, which have N conservation laws. There is no natural example known with, say, seven conservation laws. But the mere fact that there are exceptions implies that close to integrability the predictions from fluctuating hydrodynamics could be on time scales which are not accessible. The parameters entering in fluctuating hydrodynamics depend smoothly on the potential. Thus as one approaches, for example, the Toda potential no abrupt changes will be detected. In this sense, the theory cannot distinguish the Toda chain from a FPU chain both at moderate temperatures. There is another limitation which can be addressed more quantitatively. If $V(x) + Px$ has a unique minimum, then at very low temperatures the potential is close to a harmonic one. This feature will be properly reflected by fluctuating hydrodynamics through the temperature dependence of the coupling coefficients and the noise strength.

The microcanonical equilibrium state is defined by the Lebesgue measure constrained to a particular value of the conserved fields as

$$\sum_{j=1}^{N} r_j = \ell N\,, \quad \sum_{j=1}^{N} p_j = \mathsf{u}N\,, \quad \sum_{j=1}^{N}\big(\tfrac{1}{2}p_j^2 + V(r_j)\big) = \mathfrak{e}N \tag{3.36}$$

with ℓ the stretch, u the momentum, and $\mathfrak{e}$ the total energy per particle. In our context the equivalence of ensembles holds and computationally it is of advantage to switch to the canonical ensemble with respect to all three constraints. Then the dual variable for the stretch ℓ is the pressure P, for the momentum the average momentum, again denoted by u, and for the total energy $\mathfrak{e}$ the inverse temperature β. For the limit of infinite volume the symmetric choice $j \in [-N, \dots, N]$ is more convenient. In the limit $N \to \infty$ either under the canonical equilibrium state, trivially, or under the microcanonical ensemble, by the equivalence of ensembles, the collection $(r_j, p_j)_{j\in\mathbb{Z}}$ are independent random variables. Their single site probability density is given by

$$Z(P,\beta)^{-1}\mathrm{e}^{-\beta(V(r_j)+Pr_j)}(2\pi/\beta)^{-1/2}\mathrm{e}^{-\frac{1}{2}\beta(p_j-\mathsf{u})^2}\,. \tag{3.37}$$

Averages with respect to (3.37) are denoted by $\langle\cdot\rangle_{P,\beta,\mathsf{u}}$. The dependence on the average momentum can be removed by a Galilei transformation. Hence we mostly work with $\mathsf{u} = 0$, in which case we merely drop the index u. We also introduce the internal energy, e, through $\mathfrak{e} = \frac{1}{2}\mathsf{u}^2 + \mathsf{e}$, which agrees with the total energy at $\mathsf{u} = 0$. The canonical free energy, at $\mathsf{u} = 0$, is defined by

$$G(P,\beta) = -\beta^{-1}\big(-\tfrac{1}{2}\log\beta + \log Z(P,\beta)\big)\,. \tag{3.38}$$

Then

$$\ell = \langle r_0\rangle_{P,\beta}\,, \quad \mathsf{e} = \partial_\beta\big(\beta G(P,\beta)\big) - P\ell = \frac{1}{2\beta} + \langle V(r_0)\rangle_{P,\beta}\,. \tag{3.39}$$

The relation (3.39) defines $(P,\beta) \mapsto (\ell(P,\beta), \mathsf{e}(P,\beta))$, thereby the inverse map $(\ell,\mathsf{e}) \mapsto (P(\ell,\mathsf{e}),\ \beta(\ell,\mathsf{e}))$, and thus accomplishes the switch between the microcanonical thermodynamic variables ℓ, e and the canonical thermodynamic variables P, β.

It is convenient to collect the conserved fields as the 3-vector $\vec{g} = (g_1, g_2, g_3)$,

$$\vec{g}(j,t) = \big(r_j(t), p_j(t), e_j(t)\big)\,, \tag{3.40}$$

$\vec{g}(j,0) = \vec{g}(j)$. Then the conservation laws are combined as

$$\frac{d}{dt}\vec{g}(j,t) + \vec{\mathcal{J}}(j+1,t) - \vec{\mathcal{J}}(j,t) = 0 \tag{3.41}$$

with the local current functions

$$\vec{\mathcal{J}}(j) = \big(-p_j, -V'(r_{j-1}), -p_jV'(r_{j-1})\big)\,. \tag{3.42}$$

Our prime interest are the equilibrium time correlations of the conserved fields, which are defined by

$$S_{\alpha\alpha'}(j,t) = \langle g_\alpha(j,t)g_{\alpha'}(0,0)\rangle_{P,\beta} - \langle g_\alpha(0,0)\rangle_{P,\beta}\langle g_{\alpha'}(0,0)\rangle_{P,\beta}\,, \tag{3.43}$$

$\alpha, \alpha' = 1, 2, 3$. The infinite volume limit has been taken already and the average is with respect to thermal equilibrium at $\mathsf{u} = 0$. It is known that such a limit exists [28, 29]. Also the decay in j is exponentially fast, but with a correlation length increasing in time. Often it is convenient to regard $S(j,t)$, no indices, as a 3×3 matrix. In general, $S(j,t)$ has certain symmetries, the first set resulting from space-time stationarity and the second set from time reversal, even for $\alpha = 1, 3$, odd for $\alpha = 2$,

$$S_{\alpha\alpha'}(j,t) = S_{\alpha'\alpha}(-j,-t)\,, \quad S_{\alpha\alpha'}(j,t) = (-1)^{\alpha+\alpha'} S_{\alpha\alpha'}(j,-t)\,. \tag{3.44}$$

At $t = 0$ the average (3.43) reduces to a static average, which is easily computed. The fields are uncorrelated in j, i.e.

$$S(j,0) = \delta_{j0} C \tag{3.45}$$

with the static susceptibility matrix

$$C = \begin{pmatrix} \langle r_0; r_0 \rangle_{P,\beta} & 0 & \langle r_0; V_0 \rangle_{P,\beta} \\ 0 & \beta^{-1} & 0 \\ \langle r_0; V_0 \rangle_{P,\beta} & 0 & \frac{1}{2}\beta^{-2} + \langle V_0; V_0 \rangle_{P,\beta} \end{pmatrix}. \tag{3.46}$$

Here, for X,Y arbitrary random variables, $\langle X; Y \rangle = \langle XY \rangle - \langle X \rangle \langle Y \rangle$ denotes the second cumulant and $V_0 = V(r_0)$, following the same notational convention as for e_0. The conservation law implies the zeroth moment sum rule

$$\sum_{j \in \mathbb{Z}} S_{\alpha\alpha'}(j,t) = \sum_{j \in \mathbb{Z}} S_{\alpha\alpha'}(j,0) = C_{\alpha\alpha'}\,. \tag{3.47}$$

An explicit computation of $S(j,t)$ is utterly out of reach. But with the current computer power MD simulations have become an essential source of information. A broader coverage will be provided in Sect. 3.5. Just to have a first impression I show in Fig. 3.1 a recent MD simulation of the correlator for a FPU chain. One notes the central peak, called heat peak, which is standing still, and two symmetrically located peaks, called sound peaks which move outwards with the speed of sound. The peaks broaden with a certain power law which will have to be discussed. One expects, better hopes for, self-similar shape functions, at least for sufficiently long times. We do not know yet their form. But the central peak seems to have fat tails while the sound peaks fall off more rapidly, at least towards the outside of the sound cone. The area under each peak is preserved in time and normalized to 1 in our plot. If the chain is initially perturbed near 0 and the response in one of the conserved fields is observed at (j,t), then one records a signal, which is a linear combination of the peaks in Fig. 3.1, the coefficients depending on the initial perturbation. It might happen that one peak is missing. If the perturbation is orthogonal to all three physical fields, then there is no peak at all, only low amplitude noise.

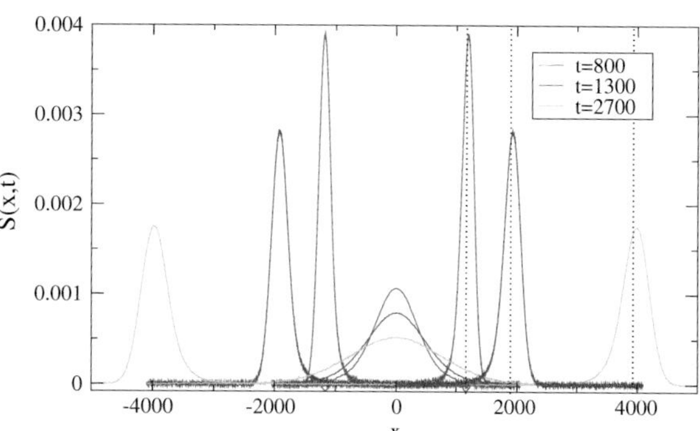

Fig. 3.1 Heat peak and sound peaks, area normalized to 1, at times $t = 800, 1300, 2700$, for a FPU chain with $N = 8192$, potential parameters $\alpha = 2$, $\beta = 1$, pressure $P = 1$, and temperature $\beta^{-1} = 0.5$. The speed of sound is $c = 1.45$

Our goal is to predict the large scale structure of the correlator $S_{\alpha\alpha'}(j,t)$ encoding the propagation of local perturbations of the equilibrium state. On the crudest level, they should be captured by linearized hydrodynamics, to which we turn next.

Linearized Hydrodynamics We start the dynamics from a product measure of the form (3.37), but replace the uniform P, u, β by slowly varying spatial fields $P(\varepsilon j), \mathsf{u}(\varepsilon j), \beta(\varepsilon j)$ with $\varepsilon \ll 1$. ε^{-1} is the macroscopic scale measured in lattice units. Equivalently we may regard ε as the lattice spacing. The relation (3.39) then defines also the slowly varying fields $\ell(\varepsilon j), \mathsf{u}(\varepsilon j), \mathfrak{e}(\varepsilon j)$. Because of the conservation laws the time change of such a state is slow and varies only over microscopic times of order $\varepsilon^{-1}t$ with macroscopic t of order 1. We average Eq. (3.41) over the slowly varying initial state, which gives then the time change of the average locally conserved fields. The expectation of the currents is more difficult. The difference in j becomes $\varepsilon\partial_x$ on the macroscopic scale. Since the currents are functions depending only on at most two neighboring lattice sites, to lowest approximation their average can be computed in the equilibrium state with the corresponding local values of the fields, i.e. $\ell(\varepsilon j, \varepsilon^{-1}t), \mathsf{u}(\varepsilon j, \varepsilon^{-1}t), \mathfrak{e}(\varepsilon j, \varepsilon^{-1}t)$. Therefore we define the hydrodynamic Euler currents by the equilibrium averages

$$\langle \vec{\mathscr{J}}(j) \rangle_{\ell,\mathsf{u},\mathfrak{e}} = \big(-\mathsf{u}, P(\ell, \mathfrak{e} - \tfrac{1}{2}\mathsf{u}^2), \mathsf{u}P(\ell, \mathfrak{e} - \tfrac{1}{2}\mathsf{u}^2) \big) = \vec{\mathsf{j}}(\ell, \mathsf{u}, \mathfrak{e}) \tag{3.48}$$

with $P(\ell, \mathsf{e})$ defined implicitly through (3.39). Our argument then leads to the macroscopic Euler equations

$$\partial_t \ell - \partial_x \mathsf{u} = 0\,, \quad \partial_t \mathsf{u} + \partial_x P(\ell, \mathfrak{e} - \tfrac{1}{2}\mathsf{u}^2) = 0\,, \quad \partial_t \mathfrak{e} + \partial_x\big(\mathsf{u}P(\ell, \mathfrak{e} - \tfrac{1}{2}\mathsf{u}^2)\big) = 0\,. \tag{3.49}$$

We refer to a forthcoming monograph [29], where the validity of the Euler equations is proved up to the first shock. Since, as emphasized already, it is difficult to deal with deterministic chaos, the authors add random velocity exchanges between neighboring particles which ensure that the dynamics locally enforces the microcanonical state.

We are interested here only in small deviations from equilibrium and therefore linearize the Euler equations as $\ell + u_1(x)$, $0 + u_2(x)$, $\mathsf{e} + u_3(x)$ to linear order in the deviations $\vec{u}(x)$. This leads to the linear equation

$$\partial_t \vec{u}(x,t) + \partial_x A\vec{u}(x,t) = 0 \tag{3.50}$$

with

$$A = \begin{pmatrix} 0 & -1 & 0 \\ \partial_\ell P & 0 & \partial_{\mathsf{e}} P \\ 0 & P & 0 \end{pmatrix}. \tag{3.51}$$

Here, and in the following, the dependence of A, C and similar quantities on the background values $\ell, \mathsf{u} = 0, \mathsf{e}$, hence on P, β, is suppressed from the notation. Beyond (3.47) there is the first moment sum rule which states that

$$\sum_{j\in\mathbb{Z}} jS(j,t) = ACt\,. \tag{3.52}$$

A proof, which in essence uses only the conservation laws and space-time stationarity of the correlations, is given in [30], see also see [31, 32]. Microscopic properties enter only minimally. However, since $C = C^{\mathrm{T}}$ and $S(j,t)^{\mathrm{T}} = S(-j,-t)$, Eq. (3.52) implies the important relation

$$AC = (AC)^{\mathrm{T}} = CA^{\mathrm{T}}, \tag{3.53}$$

with ^{T} denoting transpose. Of course, (3.53) can be checked also directly from the definitions. Since $C > 0$, A is guaranteed to have real eigenvalues and a nondegenerate system of right and left eigenvectors. For A one obtains the three eigenvalues $0, \pm c$ with

$$c^2 = -\partial_\ell P + P\partial_{\mathsf{e}} P > 0\,. \tag{3.54}$$

Thus the solution to the linearized equation has three modes, one standing still, one right moving with velocity c and one left moving with velocity $-c$. Hence we have identified the adiabatic sound speed as being equal to c.

Equation (3.50) is a deterministic equation. But the initial data are random such that within our approximation

$$\langle u_\alpha(x,0)u_{\alpha'}(x',0)\rangle = C_{\alpha\alpha'}\delta(x-x')\,. \tag{3.55}$$

To determine the correlator $S(x,t)$ with such initial conditions is most easily achieved by introducing the linear transformation R satisfying

$$RAR^{-1} = \mathrm{diag}(-c,0,c)\,, \quad RCR^{\mathrm{T}} = 1\,. \tag{3.56}$$

Up to trivial phase factors, R is uniquely determined by these conditions. Explicit formulas are found in [30]. Setting $\vec{\phi} = A\vec{u}$, one concludes

$$\partial\phi_\alpha + c_\alpha \partial_x \phi_\alpha = 0\,, \quad \alpha = -1, 0, 1\,, \tag{3.57}$$

with $\vec{c} = (-c, 0, c)$. By construction, the random initial data have the correlator

$$\langle \phi_\alpha(x,0)\phi_{\alpha'}(x',0)\rangle = \delta_{\alpha\alpha'}\delta(x - x')\,. \tag{3.58}$$

Hence

$$\langle \phi_\alpha(x,t)\phi_{\alpha'}(0,0)\rangle = \delta_{\alpha\alpha'}\delta(x - c_\alpha t)\,. \tag{3.59}$$

We transform back to the physical fields. Then in the continuum approximation, at the linearized level,

$$S(x,t) = R^{-1}\mathrm{diag}\big(\delta(x+ct), \delta(x), \delta(x-ct)\big)R^{-\mathrm{T}} \tag{3.60}$$

with $R^{-\mathrm{T}} = (R^{-1})^{\mathrm{T}}$.

Rather easily we have gained a crucial insight. $S(j,t)$ has three peaks which separate linearly in time. For example, $S_{11}(j,t)$ has three sharp peaks moving with velocities $\pm c, 0$. Physically, one expects such peaks not to be strictly sharp, but to broaden in the course of time because of dissipation. This issue will have to be explored in great detail. It follows from the zeroth moment sum rule that the area under each peak is preserved in time and thus determined through (3.60). Hence the weights can be computed from the matrix R^{-1}, usually called Landau-Plazcek ratios. A Landau-Placzek ratio could vanish, either accidentally or by a particular symmetry. An example is the momentum correlation $S_{22}(j,t)$. Since $(R^{-1})_{20} = 0$ always, its central peak is absent.

For integrable chains each conservation law generates a peak. Thus, e.g., $S_{11}(j,t)$ of the Toda chain is expected to have a broad spectrum expanding ballistically, rather than consisting of three sharp peaks.

3.3 Nonlinear Fluctuating Hydrodynamics

Euler Currents to Second Order The broadening of the peaks results from random fluctuations in the currents, which tend to be uncorrelated in space-time. Therefore the crudest model would be to assume that the current statistics is space-time Gaussian white noise. In principle, the noise components could be correlated. But since the stretch current is itself conserved, its fluctuations will be taken care of by the momentum equation. Momentum and energy currents have different signature under time reversal, hence their cross correlation vanishes. As a result, there is a fluctuating momentum current of strength σ_u and an independent energy current of strength σ_{e}. According to Onsager, noise is linked to dissipation as modeled by a diffusive term. Thus the linearized equations (3.50) are extended to

$$\partial_t \vec{\mathsf{u}}(x,t) + \partial_x\big(A\vec{\mathsf{u}}(x,t) - \partial_x D\vec{\mathsf{u}}(x,t) + B\vec{\xi}(x,t)\big) = 0\,. \tag{3.61}$$

Here $\vec{\xi}(x,t)$ is standard white noise with covariance

$$\langle \xi_\alpha(x,t)\xi_{\alpha'}(x',t')\rangle = \delta_{\alpha\alpha'}\delta(x-x')\delta(t-t') \tag{3.62}$$

and, as argued, the noise strength matrix is diagonal as

$$B = \mathrm{diag}(0,\sigma_{\mathsf{u}},\sigma_{\mathsf{e}})\,. \tag{3.63}$$

To distinguish the linearized Euler equations (3.50) from the Langevin equations (3.61), we use $\vec{\mathsf{u}} = (\mathsf{u}_1,\mathsf{u}_2,\mathsf{u}_3)$ for the fluctuating fields.

From the introduction, we know already that a Gaussian fluctuation theory will fail. But still, it is useful to first explore the structure of the Langevin equation (3.61). The stationary measures for (3.61) are spatial white noise with arbitrary mean. Since small deviations from uniformity are considered, we always impose mean zero. Then the components are correlated as

$$\langle \mathsf{u}_\alpha(x)\mathsf{u}_{\alpha'}(x')\rangle = C_{\alpha\alpha'}\delta(x-x')\,. \tag{3.64}$$

Stationarity relates the linear drift and the noise strength through the steady state covariance as

$$-(AC - CA^{\mathrm{T}})\partial_x + (DC + CD^{\mathrm{T}})\partial_x^2 = BB^{\mathrm{T}}\partial_x^2\,. \tag{3.65}$$

The first term vanishes by (3.53) and the diffusion matrix is uniquely determined as

$$D = \begin{pmatrix} 0 & 0 & 0 \\ 0 & D_{\mathsf{u}} & 0 \\ \tilde{D}_{\mathsf{e}} & 0 & D_{\mathsf{e}} \end{pmatrix}. \tag{3.66}$$

with $\tilde{D}_{\mathsf{e}} = -\langle r_0; V_0\rangle_{P,\beta}\langle r_0; r_0\rangle_{P,\beta}^{-1} D_{\mathsf{e}}$. Here $D_{\mathsf{u}} > 0$ is the momentum and $D_{\mathsf{e}} > 0$ the energy diffusion coefficient, which are related to the noise strength as

$$\sigma_{\mathsf{u}}^2 = \langle p_0; p_0\rangle_{P,\beta} D_{\mathsf{u}}\,, \quad \sigma_{\mathsf{e}}^2 = \langle e_0; e_0\rangle_{P,\beta} D_{\mathsf{e}}\,. \tag{3.67}$$

We still have to establish that the stationary measure (3.64) is unique and is approached in the limit $t \to \infty$. For this purpose it suffices that the 3×3 matrix $\mathrm{i}2\pi kA - (2\pi k)^2 D$ has its eigenvalues in the left hand complex plane, where for convenience we have switched to Fourier space with respect to x. If one drops D, then $\mathrm{i}2\pi kA$ has the eigenvalues $\mathrm{i}2\pi kc(-1,0,1)$. Hence one can use first order perturbation theory with respect to $-(2\pi k)^2 D$, which is given by $\langle \tilde{\psi}_\alpha, D\psi_\alpha \rangle$, where $A\psi_\alpha = c_\alpha \psi_\alpha$ and $A^{\mathrm{T}}\tilde{\psi}_\alpha = c_\alpha \tilde{\psi}_\alpha$ are the right and left eigenvectors of A, as listed in [30]. One simply has to follow the definitions and express every term through the cumulants of r_0 and V_0. As to be expected, the matrix elements from above have a definite value and the eigenvalues are shifted into the left hand complex plane. Similarly, $-D$ has eigenvalues $0, -D_{\mathsf{u}}, -D_{\mathsf{e}}$ and the zero eigenvalue is shifted to the left by second order perturbation in $\mathrm{i}(2\pi k)^{-1}A$. The only condition is the strict positivity of $D_{\mathsf{u}}, D_{\mathsf{e}}$.

Based on (3.61) one computes the stationary space-time covariance, which most easily is written in Fourier space,

$$S_{\alpha\alpha'}(x,t) = \langle \mathsf{u}_\alpha(x,t)\mathsf{u}_{\alpha'}(0,0)\rangle = \int dk\, \mathrm{e}^{\mathrm{i}2\pi kx}\big(\mathrm{e}^{-\mathrm{i}t2\pi kA-|t|(2\pi k)^2 D}C\big)_{\alpha\alpha'}. \tag{3.68}$$

To extract the long time behavior it is convenient to transform to normal modes. But before, we have to introduce a more systematic notation. We will use the superscript $\sharp$ for a normal mode quantity. Thus for the anharmonic chain

$$S^\sharp(j,t) = RS(j,t)R^{\mathrm{T}}, \quad S^\sharp_{\alpha\alpha'}(j,t) = \langle (R\vec{g})_\alpha(j,t)(R\vec{g})_{\alpha'}(0,0)\rangle_{P,\beta}. \tag{3.69}$$

The hydrodynamic fluctuation fields are defined on the continuum, thus functions of x, t, and we write

$$S_{\alpha\alpha'}(x,t) = \langle \mathsf{u}_\alpha(x,t)\mathsf{u}_{\alpha'}(0,0)\rangle, \quad S^\sharp(x,t) = RS(x,t)R^{\mathrm{T}}. \tag{3.70}$$

Correspondingly $A^\sharp = RAR^{-1} = \mathrm{diag}(-c, 0, c)$, $D^\sharp = RDR^{-1}$, $B^\sharp = RB$. Note that $\vec{\mathsf{u}}(x,t)$ will change its meaning when switching from linear to nonlinear fluctuating hydrodynamics.

In normal mode representation Eq. (3.68) becomes

$$S^\sharp(x,t) = \int dk\, \mathrm{e}^{\mathrm{i}2\pi kx}\mathrm{e}^{-\mathrm{i}t2\pi kA^\sharp-|t|(2\pi k)^2 D^\sharp}. \tag{3.71}$$

The leading term, $\mathrm{i}t2\pi kA^\sharp$, is diagonal, while the diffusion matrix $D^\sharp$ couples the components. But for large t the peaks are far apart and the cross terms become small. More formally we split $D^\sharp = D_{\mathrm{dia}} + D_{\mathrm{off}}$ and regard the off-diagonal part D_{off} as perturbation. When expanding, one notes that the off-diagonal terms carry an oscillating factor with frequency $c_\alpha - c_{\alpha'}$, $\alpha \neq \alpha'$. Hence these terms decay quickly and

$$S^\sharp_{\alpha\alpha'}(x,t) \simeq \delta_{\alpha\alpha'}\int dk\, \mathrm{e}^{\mathrm{i}2\pi kx}\mathrm{e}^{-\mathrm{i}t2\pi kc_\alpha-|t|(2\pi k)^2 D^\sharp_{\alpha\alpha}} \tag{3.72}$$

for large t. Each peak has a Gaussian shape function which broadens as $(D^\sharp_{\alpha\alpha}|t|)^{1/2}$.

Besides the peak structure, we have gained a second important insight. Since the peaks travel with distinct velocities, on the linearized level the three-component system decouples into three scalar equations, provided it is written in normal modes.

The linear fluctuation theory should be tested against adding nonlinear terms. The computation of the Introduction indicates that in one dimension the quadratic part of the Euler current is a relevant perturbation. This can be seen even more directly by rescaling $\vec{\mathsf{u}}(x,t)$ to large space-time scales as $\vec{\mathsf{u}}_\varepsilon(x,t) = \varepsilon^{-b}\vec{\mathsf{u}}_\varepsilon(\varepsilon^{-1}x, \varepsilon^{-z}t)$ and counting the powers of the nonlinear terms. To have the correct $t = 0$ covariance requires $b = 1/2$. The quadratic terms of the Euler currents are relevant, yielding $z = 3/2$, and the scaling exponents of cubic terms are only marginally relevant, while a possible dependence of D, B on $\vec{\mathsf{u}}(x,t)$ can be ignored. Thus, we retain the linear part (3.61) but expand the Euler currents including second order in $\vec{u}$, which turns (3.61) into the equations of nonlinear fluctuating hydrodynamics,

$$\begin{aligned}
&\partial_t \mathsf{u}_1 - \partial_x \mathsf{u}_2 = 0\,,\\
&\partial_t \mathsf{u}_2 + \partial_x\big((\partial_\ell P)\mathsf{u}_1 + (\partial_\mathsf{e} P)\mathsf{u}_3 + \tfrac{1}{2}(\partial_\ell^2 P)\mathsf{u}_1^2 - \tfrac{1}{2}(\partial_\mathsf{e} P)\mathsf{u}_2^2\\
&\qquad + \tfrac{1}{2}(\partial_\mathsf{e}^2 P)\mathsf{u}_3^2 + (\partial_\ell\partial_\mathsf{e} P)\mathsf{u}_1\mathsf{u}_3 D_\mathsf{u}\partial_x\mathsf{u}_2 + \sigma_\mathsf{u}\xi_2\big) = 0\,, \qquad (3.73)\\
&\partial_t \mathsf{u}_3 + \partial_x\big(P\mathsf{u}_2 + (\partial_\ell P)\mathsf{u}_1\mathsf{u}_2 + (\partial_\mathsf{e} P)\mathsf{u}_2\mathsf{u}_3 - \tilde{D}_\mathsf{e}\partial_x\mathsf{u}_1 - D_\mathsf{e}\partial_x\mathsf{u}_3 + \sigma_\mathsf{e}\xi_3\big) = 0\,.
\end{aligned}$$

To explore their consequences is a more demanding task than solving the linear Langevin equation and the results of the analysis will be more fragmentary.

Stationary Measure for the Physical Fields Adding quadratic terms could change drastically the character of the solution. To find out we first attempt to investigate the time-stationary measure. The vector field of the nonlinear part of (3.73) reads

$$\begin{aligned}
\vec{F} = -\partial_x\big(0, &\tfrac{1}{2}(\partial_\ell^2 P)\mathsf{u}_1^2 - \tfrac{1}{2}(\partial_\mathsf{e} P)\mathsf{u}_2^2\\
&+ \tfrac{1}{2}(\partial_\mathsf{e}^2 P)\mathsf{u}_3^2 + (\partial_\ell\partial_\mathsf{e} P)\mathsf{u}_1\mathsf{u}_3, (\partial_\ell P)\mathsf{u}_1\mathsf{u}_2 + (\partial_\mathsf{e} P)\mathsf{u}_2\mathsf{u}_3\big)\,. \qquad (3.74)
\end{aligned}$$

Formally the drift is divergence free, since

$$\sum_{\alpha=1}^{3}\int dx \frac{\delta F_\alpha(x)}{\delta \mathsf{u}_\alpha(x)} = \partial_\mathsf{e} P \int dx(-\partial_x\mathsf{u}_2 + \partial_x\mathsf{u}_2) = 0 \qquad (3.75)$$

and the infinite dimensional Lebesgue measure is invariant under the flow generated by $\vec{F}$. Since the equilibrium measure is a product, a natural ansatz for the invariant measure is Gaussian white noise retaining the physical susceptibility,

$$\exp\Big(-\sum_{\alpha,\alpha'=1}^{3}\tfrac{1}{2}(C^{-1})_{\alpha\alpha'}\int dx \mathsf{u}_\alpha(x)\mathsf{u}_{\alpha'}(x)\Big)\prod_{\alpha,x} d\mathsf{u}_\alpha(x)\,. \qquad (3.76)$$

As established before, for the linear Langevin equation this measure is stationary. Thus, to find out whether it is also stationary for (3.73), we only have to check the invariance under the nonlinear flow

$$\begin{aligned}
&\frac{d}{dt}\sum_{\alpha,\alpha'=1}^{3}(C^{-1})_{\alpha\alpha'}\int dx\mathsf{u}_{\alpha}(x)\mathsf{u}_{\alpha'}(x)\\
&=2\sum_{\alpha=1}^{3}\int dx(C^{-1}\vec{\mathsf{u}})_{\alpha}(x)\partial_t\mathsf{u}_{\alpha}(x)\\
&=\int dx a_0\mathsf{u}_2\partial_x\big((\partial_\ell^2P)\mathsf{u}_1^2-(\partial_{\mathsf{e}}P)\mathsf{u}_2^2+(\partial_{\mathsf{e}}^2P)\mathsf{u}_3^2+2(\partial_\ell\partial_{\mathsf{e}}P)\mathsf{u}_1\mathsf{u}_3\big)\\
&\quad+2\int dx(a_2\mathsf{u}_1+a_3\mathsf{u}_3)\partial_x\big((\partial_\ell P)\mathsf{u}_1\mathsf{u}_2+(\partial_{\mathsf{e}}P)\mathsf{u}_2\mathsf{u}_3\big)=0\,,
\end{aligned}\tag{3.77}$$

where $a_0=(C^{-1})_{22}$, $a_2=(C^{-1})_{13}$, $a_3=(C^{-1})_{33}$. The term cubic in u_2 vanishes by the same argument as for the one-component case. All other terms are linear in u_2, thus their sum has to vanish point-wise,

$$\begin{aligned}
&(a_0\partial_\ell^2P-a_2\partial_\ell P)\partial_x\mathsf{u}_1^2+(a_0\partial_{\mathsf{e}}^2P-a_3\partial_{\mathsf{e}}P)\partial_x\mathsf{u}_3^2\\
&\quad+2\big(a_0(\partial_\ell\partial_{\mathsf{e}}P)\partial_x(\mathsf{u}_1\mathsf{u}_3)-a_2(\partial_{\mathsf{e}}P)\mathsf{u}_3\partial_x\mathsf{u}_1-a_3(\partial_\ell P)\mathsf{u}_1\partial_x\mathsf{u}_3\big)=0\,.
\end{aligned}\tag{3.78}$$

This leads to the constraints on the coefficients as

$$a_0\partial_\ell^2P=a_2\partial_\ell P\,,\quad a_0\partial_{\mathsf{e}}^2P=a_3\partial_{\mathsf{e}}P\,,\quad a_0\partial_\ell\partial_{\mathsf{e}}P=a_3\partial_\ell P\,,\quad a_0\partial_\ell\partial_{\mathsf{e}}P=a_2\partial_{\mathsf{e}}P\,.\tag{3.79}$$

There are four constraints and five partial derivatives which may be regarded as independent parameters. Thus one would expect that there is a sub-manifold in V,P,β space for which (3.79) can be satisfied. We will come back to another representation of these constraints below. Away from the special subset of invariant Gaussian measures, we have no tools, but one would hope that the invariant measure still has a finite correlation length and exponential mixing. Based on the mechanical model, it is suggestive to assume u_2 to be independent of $\mathsf{u}_1,\mathsf{u}_3$ and to have white noise statistics. But this forces again the constraints (3.79) and results in the same Gaussian measure as before.

A basic property of the mechanical model is invariance under time reversal. On the level of fluctuating hydrodynamics this translates to the following property: We fix a time window $[0,T]$. Then, in the stationary process, the trajectories

$$(\mathsf{u}_1(t),\mathsf{u}_2(t),\mathsf{u}_3(t))\ \text{and}\ (\mathsf{u}_1(T-t),-\mathsf{u}_2(T-t),\mathsf{u}_3(T-t))\tag{3.80}$$

with $0\le t\le T$ have the same probability. To check (3.80) requires some information on the invariant measure. But under the Gaussian measure (3.76), hence assuming the validity of the constraints, it can be shown that time reversal invariance indeed holds.

Transformation to Normal Modes To proceed further, it is convenient to write (3.73) in vector form,

$$\partial_t \vec{\mathsf{u}}(x,t) + \partial_x\big(A\vec{\mathsf{u}}(x,t) + \tfrac{1}{2}\langle\vec{\mathsf{u}}, \vec{H}\vec{\mathsf{u}}\rangle - \partial_x \tilde{D}\vec{\mathsf{u}}(x,t) + \tilde{B}\vec{\xi}(x,t)\big) = 0\,, \tag{3.81}$$

where $\vec{H}$ is the vector consisting of the Hessians of the currents with derivatives evaluated at the background values $(\ell, 0, \mathsf{e})$,

$$H^{\alpha}_{\gamma\gamma'} = \partial_{\mathsf{u}_\gamma}\partial_{\mathsf{u}_{\gamma'}}\mathsf{j}_\alpha\,, \qquad \langle\vec{\mathsf{u}}, \vec{H}\vec{\mathsf{u}}\rangle = \sum_{\gamma,\gamma'=1}^{3} \vec{H}_{\gamma\gamma'}\mathsf{u}_\gamma \mathsf{u}_{\gamma'}\,. \tag{3.82}$$

As for the linear Langevin equation we transform to normal modes through

$$\vec{\phi} = R\vec{\mathsf{u}}\,. \tag{3.83}$$

Then

$$\partial_t\phi_\alpha + \partial_x\big(c_\alpha\phi_\alpha + \langle\vec{\phi}, G^\alpha\vec{\phi}\rangle - \partial_x(D^\sharp\vec{\phi})_\alpha + (B^\sharp\vec{\xi})_\alpha\big) = 0\,. \tag{3.84}$$

By construction $B^\sharp B^{\sharp\mathrm{T}} = 2D^\sharp$. The nonlinear coupling constants, denoted by $\vec{G}$, are defined by

$$G^\alpha = \tfrac{1}{2}\sum_{\alpha'=1}^{3} R_{\alpha\alpha'} R^{-\mathrm{T}} H^{\alpha'} R^{-1} \tag{3.85}$$

with the notation $R^{-\mathrm{T}} = (R^{-1})^{\mathrm{T}}$.

Since derived from a chain, the couplings are not completely arbitrary, but satisfy the symmetries

$$\begin{aligned} &G^\alpha_{\beta\gamma} = G^\alpha_{\gamma\beta}\,, \quad G^\sigma_{\alpha\beta} = -G^{-\sigma}_{-\alpha-\beta}\,, \quad G^\sigma_{-10} = G^\sigma_{01}\,,\\ &G^0_{\sigma\sigma} = -G^0_{-\sigma-\sigma}\,, \quad G^0_{\alpha\beta} = 0 \ \text{ otherwise}\,. \end{aligned} \tag{3.86}$$

In particular note that

$$G^0_{00} = 0\,, \tag{3.87}$$

always, while $G^1_{11} = -G^{-1}_{-1-1}$ are generically different from 0. This property signals that the heat peak will behave differently from the sound peaks. The $\vec{G}$-couplings are listed in [30] and as a function of P, β expressed in cumulants up to third order in r_0, V_0. The algebra is somewhat messy. But there is a short MATHEMATICA program available [33] which, for given P, β, V, computes all coupling constants, including the matrices C, A, R.

We return to the issue of Gaussian time-stationary measures, where we regard the coefficients $\vec{G}, D^\sharp, B^\sharp$ as arbitrary, up to $2D^\sharp = B^\sharp B^{\sharp\mathrm{T}}$, ignoring for a while their particular origin. The Langevin equation (3.84) is slightly formal. To have a well-defined evolution, we discretize space by a lattice of N sites. The field $\vec{\phi}(x,t)$ then becomes $\vec{\phi}_j(t)$ with components $\phi_{j,\alpha}(t), j = 1,\dots,N, \alpha = 1,2,3$. The spatial finite difference operator is denoted by ∂_j, $\partial_j f_j = f_{j+1} - f_j$, with transpose $\partial_j^{\mathrm{T}} f_j = f_{j-1} - f_j$. Then the discretized equations of fluctuating hydrodynamics read

$$\partial_t \phi_{j,\alpha} + \partial_j\big(c_\alpha \phi_{j,\alpha} + \mathscr{N}_{j,\alpha} + \partial_j^{\mathrm{T}} D^\sharp \phi_{j,\alpha} + B^\sharp \xi_{j,\alpha}\big) = 0 \tag{3.88}$$

with $\vec{\phi}_j = \vec{\phi}_{N+j}$, $\vec{\xi}_0 = \vec{\xi}_N$, where $\xi_{j,\alpha}$ are independent Gaussian white noises with covariance

$$\langle \xi_{j,\alpha}(t)\xi_{j',\alpha'}(t')\rangle = \delta_{jj'}\delta_{\alpha\alpha'}\delta(t-t')\,. \tag{3.89}$$

The diffusion matrix $D^\sharp$ and noise strength $B^\sharp$ act on components, while the difference operator ∂_j acts on the lattice site index j.

$\mathscr{N}_{j,\alpha}$ is quadratic in ϕ. But let us first consider the case $\mathscr{N}_{j,\alpha} = 0$. Then $\phi_{j,\alpha}(t)$ is a Gaussian process. The noise strength has been chosen such that one invariant measure is the Gaussian

$$\prod_{j=1}^{N}\prod_{\alpha=1}^{3} \exp[-\tfrac{1}{2}\phi_{j,\alpha}^2](2\pi)^{-1/2} d\phi_{j,\alpha} = \rho_{\mathrm{G}}(\phi)\prod_{j=1}^{N}\prod_{\alpha=1}^{3} d\phi_{j,\alpha}\,. \tag{3.90}$$

Because of the conservation laws, the hyperplanes

$$\sum_{j=1}^{N} \phi_{j,\alpha} = N\rho_\alpha\,, \tag{3.91}$$

are invariant and on each hyperplane there is a Gaussian process with a unique invariant measure given by (3.90) conditioned on that hyperplane. For large N it would become independent Gaussians with mean ρ_α, our interest being the case of zero mean, $\rho_\alpha = 0$.

The generator of the diffusion process (3.88) with $\mathscr{N}_{j,\alpha} = 0$ is given by

$$L_0 = \sum_{j=1}^{N}\Big(-\sum_{\alpha=1}^{3} \partial_j\big(c_\alpha\phi_{j,\alpha} + \partial_j^{\mathrm{T}} D^\sharp \phi_{j,\alpha}\big)\partial_{\phi_{j,\alpha}} + \sum_{\alpha,\alpha'=1}^{3} (B^\sharp B^{\sharp\mathrm{T}})_{\alpha\alpha'}\partial_j\partial_{\phi_{j,\alpha}}\partial_j\partial_{\phi_{j,\alpha'}}\Big)\,. \tag{3.92}$$

The invariance of $\rho_{\rm G}(\phi)$ can be checked through

$$L_0^* \rho_{\rm G}(\phi) = 0\,, \tag{3.93}$$

where * is the adjoint with respect to the flat volume measure. Furthermore linear functions evolve to linear functions according to

$$\mathrm{e}^{L_0 t}\phi_{j,\alpha} = \sum_{j'=1}^{N}\sum_{\alpha'=1}^{3} (\mathrm{e}^{\mathscr{A} t})_{j\alpha,j'\alpha'}\phi_{j',\alpha'}\,, \tag{3.94}$$

where the matrix $\mathscr{A} = -\partial_j \otimes \mathrm{diag}(c_1, c_2, c_3) - \partial_j\partial_j^{\mathrm{T}} \otimes D^\sharp$, the first factor acting on j and the second on α.

We now add the nonlinearity $\mathscr{N}_{j,\alpha}$. In general, this will modify the time-stationary measure and we have little control how. Therefore we propose to choose $\mathscr{N}_{j,\alpha}$ such that the corresponding vector field $\partial_j \mathscr{N}_{j,\alpha}$ is divergence free [10]. If $\mathscr{N}_{j,\alpha}$ depends only on the field at sites j and $j+1$, then the unique solution reads

$$\mathscr{N}_{j,\alpha} = \tfrac{1}{3}\sum_{\gamma,\gamma'=1}^{3} G^{\alpha}_{\gamma\gamma'}\big(\phi_{j,\gamma}\phi_{j,\gamma'} + \phi_{j,\gamma}\phi_{j+1,\gamma'} + \phi_{j+1,\gamma}\phi_{j+1,\gamma'}\big)\,. \tag{3.95}$$

For $\rho_{\rm G}$ to be left invariant under the deterministic flow generated by the vector field $-\partial_j\mathscr{N}$ requires

$$L_1\rho_{\rm G} = 0\,, \qquad L_1 = -\sum_{j=1}^{N}\sum_{\alpha=1}^{3} \partial_j\mathscr{N}_{j,\alpha}\partial_{\phi_{j,\alpha}}\,, \tag{3.96}$$

which implies

$$\sum_{j=1}^{N}\sum_{\alpha=1}^{3} \phi_{j,\alpha}\partial_j\mathscr{N}_{j,\alpha} = 0 \tag{3.97}$$

and thus the constraints

$$G^{\alpha}_{\beta\gamma} = G^{\beta}_{\alpha\gamma}\ \big(= G^{\alpha}_{\gamma\beta}\big) \tag{3.98}$$

for all $\alpha, \beta, \gamma = 1, 2, 3$, where in brackets we added the symmetry which holds by definition. Denoting the generator of the Langevin equation (3.88) by

$$L = L_0 + L_1\,, \tag{3.99}$$

one concludes $L^*\rho_{\rm G} = 0$, i.e. the time-invariance of $\rho_{\rm G}$.

In the continuum limit the condition (3.97) reads

$$\sum_{\alpha,\beta,\gamma=1}^{3} G^{\alpha}_{\beta\gamma} \int dx \phi_\alpha(x)\partial_x\big(\phi_\beta(x)\phi_\gamma(x)\big) = 0\,, \tag{3.100}$$

where $G^{\alpha}_{\beta\gamma} = G^{\alpha}_{\gamma\beta}$. By partial integration

$$2\sum_{\alpha,\beta,\gamma=1}^{3} G^{\alpha}_{\beta\gamma} \int dx \phi_\alpha(x)\phi_\beta(x)\partial_x\phi_\gamma(x)$$
$$= -\sum_{\alpha,\beta,\gamma=1}^{3} G^{\alpha}_{\beta\gamma} \int dx \phi_\beta(x)\phi_\gamma(x)\partial_x\phi_\alpha(x)\,. \tag{3.101}$$

Hence (3.100) is satisfied only if $G^{\gamma}_{\beta\alpha} = G^{\alpha}_{\beta\gamma}$, which is the condition (3.98) claimed for the discrete setting.

Equation (3.98) is the generalization of (3.79), which is specific for the anharmonic chain. In fact, while abstractly true, it is not so easy to verify directly. But now we can argue more convincingly why one should be allowed to continue with assuming the validity of the constraints (3.98). As we will discuss in the next section, the leading coupling constants are of the form $G^{\alpha}_{\alpha\alpha}$, while the sub-leading couplings have equal lower indices, $G^{\alpha}_{\gamma\gamma}$, $\gamma \neq \alpha$. The off-diagonal matrix elements are irrelevant for the large scale behavior. When one does the counting, all leading and sub-leading couplings can be chosen freely and the irrelevant couplings can be adjusted so that the constraint (3.98) is satisfied. Appealing to universality, the large space-time behavior should not depend on that particular choice. We expect that for general $\vec{G}$ the true time-stationary measure will have short range correlations and nonlinear fluctuating hydrodynamics remains a valid approximation to the dynamics of the anharmonic chain.

In related problem settings, a different point of view has been suggested [34, 35]. Firstly one notes that the Gaussian stationary measure (3.76), hence also (3.90), is simply inherited from the canonical equilibrium measure. In this respect there is no choice. Also the nonlinear Euler currents are on the safe side. But the remaining terms are phenomenological to some extent. $D^\sharp, B^\sharp$ could depend on $\vec{\phi}$ itself. One could also include higher derivative terms. In fact, one could try to choose the nonlinearities precisely in such a way that the dynamics is invariant under time-reversal and leaves the Gaussian measure invariant. The program as such may be easily endorsed, but so far I have not seen a convincing handling of the details.

3.4 Mode-Coupling Theory

Decoupling Hypothesis For the linear equations the normal modes decouple for long times. The hypothesis claims that such property persists when adding the quadratic nonlinearities. For the precise phrasing, we have to be somewhat careful. We consider a fixed component, α, in normal mode representation. It travels with velocity c_α, which is assumed to be distinct from all other mode velocities. If $G^\alpha_{\alpha\alpha} \neq 0$, then for the purpose of computing correlations of mode α at large scales, one can use the scalar conservation law

$$\partial_t \phi_\alpha + \partial_x\big(c_\alpha \phi_\alpha + G^\alpha_{\alpha\alpha}\phi_\alpha^2 - D^\sharp_{\alpha\alpha}\partial_x\phi_\alpha + B^\sharp_{\alpha\alpha}\xi_\alpha\big) = 0\,, \tag{3.102}$$

which coincides with the stochastic Burgers equation (3.19). If decoupling holds, one has the exact asymptotics as stated in (3.21) with $\lambda = 2\sqrt{2}|G^\alpha_{\alpha\alpha}|$. The universal scaling function f_{KPZ} is tabulated in [36], denoted there by f. $f_{\mathrm{KPZ}} \geq 0$, $\int dx f_{\mathrm{KPZ}}(x) = 1, f_{\mathrm{KPZ}}(x) = f_{\mathrm{KPZ}}(-x)$, $\int dx f_{\mathrm{KPZ}}(x)x^2 \simeq 0.510523$. f_{KPZ} looks like a Gaussian with a large $|x|$ decay as $\exp[-0.295|x|^3]$. Plots are provided in [18, 36].

For an anharmonic chain, $G^0_{00} = 0$ always and the decoupling hypothesis applies only to the sound peaks, provided $G^1_{11} = -G^{-1}_{-1-1} \neq 0$ which generically is the case. If $G^1_{11} \neq 0$, then the *exact* scaling form is

$$S^\sharp_{\sigma\sigma}(x,t) \cong (\lambda_s t)^{-2/3} f_{\mathrm{KPZ}}\big((\lambda_s t)^{-2/3}(x - \sigma c t)\big)\,, \quad \lambda_s = 2\sqrt{2}|G^\sigma_{\sigma\sigma}|\,, \tag{3.103}$$

$\sigma = \pm 1$. To find out about the scaling behavior of the heat mode other methods have to be developed.

For one-dimensional fluids, van Beijeren [37] follows the scheme developed in [6] and arrived first at the prediction (3.103) together with the Lévy 5/3 heat peak to be discussed below. In [37] no Langevin equations appear. I regard them as a useful intermediate step valid on a mesoscopic scale. In the Langevin form the theory can be applied to a large class of one-dimensional systems. As a tool, fluctuating hydrodynamics has been proposed considerably earlier [38] and used to predict the $t^{-2/3}$ decay of the total energy current correlation.

One-Loop, Diagonal, and Small Overlap Approximations We return to the Langevin equation (3.88) and consider the mean zero, stationary $\phi_{j,\alpha}(t)$ process with ρ_{G} as $t = 0$ measure. The stationary covariance reads

$$S^\sharp_{\alpha\alpha'}(j,t) = \langle \phi_{j,\alpha}(t)\phi_{0,\alpha'}(0)\rangle = \langle \phi_{0,\alpha'} \mathrm{e}^{Lt}\phi_{j,\alpha}\rangle_{\mathrm{eq}}\,, \quad t \geq 0\,. \tag{3.104}$$

On the left, $\langle\cdot\rangle$ denotes the average with respect to the stationary $\phi_{j,\alpha}(t)$ process and on the right $\langle\cdot\rangle_{\mathrm{eq}}$ refers to the average with respect to ρ_{G}. By construction

$$S^\sharp_{\alpha\alpha'}(j,0) = \delta_{\alpha\alpha'}\delta_{j0}\,. \tag{3.105}$$

The time derivative reads

$$\frac{d}{dt}S^{\sharp}_{\alpha\alpha'}(j,t) = \langle \phi_{0,\alpha'}(\mathrm{e}^{Lt}L_0\phi_{j,\alpha})\rangle_{\mathrm{eq}} + \langle \phi_{0,\alpha'}(\mathrm{e}^{Lt}L_1\phi_{j,\alpha})\rangle_{\mathrm{eq}}\,. \tag{3.106}$$

We insert

$$\mathrm{e}^{Lt} = \mathrm{e}^{L_0 t} + \int_0^t ds\, \mathrm{e}^{L_0(t-s)}L_1\mathrm{e}^{Ls} \tag{3.107}$$

in the second summand of (3.106). The term containing only $\mathrm{e}^{L_0 t}$ is cubic in the time zero fields and hence its average vanishes. Therefore one arrives at

$$\frac{d}{dt}S^{\sharp}_{\alpha\alpha'}(j,t) = \mathscr{A}S_{\alpha\alpha'}(j,t) + \int_0^t ds\langle \phi_{0,\alpha'}\mathrm{e}^{L_0(t-s)}L_1(\mathrm{e}^{Ls}L_1\phi_{j,\alpha})\rangle_{\mathrm{eq}}\,. \tag{3.108}$$

For the adjoint of $\mathrm{e}^{L_0(t-s)}$ we use (3.94) and for the adjoint of L_1 we use

$$\langle \phi_{j,\alpha}L_1F(\phi)\rangle_{\mathrm{eq}} = -\langle (L_1\phi_{j,\alpha})F(\phi)\rangle_{\mathrm{eq}}\,, \tag{3.109}$$

which both rely on $\langle\cdot\rangle_{\mathrm{eq}}$ being the average with respect to ρ_{G}. Furthermore

$$L_1\phi_{j,\alpha} = -\partial_j\mathscr{N}_{j,\alpha}\,. \tag{3.110}$$

Inserting in (3.108) one arrives at the identity

$$\frac{d}{dt}S^{\sharp}_{\alpha\alpha'}(j,t) = \mathscr{A}S^{\sharp}_{\alpha\alpha'}(j,t) - \int_0^t ds\langle (\mathrm{e}^{\mathscr{A}^{\mathrm{T}}(t-s)}\partial_j\mathscr{N}_{0,\alpha'})(\mathrm{e}^{Ls}\partial_j\mathscr{N}_{j,\alpha})\rangle_{\mathrm{eq}}\,. \tag{3.111}$$

To obtain a closed equation for $S^{\sharp}$ we note that the average

$$\langle \partial_{j'}\mathscr{N}_{j',\alpha'}\mathrm{e}^{L_s}\partial_j\mathscr{N}_{j,\alpha}\rangle_{\mathrm{eq}} = \langle \partial_j\mathscr{N}_{j,\alpha}(s)\partial_{j'}\mathscr{N}_{j',\alpha'}(0)\rangle \tag{3.112}$$

is a four-point correlation. We invoke the Gaussian factorization as

$$\langle \phi(s)\phi(s)\phi(0)\phi(0)\rangle \cong \langle \phi(s)\phi(s)\rangle\langle \phi(0)\phi(0)\rangle + 2\langle \phi(s)\phi(0)\rangle\langle \phi(s)\phi(0)\rangle\,. \tag{3.113}$$

The first summand vanishes because of the difference operator ∂_j. Secondly we replace the bare propagator $\mathrm{e}^{\mathscr{A}(t-s)}$ by the interacting propagator $S^{\sharp}(t-s)$, which corresponds to a partial resummation of the perturbation series in $\vec{G}$. Finally we take a limit of zero lattice spacing. This step could be avoided, and is done so in our numerical scheme for the mode-coupling equations. We could also maintain the ring geometry which, for example, would allow to investigate collisions between the moving peaks. Universality is only expected for large j, t, hence in the limit of zero lattice spacing. The continuum limit of $S^\sharp(j,t)$ is denoted by $S^\sharp(x,t)$, $x \in \mathbb{R}$. With these steps we arrive at the mode-coupling equation

$$\partial_t S^\sharp_{\alpha\beta}(x,t) = \sum_{\alpha'=1}^{3} \Big(\big(- c_\alpha \delta_{\alpha\alpha'} \partial_x + D^\sharp_{\alpha\alpha'} \partial_x^2 \big) S^\sharp_{\alpha'\beta}(x,t) + \int_0^t ds \int_{\mathbb{R}} dy M_{\alpha\alpha'}(y,s) \partial_x^2 S^\sharp_{\alpha'\beta}(x-y,t-s) \Big) \tag{3.114}$$

with the memory kernel

$$M_{\alpha\alpha'}(x,t) = 2 \sum_{\beta',\beta'',\gamma',\gamma''=1}^{3} G^\alpha_{\beta'\gamma'} G^{\alpha'}_{\beta''\gamma''} S^\sharp_{\beta'\beta''}(x,t) S^\sharp_{\gamma'\gamma''}(x,t)\,. \tag{3.115}$$

In numerical simulations of both, the mechanical model of anharmonic chains and the mode-coupling equations, it is consistently observed that $S^\sharp_{\alpha\alpha'}(j,t)$ becomes approximately diagonal fairly rapidly. To analyse the long time asymptotics on the basis of (3.114) we therefore rely on the diagonal approximation

$$S^\sharp_{\alpha\alpha'}(x,t) \simeq \delta_{\alpha\alpha'} f_\alpha(x,t)\,. \tag{3.116}$$

Then $f_\alpha(x,0) = \delta(x)$ and the f_α's satisfy

$$\partial_t f_\alpha(x,t) = (-c_\alpha \partial_x + D^\sharp_{\alpha\alpha} \partial_x^2) f_\alpha(x,t) + \int_0^t ds \int_{\mathbb{R}} dy \partial_x^2 f_\alpha(x-y,t-s) M_{\alpha\alpha}(y,s)\,, \tag{3.117}$$

$\alpha = -1, 0, 1$, with memory kernel

$$M_{\alpha\alpha}(x,t) = 2 \sum_{\gamma,\gamma'=0,\pm 1} (G^\alpha_{\gamma\gamma'})^2 f_\gamma(x,t) f_{\gamma'}(x,t)\,. \tag{3.118}$$

The solution to (3.117) has two sound peaks centered at $\pm ct$ and the heat peak sitting at 0. All three peaks have a width much less than ct. But then, in case $\gamma \neq \gamma'$, the product $f_\gamma(x,t) f_{\gamma'}(x,t) \simeq 0$ for large t. Hence for the memory kernel (3.118) we invoke a small overlap approximation as

$$M_{\alpha\alpha}(x,t) \simeq M^{\mathrm{dg}}_\alpha(x,t) = 2 \sum_{\gamma=0,\pm 1} (G^\alpha_{\gamma\gamma})^2 f_\gamma(x,t)^2\,, \tag{3.119}$$

which is to be inserted in Eq. (3.117).

Numerical Simulations of the Mode-Coupling Equations When starting this project together with Christian Mendl, in the summer of 2012 we spent many days in numerically simulating the mode coupling equations with initial conditions $S^{\sharp}_{\alpha\alpha'}(j,0) = \delta_{\alpha\alpha'}\delta_{0j}$. Only a few plots are in print [39], simply because there is such a large parameter space and it is not clear where to start and where to end. Still, for our own understanding this period was extremely helpful. Mostly we simulated in Fourier space. System size was up to 400. Speeds were of order 1, thereby limiting the simulation time to about 200, the time of the first peak collision. For such sizes the simulations are fast and many variations could be explored. We started from the scalar equation, to be discussed below, moved up to two modes, and eventually to three modes with parameters taken from an actual anharmonic chain. $|\vec{G}^{\alpha}_{\alpha'\alpha'}|$ was either 0 or somewhere in the range 0.3–2.5. $D^{\sharp}$ is a free parameter which was varied from 0 to $|\vec{G}^{\alpha}_{\alpha'\alpha'}|/2$. We always simulated the complete matrix-valued mode-coupling equations (3.114). Our main findings can be summarized as:

(a) For a large range of parameters, the diagonal approximation in generally failed for short times, but was quickly restored with the off-diagonal elements being at most 10 % of the diagonal ones.
(b) The results were fairly insensitive to the choice of $D^{\sharp}$. In fact, $D^{\sharp} = 0$ works also well. Apparently the memory term generates already enough dissipation.
(c) We varied the overlap coefficients $G^{\alpha}_{\gamma\gamma'}$ with $\gamma \neq \gamma'$. Over the time scale of the simulation no substantial changes were observed.

All these findings confirm the approximations proposed.

As our biggest surprise, except for trivial cases we never reached the asymptotic regime. The peak structure develops fairly rapidly. The peak shape then changes slowly, roughly consistent with the predicted scaling exponents, but it does not reach a self-similar form. For example in the case of the sound peak, on the scale $t^{2/3}$, rather than being symmetric, as claimed by (3.103), it is still badly distorted, tilted away from the central peak with rapid decay outside the sound cone but rather slow power law type of decay towards the central peak. To improve one would have to simulate larger system sizes and longer times. But then numerical simulations become heavy and the fun evaporates. More attention can be achieved by molecular dynamics (MD) simulation of the mechanical chain.

For given parameters V, P, β one easily computes all the required coefficients. So one goal was to run a MD and put the results on top of the ones from a simulation of the mode coupling equations. For this to be a reasonable program, one would have to simulate the mode-coupling equations for sizes of $N = 4000$ and more, which we never attempted.

For the scalar case the situation is much simpler. The mode-coupling equation takes the form

$$\partial_t f(x,t) = D\partial_x^2 f(x,t) + 2G^2 \int_0^t ds \int_{\mathbb{R}} dy \partial_x^2 f(x-y,t-s) f(y,s)^2 \,, \qquad (3.120)$$

which is the one-loop approximation for the stochastic Burgers equation [40] . For large x, t, its solution with initial condition $f(x,0) = \delta(x)$ takes the scaling form

$$f(x,t) \cong (\lambda_s t)^{-2/3} f_{\mathrm{mc}}\big((\lambda_s t)^{-2/3} x\big) . \tag{3.121}$$

Inserting in (3.120), one first finds the non-universal scaling coefficient

$$\lambda_s = 2\sqrt{2} |G^{\sigma}_{\sigma\sigma}| . \tag{3.122}$$

Secondly $\hat{f}_{\mathrm{mc}}$, the Fourier transform of f_{mc}, is defined as solution of the fixed point equation

$$\tfrac{2}{3}\hat{f}'_{\mathrm{mc}}(w) = -\pi^2 w \int_0^1 ds \hat{f}_{\mathrm{mc}}((1-s)^{2/3} w) \int_{\mathbb{R}} dq \hat{f}_{\mathrm{mc}}(s^{2/3}(w-q)) \hat{f}_{\mathrm{mc}}(s^{2/3} q) \tag{3.123}$$

with $w \geq 0$ and $\hat{f}_{\mathrm{mc}}(0) = 1, \hat{f}'_{\mathrm{mc}}(0) = 0$.

Equation (3.123) is based on the closure assumption (3.113) and there is no reason to infer that it is exact. However from our numerical simulations we conclude that f_{KPZ} differs from f_{mc} by a few percent only. We regard the scalar case as a strong support for the entire approach. But for several components the large finite size effects prohibit one to arrive at a similarly simple claim.

Asymptotic Self-similarity Within mode-coupling the asymptotic shape function for the sound peaks is given by

$$f_\sigma(x,t) \cong (\lambda_s t)^{-2/3} f_{\mathrm{mc}}\big((\lambda_s t)^{-2/3}(x - \sigma c t)\big) , \tag{3.124}$$

$\sigma = \pm 1$. For the heat peak we employ (3.117) together with (3.119), using as an input that the asymptotic form of f_σ is known already. In fact the scaling exponent for f_σ is crucial, but the precise shape of f_σ enters only mildly. Hence, again switching to Fourier space, one has to solve

$$\begin{aligned} \partial_t \hat{f}_0(k,t) &= -D^\sharp_0 (2\pi k)^2 \hat{f}_0(k,t) \\ &\quad -2 \sum_{\sigma=\pm 1} (G^0_{\sigma\sigma})^2 (2\pi k)^2 \int_0^t ds \hat{f}_0(k, t-s) \int_{\mathbb{R}} dq \hat{f}_\sigma(k-q, s) \hat{f}_\sigma(q,s) , \end{aligned} \tag{3.125}$$

$\hat{f}_0(k,0) = 1$. For $\hat{f}_\sigma$ one inserts the asymptotic result (3.121). Equation (3.125) is a linear equation which is solved through Laplace transform with the result

$$\hat{f}_0(k,t) \cong \mathrm{e}^{-|k|^{5/3} \lambda_{\mathrm{h}} t} , \tag{3.126}$$

where

$$\lambda_{\mathrm{h}} = \lambda_{\mathrm{s}}^{-2/3}(G^0_{\sigma\sigma})^2(4\pi)^2 \int_0^{\infty} dt t^{-2/3} \cos(2\pi ct) \int_{\mathbb{R}} dx f_{\mathrm{mc}}(x)^2$$
$$= \lambda_{\mathrm{s}}^{-2/3}(G^0_{\sigma\sigma})^2(4\pi)^2(2\pi c)^{-1/3}\tfrac{1}{2}\pi \frac{1}{\Gamma(\frac{2}{3})}\frac{1}{\cos(\frac{\pi}{3})}\int_{\mathbb{R}} dx f_{\mathrm{mc}}(x)^2 \qquad (3.127)$$

and we used the symmetry $G^0_{\sigma\sigma} = -G^0_{-\sigma-\sigma}$, see [30] for details. Equation (3.126) is the Fourier transform of the symmetric α-stable distribution with exponent $\alpha = 5/3$, also known as Lévy distribution. In real space the asymptotics reads, for $|x| \geq (\lambda_{\mathrm{h}} t)^{3/5}$,

$$f_0(x,t) \simeq \pi^{-1}\lambda_{\mathrm{h}} t|x|^{-8/3}\,. \qquad (3.128)$$

f_{mc} is a smooth function with rapid decay. On the other hand, f_0 has fat tails and its variance is divergent. According to (3.128), at $x = \sigma ct$ the heat peak $f_0(\sigma ct, t) \cong \pi^{-1}\lambda_{\mathrm{h}} c^{-8/3} t^{-5/3}$. This explains why there is still coupling between f_0 and f_σ, despite the large spatial separation. In fact, numerically one observes that beyond the sound cone, $x = \pm ct$, the solution decays exponentially fast. As t becomes large the tails of f_0 are build up between the two sound peaks, so to speak they unveil the Lévy distribution.

In (3.125) we could also insert for f_σ the exact scaling function f_{KPZ}, which would slightly modify λ_{h}. $f_{\mathrm{Lévy}}$ is an approximation, just as f_{mc}. But the MD simulations display so convincingly the Lévy distribution that one might be willing to regard it as exact. If so, the exact λ_{h} must be based on f_{KPZ}. To obtain the correlations of the physical fields, one has to use

$$S(j,t) = R^{-1} S^{\sharp}(j,t) R^{-\mathrm{T}}\,. \qquad (3.129)$$

In particular the correlations of the physical fields are given through

$$S_{\alpha\alpha}(j,t) = \sum_{\sigma=0,\pm 1} |(R^{-1})_{\alpha\sigma}|^2 f_\sigma(j,t)\,, \qquad (3.130)$$

where for f_σ the asymptotic scaling form is inserted. Then asymptotically the ℓ-ℓ and the e-e correlations show generically all three peaks. However, for the u-u correlations the central peak is missing asymptotically, since $(R^{-1})_{20} = 0$ independently of the interaction potential V.

We note that the coefficient $D^{\sharp}$ does not appear in the asymptotic scaling form, of course neither $B^{\sharp}B^{\sharp\mathrm{T}} = 2D^{\sharp}$. This result is consistent with the picture that noise and dissipation are required to maintain the correct local stationary measure with susceptibility C. The long time asymptotics is however governed by the nonlinearities.

No Signal Beyond the Sound Cone Physically the sound speed is an upper limit for the propagation of small disturbances. Since the initial state has a finite correlation length, one would expect that towards the outside of the sound cone correlations decay exponentially, while inside the sound cone there seems to be no particular restriction. As a consistency check, one would hope that such a general feature is properly reproduced by mode-coupling. Their numerical solutions conform with this expectation, at least for the small system sizes explored. But for the scaling limit one has to let $t \to \infty$ and the decay information seems to be lost. However there is still a somewhat subtle trace.

To explain, we have to first recall some properties of Lévy stable distributions. Except for trivial rescalings, they are characterized by two parameters, traditionally called α, β, where for simplicity we momentarily stick to this convention, without too much risk of confusion. The probability density has a simple form in Fourier space,

$$\hat{f}_{\text{Lévy},\alpha,\beta}(k) = \exp\big(-|k|^{\alpha}\big[1 - \mathrm{i}\beta\tan(\tfrac{1}{2}\pi\alpha)\mathrm{sgn}(k)\big]\big). \tag{3.131}$$

The parameter α controls the steepness, $0 < \alpha < 2$, while β controls the asymmetry, $|\beta| \le 1$. For $|\beta| > 1$ the Fourier integral no longer defines a non-negative function. At the singular point $\alpha = 2$ only $\beta = 0$ is admitted and the probability density is a Gaussian. If $|\beta| < 1$, the asymptotic decay of $f_{\text{Lévy},\alpha,\beta}(x)$ is determined by α and is given by $|x|^{-\alpha-1}$ for $|x| \to \infty$. At $|\beta| = 1$ the two tails show different decay. The functions corresponding to $\beta = 1$ and $\beta = -1$ are mirror images, for $\beta = 1$ the slow decay being for $x \to -\infty$ and still as $|x|^{-\alpha-1}$. For $0 < \alpha \le 1$, $f_{\text{Lévy},\alpha,1}(x) = 0$ for $x > 0$, while for $1 < \alpha < 2$ the decay becomes stretched exponential as $\exp(-c_0 x^{\alpha/(1-\alpha)})$ with known constant c_0. We refer to [41] for more details.

For the heat peak we obtained the symmetric Lévy distribution because the sound peaks are reflection symmetric, implying $c_1 = -c_{-1}$ and $(G^0_{11})^2 = (G^0_{-1-1})^2$. If hypothetically we would choose distinct couplings, or $c_1 \neq -c_{-1}$, then this imbalance would produce a $\beta \neq 0$. If one of the sound peaks would be completely missing, as in the case for a system with only two conserved fields, then necessarily $|\beta| = 1$. In accordance with the physical principle, the sign of β is such that the fast decay of $f_{\text{Lévy},\alpha,\pm 1}(x)$ is towards the outside of the sound cone, while the slow decay is towards the single sound peak. For finite t, this slow decay will be cut by the sound peak. Thus the scaling solution of the mode-coupling equations reproduces the rapid decay towards the exterior of the sound cone. This is a completely general fact, any number of components and any $\vec{G}$ (Schütz, Formulas for mode-coupling calculations, 2015, private communication).

Dynamical Phase Diagram As already indicated through the particular case $G^0_{00} = 0$, the large scale structure of the solution depends on whether $G^{\alpha}_{\gamma\gamma} = 0$ or not. One extreme case would be $G^{\alpha}_{\alpha\alpha} \neq 0$ for all α, implying that the three peaks have KPZ scaling behavior. The other extreme is $G^{\alpha}_{\gamma\gamma} = 0$ for all α, γ, resulting in all peaks to have diffusive broadening. For the case of only two modes, the full phase diagram has seven distinct phases, with unexpected details worked out in [42]. For the general case of n components the long time asymptotics is completely classified in [43] (Schütz, 2015, Formulas for mode-coupling calculations, private communication). Anharmonic chains have special symmetries and not all possible couplings $\vec{G}$ can be realized. Given that $G^0_{00} = 0$ and because the sound peaks are symmetric, to have a distinct scaling requires

$$G^1_{11} = 0\,, \tag{3.132}$$

which can be realized. The behavior is then determined by the value of the remaining diagonal matrix elements. For the central peak one finds

$$\sigma G^0_{\sigma\sigma} > 0\,, \tag{3.133}$$

while for the sound peak diagonals, $G^1_{-1-1} = -G^{-1}_{11}$, $G^1_{00} = -G^{-1}_{00}$, there seems to be no particular restriction. In principle, there could be sort of accidental zeros of $G^\alpha_{\gamma\gamma}$ which are then difficult to locate. A more direct approach starts from the observation that the $\vec{G}$ coefficients are expressed through cumulants in r_0, V_0 . If the integrands are antisymmetric under reflection, many terms vanish. The precise condition on the potential is to have some a_0, P_0 such that

$$V(x - a_0) + P_0 x = V(-x - a_0) - P_0 x \tag{3.134}$$

for all x. Then for arbitrary β and $P = P_0$, one finds

$$G^1_{11} = 0\,, \quad G^1_{-1-1} = -G^{-1}_{11} = 0\,, \quad G^1_{00} = -G^{-1}_{00} = 0\,, \tag{3.135}$$

while $G^\sigma_{0\sigma'} \neq 0$, generically. The standard examples for (3.134) to hold are the FPU chain with no cubic interaction term, the β-chain, and the square well potential with alternating masses, both at zero pressure.

Under (3.135) the heat mode is coupled to the sound mode, but there is no back reaction from the sound mode. Hence the sound peak is diffusive with scaling function

$$f_\sigma(x,t) = \frac{1}{\sqrt{4\pi D_s t}} \mathrm{e}^{-(x-\sigma c t)^2/4D_s t}\,. \tag{3.136}$$

D_s is a transport coefficient. It can defined through a Green-Kubo formula, which also means that no reasonably explicit answer can be expected. The feed back of the sound peak to the central peak follows by the same computation as before, with the result

$$\hat{f}_0(k,t) = \mathrm{e}^{-|k|^{3/2}\lambda_\mathrm{h} t}\,, \tag{3.137}$$

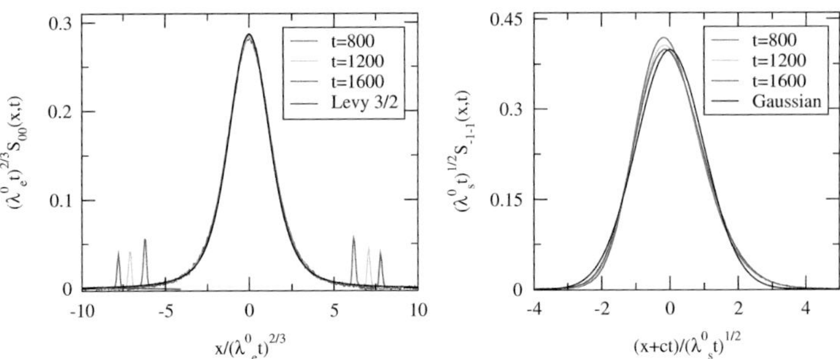

Fig. 3.2 Scaling plot of heat and sound peak for a FPU chain with $N = 8192$, potential parameters $\alpha = 0$, $\beta = 1$, pressure $P = 1$, and temperature $\beta^{-1} = 1$

where

$$\lambda_{\mathrm{h}} = (D_\sigma)^{-1/2}(G^0_{\sigma\sigma})^2(4\pi)^2(2\pi c)^{-1/2}\int_0^\infty dt\, t^{-1/2}\cos(t)(2\sqrt{\pi})^{-1}\,. \tag{3.138}$$

Since $3/2 < 5/3$, the density $f_0(x,t)$ turns out to be broader than the Lévy $5/3$ from the dynamical phase with $G^1_{11} \neq 0$.

In testing nonlinear fluctuating hydrodynamics almost subconsciously one tries to confirm (or not) the scaling exponents, resp. functions. This can be difficult because of limited size. The dynamical phase diagram offers a different option. For exceptional points in the phase diagram, without too precise a verification of the scaling, one should find that the standard scaling exponent does not properly fit the data. Such qualitative property is possibly more easy to access. In Fig. 3.2 we display heat and sound peak for a FPU chain with $G^1_{11} = 0$.

3.5 Molecular Dynamics Simulations

In 1953 Fermi, Pasta and Ulam, technically supported by Tsingou, simulated 32 particles interacting through a quartic potential at the extremely low energy of $\mathsf{e} = 5 \times 10^{-4}$ per particle (above the ground state energy) [26]. They injected energy in the highest Fourier mode and were looking for equipartition of the modes at long times. However they found quasi-periodic motion with time averages settling to some definite value different from equipartition. The observed quasi-periodicity triggered the connection to KAM tori, the discovery of integrable systems with many degrees of freedom, and to the development of the theory of solitons and breathers [44, 45]. For sure, a rich harvest, see [46] for a historical perspective. Later on Izrailev and Chirikov [47] repeated the simulation at the higher energy $\mathsf{e} = 5 \times 10^{-2}$ and observed equipartition.

Anomalous transport surfaced much later [48], see the reviews [49, 50]. One connects the two ends of the chain to thermal reservoirs. To explain, the end particles are tied down as $q_0 = 0$, $q_{N+1} = 0$, and to the equations of motion for the boundary particles one adds Langevin terms as

$$\ddot{q}_1 = -\gamma p_1 + \sqrt{2\gamma T_-}\xi_- , \quad \ddot{q}_N = -\gamma p_N + \sqrt{2\gamma T_+}\xi_+ , \tag{3.139}$$

where γ is a friction constant, $T_\pm$ are the boundary temperatures, and $\xi_\pm(t)$ are independent standard Gaussian white noises. For $T_- = T_+$ the system settles in the canonical equilibrium state. But for $T_- \neq T_+$ there is a non-trivial steady state with a non-zero energy flux $j_\mathsf{e}(N)$ depending on the length, N, of the chain. For regular heat transport, Fourier's law implies $j_\mathsf{e}(N) \simeq c_0 N^{-1}$. However for FPU chains one finds an enhanced transport as

$$j_\mathsf{e}(N) \simeq c_0 N^{-1+\alpha}(T_- - T_+) \tag{3.140}$$

with an exponent α characterizing the anomaly. [A further α, but better to stick to standard conventions.] Over the last two decades many MD simulations have been implemented for a wide variety of one-dimensional systems. Early results indicated $\alpha = 2/5$, but since about 2003 the common evidence pointed towards $\alpha = 1/3$ or at least close to it.

Nonlinear fluctuating dynamics can also deal with such open chains, at least in principle. One would impose energy imbalance boundary conditions as $\ell(0,t) = 0 = \ell(L,t)$, $\mathsf{u}(0,t) = 0 = \mathsf{u}(L,t)$, but $\mathfrak{e}(0,t) = \mathfrak{e}_-$ and $\mathfrak{e}(L,t) = \mathfrak{e}_+$ and tries to investigate the steady state. Unfortunately, at least for the moment, we have no powerful techniques to deal with this problem. On the other hand it is argued [49] that the energy flux is related to a Green-Kubo formula by

$$j_\mathsf{e}(N) \sim \int_0^{N/c} dt\big(\langle \mathscr{J}_3(t); \mathscr{J}_3(0)\rangle - \langle \mathscr{J}_3(\infty); \mathscr{J}_3(0)\rangle\big) . \tag{3.141}$$

Under the time integral appears the total energy current correlation in thermal equilibrium with its possibly non-zero value at $t = \infty$ subtracted, see Sect. 3.6 for more explanations. The decay of such correlation can be predicted by mode-coupling. In fact, we will confirm the value $\alpha = 1/3$. But the argument is subtle because it is only indirectly related to the spreading of the heat peak which is on scale $t^{3/5}$.

Because of the relation (3.141), in many MD simulations the total energy current correlation $\langle \mathscr{J}_3(t); \mathscr{J}_3(0)\rangle$ is measured as an addition to steady state transport [51, 52]. There are also MD simulations exclusively focussed on momentum and energy current correlations [53]. However simulations of correlations of the conserved fields have been fairly scarce until recently. The peak structure was noted already in [54], see also [55, 56]. But surprisingly enough, even such a basic issue as the quantitative comparison between the measured speed of sound and formula (3.54) is apparently not a routine check. So far there have been three independent sets of simulations with the specific aim to check the predictions from mode-coupling. For the details the reader is encouraged to look at the original papers. I will try to compare so to reach some sort of conclusion.

The first and second set are FPU chains with either symmetric or asymmetric potential, both the $\alpha\beta$ and the pure β chain [57, 58]. In this case one has to integrate numerically the differential equations governing the evolution, for which both a velocity-Verlet algorithm and a fourth order symplectic Runge-Kutta algorithm are used. The third set consists of chains with a piecewise constant potential [59]. We call them hard-collision, since the force is zero except for δ-spikes and the dynamics proceeds from collision to collision. Now one has to develop an efficient algorithm by which one finds the time-wise next collision. Except for rounding, there is no discretization of time. The simplest example is the hard-point potential, $V_{\mathrm{hc}}(x) = \infty$ for $x < 0$ and $V_{\mathrm{hc}}(x) = 0$ for $x \geq 0$. A variant is the infinite square well potential, $V_{\mathrm{sw}}(x) = \infty$ for $x < 0$, $x > a$ and $V_{\mathrm{sw}}(x) = 0$ for $0 \leq x \leq a$ [59, 60]. In this case two neighboring particles at the maximal distance a are reflected inwards as if connected by a massless string of finite length a. For both models the dynamics remembers the initial velocities. To have only the standard conservation laws, one imposes alternating masses, say $m_j = m_0$ for even j and $m_j = m_1$ for odd j. In both models the unit cell then contains two particles and the scheme explained before has to be extended. But at the very end the difference is minimal. A further variant is the shoulder potential $V_{\mathrm{sh}}(x)$, for which $V_{\mathrm{sh}}(x) = \infty$ for $|x| < \frac{1}{2}$, $V_{\mathrm{sh}}(x) = \varepsilon_0$ for $\frac{1}{2} \leq |x| \leq 1$, and $V_{\mathrm{sh}}(x) = 0$ for $|x| > 1$. The potential is either repulsive, $\varepsilon_0 > 0$, or attractive $\varepsilon_0 < 0$. Exploratory studies of the latter case indicate that the convergence is slower than for the extensively studied attractive case. Particles interacting with such a potential can be viewed also as a hard-core fluid with a short range potential part and thus serves as a bridge between one-dimensional fluids and anharmonic solids. The collisions resulting from the potential step make the model non-integrable.

FPU chains with an even potential at $P = 0$ constitute a distinct dynamical phase. Such phase is absent for the hard-point and the square shoulder potentials. But the square well at $P = 0$ has the same properties as can be seen from taking $a_0 = a/2$ in (3.134).

Current system sizes are $N = 2^{11}$ to 2^{13}, even size $2^{16} = 65{,}536$ has been attempted [61]. Periodic boundary conditions are imposed. Time is restricted to $t \leq t_{\max} = N/2c$, the time of the first collision between sound peaks. For given potential, one has to decide on the thermodynamic parameters, P, β. One constraint is to have c approximately in the range $1 \ldots 2$ in order to have a sufficiently long time span available. Secondly one would like to be well away from integrability. This leads to an energy per particle of order 1 in the models from above. A related issue are the coupling matrices $\vec{G}$, which should not be too small, at least for the relevant couplings. For the models from above they are tabulated. The relevant couplings show quite some variability taking values in the range $0.1 \ldots 3.4$. At the very end, one has to make a physically reasonable choice, perhaps use the same parameters as previous MD simulations so to have the possibility to compare.

A systematic study of the dependence on P, β seems to be too costly and most likely not so interesting. But it does make sense to probe values at the border. For some of the FPU simulations the energy per particle is chosen as $\mathsf{e} = 0.1$, at which point nonlinearities are small [57]. There are also very extensive simulations at the even lower energy $\mathsf{e} = 5\times 10^{-4}$, with the goal to explore the route to equipartition, which is a somewhat distinct issue [62].

Once all parameters are fixed, there are several options to run the simulation. Since in canonical equilibrium $\{r_j, p_j, j = 1, \ldots, N\}$ are independent random variables, one can sample the initial conditions through a random number generator. For the hard-collision potentials the geometric constraints are still simple enough for allowing one to generate the microcanonical ensemble by Monte Carlo methods. In our simulations the correlator S hardly depends on the choice of the ensemble. With such generated random initial data the equations of motion are simulated up to $t_{\max}$. A single run is noisy and one has to repeat many times, order 10^7. The much more common choice is to start from a reasonable nonequilibrium configuration and to equilibrate before measuring correlations. Usually one then simulates very long trajectories, up to times of order 2^{15}, over which the time lag $g_\alpha(j, t+\tau)g_{\alpha'}(0, t)$ is sampled, $\tau \leq t_{\max}$. In addition one averages over a small number of runs, of order 10^2. The total number of samples is roughly the same in both methods. The random number generator method produces the thermal average with a higher reliability.

The sampled $S_{\alpha\alpha'}(j, t)$ can be Fourier transformed in the spatial variable and/or in the time variable. One can also transform to normal coordinates. These are linear operations which can be done for each sample or only after averaging. Should one keep the full resolution or only some data points? Of course it depends somewhat on the goals. In [58, 59], the full 3×3 matrix is sampled and subsequently transformed via the theoretically computed R matrix to obtain $S^\sharp(j, t)$. This approach allows to test diagonality. Because the peak structure is most easily seen in the space-like j coordinate, maximal resolution for j is retained and only three times ($t = 250, 500, 1000$) are recorded for the purpose of making a scaling plot. In van Beijeren and Posch (TU München, June 2013, private communication), the lowest Fourier modes are measured as a function of t. In [57] the lowest Fourier modes are plotted as a function of the frequency ω. A separate issue are the much simulated total current correlations. The total currents are sampled directly, in the most complete version momentum, energy, and cross correlations, and then plotted as a function of t or ω.

We reproduce only a few figures. Many more details can be found in the original papers. In Fig. 3.3 we display the data for the FPU chain [58] with $V_{\mathrm{FPU}}(x) = \frac{1}{2}x^2 + \frac{2}{3}x_3 + \frac{1}{4}x^4$, $\beta = 2$, $P = 1$, $N = 8192$, and $t_{\max} = 2700$. The sound speed is $c = 1.45$. Note that the sound peak is somewhat distorted, not symmetric relative to ct, but has a rapid fall-off away from the sound cone. As only fit parameter one uses λ_{h}, resp. λ_{s}. The optimal fit at the longest available time is denoted by $\lambda_{\mathrm{h}}^{\mathrm{emp}}$, resp. $\lambda_{\mathrm{s}}^{\mathrm{emp}}$, standing for empirical value. In most cases there is also a theoretical value based on decoupling and/or mode-coupling, which is indicated in square brackets. For the FPU simulations the results are for the heat peak $\lambda_{\mathrm{h}}^{\mathrm{emp}} = 13.8$ [1.97], and

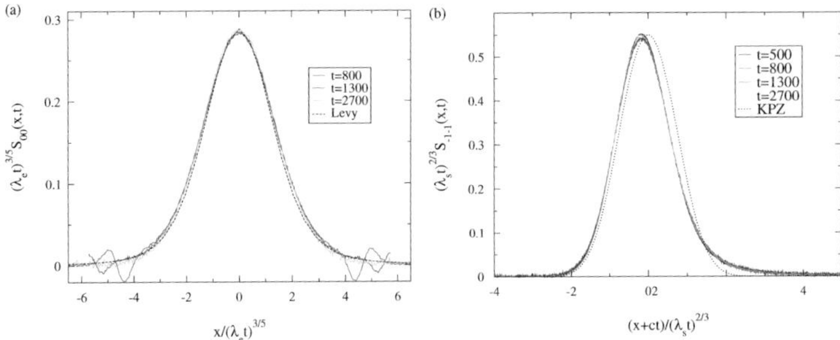

Fig. 3.3 Scaling plot of (**a**) heat and (**b**) sound peak for a FPU chain with $N = 8192$, potential parameters $\alpha = 2$, $\beta = 1$, pressure $P = 1$, and temperature $\beta^{-1} = 0.5$

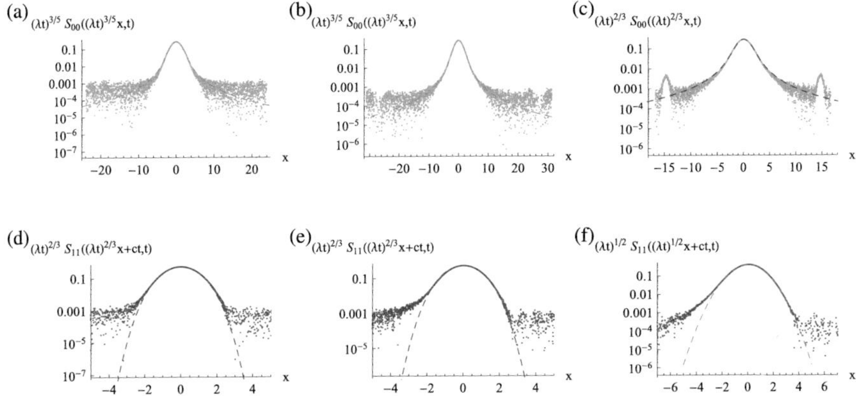

Fig. 3.4 Heat and sound peak for shoulder, hard-point, and square well potential in logarithmic scale: (**a**) shoulder, heat; (**b**) hard-point gas, heat; (**c**) square-well, heat; (**d**) shoulder, sound; (**e**) hard-point gas, sound; (**f**) square-well, sound

for the sound peak $\lambda_{\mathrm{s}}^{\mathrm{emp}} = 2.05$ [0.68]. From the visual appearance, one might have guessed the theoretical values to be just the other way round. So maybe the system tries to generate the optimal Lévy peak at non-relaxed sound peaks.

In Fig. 3.4 we reproduce the plot of heat and sound peak for the hard-collision models with shoulder, hard-point, and square well potential, in the latter two cases with alternating masses [59]. To better judge the quality of the fit we provide a logarithmic plot. In all cases $N = 4096$ with $t_{\max} = 1024$. The following parameters have been chosen, shoulder: $\varepsilon_0 = 1$, $P = 1.2$, $\beta = 2$, $c = 1.74$, hard-point: $m_1/m_0 = 3$, $P = 2$, $\beta = 0.5$, $c = 1.73$, square well: $m_1/m_0 = 3$, $a = 1$, $P = 0$, $\beta = 2$, $c = 1.73$. The fit to the predicted scaling function has an error less than 5 %. For the hard-collision models the simulation results are, shoulder: $\lambda_{\mathrm{s}}^{\mathrm{emp}} = 1.62$ [1.71], $\lambda_{\mathrm{h}}^{\mathrm{emp}} = 1.44$ [1.04], hard-point: $\lambda_{\mathrm{s}}^{\mathrm{emp}} = 1.42$ [2.00], $\lambda_{\mathrm{h}}^{\mathrm{emp}} = 1.04$ [0.95], square well: $\lambda_{\mathrm{h}}^{\mathrm{emp}} = 0.95$ [1.04]. Recall that the square well potential at $P = 0$ is in a distinct dynamical universality class, hence the different scaling exponents. For this model λ_s is related to a diffusion constant, which can be obtained only numerically.

The MD simulation [57] is at low temperatures and considers the positional correlations, which in Fourier space differ from the stretch correlations only by a k-dependent prefactor. Thus sampled is the correlator

$$(1 - \cos(2\pi k))^{-1} \int dt\, \mathrm{e}^{\mathrm{i}\omega t} \hat{S}_{11}(k,t) = (1 - \cos(2\pi k))^{-1} \hat{S}_{11}(k,\omega)\,. \tag{3.142}$$

For low temperatures the area under the central peak is a factor 10 smaller than the one under the sound peak. Hence only the sound peak is explored. Its asymptotic scaling form is

$$\hat{S}_{11}(k,t) = \cos(2\pi \mathrm{i}kct) C_{11} \hat{f}_{\mathrm{KPZ}}(k(\lambda_{\mathrm{h}}|t|)^{2/3})\,. \tag{3.143}$$

Considering only the right moving sound peak by setting $\omega_{\max} = 2\pi kc$,

$$\begin{aligned}\hat{S}_{11}(k,\omega + \omega_{\max}) &= \int dt\, \mathrm{e}^{\mathrm{i}\omega t} \tfrac{1}{2} C_{11} \hat{f}_{\mathrm{KPZ}}(k(\lambda_{\mathrm{h}}|t|)^{2/3}) \\ &= \int dt\, \mathrm{e}^{\mathrm{i}(\omega/\lambda_{\mathrm{h}}|k|^{3/2})t} (\lambda_{\mathrm{h}}|k|^{3/2})^{-1} \tfrac{1}{2} C_{11} \hat{f}_{\mathrm{KPZ}}(|t|^{2/3})\,.\end{aligned} \tag{3.144}$$

Thus defining

$$h_{\mathrm{KPZ}}(\omega) = \int dt\, \mathrm{e}^{\mathrm{i}\omega t} \hat{f}_{\mathrm{KPZ}}(|t|^{2/3})\,, \tag{3.145}$$

one arrives at

$$\hat{S}_{11}(k,\omega + \omega_{\max}) = \tfrac{1}{2} C_{11} (\lambda|k|^{3/2})^{-1} h_{\mathrm{KPZ}}(\omega/\lambda_{11}|k|^{3/2})\,. \tag{3.146}$$

If one normalizes the maximum of $\hat{S}$ to 1, then $h_{\mathrm{KPZ}}(\omega)/h_{\mathrm{KPZ}}(0)$ replaces $h_{\mathrm{KPZ}}(\omega)$, which amounts to setting the prefactor in (3.146) equal to 1. In Fig. 3.5 the spectrum at two different choices of the asymmetry parameter is displayed.

In the simulation the potential is chosen as $V(x) = \frac{1}{2}x^2 + \alpha\frac{1}{3}x^3 + \frac{1}{4}x^4$, where the asymmetry varies from 0 to 2. Increasing α, the pressure P increases from 0 to -0.2 and the inverse temperature from 9.55 to 9.75. The sound speed $c \simeq 1.1$. In frequency space the peak moves linearly in k at around $\omega = 0.01$. For the shape function one uses $\hat{S}(k,\omega)$ for $k = 1, 2, 4, 8, 16$ to generate a scaling plot. Over the whole range of α's the fit with the scaling function (3.145) is fairly convincing. The optimal fit parameter starts from $\lambda_{\mathrm{s}}^{\mathrm{emp}} = 0.02$ [0.04] at $\alpha = 0.2$ to $\lambda_{\mathrm{s}}^{\mathrm{emp}} = 0.08$ [0.37] at $\alpha = 1.6$ and to $\lambda_{\mathrm{s}}^{\mathrm{emp}} = 0.07$ [0.53] at $\alpha = 2.0$.

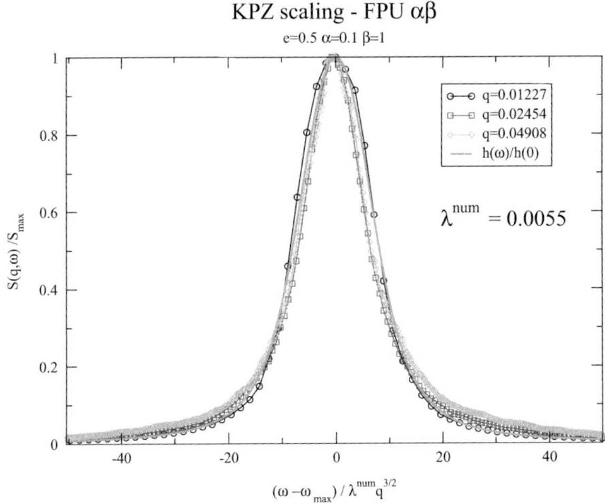

Fig. 3.5 Scaling plot of the sound peak in $(k = q, \omega)$ variables of a FPU chain with $N = 1024$, potential parameters $\alpha = 0.1$, $\beta = 1$, pressure $P = -0.04$, and temperature $\beta^{-1} = 0.105$. The speed of sound is $c = 1.11$

Given the diversity of models, parameters, and numerical schemes only tentative conclusions can be drawn.

(a) The separation into three peaks is a fast process. In normal mode representation the off-diagonal matrix elements are indeed small. There are no correlations beyond the sound cone.
(b) The distinction between $G^1_{11} \neq 0$ and $G^1_{11} = 0$ is seen very convincingly.
(c) The central peak is on the $t^{3/5}$ scale and adjusts well to the predicted Lévy distribution with one caveat. As seen from the center, the shape function fairly rapidly switches into the power law decay. However at the location of the sound peaks there are some wiggles and beyond only small amplitude noise is observed. In this sense, the Lévy distribution gets uncovered as time progresses. The Lévy distribution is a fairly direct consequence of $G^0_{00} = 0$ and thus one of the strongest supports for the theory.
(d) The sound peaks are mirror images of each other. One plots them on the $t^{2/3}$ scale, but then there is still a slow change in time. For the hard-collision models the shape is almost perfect, but the λ_s parameter is dropping in time. For the FPU models the peak is distorted and still away from the symmetric shape predicted by the theory. To the outside of the sound cone there is the rapid fall-off in accordance with the KPZ scaling function. But towards the heat peak there is slow decay. The sound peak is tilted away from the central peak. Apparently there is still a strong interaction between the peaks.
(e) In the majority of the simulations the peaks vary slowly on the scale $t^{2/3}$ for sound, resp. $t^{3/5}$ for the heat peak. Thus it becomes meaningful to use as a fit the theoretical scaling function with λ_s, resp. λ_h, as only free parameter.

This optimal choice has been denoted above as empirical value $\lambda_{s/h}^{emp}$. The error between the measured and theoretical shape function is less than 5 %. However one observes that λ_s^{emp} and λ_h^{emp} are still changing in time signaling that the simulation has not yet reached the truly asymptotic regime. In some simulations $\lambda_{s/h}^{emp}$ drops monotonically in time and differs not too strongly from $\lambda_{s/h}$. One is then willing to believe that for even longer simulation times the asymptotic value is reached. But in other simulations there is a much stronger discrepancy, which asks for more explanations.

Mode-coupling is not specific to anharmonic chains. In principle any one-dimensional system with conserved fields can be handled by the same scheme. This offers the possibility to test the theory through other models, possibly finding systems with less strong finite time effects. One obvious choice are stochastic lattice gases with several type of particles like several lane TASEP [43, 63] and the AHR model [64, 65]. For them the couplings can be more easily adjusted than for anharmonic chains, which offers the possibility to test the dynamical phase diagram. Also anharmonic chains with a stochastic collision mechanism, respecting the conservation laws, have been studied in considerable detail [42, 66, 67].

3.6 Total Current Correlations

The total current is a fluctuation observable, in contrast to $S_{\alpha\alpha'}(j,t)$ which refers to the average of the product of two local observables. Thus we need some additional considerations to establish the link to nonlinear fluctuating hydrodynamics. For a ring of size N, $\Lambda_N = [1,\dots,N]$, the total currents are defined by

$$\vec{\mathscr{J}}_{\mathrm{tot},\Lambda_N}(t) = \frac{1}{\sqrt{N}}\sum_{j=1}^{N}\vec{\mathscr{J}}(j,t) \tag{3.147}$$

and the total current covariance reads

$$\begin{aligned}\Gamma_{\Lambda_N,\alpha\alpha'}(t) &= \langle \mathscr{J}_{\mathrm{tot},\Lambda_N,\alpha}(t);\, \mathscr{J}_{\mathrm{tot},\Lambda_N,\alpha'}(0)\rangle_{P,\beta,\Lambda_N}\\ &= \sum_{j=1}^{N}\langle \mathscr{J}_\alpha(j,t);\, \mathscr{J}_{\alpha'}(0,0)\rangle_{P,\beta,\Lambda_N}\,. \end{aligned}\tag{3.148}$$

The cumulant $\langle\cdot;\cdot\rangle$ means that the static average is subtracted and system size is indicated explicitly. In the limit $N\to\infty$

$$\Gamma_{\alpha\alpha'}(t) = \sum_{j\in\mathbb{Z}}\langle \mathscr{J}_\alpha(j,t);\, \mathscr{J}_{\alpha'}(0,0)\rangle_{P,\beta}\,. \tag{3.149}$$

For fixed t the integrand decays exponentially in j, but with a correlation length increasing in time.

Before we emphasized the transformation to normal coordinates. For the currents we stick to the physical fields. The current $\mathscr{J}_{\mathrm{tot},\Lambda_N,1}(t)$ is itself conserved. Thus only the $(2,3)$ block has a time variation. Using stationarity and time-reversal, the diagonal elements are even, $\Gamma_{\alpha\alpha}(t) = \Gamma_{\alpha\alpha}(-t)$, while the off-diagonal elements are odd and satisfy

$$\Gamma_{23}(t) = -\Gamma_{23}(-t) = \Gamma_{32}(-t) = -\Gamma_{32}(t)\,. \tag{3.150}$$

Mode-coupling predicts that this matrix element vanishes. In our simulations we observe an exponential decay with a decay time of order 20–30. Thus the correlations of real interest are $\Gamma_{22}(t)$ and $\Gamma_{33}(t)$. For chains the latter is the most frequently simulated equilibrium time correlation. The momentum current correlations have been measured in [53] (van Beijeren and Posch, June 2014, GGI, Firenze, private communication) and the momentum-energy cross correlations only recently in [68].

$\Gamma_{\Lambda_N,\alpha\alpha'}(t)$ is a fluctuation observable, for which the equivalence of ensembles does not hold. But the time-independent difference between microcanonical and canonical average can be computed explicitly. For the microcanonical ensemble, $\lim_{t\to\infty}\Gamma_{\Lambda_N,\mathrm{micro}}(t) = 0$. On the other hand, there is no reason for $\Gamma_{\Lambda_N}(t)$ to vanish asymptotically if the canonical average, as in (3.148), is used. In fact, for infinite volume,

$$\lim_{t\to\infty}\Gamma_{22}(t) = \beta^{-1}c^2\,, \qquad \lim_{t\to\infty}\Gamma_{33}(t) = \beta^{-1}P^2\,. \tag{3.151}$$

The asymptotic values in (3.151) are called Drude weight, which has received a lot of attention in the context of current correlations for integrable quantum chains. A non-zero Drude weight indicates that the correlator of the corresponding conserved field has a ballistically moving component, but it cannot resolve the structure of this component. For non-integrable anharmonic chains, the ballistic pieces are just the sharply concentrated sound peaks, while in the integrable case one expects to have a broad spectrum which expands ballistically.

The link to mode-coupling is achieved through the general observation that the current correlations are proportional to the memory kernel. We use its diagonal approximation and insert the asymptotic form of f_α. The memory kernel is in normal mode representation. Thus we still have to transform back to the physical fields through the R matrix. The computation can be found in [68] with the result

$$\begin{aligned}\Gamma^{\Delta}_{22}(t) &= \Gamma_{22}(t) - \beta^{-1}c^2 \simeq \tfrac{1}{2}(\lambda_{\mathrm{h}}t)^{-3/5}\langle\psi_0, H^{\mathrm{u}}\psi_0\rangle^2\int dx f_{\mathrm{Lévy},5/3}(x)^2\\ &\quad+(\lambda_{\mathrm{s}}t)^{-2/3}\langle\psi_1, H^{\mathrm{u}}\psi_1\rangle^2\int dx f_{\mathrm{KPZ}}(x)^2\\ \Gamma^{\Delta}_{33}(t) &= \Gamma_{33}(t) - \beta^{-1}P^2 \simeq c^2\beta^{-2}(\lambda_{\mathrm{s}}t)^{-2/3}\int dx f_{\mathrm{KPZ}}(x)^2\,,\end{aligned} \tag{3.152}$$

where $\{\psi_\alpha\}$ are the eigenvectors of A, $A\psi_0 = 0$, $A\psi_1 = c\psi_1$, see [30]. If (3.134) is satisfied, e.g. an even potential at $P = 0$, then the sound peak is diffusive, see Eq. (3.136), and the central peak is Lévy 3/2. Furthermore $\langle\psi_0, H^{\mathrm{u}}\psi_0\rangle = 0 = \langle\psi_1, H^{\mathrm{u}}\psi_1\rangle$. Hence $\Gamma_{22}(t) - \beta^{-1}c^2$ is expected to decay integrably, while

$$\Gamma_{33}(t) - \beta^{-1}P^2 \simeq c^2\beta^{-2}(8\pi D_s t)^{-1/2}\,. \tag{3.153}$$

The energy current correlation is predicted to decay as $t^{-2/3}$ which has been reported in MD simulations already more than 15 years ago [51, 52]. The true mechanism behind the decay is actually somewhat subtle. From the conservation law it follows that the second moment of the heat peak is related to the second time derivate of the current correlation. Using the asymptotic form (3.128), including the cut-off at the sound peak, one arrives at

$$\frac{d^2}{dt^2}\int_{-ct}^{ct} dx\, x^2 f_0(x,t) \simeq \tfrac{8}{3\pi}(\lambda_{\mathrm{h}})^{5/3}c^{1/3}(\lambda_{\mathrm{h}}t)^{-2/3}\,. \tag{3.154}$$

This argument overlooks that the scaling form is for the normal mode representation, while $\Gamma_{33}(t)$ refers to the physical energy current. The complete computation leads however to the same power law except for a different prefactor [68].

The momentum current correlation should decay as $t^{-3/5}$, which is a recent finding. However, its prefactor could vanish, in principle. Mode-coupling with the currently available precision would not provide an answer, then. In contrast the prefactor for $\Gamma_{33}(t)$ is strictly positive.

Our simulation results [68] are shown in Figs. 3.6 and 3.7. The parameters are as before, $N = 4096$, $t_{\max} = 1024$, shoulder: $\varepsilon_0 = 1$, $P = 1.2$, $\beta = 2$, $c = 1.74$, and hard-point: $m_1/m_0 = 3$, $P = 2$, $\beta = 0.5$, $c = 1.73$. The red lines indicate the predictions based on mode-coupling. It is interesting to note that for the shoulder potential the evidence for a $t^{-2/3}$ decay is not so overwhelming as one might have anticipated and, by looking at a different time window, one could as well fit to a slightly different exponent. On the other hand the hard-point potential with alternating masses shows a very clean power law decay. Through MD with shoulder potential the predicted decay of $\Gamma_{22}(t)$ is well confirmed. However, for the hard-point potential it so happens that both prefactors, $\langle\psi_0, H^{\mathrm{u}}\psi_0\rangle$ and $\langle\psi_1, H^{\mathrm{u}}\psi_1\rangle$, vanish. Numerically we estimate a decay as t^{-1}. Again this is a strong qualitative support of nonlinear fluctuating hydrodynamics. One might have thought that all hard-collision potentials have the same asymptotic power law for the momentum current correlation. This expectation is born out under the proviso that the respective prefactors do not vanish. Matrix elements, as $\langle\psi_0, H^{\mathrm{u}}\psi_0\rangle$, must come from a microscopic computation and cannot be deduced by a mere inspection of the potential.

An additional confirmation of (3.151), (3.153) has been accomplished recently [69]. For the potential $V(x) = \frac{1}{3}ax^3 + \frac{1}{4}x^4$, $\beta = 1$ and at $P = 0.59$ with $a = -2$, resp. at $P = -0.5$, $a = 1.89$, up to very small errors the signature of the $\vec{G}$ matrices

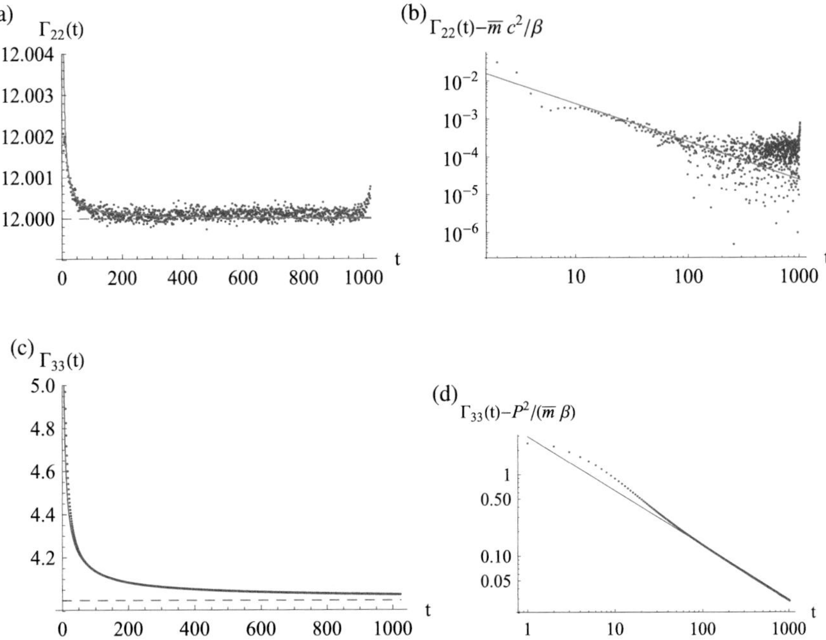

Fig. 3.6 Total momentum and energy current correlations for hard-point particles with alternating masses. (**a**) Momentum current correlations; (**b**) logarithmic plot of $\Gamma^{\Delta}_{22}(t)$; (**c**) energy current correlations; (**d**) logarithmic plot of $\Gamma^{\Delta}_{33}(t)$

is identical to an even potential at $P = 0$. In the MD simulation the energy current is found to decay as $t^{-1/2}$ and the momentum current seems to be integrable. Keeping all parameters fixed and shifting slightly to $a = -2.7$, resp. $a = 2$, the decay as stated in (3.152) is restored.

3.7 Other 1D Hamiltonian Systems

For the anharmonic chains studied so far, the potential depends only on $q_{j+1} - q_j$ and hence remains without change under spatial translations. Physically this property is obvious and seems hard to avoid. However there could be a substrate potential which forces the particles preferentially to particular locations. One could consider a two-component system, which then has an acoustic and an optical mode. The latter would be comparable to a one-component system with an on-site potential. Such considerations lead to the more general class of Hamiltonians

$$H_{\mathrm{os}} = \sum_{j=1}^{N} \big(\tfrac{1}{2}p_j^2 + V_{\mathrm{os}}(q_j)\big) + \sum_{j=1}^{N-1} V(q_{j+1} - q_j) \tag{3.155}$$

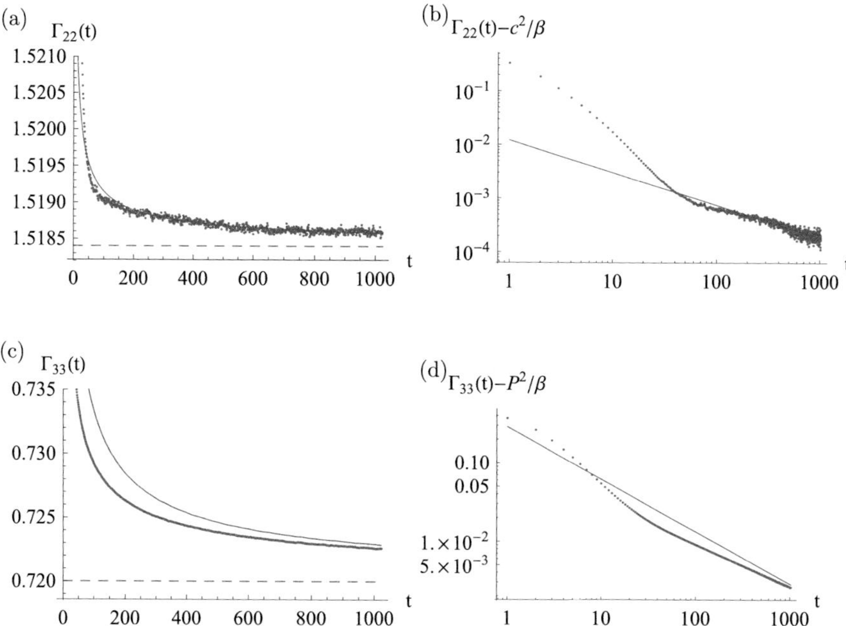

Fig. 3.7 Total momentum and energy current correlations for a hard collision model with shoulder potential

with some confining on-site potential V_{os}. The only conserved field is the energy. There is a unique equilibrium measure. The Euler currents vanish. From the perspective of fluctuating hydrodynamics all evidence points towards diffusive energy transport. For the case of a quadratic V and for $V_{\text{os}} = V_{\text{FPU}}$ very detailed MD simulation confirm diffusive transport [70].

More interesting are models with two conservation laws. We discuss separately coupled rotators, which can be thought of as a classical limit of a quantum Heisenberg chain, and the discrete nonlinear Schrödinger equation on a lattice, which is the classical field theory for lattice bosons.

Coupled Rotators The Hamiltonian of the rotator chain reads

$$H_{\text{CR}} = \sum_{j=1}^{N} \left(\tfrac{1}{2}p_j^2 + V(\varphi_{j+1} - \varphi_j)\right) \tag{3.156}$$

with periodic boundary conditions, $\varphi_{N+1} = \varphi_1$. At first glance we have only rebaptized q_j as φ_j. But the φ_j's are angles and the p_j's angular momenta. Hence the phase space is $(S^1 \times \mathbb{R})^N$ with S^1 denoting the unit circle. The standard choice for V is $V(\vartheta) = -\cos\vartheta$, but in our context any 2π-periodic potential is admitted.

The equations of motion are

$$\frac{d}{dt}\varphi_j = p_j\,, \quad \frac{d}{dt}p_j = V'(\varphi_{j+1} - \varphi_j) - V'(\varphi_j - \varphi_{j-1})\,. \tag{3.157}$$

Obviously angular momentum is locally conserved with the angular momentum current

$$\mathscr{J}_1(j) = -V'(\varphi_j - \varphi_{j-1})\,. \tag{3.158}$$

As local energy we define $e_j = \frac{1}{2}p_j^2 + V(\varphi_{j+1} - \varphi_j)$. Then e_j is locally conserved, since

$$\frac{d}{dt}e_j = p_{j+1}V'(\varphi_{j+1} - \varphi_j) - p_j V'(\varphi_j - \varphi_{j-1})\,, \tag{3.159}$$

from which one reads off the energy current

$$\mathscr{J}_2(j) = -p_j V'(\varphi_j - \varphi_{j-1})\,. \tag{3.160}$$

For the angles, in analogy to the stretch, one defines the phase difference $\tilde{r}_j = \Theta(\varphi_{j+1} - \varphi_j)$, where Θ is 2π-periodic and $\Theta(x) = x$ for $|x| \leq \pi$. Because of the jump discontinuity the stretch is not conserved. A rotator chain has only two conserved fields.

To apply nonlinear fluctuating hydrodynamics one has to compute the Euler currents in local equilibrium. Since there are two conserved fields the canonical equilibrium state reads

$$\frac{1}{Z_N}\prod_{j=1}^{N}\exp\big[-\beta\big(\tfrac{1}{2}(p_j - u)^2 + V(\varphi_{j+1} - \varphi_j)\big)\big]d\varphi_j dp_j \tag{3.161}$$

with u the average angular momentum. Now

$$\langle \mathscr{J}_1(j)\rangle_N = -\langle V'(\varphi_j - \varphi_{j-1})\rangle_N\,, \quad \langle \mathscr{J}_2(j)\rangle_N = -u\langle V'(\varphi_j - \varphi_{j-1})\rangle_N\,, \tag{3.162}$$

average with respect to the canonical ensemble (3.161). We claim that

$$\lim_{N\to\infty}\langle V'(\varphi_j - \varphi_{j-1})\rangle_N = 0\,. \tag{3.163}$$

For this purpose we expand in Fourier series as

$$e^{-V(\vartheta)} = \sum_{m\in\mathbb{Z}} a(m)e^{-im\vartheta}\,, \quad f(\vartheta) = \sum_{m\in\mathbb{Z}}\hat{f}(m)e^{-im\vartheta}\,. \tag{3.164}$$

Then, working out all Kronecker deltas from the integration over the φ_j's, one arrives at

$$\langle f(\varphi_{j+1}-\varphi_j)\rangle_N = \Big(\sum_{m\in\mathbb{Z}} a(m)^N\Big)^{-1} \sum_{m\in\mathbb{Z}} a(m)^N \Big(a(m)^{-1}\sum_{\ell\in\mathbb{Z}} \hat{f}(\ell-m)a(\ell)\Big). \tag{3.165}$$

Since $a(0) > |a(m)|$ for all $m \neq 0$,

$$\lim_{N\to\infty} \langle f(\varphi_{j+1}-\varphi_j)\rangle_N = a(0)^{-1}\sum_{\ell\in\mathbb{Z}} \hat{f}(\ell)a(\ell) = \frac{1}{Z_1}\int_{-\pi}^{\pi} d\vartheta f(\vartheta)e^{-\beta V(\vartheta)}. \tag{3.166}$$

For $f(\vartheta) = V'(\vartheta)$, the latter integral vanishes because of periodic boundary conditions in ϑ. We conclude that both currents vanish on average.

As before we consider the infinite lattice with thermal expectation $\langle\cdot\rangle_{u,\beta}$ and form the equilibrium time correlations as

$$S_{\alpha\alpha'}(j,t) = \langle g_\alpha(j,t)g_{\alpha'}(0,0)\rangle_{u,\beta} - \langle g_\alpha(0)\rangle_{u,\beta}\langle g'_\alpha(0)\rangle_{u,\beta}\,, \tag{3.167}$$

$\vec{g}(j,t) = \big(p_j(t), e_j(t)\big)$. p_j is odd and e_j is even under time reversal. Hence in the Green-Kubo formula the cross term vanishes. Thus, for large j, t, fluctuating hydrodynamics predicts

$$S_{\alpha\alpha'}(j,t) = \delta_{\alpha\alpha'}(4\pi D_\alpha t)^{-1/2} f_{\mathrm{G}}((4\pi D_\alpha t)^{-1/2} j)\,, \tag{3.168}$$

where f_{G} is the unit Gaussian. D_α is the diffusion coefficient of mode α. Of course, it can be written as a time-integral over the corresponding total current-current correlation, but its precise value has to be determined numerically. This has been done for the standard choice $V(x) = -\cos x$, to which we specialize now. For $\beta = 1$, with a lattice size $N = 500$, the diffusive peaks are well established at $t = 2000$ [71, 72]. Energy diffusion has been confirmed much earlier [73, 74].

At low temperatures one finds a different, perhaps more interesting scenario. At zero temperature, there is the one-parameter family of ground states with $\varphi_j = \bar{\varphi}$, $p_j = 0$. When heating up, under the canonical equilibrium measure, the phase φ_j jumps to φ_{j+1} with a jump size $\mathcal{O}(1/\sqrt{\beta})$. Next we have to understand how the conservation of $\tilde{r}_j$ field is broken. In a pictorial language, the event that $|\varphi_{j+1}(t) - \varphi_j(t)| = \pi$ is called an umklapp for phase difference $\tilde{r}_j$ or an umklapp process to emphasize its dynamical character. At low temperatures a jump of size π has a small probability of order $e^{-\beta\Delta V}$ with $\Delta V = 2$ the height of the potential barrier. Hence $\tilde{r}_j$ is locally conserved up to umklapp processes occurring at a very small frequency only. This can be measured more quantitatively by considering the average

$$\Gamma_{\mathrm{uk}}(t) = \sum_{j\in\mathbb{Z}} \big(\langle\tilde{r}_j(t)\tilde{r}_0(0)\rangle_{u,\beta} - \langle\tilde{r}_0\rangle_{u,\beta}^2\big)\,. \tag{3.169}$$

At $\beta = 1$, $\Gamma_{\mathrm{uk}}(t)$ decays exponentially due to umklapp. But at $\beta = 5$ the decay rate is already very much suppressed, see [72].

In the low temperature regime it is tempting to use an approximation, where the potential $V(x) = -\cos x$ is Taylor expanded at the minimum $x = 0$. But such procedure would underestimate the regime of low temperatures, as can be seen from the example of a potential, still with $\Delta V = 2$, but several shallow minima. The proper small parameter is β^{-1} such that $\beta\Delta V > 1$. To arrive at an optimal low temperature hamiltonian, we first parametrize the angles $\varphi_1, \ldots, \varphi_N$ through $r_j = \phi_{j+1} - \phi_j$ with $r_j \in [-\pi, \pi]$. To distinguish, we denote the angles in this particular parametrization by ϕ_j. The dynamics governed by H_{CR} corresponds to periodic boundary conditions at $r_j = \pm\pi$. For a low temperature description we impose instead specular reflection, i.e., if $r_j = \pm\pi$, then p_j, p_{j+1} are scattered to $p'_j = p_{j+1}$, $p'_{j+1} = p_j$. By fiat all umklapp processes are now suppressed, while between two umklapp events the CR dynamics and the low temperature dynamics are identical. The corresponding hamiltonian reads

$$H_{\mathrm{CR,lt}} = \sum_{j=1}^{N} \left(\tfrac{1}{2}p_j^2 + \tilde{V}(\phi_{j+1} - \phi_j)\right) \tag{3.170}$$

with

$$\tilde{V}(x) = -\cos x \ \text{ for } |x| \le \pi\,, \qquad \tilde{V}(x) = \infty \ \text{ for } |x| > \pi\,, \tag{3.171}$$

periodic boundary conditions $\phi_{N+1} = \phi_1$ being understood. The pair (ϕ_j, p_j) are canonically conjugate variables. Note that as weights $\exp[-\beta H_{\mathrm{CR}}] = \exp[-\beta H_{\mathrm{CR,lt}}]$. Thus all equilibrium properties of the coupled rotators remain untouched.

The hamiltonian $H_{\mathrm{CR,lt}}$ is a variant of the hard collision model with square well potential as discussed before, see [59]. The dynamics governed by $H_{\mathrm{CR,lt}}$ has three conserved fields, the stretch $r_j = \phi_{j+1} - \phi_j$, the momentum p_j, and the energy $e_j = \frac{1}{2}p_j^2 + \tilde{V}(r_j)$. Because of $\phi_1 = \phi_{N+1}$, one has $\sum_{j=1}^{N} r_j = 0$. The model is in the dynamical phase characterized by an even potential at zero pressure.

We claim that, for $\beta\Delta V > 1$, the CR equilibrium time correlations are well approximated by those of $H_{\mathrm{CR,lt}}$, provided the time of comparison is not too long. The latter correlations can be obtained within the framework of nonlinear fluctuating hydrodynamics. Thereby one arrives at fairly explicit dynamical predictions for the low temperature regime of the CR model.

One physically interesting information concerns the Landau-Placzek ratio at low temperatures. We use (3.130) and expand R^{-1} in $1/\beta$. To lowest order it suffices to use the harmonic approximation, $\tilde{V}(x) = \frac{1}{2}ax^2$, $a = 1$ for the cosine potential.

Using the formulas in Appendix A of [30] one arrives at the following value for the Landau-Placzek ratios,

$$r\text{-}r: \ (2a\beta)^{-1}(1,0,1)\,, \quad p\text{-}p: \ (2\beta)^{-1}(1,0,1)\,, \quad e\text{-}e: \ (2\beta)^{-1}(a\ell^2,\beta^{-1},a\ell^2)\,. \tag{3.172}$$

The correlations are small, order β^{-1}. For the stretch correlations there is no central peak, to this order, and for the energy correlations the central peak is down by a factor β^{-1} relative to the sound peaks.

To have a unified picture we add $\tilde{r}_j$ to the list of fields of physical interest. At high temperatures $\tilde{r}_j$ is not conserved and one has diffusive spreading of the conserved fields. At low temperatures $\tilde{r}_j$ is conserved up to small errors and the conventional three-peak structure, including universal shape functions, results. For extremely long times umklapp processes will happen and one expects that they force a cross over to the Gaussian scaling (3.168). The precise dynamical structure of such cross over still needs to be investigated.

Nonlinear Schrödinger Equation on a Lattice A further example with a dynamically distinct low temperature phase is the nonlinear Schödinger equation on the one-dimensional lattice. In this case the lattice field is $\psi_j \in \mathbb{C}$, for which real and imaginary part are the canonically conjugate fields. The Hamiltonian reads

$$H = \sum_{j=1}^{N} \big(\tfrac{1}{2}|\psi_{j+1} - \psi_j|^2 + \tfrac{1}{2}g|\psi_j|^4\big) \tag{3.173}$$

with periodic boundary conditions and coupling $g > 0$. The sign of the hopping term plays no role, since it can be switched through the gauge transformation $\psi_j \rightsquigarrow \mathrm{e}^{\mathrm{i}\pi j}\psi_j$. The chain is non-integrable and the locally conserved fields are the number density $\rho_j = |\psi_j|^2$ and the local energy $e_j = \frac{1}{2}|\psi_{j+1} - \psi_j|^2 + \frac{1}{2}g|\psi_j|^4$. Hence the canonical equilibrium state is given by

$$Z^{-1}\mathrm{e}^{-\beta(H-\mu\mathsf{N})}\prod_{j=1}^{N} d\psi_j d\psi_j^*\,, \quad \mathsf{N} = \sum_{j=1}^{N}|\psi_j|^2\,, \tag{3.174}$$

with the chemical potential μ. We assume $\beta > 0$. But also negative temperature states, in the microcanonical ensemble, have been studied [75, 76]. Then the dynamics is dominated by a coarsening process mediated through breathers. In equilibrium, the ψ-field has high spikes at random locations embedded in a low noise background, which is very different from the positive temperature states considered here. For them the density and energy currents are symbolically of the form $\mathrm{i}(z - z^*)$, hence their thermal average vanishes. Both fields are expected to have diffusive transport. In fact, this is confirmed by MD simulations [77]. They also show Gaussian cross-correlations, which is possible since density and energy are both even under time reversal. In the previous studies [78] transport coefficients have been measured in the steady state set-up.

To understand the low temperature phase, it is convenient to transform to the new canonical pairs ρ_j, φ_j through

$$\psi_j = \sqrt{\rho_j}\, e^{i\varphi_j} . \tag{3.175}$$

In these variables the Hamiltonian becomes

$$H = \sum_{j=1}^{N} \big(- \sqrt{\rho_{j+1}\, \rho_j} \cos(\varphi_{j+1} - \varphi_j) + \rho_j + \tfrac{1}{2} g \rho_j^2 \big) . \tag{3.176}$$

The equations of motion read then

$$\partial_t \varphi_j = -\partial_{\rho_j} H , \quad \partial_t \rho_j = \partial_{\varphi_j} H . \tag{3.177}$$

φ_j takes values on the circle S^1 and $\rho_j \geq 0$. From the continuity of $\psi_j(t)$ when moving through the origin, one concludes that at $\rho_j(t) = 0$ the phase jumps from $\varphi_j(t)$ to $\varphi_j(t) + \pi$.

One recognizes the similarity to the coupled rotators (3.156). But now the equilibrium measure carries a nearest neighbor coupling. For $\mu > 0$, in the limit $\beta \to \infty$ the canonical measure converges to the one-parameter family of ground states with $\rho_j = \bar{\varphi} = \mu/g$, $\varphi_j = \bar{\varphi}$ with $\bar{\varphi}$ uniformly distributed on S^1. At low temperatures the field of phase differences $\tilde{r}_j$ is approximately conserved. The low temperature hamiltonian is constructed in such a way that the equilibrium ensemble remains unchanged while all umklapp processes are suppressed. To achieve our goal we follow verbatim the CR blueprint. The phases are parametrized such that $\varphi_{j+1} - \varphi_j$ lies in the interval $[-\pi, \pi]$ and this particular parametrization denoted by ϕ_j. Umklapp is a point at the boundary of this interval. Now (ϕ_j, ρ_j) are a pair of canonically conjugate variables, only $\rho_j \geq 0$ instead of $p_j \in \mathbb{R}$. Thus the proper low temperature hamiltonian reads

$$H_{\mathrm{lt}} = \sum_{j=0}^{N-1} \Big(\sqrt{\rho_{j+1}\, \rho_j}\, U(\phi_{j+1} - \phi_j) + V(\rho_j) \Big) , \tag{3.178}$$

where

$$U(x) = -\cos(x) \ \text{ for } \ |x| \leq \pi , \qquad U(x) = \infty \ \text{ for } \ |x| > \pi , \tag{3.179}$$

and

$$V(x) = x + \tfrac{1}{2} g x^2 \ \text{ for } \ x \geq 0 , \qquad V(x) = \infty \ \text{ for } \ x < 0 . \tag{3.180}$$

The low temperature Hamiltonian has a nearest neighbor coupling, which complicates the scheme through which the $\vec{G}$ matrices are determined [77]. Progress is achieved through the miraculous identity $\vec{\mathrm{j}} = (\mu, P, \mu P)$ for the Euler currents. The $\vec{G}$ coefficients are evaluated at $P = 0$. As for a generic anharmonic chain, one finds that $G^0_{00} = 0$ and $G^1_{11} \neq 0$. Thus the heat peak is predicted to be Lévy 5/3 and the sound peaks to be KPZ. The sound peak for the density-density correlation was first observed in [79] using k, ω space, see also [80]. In [77] we use normal mode representation, as explained in this article. The sound peaks fit nicely with KPZ, but the normalized heat peak is very broad and noisy, still with a shape not unlikely Lévy 5/3.

Acknowledgements First of all I owe my thanks to Christian Mendl. Only through his constant interest, his help in checking details, and his superb numerical efforts my project could advance to its current level. Various collaborations developed and I am most grateful to my coauthors A. Dhar, P. Ferrari, D. Huse, M. Kulkarni, T. Sasamoto, and G. Stoltz. At various stages of the project, I greatly benefited from discussions with H. van Beijeren, C. Bernardin, J. Krug, J.L. Lebowitz, S. Lepri, R. Livi, S. Olla, H. Posch, and G. Schütz. This work was supported by a Research Chair of the Foundation Sciences Mathématiques de Paris.

References

1. Green, M.S.: Markoff random processes and the statistical mechanics of time-dependent phenomena. II. Irreversible processes in fluids. J. Chem. Phys. **22**, 398–413 (1954)
2. Kubo, R.: Statistical-mechanical theory of irreversible processes. I. General theory and simple applications to magnetic and conduction problems. J. Phys. Soc. Jpn. **12**, 570–586 (1957)
3. Kubo, R.: Some aspects of the statistical-mechanical theory of irreversible processes. In: Brittin, W.E., Dunham, L.G. (eds.) Lecture Notes in Theoretical Physics, pp. 120–203. Interscience, New York (1959)
4. Alder, B.J., Wainwright, T.E.: Decay of the velocity autocorrelation function. Phys. Rev. A **1**, 18–21 (1970)
5. Pomeau, Y., Résibois, P.: Time-dependent correlation functions and mode-mode coupling theories. Phys. Rep. C **19**, 63–139 (1975)
6. Ernst, M.H., Hauge, E.H., van Leeuwen, J.M.J.: Asymptotic time behavior of correlation functions. II. Kinetic and potential terms. J. Stat. Phys. **15**, 7–22 (1976)
7. Cohen, E.G.D.: Fifty years of kinetic theory. Physica A **194**, 229–257 (1993)
8. Ortiz de Zarate, J.M., Sengers, J.V.: Hydrodynamic Fluctuations in Fluids and Fluid Mixtures. Elsevier, New York (2006)
9. Résibois, P., De Leener, M.: Classical Kinetic Theory of Fluids. Wiley, New York (1977)
10. Sasamoto, T., Spohn, H.: Superdiffusivity of the 1D lattice Kardar-Parisi-Zhang equation. J. Stat. Phys. **137**, 917–935 (2009)
11. Kardar, M., Parisi, G., Zhang Y.-C.: Dynamic scaling of growing interfaces. Phys. Rev. Lett. **56**, 889–892 (1986)
12. Kupiainen, A.: Renormalization group and stochastic PDE's. Ann. H. Poincaré online (2015). arXiv:1410.3094
13. Bertini, L., Cancrini, N.: The stochastic heat equation: Feynman-Kac formula and intermittence. J. Stat. Phys. **78**, 1377–1401 (1995)
14. Bertini, L., Giacomin, G.: Stochastic Burgers and KPZ equations from particle systems. Commun. Math. Phys. **183**, 571–607 (1997)

15. Hairer, M.: Solving the KPZ equation. Ann. Math. **178**, 559–664 (2013)
16. Hairer, M.: A theory of regularity structures. Invent. Math. **198**, 269–504 (2014)
17. Forster, D., Nelson, D.R., Stephen, M.J.: Large-distance and long-time properties of a randomly stirred fluid. Phys. Rev. A **16**, 732–749 (1977)
18. Prähofer, M., Spohn, H.: Exact scaling functions for one-dimensional stationary KPZ growth. J. Stat. Phys. **115**, 255–279 (2004)
19. Ferrari, P.L., Spohn, H.: Scaling limit for the space-time covariance of the stationary totally asymmetric simple exclusion process. Commun. Math. Phys. **265**, 1–44 (2006)
20. Baik, J., Ferrari, P.L., Péché, S.: Limit process of stationary TASEP near the characteristic line. Commun. Pure Appl. Math. **63**, 1017–1070 (2010)
21. Ferrari, P.L., Spohn, H., Weiss, T.: Brownian motions with one-sided collisions: the stationary case. Electron. J. Probab. **20**, 69 (2015)
22. Imamura, T., Sasamoto, T.: Exact solution for the stationary KPZ equation. Phys. Rev. Lett. **108**, 190603 (2012)
23. Imamura, T., Sasamoto, T.: Stationary correlations for the 1D KPZ equation. J. Stat. Phys. **150**, 908–939 (2013)
24. Borodin, A., Corwin, I., Ferrari, P.L., Vető, B.: Height fluctuations for the stationary KPZ equation. Math. Phys. Anal. Geom. **18**, 20 (2015)
25. Miloshevich, G., Nguenang, J.-P., Dauxois, T., Khomeriki, R., Ruffo, S.: Instabilities in long-range oscillator chains. Phys. Rev. E **91**, 032927 (2014)
26. Fermi, E., Pasta, J., Ulam, S.: Studies of nonlinear problems. Los Alamos Report LA-1940 (1955). Published in Collected Papers of Enrico Fermi, E. Segré (ed.), University of Chicago Press (1965)
27. Toda, M.: Vibration of a chain with a non-linear interaction. J. Phys. Soc. Jpn. **22**, 431–436 (1967)
28. Even, N., Olla, S.: Hydrodynamic limit for an Hamiltonian system with boundary conditions and conservative noise. Arch. Ration. Mech. Appl. **213**, 561–585 (2014)
29. Bernardin, C., Olla, S.: Non-equilibrium macroscopic dynamics of chains of anharmonic oscillators (2014). www.ceremade.dauphine.fr/~olla/springs13.pdf
30. Spohn, H.: Nonlinear fluctuating hydrodynamics for anharmonic chains. J. Stat. Phys. **154**, 1191–1227 (2014)
31. Tóth, B., Valkó, B.: Onsager relations and Eulerian hydrodynamic limit for systems with several conservation laws. J. Stat. Phys. **112**, 497–521 (2003)
32. Grisi, R., Schütz, G.: Current symmetries for particle systems with several conservation laws. J. Stat. Phys. **145**, 1499–1512 (2011)
33. Mendl, C.B.: Department of Mathematics, TU München (2014). www.github.com/cmendl/fluct-hydro-chains
34. van Saarloos, W., Bedeaux, D., Mazur, P.: Non-linear hydrodynamic fluctuations around equilibrium. Physica A **110**, 147–170 (1982)
35. Zubarev, D.N., Morozov, V.G.: Statistical mechanics of nonlinear hydrodynamic fluctuations. Physica A **120**, 411–467 (1983)
36. Prähofer, M.: Exact scaling functions for one-dimensional stationary KPZ growth (2006). www-m5.ma.tum.de/KPZ
37. van Beijeren, H.: Exact results for anomalous transport in one-dimensional Hamiltonian systems. Phys. Rev. Lett. **108**, 180601 (2012)
38. Narayan, O., Ramaswamy, S.: Anomalous heat conduction in one-dimensional momentum conserving systems. Phys. Rev. Lett. **89**, 200601 (2002)
39. Mendl, C.B., Spohn, H.: Dynamic correlators of FPU chains and nonlinear fluctuating hydrodynamics. Phys. Rev. Lett. **111**, 230601 (2013)
40. van Beijeren, H., Kutner, R., Spohn, H.: Excess noise for driven diffusive systems. Phys. Rev. Lett. **54**, 2026–2029 (1985)
41. Uchaikin, V., Zolotarev, V.: Chance and Stability, Stable Distributions and Applications. W. de Gruyter, Berlin (1999)

42. Spohn, H., Stoltz, G.: Nonlinear fluctuating hydrodynamics in one dimension: the case of two conserved fields. J. Stat. Phys. **160**, 861–884 (2015)
43. Popkov, V., Schadschneider, A., Schmidt, J., Schütz, G.M.: The Fibonacci family of dynamical universality classes. Proc. Natl. Acad. Sci. **112**, 12645–12650 (2015)
44. Campbell, D.K., Rosenau, P., Zaslavsky, G.: Introduction: the Fermi-Pasta-Ulam problem – the first fifty years. Chaos **1**, 015101 (2005)
45. Gallavotti, G.: The Fermi-Pasta-Ulam Problem: A Status Report. Lecture Notes in Physics, vol. 728. Springer, Berlin (2008)
46. Dauxois, T., Peyrard, M., Ruffo, S.: The Fermi-Pasta-Ulam "numerical experiment": history and pedagogical perspectives. Eur. J. Phys. **26**, S3–S11 (2005)
47. Izrailev, F.M., Chirikov, B.V.: Statistical properties of a nonlinear string. Sov. Phys. Dokl. **11**, 30–31 (1966)
48. Lepri, S., Livi, R., Politi, A.: Heat conduction in chains of nonlinear oscillators. Phys. Rev. Lett. **78**, 1896–1899 (1997)
49. Lepri, S., Livi, R., Politi, A.: Thermal conduction in classical low-dimensional lattices. Phys. Rep. **377**, 1–80 (2003)
50. Dhar, A.: Heat transport in low-dimensional systems. Adv. Phys. **57**, 457–537 (2008)
51. Hatano, T.: Heat conduction in the diatomic Toda lattice revisited. Phys. Rev. E **59**, R1–R4 (1999)
52. Grassberger, P., Nadler, W., Yang, L.: Heat conduction and entropy production in a one-dimensional hard-particle gas. Phys. Rev. Lett. **89**, 180601 (2002)
53. Lee-Dadswell, G.R., Nickel, B.G., Gray, C.G.: Thermal conductivity and bulk viscosity in quartic oscillator chains. Phys. Rev. E **72**, 031202 (2005)
54. Prosen, T., Campbell, D.K.: Normal and anomalous heat transport in one-dimensional classical lattices. Chaos **15**, 015117 (2005)
55. Zhao, H.: Identifying diffusion processes in one-dimensional lattices in thermal equilibrium. Phys. Rev. Lett. **96**, 140602 (2006)
56. Chen, S., Zhang,Y., Wang, J., Zhao, H.: Diffusion of heat, energy, momentum and mass in one-dimensional systems. Phys. Rev. E **87**, 032153 (2013)
57. Straka, M.: KPZ scaling in the one-dimensional FPU $\alpha-\beta$ model. Master's thesis, University of Florence (2013)
58. Das, S.G., Dhar, A., Saito, K., Mendl, C.B., Spohn, H.: Numerical test of hydrodynamic fluctuation theory in the Fermi-Pasta-Ulam chain. Phys. Rev. E **90**, 012124 (2014)
59. Mendl, C.B., Spohn, H.: Equilibrium time-correlation functions for one-dimensional hard-point systems. Phys. Rev. E **90**, 012147 (2014)
60. Delfini, L., Denisov, S., Lepri, S., Livi, R., Mohanty, P.K.: Energy diffusion in hard-point systems. Eur. Phys. J. **146**, 21–35 (2007)
61. Das, S.G., Dhar, A., Narayan, O.: Heat conduction in the α-β-Fermi-Pasta-Ulam chain. J. Stat. Phys. **154**, 204–213 (2013)
62. Benettin, G., Ponno, A., Christodoulidi, H.: The Fermi–Pasta–Ulam problem and its underlying integrable dynamics. J. Stat. Phys. **152**, 195–212 (2013)
63. Popkov, V., Schmidt, J., Schütz, G.: Non-KPZ modes in two-species driven diffusive systems. Phys. Rev. Lett. **112**, 200602 (2014)
64. Arndt, P.F., Heinzel, T., Rittenberg, V.: Spontaneous breaking of translational invariance and spatial condensation in stationary states on a ring. I. The neutral system. J. Stat. Phys. **97**, 1–65 (1999)
65. Ferrari, P.L., Sasamoto, T., Spohn, H.: Coupled Kardar-Parisi-Zhang equations in one dimension. J. Stat. Phys. **153**, 377–399 (2013)
66. Bernardin, C., Gonçalves, P., Jara, M.: $3/4$-superdiffusion in a system of harmonic oscillators perturbed by a conservative noise. Arch. Rat. Mech. Anal. online (2015). arXiv:1402.1562
67. Jara, M., Komorowski, T., Olla, S.: Superdiffusion of energy in a chain of harmonic oscillators with noise. Commun. Math. Phys. **339**, 407–453 (2015)
68. Mendl, C.B., Spohn, H.: Current fluctuations for anharmonic chains in thermal equilibrium. J. Stat. Mech. **2015**, 03007 (2015)

69. Lee-Dadswell, G.R.: Universality classes for thermal transport in one-dimensional oscillator chains. Phys. Rev. E **91**, 032102 (2015)
70. Aoki, K., Kusnezov, D.: Non-equilibrium statistical mechanics of classical lattice ϕ^4 field theory. Ann. Phys. **295**, 50–80 (2002)
71. Li, Y., Liu, S., Li, N., Hänggi, P., Li, B.: 1D momentum-conserving systems: the conundrum of anomalous versus normal heat transport. New J. Phys. **17**, 043064 (2015)
72. Das, S.G., Dhar, A.: Role of conserved quantities in normal heat transport in one dimension (2014). arXiv:1411.5247
73. Giardinà, C., Livi, R., Politi, A., Vassalli, M.: Finite thermal conductivity in 1d lattices. Phys. Rev. Lett. **84**, 2144–2147 (2000)
74. Gendelman, O.V., Savin, A.V.: Normal heat conductivity of the one-dimensional lattice with periodic potential of nearest-neighbor interaction. Phys. Rev. Lett. **84**, 2381–2384 (2000)
75. Iubini, S., Lepri, S., Livi, R., Politi, A.: Off-equilibrium dynamics of the discrete nonlinear Schrödinger chain. J. Stat. Mech. **2013**, 08017 (2013)
76. Iubini, S., Lepri, S., Politi, A.: Coarsening dynamics in a simplified DNLS model (2013). J. Stat. Phys. **154**, 1057–1073 (2014)
77. Mendl, C.B., Spohn, H.: The low temperature dynamics of the one-dimensional discrete nonlinear Schrödinger equation. J. Stat. Mech. **2015**, 08028 (2015)
78. Iubini, S., Lepri, S., Politi, A.: Nonequilibrium discrete nonlinear Schrödinger equation. Phys. Rev. E **86**, 011108 (2012)
79. Kulkarni, M., Lamacraft, A.: Finite-temperature dynamical structure factor of the one-dimensional Bose gas: from the Gross-Pitaevskii equation to the Kardar-Parisi-Zhang universality class of dynamical critical phenomena. Phys. Rev. A **88**, 021603(R) (2013)
80. Kulkarni, M., Huse, D., Spohn, H.: Fluctuating hydrodynamics for a discrete Gross-Pitaevskii equation: mapping to Kardar-Parisi-Zhang universality class. Phys. Rev. A **92**, 043612 (2015)

Chapter 4
Kinetic Theory of Phonons in Weakly Anharmonic Particle Chains

Jani Lukkarinen

Abstract The aim of the chapter is to develop the kinetic theory of phonons in classical particle chains to a point which allows comparing the kinetic theory of normally conducting chains, with an anharmonic pinning potential, to the kinetic theory of the anomalously conducting FPU chains. In addition to reviewing the related literature, the chapter contains a streamlined derivation of the phonon Boltzmann collision operators using Wick polynomials, as well as details about the estimates which are needed to study the effect of the collision operator. This includes explicit solutions of the collisional constraints, both with and without harmonic pinning. We also recall in detail the derivation of the Green–Kubo formula for thermal conductivity in these systems, and the relation between entropy and the Boltzmann H-theorem for the phonon Boltzmann equations. The focus is in systems which are spatially translation invariant perturbations of thermal equilibrium states. We apply the results to obtain detailed predictions from kinetic theory for the Green–Kubo correlation functions, and hence the thermal conductivities, of the chain with a quartic pinning potential as well as the standard FPU-β chain.

4.1 Kinetic Scaling Limit for Weakly Anharmonic Chains

4.1.1 Introduction

Kinetic theory describes motion which is *transport dominated* in the sense that typically the solutions to the kinetic equations correspond to constant velocity, i.e., *ballistic*, motion intercepted by *collisions* whose frequency is order one on the kinetic space-time scales. The constant velocity part of the transport, the *free streaming motion*, can arise via several different microscopic mechanisms, the obvious case being free motion of classical particles. Here, we are interested in evolution of energy density in chains composed out of locally interacting classical

J. Lukkarinen (✉)
Department of Mathematics and Statistics, University of Helsinki, P.O. Box 68, FI-00014 Helsingin yliopisto, Finland
e-mail: jani.lukkarinen@helsinki.fi

S. Lepri (ed.), *Thermal Transport in Low Dimensions*, Lecture Notes in Physics 921, DOI 10.1007/978-3-319-29261-8_4

particles, in the limit where the interactions are dominated by a stable harmonic potential. For such models, the energy transport is dominated by the free streaming of *phonons* which correspond to eigenmodes of the motion generated by the harmonic potential.

Our goal is to derive and study the properties of kinetic theory given by a phonon Boltzmann equation. Even with most generous estimates for the terms neglected in its derivation, the Boltzmann equation can become an exact description of the energy transport only in a *kinetic scaling limit* which produces a total scale separation between the free streaming motion of phonons and the collisions induced by the perturbation. For the above particle chains this amounts to studying lengths and time intervals both of which are proportional to λ^{-2}, where λ is the strength of the anharmonic perturbation, and then taking the limit $\lambda \to 0$.

In particular in one dimension, i.e., for particle chains, it is difficult to estimate reliably the error terms associated with the approximation of the original motion by the phonon Boltzmann equation. At present, no such full mathematical analysis has been achieved. However, there are instances in which the one-dimensional phonon Boltzmann equation has proven to be useful: in Sect. 4.3.4, the thermal conductivity predicted by the Boltzmann equation of systems with an anharmonic *onsite* potential is compared to numerical simulations, and the results are found to agree within numerical accuracy for sufficiently weak couplings.

Although just one example, the agreement is significant since it shows that, under proper conditions, the kinetic theory from the phonon Boltzmann equation can produce meaningful predictions even for one-dimensional transport problems. As will be apparent later, computing these predictions requires some additional effort compared to the standard rarefied gas Boltzmann equation. The effort is rewarded in a prediction which has *no adjustable parameters* and hence can be compared directly with experiments and numerical simulations.

The drawback of the phonon Boltzmann equation is that, a priori, it only describes the lowest order effects of the perturbation. Hence, it might not capture all processes relevant to the energy transport. In addition, since it is not known which precise conditions guarantee that the corrections are truly vanishing in the kinetic scaling limit, one has to leave the option open that even the lowest order effects are not being described accurately by the equation.

A discussion about the physical and mathematical conditions under which the phonon Boltzmann equation should be an accurate approximation is given in [21], while [23] could serve as a textbook reference on physics of phonons. The present rigorous results on linear and nonlinear perturbations of wave-like evolution equations support the basic principle that for sufficiently dispersive systems, in particular for crystals in three or higher dimensions, the derivation of the phonon Boltzmann equation yields an accurate approximation to the behaviour of the system at kinetic time-scales with weak coupling. The argument is particularly convincing for spatially homogeneous systems and for linear perturbations such as when the particle masses have a weak random disorder [4, 14], and for the closely related quantum models, random Schrödinger equation [6, 7] and the Anderson model [5]. In addition, it has been proven that the Wigner transforms of unperturbed wave-equations do, in great generality, converge to the collisionless Boltzmann equation [8, 18], although acoustic modes may require somewhat more refined treatment [9]. Nonlinear perturbations are more intricate, and we refer to [2, 16] for recent results on the rigorous analysis of their kinetic scaling limits.

Quite often the addition of the collision operator to the transport equation leads to loss of "memory" at each collision and eventual equilibration via diffusion: this would correspond to normal energy transport by phonons. However, if the transport is anomalous, it is highly likely to be reflected also in the solutions of the corresponding kinetic equation. An example of such behaviour is given by the FPU-β chain. The aim of this contribution is to develop the general kinetic theory of classical particle chains to the point which allows to compare the kinetic theory of normally conducting chains with an anharmonic pinning potential and the anomalously conducting FPU chains.

The first section concerns the derivation of the phonon Boltzmann equation in the one-dimensional case, for general dispersion relations and third and fourth order perturbations of the harmonic potential. We begin with the infinite volume harmonic model in Sect. 4.1.2, and make a brief comment on the finite volume case and the relevance of boundary conditions in Sect. 4.1.2.1. The anharmonic perturbations are discussed in Sect. 4.1.3 where we also explain our choice for the definition of the related energy density and derive the related energy current observables.

The role of the proper choice of initial data for the derivation of the Boltzmann equation cannot be stressed enough. We describe our choices in Sect. 4.1.4, and introduce the notations needed in the analysis of the cumulants of the phonon fields. We conclude the first section with a derivation of the Green–Kubo formula for the particle chains, under the assumption that energy density satisfies a closed diffusion equation. (This would correspond to a case with only one locally conserved field.)

The first section concerns fairly general particle chains, but in order to derive the phonon Boltzmann equation, further restrictions are required. In Sect. 4.2.1, we do the derivation in detail for the FPU chains, and sketch it for the onsite anharmonic perturbations. The rest of the section concerns only chains with stable *nearest neighbour* harmonic couplings. We explain why the nearest neighbour couplings do not produce any collisions at the kinetic time-scale from the third order terms of the potential (Sect. 4.2.2.1), and show that only phonon number conserving collisions appear from the fourth order terms (Sect. 4.2.2.2). To use the collision operator, one needs to resolve the related collisional constraints. This is done for FPU-chains in Sect. 4.2.2.3 and for the chains with a harmonic pinning potential in Sect. 4.2.2.4.

The final section is devoted to the analysis of the resulting class of phonon Boltzmann equations. We first show how an assumption of an increase of microscopic entropy in homogeneous models leads to an H-theorem for the phonon Boltzmann equation. In Sect. 4.3.2, it is explained how the H-theorem can be used to classify all steady states of the Boltzmann equation. The kinetic description derived in Sect. 4.1.5 for the time correlation function in the Green–Kubo formula is applied to these models in Sect. 4.3.3. This results in explicit predictions for the leading asymptotics of the decay of the correlations. The decay is found to be integrable for the onsite perturbation, and we make a comparison of the resulting thermal conductivities in Sect. 4.3.4. We discuss the corresponding prediction in FPU models in Sect. 4.3.5. The result is decay $O(t^{-3/5})$, which predicts anomalous conduction in the FPU chains.

More comments and discussion about the results are given in Sect. 4.4, before the Acknowledgements and References.

4.1.2 Free Motion of Phonons

Phonons correspond to eigenmodes of harmonic Bravais lattices. The general case includes multi-component phonon fields which may arise both from the dimensionality of the lattice—the evolution equation could describe *displacements* from some reference lattice positions and thus have d components in d dimensions—or from the reference cell of periodicity containing more than one particle. An example of the latter is provided by the one-dimensional chain with alternating masses: then the reference system remains invariant only under shifts over two lattices sites which corresponds to a Bravais lattice with two particles per cell of periodicity. For notational simplicity, let us here only consider the case of one-component classical phonon fields and assume that all particles have a unit mass (this is always possible to achieve for one-component fields by choosing a suitable time-scale).

Then the state of the system at time t is determined by $(q_x(t), p_x(t))$, $x \in \mathbb{Z}$, which satisfy the evolution equations

$$\dot{q}_x(t) = p_x(t), \quad \dot{p}_x(t) = -\sum_y \alpha(x-y)q_y(t) . \tag{4.1}$$

The equations are of a Hamiltonian form, corresponding to the Hamiltonian function

$$H(p,q) = \sum_{x\in\mathbb{Z}} \tfrac{1}{2}p_x^2 + \sum_{x,y\in\mathbb{Z}} \tfrac{1}{2}q_x\alpha(x-y)q_y . \tag{4.2}$$

We suppose that the harmonic interactions have a *short range*. For instance, only finitely many $\alpha(z)$ are different from zero, or $|\alpha(z)|$ decreases exponentially fast to zero as $|z| \to \infty$.

The evolution equation (4.1) can be solved by using Fourier-transform. We define the Fourier-fields as $\hat{q}(t,k) = \sum_{x\in\mathbb{Z}} \mathrm{e}^{-\mathrm{i}2\pi kx}q_x(t)$ and $\hat{p}(t,k) = \sum_{x\in\mathbb{Z}} \mathrm{e}^{-\mathrm{i}2\pi kx}p_x(t)$. Here $k \in \mathbb{T}$ where $\mathbb{T}$ denotes the one-torus which we parametrize using the interval $[-\frac{1}{2}, \frac{1}{2}]$ and then identify its endpoints, $-\frac{1}{2}$ and $\frac{1}{2}$. The evolution equation for the Fourier-fields then reads, in a matrix form,

$$\frac{\mathrm{d}}{\mathrm{d}t}\begin{pmatrix}\hat{q}(t,k)\\ \hat{p}(t,k)\end{pmatrix} = \begin{pmatrix}0 & 1\\ -\hat{\alpha}(k) & 0\end{pmatrix}\begin{pmatrix}\hat{q}(t,k)\\ \hat{p}(t,k)\end{pmatrix} . \tag{4.3}$$

For each k, the equation is easily solved by Jordan decomposition of the matrix on the right hand side. This provides two independently evolving eigenmodes $a_t(k,\sigma)$, $\sigma = \pm 1$, as long as $\hat{\alpha}(k) \neq 0$. For those k with $\hat{\alpha}(k) = 0$, the solution is given by constant speed increase: $\hat{p}(t,k) = \hat{p}(0,k)$ and $\hat{q}(t,k) = \hat{q}(0,k) + t\hat{p}(0,k)$. Typically, this can happen at most at a finite number of points whose neighbourhoods might require special treatment in the kinetic theory of phonons.

The eigenmode fields $a_t(k,\sigma)$, which we call *phonon modes* of the harmonic chain, satisfy the evolution equation

$$\frac{\mathrm{d}}{\mathrm{d}t} a_t(k,\sigma) = -\mathrm{i}\sigma\omega(k) a_t(k,\sigma)\,, \tag{4.4}$$

where $\omega(k) = \sqrt{\hat{\alpha}(k)}$. Thus the square root of the Fourier-transform of α determines the *dispersion relation* ω of the phonons, and we arbitrarily choose the principal branch of the square root so that always $\mathrm{Re}\,\omega(k) \geq 0$. There are several possible ways to normalize the eigenmodes. Here we employ the following choice, standard to phonon physics,

$$a_t(k,\sigma) = \frac{1}{\sqrt{2\omega(k)}} \left(\omega(k)\hat{q}(t,k) + \mathrm{i}\sigma\hat{p}(t,k)\right), \quad \sigma \in \{\pm 1\},\ k \in \mathbb{T}\,. \tag{4.5}$$

It is straightforward to check that these fields indeed satisfy the evolution equation (4.4) whenever $\omega(k) \neq 0$. In addition, if we choose a solution to (4.4) and then define

$$\hat{q}(t,k) = \frac{1}{\sqrt{2\omega(k)}} \sum_\sigma a_t(k,\sigma) \quad \text{and} \quad \hat{p}(t,k) = \frac{\sqrt{2\omega(k)}}{2} \sum_\sigma (-\mathrm{i}\sigma) a_t(k,\sigma)\,, \tag{4.6}$$

then $(\hat{q}(t,k), \hat{p}(t,k))$ yields a solution to (4.3). Given some initial field a_0, the solution of (4.4) is straightforward, yielding $a_t(k,\sigma) = \mathrm{e}^{-\mathrm{i}\sigma\omega(k)t} a_0(k,\sigma)$. Therefore, if we let a_0 to be determined by the initial data $(\hat{q}(0,k), \hat{p}(0,k))$, we can now obtain the solution to the original evolution problem (4.3) for every k for which $\hat{\alpha}(k) \neq 0$, just by inserting $a_t(k,\sigma) = \mathrm{e}^{-\mathrm{i}\sigma\omega(k)t} a_0(k,\sigma)$ in (4.6).

For later use, it is also important to assume that the harmonic interactions are *stable*, i.e., they do not have any solutions which increase exponentially in time. By the above discussion, this is equivalent to assuming that the Fourier-transform of α is pointwise non-negative, $\hat{\alpha}(k) \geq 0$ for all $k \in \mathbb{T}$. A standard example is given by the *nearest neighbour interactions* for which $\alpha(z) = 0$ if $|z| > 1$. A short computation reveals that the only stable nearest neighbour dispersion relations are given by $\omega(k) = \overline{\omega}(1 - 2\delta\cos(2\pi(k - k_0)))^{1/2}$ for some $\overline{\omega} > 0$, $|\delta| \leq \frac{1}{2}$ and $k_0 \in \mathbb{T}$. Since for phonons we also require that $\alpha(z)$ are real, we can also always choose $k_0 = 0$ above.

The parameter $\overline{\omega}$ determines the average angular frequency of the oscillations, the maximum value being $\overline{\omega}\sqrt{1+2|\delta|}$ and the minimum $\overline{\omega}\sqrt{1-2|\delta|}$. If $|\delta| < \frac{1}{2}$, the function $\omega(k)$ is thus an even, real analytic, function. These systems describe *optical phonon modes*, and one can also think of the system as having *harmonic pinning*. If $\delta < 0$, it is possible to make a change of variables which reverses its sign to positive: this is achieved with a shift by $\frac{1}{2}$ in the Fourier variable, in other words, by using $((-1)^x q_x, (-1)^x p_x)$ as the new lattice fields. Enforcing this convention has the benefit that then the slowest evolving fields are also spatially slowly varying, corresponding to small values of k. Therefore, we only consider the cases $k_0 = 0$ and $0 < \delta \leq \frac{1}{2}$ in the following, excluding also the degenerate constant dispersion relation corresponding to $\delta = 0$.

If $\delta = \frac{1}{2}$ one has $\omega(k) = \overline{\omega}\sqrt{2}|\sin(\pi k)|$, with an $|k|$-singularity near the origin. These systems describe *acoustic phonon modes* and the seemingly innocuous singularity has important implications for the energy transport properties of the system. The FPU-chains belong to this category.

Finally, let us point out that in the stable case, $\omega(k) \geq 0$, the phonon fields defined by (4.5) are directly connected to the energy of the harmonic system by $\frac{1}{2}\sum_\sigma \int \mathrm{d}k\, \omega(k)|a_t(k,\sigma)|^2 = H(p(t), q(t))$. Hence, in analogy to quantum mechanics, the complex phonon fields $a_t(k,\sigma)$ can be thought of as wave functions of phonons each of which carries an energy $\omega(k)$.

4.1.2.1 Effect of Boundary Conditions in Finite Non-periodic Chains

For notational simplicity, the above discussion uses formally an infinite lattice setup. Its proper interpretation is to consider the infinite system to be an approximation for a finite, but large system. This identification has an important mathematical consequence: the standard methods of ℓ^p-spaces do not immediately apply to the infinite system. For instance, typical samples from infinite volume Gibbs states are only logarithmically bounded at infinity [10]. Hence, should the need arise, it is better to go back to the finite systems to resolve any possible issues about existence of solutions or to handle singularities. (For discussion about the existence and properties of the infinite volume dynamics, see [3, 10].)

In a finite system, the choice of boundary conditions begins to play a role. The above computations involving Fourier-transforms can be given an exact correspondence by choosing a box with periodic boundary conditions. In this case, the wave evolution can move energy around the lattice unimpeded, as indicated by the solutions which are given by multiplication operators in the Fourier-space. One can find more details about the correspondence, for instance, in [16].

If the finite lattice does not have periodic boundary conditions, the boundaries will influence the transport of phonons. For boundary conditions which preserve energy, such as Dirichlet and Neumann, the effect can be understood as a reflection of phonons at the boundary. It is also possible to combine these with partial periodic transmission through the boundary.

The effect of standard heat baths, such as Langevin heat baths, on the kinetic equations is more difficult to analyse. These will introduce also absorption and source terms to the phonon evolution equations, but since the heat baths are typically acting only on a few particles of the chain, their coupling to the wave evolution is nontrivial and somewhat singular. (In general, it is difficult to affect the long wavelength part of the evolution by any local change to the system.) Unfortunately, a systematic analysis of these effects appears to be missing from the current literature.

4.1.3 Particle Chains with Anharmonic Perturbations

For the discussion about the effect of local nonlinear perturbations to the above wave equation, let us introduce, in addition to the harmonic coupling function $\alpha : \mathbb{Z} \to \mathbb{R}$, its second and third polynomial counterparts, $\alpha_3 : \mathbb{Z}^2 \to \mathbb{R}$ and $\alpha_4 : \mathbb{Z}^3 \to \mathbb{R}$. These are assumed to be "small" in the sense that for large microscopic times the evolution is dominated by the linear term. Explicitly, we take the perturbed evolution equations to be

$$\begin{aligned} \dot{q}_x(t) &= p_x(t)\,, \\ \dot{p}_x(t) &= -\sum_y \alpha(y) q_{x-y}(t) - \sum_{y_1,y_2} \alpha_3(y_1,y_2) q_{x-y_1}(t) q_{x-y_2}(t) \\ &\quad - \sum_{y_1,y_2,y_3} \alpha_4(y_1,y_2,y_3) q_{x-y_1}(t) q_{x-y_2}(t) q_{x-y_3}(t)\,. \end{aligned} \tag{4.7}$$

This corresponds to Hamiltonian evolution with the interaction potential

$$\begin{aligned} \mathscr{V}(q) &= \frac{1}{2}\sum_{x,y} \alpha(y) q_x q_{x-y} + \frac{1}{3}\sum_{x,y_1,y_2} \alpha_3(y_1,y_2) q_x q_{x-y_1}(t) q_{x-y_2}(t) \\ &\quad + \frac{1}{4}\sum_{x,y_1,y_2,y_3} \alpha_4(y_1,y_2,y_3) q_x q_{x-y_1}(t) q_{x-y_2}(t) q_{x-y_3}(t)\,, \end{aligned} \tag{4.8}$$

supposing, as we will do here, that the following symmetries hold:

$$\alpha(y) = \alpha(-y)\,, \tag{4.9}$$

$$\alpha_3(y_1,y_2) = \alpha_3(y_2,y_1) = \alpha_3(-y_1, y_2 - y_1)\,, \tag{4.10}$$

$$\alpha_4(y_1,y_2,y_3) = \alpha_4(y_2,y_1,y_3) = \alpha_4(y_1,y_3,y_2) = \alpha_4(-y_1, y_2 - y_1, y_3 - y_1)\,. \tag{4.11}$$

The symmetries arise from assuming that the coefficients correspond to a generic label-exchange symmetric, translation invariant potential. We have also used here the fact that any permutation can be expressed as a product of adjacent transpositions. Therefore, it suffices to check symmetry with respect to the transpositions to get full permutation invariance.

For example, the standard FPU-chain is given by

$$\mathscr{V}_{\rm FPU}(q) = \sum_x U(q_x - q_{x-1}), \quad U(r) = \frac{1}{4}\varpi^2 r^2 + \lambda_3 \frac{1}{3} r^3 + \lambda_4 \frac{1}{4} r^4, \tag{4.12}$$

and it satisfies the evolution equation $\ddot{q}_x = U'(q_{x+1} - q_x) - U'(q_x - q_{x-1})$. A brief computation, using the periodicity, shows that the FPU chain corresponds to choosing the coefficient functions above so that their nonzero values are

$$\alpha(y) = \frac{1}{2}\varpi^2 \begin{cases} 2, & y = 0 \\ -1, & y \in \{\pm 1\} \end{cases}, \tag{4.13}$$

$$\alpha_3(y_1, y_2) = \lambda_3 \begin{cases} 1, & y_1 = y_2 = 1,\ y_1 = -1, y_2 = 0, \text{or } y_1 = 0, y_2 = -1 \\ -1, & y_1 = y_2 = -1,\ y_1 = 1, y_2 = 0, \text{or } y_1 = 0, y_2 = 1 \end{cases}, \tag{4.14}$$

$$\alpha_4(\mathbf{y}) = \lambda_4 \begin{cases} 2, & y_1 = y_2 = y_3 = 0 \\ 1, & \pm\mathbf{y} \in \{(0,1,1), (1,0,1), (1,1,0)\}, \\ -1, & \pm\mathbf{y} \in \{(1,1,1), (1,0,0), (0,1,0), (0,0,1)\} \end{cases}. \tag{4.15}$$

For later use, let us also record their Fourier-transforms. Using the shorthand notations $p_i = 2\pi k_i$, $i = 1, 2, 3$, they are given by

$$\hat{\alpha}(k_1) = \varpi^2(1 - \cos p_1) = 2\varpi^2 \sin^2 \frac{p_1}{2}, \tag{4.16}$$

$$\hat{\alpha}_3(k_1, k_2) = \mathrm{i}\lambda_3 2^3 \sin \frac{p_1 + p_2}{2} \sin \frac{p_1}{2} \sin \frac{p_2}{2}, \tag{4.17}$$

$$\hat{\alpha}_4(k_1, k_2, k_3) = \lambda_4 2\, \mathrm{Re}\Big[\prod_{\ell=1}^{3}(1 - \mathrm{e}^{-\mathrm{i}p_\ell})\Big] = -\lambda_4 2^4 \sin \frac{p_1 + p_2 + p_3}{2} \prod_{\ell=1}^{3} \sin \frac{p_\ell}{2}. \tag{4.18}$$

Onsite perturbations also belong to the above category. For instance,

$$\mathscr{V}_{\rm OS}(q) = \sum_{x,y\in\mathbb{Z}} \tfrac{1}{2} q_x \alpha(x-y) q_y + \sum_x V(q_x), \quad V(q) = \lambda_3 \frac{1}{3} q^3 + \lambda_4 \frac{1}{4} q^4, \tag{4.19}$$

has

$$\alpha_3(y_1, y_2) = \lambda_3 \mathbb{1}(y_1 = y_2 = 0)\,, \tag{4.20}$$

$$\alpha_4(y_1, y_2, y_3) = \lambda_4 \mathbb{1}(y_1 = y_2 = y_3 = 0)\,. \tag{4.21}$$

Both Fourier-transforms are thus constant: $\hat{\alpha}_3(k_1, k_2) = \lambda_3$ and $\hat{\alpha}_4(k_1, k_2, k_3) = \lambda_4$.

After these examples, let us come back to the general case, Eq. (4.7). Taking a Fourier-transform yields

$$\begin{aligned}
\frac{\mathrm{d}}{\mathrm{d}t}\hat{q}(t,k) &= \hat{p}(t,k)\,,\\
\frac{\mathrm{d}}{\mathrm{d}t}\hat{p}(t,k) &= -\hat{\alpha}(k)\hat{q}(t,k) - \int_{\mathbb{T}^2} \mathrm{d}k_1'\mathrm{d}k_2'\, \delta_{\mathbb{T}}(k - k_1' - k_2')\hat{\alpha}_3(k_1', k_2')\hat{q}(t,k_1')\hat{q}(t,k_2')\\
&\quad - \int_{\mathbb{T}^3} \mathrm{d}k_1'\mathrm{d}k_2'\mathrm{d}k_3'\, \delta_{\mathbb{T}}(k - \sum_{\ell=1}^{3} k_\ell')\hat{\alpha}_4(\mathbf{k}') \prod_{\ell=1}^{3} \hat{q}(t,k_\ell')\,,
\end{aligned} \tag{4.22}$$

where we have denoted the periodic δ-function by $\delta_{\mathbb{T}}$. We will drop the subscript in the following and merely use δ. This implies that the phonon fields, defined still by (4.5), now satisfy the evolution equation

$$\begin{aligned}
\frac{\mathrm{d}}{\mathrm{d}t}a_t(k_0,\sigma_0) &= -\mathrm{i}\sigma_0\omega(k_0)a_t(k_0,\sigma_0)\\
&\quad - \mathrm{i}\sigma_0 \sum_{\sigma_1,\sigma_2\in\{\pm1\}} \int_{\mathbb{T}^2} \mathrm{d}k_1\mathrm{d}k_2\, \delta(k_0 - k_1 - k_2)\\
&\quad \times \Phi_3(k_0,k_1,k_2)a_t(k_1,\sigma_1)a_t(k_2,\sigma_2)\\
&\quad - \mathrm{i}\sigma_0 \sum_{\sigma_1,\sigma_2,\sigma_3\in\{\pm1\}} \int_{\mathbb{T}^3} \mathrm{d}k_1\mathrm{d}k_2\mathrm{d}k_3\, \delta(k_0 - \sum_{\ell=1}^{3} k_\ell)\\
&\quad \times \Phi_4(k_0,k_1,k_2,k_3) \prod_{\ell=1}^{3} a_t(k_\ell,\sigma_\ell)\,,
\end{aligned} \tag{4.23}$$

where the interaction amplitude functions are

$$\Phi_3(k_0,k_1,k_2) = \hat{\alpha}_3(k_1,k_2) \prod_{\ell=0}^{2} \frac{1}{(2\omega(k_\ell))^{\frac{1}{2}}}\,, \tag{4.24}$$

$$\Phi_4(k_0,k_1,k_2,k_3) = \hat{\alpha}_4(k_1,k_2,k_3) \prod_{\ell=0}^{3} \frac{1}{(2\omega(k_\ell))^{\frac{1}{2}}}\,. \tag{4.25}$$

Since the δ-function enforces the sum of the integration variables to be equal to k_0, we can thus employ the following definitions for the FPU-chains

$$\Phi_3(k_0, k_1, k_2) = \mathrm{i}\lambda_3 2^{\frac{3}{4}} \overline{\omega}^{-\frac{3}{2}} \prod_{\ell=0}^{2} g(k_\ell)\,, \tag{4.26}$$

$$\Phi_4(k_0, k_1, k_2, k_3) = -2\lambda_4 \overline{\omega}^{-2} \prod_{\ell=0}^{3} g(k_\ell)\,, \tag{4.27}$$

where $g(k) = \mathrm{sign}(k)\sqrt{|\sin(\pi k)|}$ for $|k| \leq \frac{1}{2}$.

To study energy transport, it is necessary to split the total energy into local components. The harmonic energy, corresponding to the energy of free phonons, is most conveniently distributed using a form which is symmetric in Fourier-space. By the definition (4.5), we have for any k_1, k_2

$$\sqrt{\omega(k_1)\omega(k_2)} \sum_{\sigma=\pm 1} a(k_1, -\sigma) a(k_2, \sigma) = \omega(k_1)\omega(k_2)\hat{q}(k_1)\hat{q}(k_2) + \hat{p}(k_1)\hat{p}(k_2)\,. \tag{4.28}$$

Thus, if we denote the inverse Fourier transform of ω by $\tilde{\omega}$, we can define

$$\begin{aligned} H_x^{(0)}(p, q) &= \frac{1}{2} p_x^2 + \frac{1}{2}\left(\sum_y \tilde{\omega}(y) q_{x-y}\right)^2 \\ &= \frac{1}{2} \sum_{\sigma=\pm 1} \int_{\mathbb{T}^2} \mathrm{d}k\, \mathrm{e}^{\mathrm{i}2\pi x(k_1+k_2)} \Phi_2(k) a(k_1, -\sigma) a(k_2, \sigma)\,, \end{aligned} \tag{4.29}$$

where $\Phi_2(k_1, k_2) = \sqrt{\omega(k_1)\omega(k_2)}$. Since ω is real and symmetric, we have here $\tilde{\omega}(x) \in \mathbb{R}$. Thus $H_x^{(0)} \geq 0$ and $\sum_x H_x^{(0)}(p, q)$ is equal to the harmonic part of the total energy, to $\frac{1}{2}\sum_x p_x^2 + \frac{1}{2}\sum_{x,y} \frac{1}{2} q_x \alpha(x-y) q_y$. Thus this choice allows distributing the positive total harmonic energy into positive contributions localized at each lattice site.

For the anharmonic terms of the potential energy we use the form suggested by the notation in (4.8), and define the local energy at site x by

$$\begin{aligned} H_x(p, q) &= H_x^{(0)}(p, q) + \frac{1}{3} \sum_{y_1, y_2} \alpha_3(y_1, y_2) q_x q_{x-y_1} q_{x-y_2} \\ &\quad + \frac{1}{4} \sum_{y_1, y_2, y_3} \alpha_4(y_1, y_2, y_3) q_x q_{x-y_1} q_{x-y_2} q_{x-y_3} \\ &= \frac{1}{2} \sum_{\sigma=\pm 1} \int_{\mathbb{T}^2} \mathrm{d}k\, \mathrm{e}^{\mathrm{i}2\pi x(k_1+k_2)} \Phi_2(k) a(k_1, -\sigma) a(k_2, \sigma) \end{aligned}$$

$$+\frac{1}{3}\sum_{\sigma\in\{\pm1\}^3}\int_{\mathbb{T}^3}\mathrm{d}k\,\mathrm{e}^{\mathrm{i}2\pi x\sum_{\ell=1}^{3}k_\ell}\Phi_3(k)\prod_{\ell=1}^{3}a(k_\ell,\sigma_\ell)$$

$$+\frac{1}{4}\sum_{\sigma\in\{\pm1\}^4}\int_{\mathbb{T}^4}\mathrm{d}k\,\mathrm{e}^{\mathrm{i}2\pi x\sum_{\ell=1}^{4}k_\ell}\Phi_4(k)\prod_{\ell=1}^{4}a(k_\ell,\sigma_\ell)\,. \tag{4.30}$$

Then clearly $\sum_x H_x(p,q) = \frac{1}{2}\sum_x p_x^2 + \mathscr{V}(q) = H(p,q)$, and H_x is a local function at x (it depends mainly on fields near the point x). In addition, H_x is translated just like the fields if these are shifted by x_0: if we choose some $x_0 \in \mathbb{Z}$ and define $\tilde{q}_x = q_{x+x_0}$ and $\tilde{p}_x = p_{x+x_0}$, then $H_x(\tilde{q},\tilde{p}) = H_{x+x_0}(q,p)$.

There are various possible choices for how to split the total energy into local energy density. The above definition of energy density is perhaps not the most standard one, but it has two appealing features for those systems which are dominated by phonon transport. First, it has simple algebraic dependence on the phonon eigenmode fields. Secondly, its dependence on the position variable x is located entirely in the Fourier-factor. This allows a definition of a current observable with a simple dependence on the particle interactions and the simplest choice for the associated discrete derivative.

Explicitly, consider some $k \neq 0$ and any $x_1, x_2 \in \mathbb{Z}$ with $x_1 \leq x_2$. Setting $y = x_2 - x_1$, we then have $y \geq 0$ and

$$\sum_{x=x_1}^{x_2}\mathrm{e}^{\mathrm{i}2\pi xk} = \mathrm{e}^{\mathrm{i}2\pi x_1 k}\sum_{y'=0}^{y}\mathrm{e}^{\mathrm{i}2\pi y'k} = \mathrm{e}^{\mathrm{i}2\pi x_1 k}\frac{1-\mathrm{e}^{\mathrm{i}2\pi(y+1)k}}{1-\mathrm{e}^{\mathrm{i}2\pi k}}$$

$$= \frac{\mathrm{i}\mathrm{e}^{-\mathrm{i}\pi k}}{2\sin(\pi k)}\left[\mathrm{e}^{\mathrm{i}2\pi x_1 k} - \mathrm{e}^{\mathrm{i}2\pi(x_2+1)k}\right]. \tag{4.31}$$

Hence, if we define $J_{x-1,x}(t)$ by replacing in (4.30) every factor "$\mathrm{e}^{\mathrm{i}2\pi x\bar{k}}\prod_\ell a(k_\ell,\sigma_\ell)$", where $\bar{k} = \sum_\ell k_\ell$, by "$\frac{\mathrm{i}\mathrm{e}^{-\mathrm{i}\pi\bar{k}}}{2\sin(\pi\bar{k})}\mathrm{e}^{\mathrm{i}2\pi x\bar{k}}\partial_t\left(\prod_\ell a_t(k_\ell,\sigma_\ell)\right)$", we have for all $x_1, x_2 \in \mathbb{Z}$ with $x_1 \leq x_2$ that

$$\partial_t\left(\sum_{x=x_1}^{x_2}H_x(q(t),p(t))\right) = J_{x_1-1,x_1}(t) - J_{x_2,x_2+1}(t)\,. \tag{4.32}$$

Since x_1, x_2 are arbitrary, we can interpret $J_{x,x+1}(t)$ as the energy flux from site x to $x+1$ at time t.

To be precise, the above construction can only be used inside the integral if $\bar{k} \neq 0$. To see why the resulting current observable still works, let us for the moment regularize the system by replacing the infinite lattice $\mathbb{Z}$ by a finite periodic lattice of length $L \gg 1$. Then $k_\ell \in \mathbb{Z}/L$, and either $\bar{k} = \sum_\ell k_\ell \in \mathbb{Z}$ or the distance of $\bar{k}$ from $\mathbb{Z}$ is at least $1/L$. Let us thus separate all terms with $\bar{k} \in \mathbb{Z}$ in the definition of H_x, and use the periodic identification. Then the terms correspond to $\bar{k} = 0$. Since for every x on the periodic lattice $\mathrm{e}^{\mathrm{i}2\pi x\bar{k}}\mathbb{1}(\bar{k}=0) = \frac{1}{L}\sum_y \mathrm{e}^{\mathrm{i}2\pi y\bar{k}}$, these terms sum to $H(p,q)/L$, i.e., to the total energy density. By conservation of energy, these terms do not contribute to $\partial_t H_x(q(t),p(t))$. The remaining terms converge to the claimed result as $L \to \infty$, provided that the integrals over $\mathbb{T}^n$ are understood as principal value integrals around the subset with $\bar{k} = 0$.

In summary, the definition

$$\begin{aligned}
J_{x-1,x} = {} & \frac{1}{2}\sum_{\sigma}\int_{\mathbb{T}^2} \mathrm{d}k\, \Phi_2(k)\, \frac{\mathrm{i}\mathrm{e}^{-\mathrm{i}\pi\bar{k}}}{2\sin(\pi\bar{k})}\mathrm{e}^{\mathrm{i}2\pi x\bar{k}}\Big|_{\bar{k}=k_1+k_2} \partial_t\left(a_t(k_1,-\sigma)a_t(k_2,\sigma)\right) \\
& + \frac{1}{3}\sum_{\sigma\in\{\pm1\}^3}\int_{\mathbb{T}^3} \mathrm{d}k\, \Phi_3(k)\, \frac{\mathrm{i}\mathrm{e}^{-\mathrm{i}\pi\bar{k}}}{2\sin(\pi\bar{k})}\mathrm{e}^{\mathrm{i}2\pi x\bar{k}}\Big|_{\bar{k}=\sum_{\ell=1}^3 k_\ell} \partial_t\left(\prod_{\ell=1}^3 a_t(k_\ell,\sigma_\ell)\right) \\
& + \frac{1}{4}\sum_{\sigma\in\{\pm1\}^4}\int_{\mathbb{T}^4} \mathrm{d}k\, \Phi_4(k)\, \frac{\mathrm{i}\mathrm{e}^{-\mathrm{i}\pi\bar{k}}}{2\sin(\pi\bar{k})}\mathrm{e}^{\mathrm{i}2\pi x\bar{k}}\Big|_{\bar{k}=\sum_{\ell=1}^4 k_\ell} \partial_t\left(\prod_{\ell=1}^4 a_t(k_\ell,\sigma_\ell)\right)
\end{aligned} \tag{4.33}$$

yields a current observable associated to the energy density (4.30) which satisfies the lattice continuity equation (4.32). Let us point out that no spurious imaginary part is created by this choice: since $a_t(k,\sigma)^* = a_t(-k,-\sigma)$, $\Phi_n(k)^* = \Phi_n(-k)$, for all n, and $\omega(-k) = \omega(k)$, we have always $J_{x,x+1}(t)^* = J_{x,x+1}(t)$ and hence the current observable is real-valued. We may also simplify the prefactor above by using $\frac{\mathrm{i}\mathrm{e}^{-\mathrm{i}\pi\bar{k}}}{2\sin(\pi\bar{k})} = \frac{1}{2}\left(1 + \mathrm{i}\cot(\pi\bar{k})\right)$.

The above current observable also behaves as expected with respect to spatial translations of the lattice. If we choose some $x_0 \in \mathbb{Z}$ and define $\tilde{q}_x = q_{x+x_0}$ and $\tilde{p}_x = p_{x+x_0}$, then the a-field of the translated configuration $(\tilde{q},\tilde{p})$ satisfies $\tilde{a}_t(k,\sigma) = \mathrm{e}^{\mathrm{i}2\pi x_0 k}a_t(k,\sigma)$. Thus each of the three terms in the sum defining the current in (4.33) will acquire a factor $\mathrm{e}^{\mathrm{i}2\pi x_0\bar{k}}$ with $\bar{k} = \sum_{\ell=1}^3 k_\ell$. Hence, the current of the translated configuration satisfies $\tilde{J}_{x,x+1} = J_{x+x_0,x+x_0+1}$, similarly to what was proven earlier for the energy density observables H_x.

4.1.4 Choice of Initial Data

We consider the above evolution equation with random initial data. The distribution of initial data plays as important a role as the choice of the coupling functions. For instance, there always are degenerate initial data which do not thermalize. Assume, for instance, that $q(0)$ is chosen as any local minimum or maximum configuration of the potential $\mathscr{V}$ and let the particles be initially at rest, $p(0) = 0$. Since then $\nabla\mathscr{V}(q(0)) = 0$, this configuration is a stationary solution of the evolution equations. However, it need not be stable, and it can well happen that even a minute perturbation of the initial data will take the asymptotic state of the system far from the original stationary state.

On the other hand, for normally conducting systems sufficiently chaotic initial data should lead to thermalization of the state. The precise mechanism and classification of what kind of initial data should have this property, and in which precise mathematical sense, is a long-standing open problem in mathematical physics. The most difficult part of the problem seems to be to control the beginning of the thermalization process, the *microscopic thermalization*, where the state of the system approaches local stationarity.

However, many important physical properties of the system do not require full understanding of the thermalization process. For instance, for systems with normal heat conduction, the Green–Kubo formula allows to compute the thermal conductivity of the system. As shown in the next subsection, the formula involves studying the time-correlations of the system *starting it from a thermal equilibrium state* or, equivalently, to study the evolution of correlation functions for small perturbations of the equilibrium state. Hence, thermal equilibrium states and their perturbations form an important class of initial data, and our assumptions should allow for at least these.

To arrive at the kinetic theory of phonons, we need to make some rather specific assumptions about the initial data. The principle here is that the states are chosen to mimic states produced by typical microscopic time-averages. Note that this concept involves a choice of scale, the time-scale over which the time-averages are taken. These questions touch upon the tough question of microscopic thermalization, and we do not wish to speculate further about them here. Instead, let us postulate a number of assumptions on the initial data based on reasonable assumptions about the physical characteristics of the time evolution.

We assume here that the initial data for the system has the following properties.

1. **Random:** We suppose that the initial data $(q_x(0), p_x(0))$ is randomly distributed. The initial randomness makes also the configurations $(q_x(t), p_x(t))$ at later times random variables.
2. **Chaotic:** We suppose that initial distribution of particles in regions which are far apart are almost independent. More precisely, we suppose that all correlation functions of the initial data decay fast (on the microscopic scale). The speed is assumed to be at least so fast that the correlation functions are ℓ_2-summable.
3. **Spatially homogenized:** We assume that there is a scale ε^{-1}, which is large in microscopic units, such that the correlation functions are nearly invariant under translations of lengths less than ε^{-1}.

The randomness in the initial data makes also the phonon fields $a_t(k, \sigma)$ random variables. In principle, the field is not defined for k such that $\omega(k) = 0$, but we assume these to have been suitably regularized. Consider for instance the case in which the interactions depend only on the differences $q_x - q_{x-1}$, such as for the FPU-β chains. Given any initial data (on a finite periodic chain), we can regularize it before defining the phonon fields without altering the evolution equations in any way. We move from (q,p) to $(\tilde{q},\tilde{p})$ by first setting $\bar{p} = \frac{1}{N}\sum_x p_x(0)$ and $\bar{q} = \frac{1}{N}\sum_x q_x(0)$, and then we define $\tilde{q}_x(t) = q_x(t) - \bar{q} - t\bar{p}$ and $\tilde{p}_x(t) = p_x(t) - \bar{p}$. Under these assumptions, the total momentum is conserved, and thus $\sum_x \tilde{p}_x(t) = 0 = \sum_x \tilde{q}_x(t)$ for all t. In addition, $(\tilde{q}_x(t), \tilde{p}_x(t))$ are a solution to the original evolution equations. We take $a_t(k,\sigma)$ to be defined via these regularized fields which guarantees that $a_t(0,\sigma) = 0$ for all t.

As we are interested in the evolution of the energy density (4.30), it suffices to study the correlation functions of $a_t(k,\sigma)$. For instance, to study the energy density averaged over the initial data, one needs the correlation functions up to order 4.

To control the correlation functions of the above type of chaotic, spatially homogenized states, it is often better to consider the cumulants of the fields rather than their moments. For instance, consider the field p_x at two points x_1, x_2 which are far apart. By the chaoticity assumption, then $e_1 = p_{x_1}^2$ and $e_2 = p_{x_2}^2$ should be nearly independent random variables, and hence $\langle e_1^n e_2^m \rangle \approx \langle e_1^n \rangle \langle e_2^m \rangle$, for $n, m \neq 0$. Unless one of the particles is frozen to its initial position, these moments are not zero. However, the corresponding cumulant is then nearly zero whenever both $n, m \neq 0$.

Hence, for two asymptotically independent regions, it is the cumulants, not moments, which will vanish in the limit of taking the regions infinitely far from each other. We consider here only initial states for which this decay is absolutely square summable in space, "ℓ_2-clustering" for short. This particular assumption allows for an easy identification of the (distribution type) singularities of the Fourier-transform for any spatially homogeneous initial data.

To make the computations manageable, we rely here on the notations and basic results used in [13]. In particular, for a random field $\psi(x)$, $x \in \mathbb{Z}$, and any sequence J of n lattice points the shorthand notations $\langle \psi(x)^J \rangle$, $\kappa[\psi(x_J)]$, and $:\psi(x)^J:$ refer to the expectation $\langle \prod_{i=1}^n \psi(x_{J_i}) \rangle$, to the cumulant $\kappa[\psi(x_{J_1}), \ldots, \psi(x_{J_n})]$, and to the Wick polynomial $:\prod_{i=1}^n \psi(x_{J_i}):$, respectively. In addition, we define $\psi(x)^J = 1$ if J is an empty sequence. Similar notations will be used for random fields on other label sets in addition to the lattice points; for instance, for the fields $a(k,\sigma)$ and any sequence $J = (k_i, \sigma_i)_{i=1}^n$ we write $\kappa[a(k,\sigma)_J] = \kappa[a(k_1,\sigma_1), \ldots, a(k_n,\sigma_n)]$.

We employ here the following basic relations between the above constructions. The proofs of these results can be found in many sources, in particular, in [13] using the above notations. The moments-to-cumulants formula states that for any sequence J

$$\langle \psi(x)^J \rangle = \sum_{\pi \in \mathscr{P}(J)} \prod_{A \in \pi} \kappa[\psi(x_A)] , \tag{4.34}$$

where $\mathscr{P}(J)$ denotes the collection of partitions of the sequence J.[1] For a partition $\pi \in \mathscr{P}(I)$, let us call the subsets $A \in \pi$ *clusters* or *blocks*. We also recall that

[1] Some care is needed in the interpretation of the moments-to-cumulants formula in order to get all combinatorial factors correctly. It is safe to think that the elements in the sequence J of length n are labelled by the set $I_n = \{1, 2, \ldots n\}$, and the collection of partitions $\mathscr{P}(J)$ refers to the collection

the cumulants are permutation invariant and multilinear, which means that they are linear in each of the arguments separately. Thus cumulants of Fourier-transforms can be easily expressed as a linear combination of the cumulants of the original field.

Wick polynomials of random variables allow simplifying the moments-to-cumulants expansions by removing all those partitions π from the sum which contain a cluster internal to the Wick polynomial part of the expectation. Explicitly, consider some $L \geq 1$ and a collection of $L+1$ index sequences J' and J_ℓ, for $\ell = 1, \ldots, L$. Then, for the merged sequence $I = J' + \sum_{\ell=1}^{L} J_\ell$ we have

$$\left\langle \psi(x)^{J'} \prod_{\ell=1}^{L} :\psi(x)^{J_\ell}: \right\rangle = \sum_{\pi \in \mathscr{P}(I)} \prod_{A \in \pi} (\kappa[\psi(x_A)] \mathbb{1}(A \not\subset J_\ell \ \forall \ell)) \, . \tag{4.35}$$

In particular, the formula implies that cumulants and Wick polynomials satisfy the following relation: If the sequence J is non-empty and has x_1 as its first element, then

$$\kappa[\psi(x_J)] = \langle \psi(x_1) :\psi_t(x)^{J \setminus x_1}: \rangle \, , \tag{4.36}$$

where $J \setminus x_1$ denotes the sequence which is obtained when x_1 is removed from J. By permutation invariance of the cumulants, the formula holds also if x_1 is replaced by any other element in J.

Let us also point out that, since the cumulants are multilinear, their time-evolution can be written down fairly easily for the present kind of hybrid dynamics, where all randomness is contained in the initial data. Namely, by using a straightforward "telescoping" argument and (4.36), it is clear that for any sequence J

$$\partial_t \kappa[\psi_t(x_J)] = \sum_{\ell \in J} \langle \partial_t \psi_t(x_\ell) :\psi_t(x)^{J \setminus x_\ell}: \rangle \, . \tag{4.37}$$

For the manipulation of terms such as $\partial_t \psi_t$ here, it is possible to move back and forth between standard and Wick polynomials by using the following identities:

$$\psi(x)^J = \sum_{U \subset J} :\psi(x)^U: \langle \psi(x)^{J \setminus U} \rangle \, , \tag{4.38}$$

and

$$:\psi(x)^J: = \sum_{U \subset J} \psi(x)^U \sum_{\pi \in \mathscr{P}(J \setminus U)} (-1)^{|\pi|} \prod_{A \in \pi} \kappa[\psi(x_A)] \, , \tag{4.39}$$

where $|\pi|$ denotes the number of clusters in the partition π.

---

of standard set partitions of I_n. Therefore, if $\ell \in A \in \pi \in \mathscr{P}(J)$, we have $\ell \in I_n$, x_ℓ denotes the value of x at the ℓ:th position of the sequence J, and x_A denotes the subsequence $(x_\ell)_{\ell \in A}$ of J.

Let us for the moment denote the inverse Fourier-transform of $\sqrt{2\omega(k)}a(k,\sigma)$ by $\psi(x,\sigma)$. By the definition in (4.5), we in fact then have $\psi(x,\sigma) = \sum_y \tilde{\omega}(x-y)q_y + \mathrm{i}\sigma p_x$. The ℓ_2-clustering assumption of the (q,p) variables then implies that the n:th cumulant $\mathbb{1}(y_1 = 0)\kappa[\psi(x+y_1,\sigma_1),\ldots,\psi(x+y_n,\sigma_n)]$ is square summable in y for every x. Thus one can safely take Fourier-transform with respect to the variables y, and the result is a function in $L^2(\mathbb{T}^n)$ for every fixed $x \in \mathbb{Z}$. Since the second assumption implies that the function is slowly varying in x, on the scale ε^{-1}, we can thus conclude that for each $\sigma \in \{\pm 1\}^n$ there is a function $F_n(x,k)$ such that it is slowly varying in the lattice position x on the scale ε^{-1}, it is L^2-integrable in $k \in \mathbb{T}^n$, and

$$\kappa[\hat{\psi}(k_1,\sigma_1),\ldots,\hat{\psi}(k_n,\sigma_n)] = \sum_x \mathrm{e}^{-\mathrm{i}2\pi x\cdot\sum_{\ell=1}^n k_\ell} F_n(x,k_1,k_2,\ldots,k_n;\sigma)\,. \tag{4.40}$$

In fact, there are many such functions: by using the periodic lattice Λ as a middle step, we find that, for instance, any convex combination of the n functions, defined for $\ell_0 \in \{1,\ldots,n\}$ by

$$\begin{aligned} &F_{n,\ell_0}(x,k_1,k_2,\ldots,k_n;\sigma) \\ &\quad = \sum_{y\in\Lambda^n} \mathrm{e}^{-\mathrm{i}2\pi\sum_{\ell=1}^n y_\ell k_\ell}\, \mathbb{1}(y_{\ell_0}=0)\kappa[\psi(x+y_1,\sigma_1),\ldots,\psi(x+y_n,\sigma_n)]\,, \end{aligned} \tag{4.41}$$

will work.

In the translation invariant case, when the scale of spatial variation $\varepsilon \to 0$, F_n is also independent of x and hence can be thought of as a proper function in $L^2(\mathbb{T}^n)$. Since $\hat{\psi}(k,\sigma) = \sqrt{2\omega(k)}a(k,\sigma)$, this yields the following structure for cumulants of the a-fields for translation invariant, ℓ_2-clustering, initial data:

$$\kappa[a(k_1,\sigma_1),\ldots,a(k_n,\sigma_n)] = \delta\Big(\sum_{\ell=1}^n k_\ell\Big) F_n(k,\sigma) \prod_{\ell=1}^n \frac{1}{(2\omega(k_\ell))^{1/2}}\,. \tag{4.42}$$

In case ω has zeroes, we recall our regularized definition of the field a and conclude that F_n must then also be zero at such points. Let $W_n(k,\sigma)$ denote the product $F_n(k,\sigma)\prod_{\ell=1}^n(2\omega(k_\ell))^{-1/2}$. Thus, if $\omega(k_\ell) = 0$ for some ℓ, then $W_n(k,\sigma) = 0$ as well.

Therefore, although the cumulants are singular, their singularity structure is simple, entirely encoded in the δ-multiplier. In contrast, by the moments-to-cumulants formula, the moments satisfy for any sequence J of labels

$$\langle a(k,\sigma)^J\rangle = \sum_{\pi\in\mathscr{P}(J)} \prod_{A\in\pi} \bigg(\delta\Big(\sum_{j\in A} k_j\Big) W_{|A|}(k_A,\sigma_A)\bigg)\,. \tag{4.43}$$

This has ever more complicated singularity structure as the order of the moment is increased. (The above discussion can be made mathematically rigorous by replacing the infinite lattice by a periodic d-dimensional lattice. See [16] for details.)

4.1.5 Green–Kubo Formula

The Green–Kubo formula gives an expression for the thermal conductivity if the system has normal heat conduction, i.e., diffusive energy transport. However, it can also be used to inspect anomalous transport since it always allows studying certain relaxation characteristics of perturbations of thermal states.

Let us recall here the argument in the diffusive case, assuming that energy is the only relevant ergodic invariant. The basic assumption is that after a relatively short thermalization period the macroscopic energy density $e(x,t)$ evolves according to the Fourier's law

$$\partial_t e = \partial_x \left(D(e) \partial_x e \right) . \tag{4.44}$$

This is a nonlinear diffusion equation with a diffusion "constant" D which depends on the value of energy density e. The steady states are then given by microcanonical ensembles whose only variable is the uniform energy density $\bar{e}$.

Consider next an infinite system with a steady state whose energy density is $\bar{e}$. Perturb the state, locally near origin, so that the perturbation has finite energy. Then, for normally conducting systems, one would expect that the energy density at "infinity" always remains equal to $\bar{e}$ and that the final state of an infinite system should be the same as the initial steady state, since for finite systems of volume V the final steady state should be of uniform energy density $\bar{e} + O(V^{-1})$. This implies that there should be a time t_0 after which the energy density is ever better approximated by a solution to the *linear* diffusion equation

$$\partial_t e = D(\bar{e}) \partial_x^2 e \, . \tag{4.45}$$

For a linear diffusion equation, it is possible to find out the diffusion constant from the leading evolution of the spatial spread of the perturbation. Namely, since then

$$e(t + t_0, x) \simeq \int \mathrm{d}y \frac{1}{\sqrt{4\pi D(\bar{e}) t}} \mathrm{e}^{-\frac{1}{4D(\bar{e})t}(x-y)^2} e(t_0, y) \, , \tag{4.46}$$

where $\int \mathrm{d}x \, x^2 |e(t_0, x) - \bar{e}| < \infty$, it follows that

$$\lim_{t\to\infty} \frac{1}{t} \int \mathrm{d}x \, x^2 (e(t,x) - \bar{e}) = D(\bar{e}) 2 \delta E \, , \tag{4.47}$$

where δE denotes the excess energy of the perturbation, which is conserved and hence $\delta E = \int \mathrm{d}x\,(e(t,x) - \bar{e})$ for all $t \geq 0$.

Translated into the present lattice setting, one analogously arrives at the formula

$$\lim_{t\to\infty} \frac{1}{t} \sum_{x\in\mathbb{Z}} x^2 \kappa[H_x(t), H_0(0)] = D(e) 2 \sum_{x\in\mathbb{Z}} \kappa[H_x(0), H_0(0)]\,, \tag{4.48}$$

where $\kappa[a,b] = \langle ab\rangle - \langle a\rangle\langle b\rangle = \langle (a - \langle a\rangle)(b - \langle b\rangle)\rangle$ denotes the second cumulant, i.e., covariance, of the random variables a, b defined by solving the time-evolution using equilibrium initial data with mean energy e. Namely, if we divide the equation by $\langle H_0\rangle$, the sum on the right hand side corresponds to the excess energy from a perturbation of the initial measure μ_0 to the probability measure $\frac{H_0}{\langle H_0\rangle}\mu_0$. The perturbation is then localized near the origin. Since the left hand side is a discrete version of the left hand side of (4.47), the limit in (4.48) should hold if the relaxation occurs via diffusion of energy, as given by (4.44).

To get the standard Green–Kubo formula, we next also assume equivalence of the infinite volume microcanonical and canonical ensembles, and use here initial data distributed according to a canonical ensemble at a temperature T. Then e in (4.48) denotes the corresponding (uniform) equilibrium energy density, $e = e(T) = \langle H_0\rangle_T$. Making this change of variables in (4.44) implies the standard Fourier's law. Explicitly, we define the temperature distribution corresponding to a given $e(x,t)$ by inverting the above function to yield $T(t,x) = T(e(t,x))$ which then satisfies

$$\frac{\partial e(T)}{\partial T} \partial_t T = \partial_x \left(\kappa(T) \partial_x T\right)\,, \tag{4.49}$$

where the *thermal conductivity* is given by $\kappa(T) = D(e(T))\frac{\partial e(T)}{\partial T}$. In the finite volume Λ canonical Gibbs state, we have $e_\Lambda(T) = \langle H_0\rangle_\Lambda = Z^{-1}\int \mathrm{d}^\Lambda q\,\mathrm{d}^\Lambda p\,\mathrm{e}^{-H/T} H_0$ where Z denotes the partition function, $Z = \int \mathrm{d}^\Lambda q\,\mathrm{d}^\Lambda p\,\mathrm{e}^{-H/T}$. Therefore,

$$\frac{\partial e_\Lambda(T)}{\partial T} = \frac{1}{T^2}\left(\langle H H_0\rangle_\Lambda - \langle H\rangle_\Lambda \langle H_0\rangle_\Lambda\right) = \frac{1}{T^2} \sum_{x\in\Lambda} \kappa_\Lambda[H_x(0), H_0(0)]\,. \tag{4.50}$$

Taking the infinite volume limit $\Lambda \to \mathbb{Z}$, and then using (4.48), thus yields

$$\kappa(T) = \frac{1}{2T^2} \lim_{t\to\infty} \frac{1}{t} \sum_{x\in\mathbb{Z}} x^2 \kappa[H_x(t), H_0(0)]\,. \tag{4.51}$$

The above argument shows that to find the numerical value of the thermal conductivity at a given temperature, it suffices to study the asymptotic increase of the spatial quadratic spread of the energy density, when starting the system with samples taken from the corresponding canonical Gibbs state. Explicitly, one should study the observable

$$S(t) = \sum_{x\in\mathbb{Z}} x^2 \kappa[H_x(t), H_0(0)]\,, \tag{4.52}$$

show that it is $O(t)$ for large t, and then solve the leading asymptotics.

For this, it is convenient to rewrite $S(t)$ in terms of current observables. Namely, for any J satisfying (4.32) we have for $t > 0$ and all $x \in \mathbb{Z}$

$$H_x(t) = H_x(0) + \int_0^t \mathrm{d}s\ (J_{x-1,x}(s) - J_{x,x+1}(s))\ , \tag{4.53}$$

$$H_x(-t) = H_x(0) - \int_{-t}^0 \mathrm{d}s\ (J_{x-1,x}(s) - J_{x,x+1}(s))\ . \tag{4.54}$$

Since the Gibbs state is invariant under both spatial translations and the Hamiltonian time-evolution, and these transform the energy observables as expected, we have here

$$S(t) = \sum_{x\in\mathbb{Z}} x^2 \kappa[H_0(0), H_{-x}(-t)] = \sum_{x\in\mathbb{Z}} x^2 \kappa[H_x(-t), H_0(0)] = S(-t)\,. \tag{4.55}$$

Hence, $S(t)$ is symmetric, and

$$\begin{aligned} 2S(t) &= 2S(0) + \sum_{x\in\mathbb{Z}} x^2 \int_0^t \mathrm{d}s\, \kappa[J_{x-1,x}(s) - J_{x,x+1}(s), H_0(0)] \\ &\quad - \sum_{x\in\mathbb{Z}} x^2 \int_{-t}^0 \mathrm{d}s\, \kappa[J_{x-1,x}(s) - J_{x,x+1}(s), H_0(0)] \\ &= 2S(0) + \sum_{x\in\mathbb{Z}} (x^2 - (x-1)^2) \int_0^t \mathrm{d}s\, \kappa[J_{x-1,x}(0), (H_0(-s) - H_0(s))]\,, \end{aligned} \tag{4.56}$$

where we have used the earlier proven simple rule for translation of the current observable. Therefore, by (4.53) and (4.54), we have

$$\begin{aligned} S(t) &= S(0) - \frac{1}{2}\sum_{x\in\mathbb{Z}} (2x-1) \int_0^t \mathrm{d}s \int_{-s}^s \mathrm{d}r\, \kappa[J_{x-1,x}(0), J_{-1,0}(r) - J_{0,1}(r)] \\ &= S(0) + \int_0^t \mathrm{d}s \int_{-s}^s \mathrm{d}r \sum_{x\in\mathbb{Z}} \kappa[J_{x-1,x}(0), J_{0,1}(r)]\,. \end{aligned} \tag{4.57}$$

We define the *current-current correlation function*, also called the *Green–Kubo correlator*, by the formula $C(t;T) = \sum_{x\in\mathbb{Z}} \kappa[J_{x-1,x}(0), J_{0,1}(t)]$. Since then $C(-t) = \sum_{x\in\mathbb{Z}} \kappa[J_{x-1,x}(t), J_{0,1}(0)] = \sum_{x\in\mathbb{Z}} \kappa[J_{0,1}(t), J_{1-x,2-x}(0)] = C(t)$, we find from (4.57)

$$S(t) = S(0) + 2\int_0^t \mathrm{d}s \int_0^s \mathrm{d}r\, C(r) = S(0) + 2t\int_0^t \mathrm{d}r \left(1 - \frac{r}{t}\right) C(r)\,. \tag{4.58}$$

If $\int_0^\infty \mathrm{d}r\, |C(r;T)| < \infty$, we can rely on dominated convergence in (4.58) and conclude that then, by (4.51),

$$\kappa(T) = \frac{1}{T^2}\int_0^\infty \mathrm{d}r\, C(r;T)\,. \tag{4.59}$$

This is the standard Green–Kubo formula, and thus, in great generality, the integrability of the correlation function $C(t;T)$ is sufficient to imply normal thermal conduction at the temperature T.

However, we can also conclude that, if $\int_0^t \mathrm{d}r \left(1 - \frac{r}{t}\right) C(r;T)$ is *not* bounded in t, then the energy relaxation cannot be diffusive at temperature T. Hence, the Green–Kubo correlation function $C(t;T)$ can also be used to prove *superdiffusion* of energy and to check at which time scale the energy spread occurs: if $\int_0^t \mathrm{d}r \left(1 - \frac{r}{t}\right) C(r;T) = O(t^p)$, with $p > 0$, then the spatial spread at time t is $O(t^{(1+p)/2})$. However, the spread from the observable $S(t)$ can be somewhat misleading for systems which have polynomial decay or multiscale behaviour. In these cases, $S(t)$ could be entirely dominated by tail behaviour, or just by one of the many scales, instead of describing the actual speed of spreading of local perturbations. We will find such an example from the kinetic theory of FPU-β chains, discussed in Sect. 4.3.5.

Let us conclude the section by computing an expression for the current correlation $C(t)$ for the particle chains. Some care needs to be taken with the sum over x and the principal value integral in the definition of the current observable. To control their behaviour, let us insert $\int_{\mathbb{T}} \mathrm{d}\bar{k}\, \delta(\bar{k} - \sum_\ell k_\ell)$ into the integrals defining J in (4.33). Then for each t there is some function f_t on the torus, such that

$$\kappa[J_{x-1,x}(0), J_{0,1}(t)] = \int_{\mathbb{T}} \mathrm{d}\bar{k}\, \mathrm{e}^{\mathrm{i}2\pi x\bar{k}} \frac{1}{2}\left(1 + \mathrm{i}\cot(\pi\bar{k})\right) f_t(\bar{k})\,, \tag{4.60}$$

as a principal value integral around zero. Hence,

$$\begin{aligned}\sum_{x\in\mathbb{Z}} \kappa[J_{x-1,x}(0), J_{0,1}(t)] &= \frac{1}{2} f_t(0) + \mathrm{i}\sum_{x\in\mathbb{Z}} \int_{\mathbb{T}} \mathrm{d}\bar{k}\, \mathrm{e}^{\mathrm{i}2\pi x\bar{k}} \frac{f_t(\bar{k}) - f_t(-\bar{k})}{4} \cot(\pi\bar{k}) \\ &= \frac{1}{2} f_t(0) + \frac{\mathrm{i}}{2\pi} \partial_{\bar{k}} f_t(0)\,. \end{aligned} \tag{4.61}$$

In fact, here $f_t(0) = 0$ due to energy conservation, as discussed above. Also, for any function F we have $\partial_{\bar{k}} \int_{\mathbb{T}^n} \mathrm{d}k\, \delta(\bar{k} - \sum_\ell k_\ell) F(k)\big|_{\bar{k}=0} = \int_{\mathbb{T}^n} \mathrm{d}k\, \delta(\sum_\ell k_\ell) \partial_{k_1} F(k)$.

Therefore, for the particle chains the Green–Kubo correlation function is given by

$$\sum_{x\in\mathbb{Z}} \kappa[J_{x-1,x}(0), J_{0,1}(t)] = \sum_{n',n=2}^{4} \frac{1}{nm} \sum_{\sigma',\sigma} w_n(\sigma) w_{n'}(\sigma') \int_{\mathbb{T}^n} \mathrm{d}k\, \delta\Big(\sum_{\ell=1}^{n} k_\ell\Big) \int_{\mathbb{T}^{n'}} \mathrm{d}k'$$
$$\times \frac{\mathrm{i}}{4\pi} \Phi_{n'}(k') \Big(-1 + \mathrm{i}\cot(\pi \sum_{\ell=1}^{n'} k'_\ell)\Big)$$
$$\times \kappa\Big[\partial_{k_1}\Big[\Phi_n(k)\partial_s\Big(\prod_{\ell=1}^{n} a_s(k_\ell,\sigma_\ell)\Big)_{s=0}\Big], \partial_t\Big(\prod_{\ell=1}^{n'} a_t(k'_\ell,\sigma'_\ell)\Big)\Big], \tag{4.62}$$

where the weight $w_n(\sigma) = 1$, for $\sigma \in \{\pm 1\}^n$, unless $n = 2$ and $\sigma_1 = \sigma_2$, in which case it is zero. By translation invariance, the final cumulant is proportional to $\delta(\sum_{\ell=1}^{n} k_\ell + \sum_{\ell=1}^{n'} k'_\ell)$, and thus the integral over k' has to be treated similarly to what has been done above for the integral over k. Following the same steps as above, we thus arrive at the simplified formula

$$\sum_{x\in\mathbb{Z}} \kappa[J_{x-1,x}(0), J_{0,1}(t)] = \sum_{n',n=2}^{4} \sum_{\sigma'\in\{\pm1\}^{n'}} \sum_{\sigma\in\{\pm1\}^{n}} w_n(\sigma) w_{n'}(\sigma')$$
$$\times \int_{\mathbb{T}^n} \mathrm{d}k\, \delta\Big(\sum_{\ell=1}^{n} k_\ell\Big) \int_{\mathbb{T}^{n'}} \mathrm{d}k'\, \kappa\Big[\mathscr{J}_n(k,\sigma;0), \mathscr{J}_{n'}(k',\sigma';t)\Big], \tag{4.63}$$

where for $k \in \mathbb{T}^n$, $\sigma \in \{\pm 1\}^n$, we have defined

$$\mathscr{J}_n(k,\sigma;t) = \frac{1}{n}\frac{\mathrm{i}}{2\pi} \partial_{k_1}\Big[\Phi_n(k)\partial_t\Big(\prod_{\ell=1}^{n} a_t(k_\ell,\sigma_\ell)\Big)\Big]. \tag{4.64}$$

Assuming that the harmonic terms dominate over the anharmonic ones, we can obtain the leading harmonic part of the above expression by considering only the term $n' = 2 = n$ above and approximating $\partial_t a_t(k,\sigma) \approx -\mathrm{i}\sigma\omega(k) a_t(k,\sigma)$. This yields an approximation

$$\mathscr{J}_2(k,\sigma;t) \approx \frac{1}{4\pi} \partial_{k_1}\Big[\Phi_n(k)(\sigma_1\omega(k_1) + \sigma_2\omega(k_2)) a_t(k_1,\sigma_1) a_t(k_2,\sigma_2)\Big]. \tag{4.65}$$

Since these terms appear in the integrand only when $\sigma_2 = -\sigma_1$ and $k_2 = -k_1$, for which $\sigma_1\omega(k_1) + \sigma_2\omega(k_2) = 0$, only the term where the factor $\sigma_1\omega(k_1) + \sigma_2\omega(k_2)$ is differentiated yields a non-zero contribution to the integral. Since then also $\Phi_2(k) = \omega(k_1)$, it is possible to substitute the above harmonic current term by

$$\omega(k_1)\frac{\sigma_1}{4\pi}\omega'(k_1)a_t(k_1,\sigma_1)a_t(k_2,\sigma_2) \tag{4.66}$$

in the integrand. Thus the harmonic part of the Green–Kubo correlation function evaluates to

$$\sum_{\sigma',\sigma\in\{\pm 1\}} \frac{\sigma'\sigma}{4}\int_{\mathbb{T}} \mathrm{d}k \int_{\mathbb{T}^2} \mathrm{d}k' \,\frac{\omega'(k)}{2\pi}\omega(k)\frac{\omega'(k_1')}{2\pi}\omega(k_1') \\ \times \kappa\Big[a_0(k,\sigma)a_0(-k,-\sigma), a_t(k_1',\sigma')a_t(k_2',-\sigma')\Big]. \tag{4.67}$$

As $\omega'(-k) = -\omega'(k)$, we can simplify the expression (4.67) further to

$$\int_{\mathbb{T}} \mathrm{d}k \int_{\mathbb{T}^2} \mathrm{d}k' \,\frac{\omega'(k)}{2\pi}\omega(k)\frac{\omega'(k_2')}{2\pi}\omega(k_2')\kappa\Big[a_0(-k,-1)a_0(k,1), a_t(k_1',-1)a_t(k_2',1)\Big]. \tag{4.68}$$

Here $\frac{\omega'(k)}{2\pi}\omega(k) = \frac{1}{4\pi}\hat{\alpha}'(k)$ which is equal to $\overline{\omega}^2\delta\sin(2\pi k)$ for chains with stable nearest neighbour interactions, using the parametrization given in Sect. 4.1.2. Thus the harmonic interaction part of the Green–Kubo correlation function for nearest neighbour chains is equal to

$$\overline{\omega}^4\delta^2\kappa\bigg[\int_{\mathbb{T}} \mathrm{d}k \sin(2\pi k)a_0(-k,-1)a_0(k,1), \int_{\mathbb{T}^2} \mathrm{d}k' \,\sin(2\pi k_2')a_t(k_1',-1)a_t(k_2',1)\bigg]. \tag{4.69}$$

This result coincides with the formula used in Sect. 3 in [1]. As explained there, the above time-correlator can also be computed by solving the evolution of the Wigner function starting from a suitably perturbed initial Gibbs state. Namely, consider the Gibbs states with perturbed Hamiltonians $H^\varepsilon(q,p) = H(q,p) - T\varepsilon\mathscr{J}(q,p)$ where

$$\mathscr{J}(q,p) = \int_{\mathbb{T}} \mathrm{d}k \sin(2\pi k)a_0(-k,-1)a_0(k,1) \tag{4.70}$$

and denote the expectation over the canonical Gibbs state at temperature T and the Hamiltonian H^ε by $\langle\cdot\rangle^{(\varepsilon)}$. For instance by using approximations by finite periodic lattices, it follows that

$$\partial_\varepsilon \left\langle \int_{\mathbb{T}^2} \mathrm{d}k' \, \sin(2\pi k_2') a_t(k_1', -1) a_t(k_2', 1) \right\rangle^{(\varepsilon)} \Bigg|_{\varepsilon=0}$$

$$= \kappa\left[\int_{\mathbb{T}} \mathrm{d}k \sin(2\pi k) a_0(-k, -1) a_0(k, 1), \int_{\mathbb{T}^2} \mathrm{d}k' \, \sin(2\pi k_2') a_t(k_1', -1) a_t(k_2', 1) \right] . \tag{4.71}$$

Note that the *time-evolution* in both expectations is the same, determined by the original Hamiltonian H. Hence, computing the left hand side involves solving the original problem with non-stationary initial data. The perturbation in the initial data can be captured by changing $\beta\omega(k)$ to $\beta\omega(k) - \varepsilon\sigma \sin(2\pi k)$ in the harmonic part multiplying $a_0(k, -\sigma) a_0(k, \sigma)$. In particular, also the perturbed initial state is translation invariant in space.

4.2 Phonon Boltzmann Equation of Spatially Homogeneous Anharmonic Chains

4.2.1 Identifying the Collision Operator

The goal of the section is to derive the form of the phonon Boltzmann collision operators for the particle chains. We do this only for spatially homogeneous initial data which suffices to cover the evolution of the Green–Kubo correlation function, as explained in Sect. 4.1.5. Allowing for inhomogeneous initial data would be an important extension, but the inhomogeneity introduces new multiscale effects which we wish to avoid in the present contribution. See the discussion in Sect. 6 of [13] for an example of the problems associated with spatially inhomogeneous initial data for a nonlinear evolution equation similar to the FPU chains.

Let us first recall the parametrization of the spatially homogeneous correlation functions $W_n(k, \sigma; t)$ satisfying (4.42),

$$\kappa[a_t(k_1, \sigma_1), \ldots, a_t(k_n, \sigma_n)] = \delta\Big(\sum_{\ell=1}^{n} k_\ell\Big) W_n(k, \sigma; t) . \tag{4.72}$$

As discussed earlier, this requirement does not fix the functions W_n completely, and here we resolve the ambiguity by requiring that they are constant in the first variable, k_1. This is equivalent to using the definition

$$W_n(k, \sigma; t) = \int_{\mathbb{T}} \mathrm{d}k_1' \, \kappa[a_t(k_1', \sigma_1), a_t(k_2, \sigma_2), \ldots, a_t(k_n, \sigma_n)] . \tag{4.73}$$

We can then also obtain the Wigner function $W(k;t)$ of the state, as used in earlier works, by setting $W(k;t) = W_2((k',k),(-1,1);t)$ for all $k',k \in \mathbb{T}$.

The time-evolution of the correlation functions W_n can now be solved from (4.37). Recalling the earlier notations involving a sequence J of labels, we have

$$\partial_t \kappa[a_t(k,\sigma)_J] = \sum_{\ell\in J} \langle \partial_t a_t(k_\ell,\sigma_\ell) :a_t(k,\sigma)^{J\setminus\ell}:\rangle . \tag{4.74}$$

Since the derivative of the phonon field satisfies (4.23), we thus obtain

$$\begin{aligned}\partial_t \kappa[a_t(k,\sigma)_J] &= -\mathrm{i}\Omega_n(k,\sigma)_J \kappa[a_t(k,\sigma)_J] \\ &\quad -\mathrm{i}\sum_{\ell\in J}\sigma_\ell \sum_{\sigma'\in\{\pm 1\}^2}\int_{\mathbb{T}^2} \mathrm{d}k'\,\delta(k_\ell - k_1' - k_2')\Phi_3(k_\ell,k') \\ &\quad \times \langle a_t(k_1',\sigma_1')a_t(k_2',\sigma_2') :a_t(k,\sigma)^{J\setminus\ell}:\rangle \\ &\quad -\mathrm{i}\sum_{\ell\in J}\sigma_\ell \sum_{\sigma'\in\{\pm 1\}^3}\int_{\mathbb{T}^3} \mathrm{d}k'\,\delta(k_\ell - \sum_{i=1}^3 k_i')\Phi_4(k_\ell,k') \\ &\quad \times \langle a_t(k_1',\sigma_1')a_t(k_2',\sigma_2')a_t(k_3',\sigma_3') :a_t(k,\sigma)^{J\setminus\ell}:\rangle ,\end{aligned} \tag{4.75}$$

where $n = |J|$ and we have used (4.36) to simplify the linear, free evolution, term into a multiplication by the function

$$\Omega_n(k,\sigma) = \sum_{\ell=1}^{n} \sigma_\ell \omega(k_\ell) . \tag{4.76}$$

The remaining expectation values can be represented in terms of cumulants using (4.35). To derive the Boltzmann equation, we will need only the first four cumulants but let us postpone the detailed expansion for later.

We next exponentiate the free evolution part by integrating the identity

$$\begin{aligned}&\partial_s \left(\mathrm{e}^{-\mathrm{i}\Omega_n(k,\sigma)_J(t-s)}\kappa[a_s(k,\sigma)_J]\right) \\ &\quad = \mathrm{e}^{-\mathrm{i}\Omega_n(k,\sigma)_J(t-s)}\left(\mathrm{i}\Omega_n(k,\sigma)_J\kappa[a_s(k,\sigma)_J] + \partial_s\kappa[a_s(k,\sigma)_J]\right) .\end{aligned} \tag{4.77}$$

This results in the following Duhamel perturbation formula

$$\begin{aligned}\kappa[a_t(k,\sigma)_J] &= \mathrm{e}^{-\mathrm{i}\Omega_n(k,\sigma)_J t}\kappa[a_0(k,\sigma)_J] \\ &\quad -\mathrm{i}\sum_{\ell\in J}\sigma_\ell\int_0^t \mathrm{d}s\,\mathrm{e}^{-\mathrm{i}\Omega_n(k,\sigma)_J(t-s)}\sum_{\sigma'\in\{\pm 1\}^2}\int_{\mathbb{T}^2}\mathrm{d}k'\,\delta(k_\ell - k_1' - k_2')\Phi_3(k_\ell,k') \\ &\quad \times \langle a_s(k_1',\sigma_1')a_s(k_2',\sigma_2') :a_s(k,\sigma)^{J\setminus\ell}:\rangle\end{aligned}$$

$$
-\mathrm{i}\sum_{\ell\in J}\sigma_\ell\int_0^t \mathrm{d}s\,\mathrm{e}^{-\mathrm{i}\Omega_n(k,\sigma)_J(t-s)}\sum_{\sigma'\in\{\pm1\}^3}\int_{\mathbb{T}^3}\mathrm{d}k'\,\delta(k_\ell-\sum_{i=1}^3 k'_i)\Phi_4(k_\ell,k')
$$
$$
\times\,\langle a_s(k'_1,\sigma'_1)a_s(k'_2,\sigma'_2)a_s(k'_3,\sigma'_3)\,{:}a_s(k,\sigma)^{J\setminus\ell}{:}\rangle\,. \tag{4.78}
$$

The case with $n=2$ and $J=((k_1,-1),(k_2,1))$ determines the Wigner function $W(k;t)$ after integration over k_1. However, in this case $\Omega_n(k,\sigma)_J=\omega(k_2)-\omega(k_1)$ which is identically zero whenever $k_1+k_2=0$. By translation invariance, all three terms on the right hand side of (4.78) are now proportional to $\delta(k_1+k_2)$, the last two since the remaining expectations are proportional to $\delta(k_{\ell'}+\sum_i k'_i)$ with $(\ell')=J\setminus\ell$. Therefore, all the phase factors involving Ω_n are equal to one in this case.

Let us next assume that the fields have been centred, i.e., that $\langle a_t(k,\sigma)\rangle=0$. In general, for a spatially homogeneous case we could have $\langle a_t(k,\sigma)\rangle=\delta(k)c_t(\sigma)$. However, for instance in the FPU model we have normalized the fields so that $a_t(0,\sigma)=0$ for all t, which implies $c_t(\sigma)=0$. Under this simplifying assumption, we obtain using (4.35) and the obvious shorthand notations for the fields a'_s that

$$
\langle a'_1a'_2\,{:}a{:}\rangle=\kappa[a'_1,a'_2,a]\,, \tag{4.79}
$$
$$
\begin{aligned}
\langle a'_1a'_2a'_3\,{:}a{:}\rangle&=\kappa[a'_1,a'_2,a'_3,a]\\
&\quad+\kappa[a'_1,a'_2]\kappa[a'_3,a]+\kappa[a'_1,a'_3]\kappa[a'_2,a]+\kappa[a'_2,a'_3]\kappa[a'_1,a]\,,
\end{aligned} \tag{4.80}
$$

where $a=a_s(k_{\ell'},\sigma_{\ell'})$. Therefore, integration of (4.78) over k_1 and writing $k=k_2$ yields

$$
\begin{aligned}
W(k;t)&=W(k;0)\\
&\quad-\mathrm{i}\int_0^t\mathrm{d}s\sum_{\sigma'\in\{\pm1\}^3}\sigma'_3\int_{\mathbb{T}^3}\mathrm{d}k'\,\delta(k'_3-\sigma'_3k)\delta(\sum_{i=1}^3 k'_i)\Phi_3(k')W_3(k',\sigma';s)\\
&\quad-\mathrm{i}\int_0^t\mathrm{d}s\sum_{\sigma'\in\{\pm1\}^4}\sigma'_4\int_{\mathbb{T}^4}\mathrm{d}k'\,\delta(k'_4-\sigma'_4k)\delta(\sum_{i=1}^4 k'_i)\Phi_4(k')W_4(k',\sigma';s)\\
&\quad-3\mathrm{i}\int_0^t\mathrm{d}s\sum_{\sigma'\in\{\pm1\}^4}\sigma'_4\int_{\mathbb{T}^4}\mathrm{d}k'\,\delta(k'_4-\sigma'_4k)\delta(\sum_{i=1}^4 k'_i)\delta(k'_1+k'_2)\Phi_4(k')\\
&\qquad\times W_2((k'_1,k'_2),(\sigma'_1,\sigma'_2);s)W_2((k'_3,k'_4),(\sigma'_3,\sigma'_4);s)\,,
\end{aligned} \tag{4.81}
$$

where we have enumerated the sum over $\ell\in J$ by using $\sigma'_n=-\sigma_\ell$, when clearly $k_\ell=-\sigma'_3k$ and $k_{\ell'}=\sigma'_3k$. We have also used here the permutation invariance of Φ_n and the obvious antisymmetries of Φ_n, $n=3,4$, under a reversal of the sign of the first component.

The last term in (4.81) can be simplified further using symmetries of Φ_4 to

$$
-3\mathrm{i}\int_0^t \mathrm{d}s \sum_{\sigma\in\{\pm 1\}^2} \sigma_2 W_2((-\sigma_2 k, \sigma_2 k), (\sigma_1, \sigma_2); s)
\times \int_{\mathbb{T}} \mathrm{d}k'\, \Phi_4(-k', k', -k, k) \sum_{\sigma'\in\{\pm 1\}^2} W_2((-k', k'), (\sigma_1', \sigma_2'); s)\,. \tag{4.82}
$$

Inserting the definition of the FPU interaction amplitude Φ_4 given in (4.27) then yields

$$
\mathrm{i}6\lambda_4\overline{\omega}^{-2}\int_0^t \mathrm{d}s\, |\sin(\pi k)| \sum_{\sigma\in\{\pm 1\}^2} \sigma_2 W_2((-\sigma_2 k, \sigma_2 k), (\sigma_1, \sigma_2); s)
\times \int_{\mathbb{T}} \mathrm{d}k'\, |\sin(\pi k')| \sum_{\sigma'\in\{\pm 1\}^2} W_2((-k', k'), (\sigma_1', \sigma_2'); s)\,. \tag{4.83}
$$

Since now $\overline{\omega}|\sin(\pi k)| = \omega(k)/\sqrt{2}$ and $\sum_{\sigma\in\{\pm 1\}^2} \sigma_2 W_2((-\sigma_2 k, \sigma_2 k), (\sigma_1, \sigma_2); s) = W_2((-k, k), (1, 1); s) - W_2((-k, k), (-1, -1); s)$, for the FPU chains we may write this term as

$$
\mathrm{i}3\lambda_4\overline{\omega}^{-4}\int_0^t \mathrm{d}s\, \omega(k) \sum_{\sigma\in\{\pm 1\}} \sigma W_2((-k, k), (\sigma, \sigma); s)
\times \int_{\mathbb{T}} \mathrm{d}k'\, \omega(k') \sum_{\sigma'\in\{\pm 1\}^2} W_2((-k', k'), (\sigma_1', \sigma_2'); s)\,. \tag{4.84}
$$

It is customary in kinetic theory to ignore the term (4.84) in the Boltzmann equation. At the first glance, it is not obvious why this should be so: the term does not involve any non-Gaussian factors and it has an apparent magnitude $O(\lambda_4 t)$. Although no full mathematical study has been made about the effect of the term on the kinetic time scales $O(\lambda_4^{-2})$, the following argument suggests that, indeed, it behaves better than suggested by its apparent magnitude which is $O(\lambda_4^{-1})$ on the kinetic time scale. In fact, the argument indicates that the term is $O(\lambda_4)$ uniformly in t and k, and hence could be safely neglected when $\lambda_4 \ll 1$.

We study the two W_2-dependent factors in (4.84) separately. The second factor is

$$
\int_{\mathbb{T}} \mathrm{d}k\, \omega(k) \sum_{\sigma\in\{\pm 1\}^2} W_2((-k, k), (\sigma_1, \sigma_2); s)
= \int_{\mathbb{T}^2} \mathrm{d}^2 k\, \sqrt{\omega(k_1)\omega(k_2)} \sum_{\sigma\in\{\pm 1\}^2} \kappa[a_s(k_1, \sigma_1), a_s(k_2, \sigma_2)]
= 2\int_{\mathbb{T}^2} \mathrm{d}^2 k\, \omega(k_1)\omega(k_2)\langle \hat{q}(s, k_1)\hat{q}(s, k_2)\rangle\,, \tag{4.85}
$$

where we have applied the inversion formula given in (4.6) and the assumption that $\langle a_s(k,\sigma)\rangle = 0$. This is equal to $2\langle(\tilde{\omega} * q(s))(0)^2\rangle$ which for a finite periodic lattice would correspond to $2|\Lambda|^{-1}\sum_x \langle(\tilde{\omega} * q(s))(x)^2\rangle = 2|\Lambda|^{-1}\langle\sum_{x,y} q_x(s)\,\alpha(x-y)q_y(s)\rangle$, by translation invariance. Therefore, the second factor is proportional to the average harmonic potential energy, and we would expect it to be slowly varying and to equilibrate to the corresponding equilibrium value as $s \to \infty$. In particular, it can be bounded by the total energy density which is a constant, so the value of this term is $O(1)$ uniformly in s.

The first factor in (4.84) depends only on the "off-diagonal" elements of the second order correlation matrix. By (4.78), they have rapidly oscillating phase factors $\mathrm{e}^{-\mathrm{i}s\Omega}$ where $\Omega = 2\sigma\omega(k)$. If we assume that this oscillatory behaviour continues for all s, it seems reasonable to expect that the value of (4.84) can be well approximated by using this oscillatory factor and assuming that the remainder is so slowly varying that it can be replaced by a constant. This yields a factor which is $O(\lambda_4)$ times the integral $\Omega\int_0^t \mathrm{d}s\,\mathrm{e}^{-\mathrm{i}s\Omega} = \mathrm{i}(1-\mathrm{e}^{-\mathrm{i}t\Omega})$ which is bounded by 2 for all t and k. Of course, the replacement of a slowly varying term by a constant is not totally accurate here, but it seems reasonable to assume that under the present conditions the resulting corrections would also be $O(\lambda_4)$, at least up to the kinetic time scales.

The rest of the argument leading to the phonon Boltzmann collision operator is fairly standard. The clusterings relevant to the third cumulant are

$$\langle a_1'a_2' :a_1a_2:\rangle = \kappa[a_1',a_2',a_1,a_2] + \kappa[a_1',a_1]\kappa[a_2',a_2] + \kappa[a_1',a_2]\kappa[a_2',a_1]\,, \quad (4.86)$$

and $\langle a_1'a_2'a_3' :a_1a_2:\rangle$ whose expansion has only terms containing higher order cumulant factors. We continue to assume that the oscillations of the higher order cumulants are sufficient to suppress any term which involves integrals of W_n, $n > 2$, over any of its arguments. The clusterings relevant to the fourth cumulant are $\langle a_1'a_2' :a_1a_2a_3:\rangle$, which produces only terms containing higher order cumulant factors, and

$$\begin{aligned}\langle a_1'a_2'a_3' :a_1a_2a_3:\rangle &= (\text{higher order terms})\\ &\quad + \sum_{\pi\in\mathrm{Perm}(3)} \kappa[a_{\pi(1)}',a_1]\kappa[a_{\pi(2)}',a_2]\kappa[a_{\pi(3)}',a_3]\,, \end{aligned}\quad (4.87)$$

where the sum goes through all permutations π of three elements.

In summary, we then obtain the following approximations from the third and fourth cumulants,

$$\begin{aligned}W_3(k,\sigma;t) &\approx \mathrm{e}^{-\mathrm{i}\Omega_3(k,\sigma)t}W_3(k,\sigma;0)\\ &\quad - 2\mathrm{i}\sum_{\ell=1}^{3}\sigma_\ell\int_0^t \mathrm{d}s\,\mathrm{e}^{-\mathrm{i}\Omega_3(k,\sigma)(t-s)}\Phi_3(k)\prod_{i\neq\ell}\sum_{\sigma'\in\{\pm1\}} W_2((-k_i,k_i),(\sigma',\sigma_i);s)\,, \end{aligned}\quad (4.88)$$

$$W_4(k,\sigma;t) \approx \mathrm{e}^{-\mathrm{i}\Omega_4(k,\sigma)t} W_4(k,\sigma;0) + 6\mathrm{i}\sum_{\ell=1}^{4}\sigma_\ell \int_0^t \mathrm{d}s\, \mathrm{e}^{-\mathrm{i}\Omega_4(k,\sigma)(t-s)} \Phi_4(k) \prod_{i\neq\ell} \sum_{\sigma'\in\{\pm1\}} W_2((-k_i,k_i),(\sigma',\sigma_i);s)\,, \tag{4.89}$$

where we have simplified the result by applying the invariance properties of the amplitude factors Φ_3 and Φ_4. We insert these approximations in (4.81). The terms depending on the cumulants at time zero are again assumed to be negligible since they involve highly oscillatory integrands.

Since $\sum_{\sigma'\in\{\pm1\}} W_2((-k_i,k_i),(\sigma',\sigma_i);s)$ is equal to $W(\sigma_i k_i, s)$ plus a rapidly oscillatory (σ_i,σ_i)-term, this results in the approximation

$$\begin{aligned} W(k;t) \approx W(k;0) &+ 2\int_0^t \mathrm{d}s \sum_{\sigma\in\{\pm1\}^3} \sigma_3 \int_{\mathbb{T}^3} \mathrm{d}k'\, \delta(k_3'-\sigma_3 k) \\ &\times \delta(\sum_{i=1}^3 k_i')(-\Phi_3(k')^2) \int_s^t \mathrm{d}t'\, \mathrm{e}^{-\mathrm{i}\Omega_3(k',\sigma)(t'-s)} \sum_{\ell=1}^3 \sigma_\ell \prod_{i\neq\ell} W(\sigma_i k_i', s) \\ &+ 6\int_0^t \mathrm{d}s \sum_{\sigma\in\{\pm1\}^4} \sigma_4 \int_{\mathbb{T}^4} \mathrm{d}k'\, \delta(k_4'-\sigma_4 k) \\ &\times \delta(\sum_{i=1}^4 k_i')\Phi_4(k')^2 \int_s^t \mathrm{d}t'\, \mathrm{e}^{-\mathrm{i}\Omega_4(k',\sigma)(t'-s)} \sum_{\ell=1}^4 \sigma_\ell \prod_{i\neq\ell} W(\sigma_i k_i', s)\,, \end{aligned} \tag{4.90}$$

where we have used Fubini's theorem to exchange the order of the time-integrals. If we swap the sign of both σ and k', clearly also $\Omega_n(k',\sigma)$, $n=3,4$, change their sign. Therefore, in the above formula, we may replace both of the terms $\int_s^t \mathrm{d}t'\, \mathrm{e}^{-\mathrm{i}(t'-s)\Omega}$ by $\frac{1}{2}\int_{-(t-s)}^{t-s} \mathrm{d}r\, \mathrm{e}^{-\mathrm{i}r\Omega} \approx \pi\delta(\Omega)$, for large t.

Inserting the explicit expressions for Φ_n thus results in the formula

$$\begin{aligned} W(k;t) \approx W(k;0) &+ 2^{\frac{5}{2}}\pi\lambda_3^2\overline{\omega}^{-3} \int_0^t \mathrm{d}s \sum_{\sigma\in\{\pm1\}^3} \int_{\mathbb{T}^3} \mathrm{d}k'\, \delta(k_3'-k) \\ &\times \delta(\sum_{i=1}^3 \sigma_i' k_i')\delta(\sum_{i=1}^3 \sigma_i\omega(k_i')) \prod_{\ell=1}^3 |\sin(\pi k_\ell')| \sum_{\ell=1}^3 \sigma_3\sigma_\ell \prod_{i\neq\ell} W(k_i', s) \\ &+ 24\pi\lambda_4^2\overline{\omega}^{-4} \int_0^t \mathrm{d}s \sum_{\sigma\in\{\pm1\}^4} \int_{\mathbb{T}^4} \mathrm{d}k'\, \delta(k_4'-k)\delta(\sum_{i=1}^4 \sigma_i' k_i')\delta(\sum_{i=1}^4 \sigma_i\omega(k_i')) \\ &\times \prod_{\ell=1}^4 |\sin(\pi k_\ell')| \sum_{\ell=1}^4 \sigma_4\sigma_\ell \prod_{i\neq\ell} W(k_i', s)\,. \end{aligned} \tag{4.91}$$

The right hand side is an integrated version of the solution to the homogeneous Boltzmann equation

$$\partial_t W(k;t) = \lambda_3^2 \mathscr{C}_3[W(\cdot;t)](k) + \lambda_4^2 \mathscr{C}_4[W(\cdot;t)](k)\,, \tag{4.92}$$

where, using shorthand notations $\omega_\ell = \omega(k_\ell)$, $W_\ell = W(k_\ell)$, $\ell = 0,1,2,3,4$,

$$\mathscr{C}_3[W](k_0) = 2^{\frac{7}{2}}\pi\overline{\omega}^{-3} \sum_{\sigma\in\{\pm 1\}^2} \int_{\mathbb{T}^2} \mathrm{d}k\, \delta(k_0 + \sigma_1 k_1 + \sigma_2 k_2)\delta(\omega_0 + \sigma_1\omega_1 + \sigma_2\omega_2) \\ \times \prod_{\ell=0}^{2} |\sin(\pi k_\ell)| \times (W_1 W_2 + \sigma_1 W_0 W_2 + \sigma_2 W_0 W_1)\,, \tag{4.93}$$

and

$$\mathscr{C}_4[W](k_0) = 48\pi\overline{\omega}^{-4} \sum_{\sigma\in\{\pm 1\}^3} \int_{\mathbb{T}^3} \mathrm{d}k\, \delta(k_0 + \sum_{i=1}^{3}\sigma_i k_i)\delta(\omega_0 + \sum_{i=1}^{3}\sigma_i\omega_i) \prod_{\ell=0}^{3} |\sin(\pi k_\ell)| \\ \times (W_1 W_2 W_3 + \sigma_1 W_0 W_2 W_3 + \sigma_2 W_0 W_1 W_3 + \sigma_3 W_0 W_1 W_2)\,. \tag{4.94}$$

Other interactions can be treated similarly. For instance, the onsite perturbations have $\hat{\alpha}_3(k_1,k_2) = \lambda_3$ and $\hat{\alpha}_4(k_1,k_2,k_3) = \lambda_4$ and, for general dispersion relations ω, yield the collision operators

$$\mathscr{C}_3[W](k_0) = 4\pi \sum_{\sigma\in\{\pm 1\}^2} \int_{\mathbb{T}^2} \mathrm{d}k\, \delta(k_0 + \sigma_1 k_1 + \sigma_2 k_2)\delta(\omega_0 + \sigma_1\omega_1 + \sigma_2\omega_2) \prod_{\ell=0}^{3} \frac{1}{2\omega_\ell} \\ \times (W_1 W_2 + \sigma_1 W_0 W_2 + \sigma_2 W_0 W_1)\,, \tag{4.95}$$

and

$$\mathscr{C}_4[W](k_0) = 12\pi \sum_{\sigma\in\{\pm 1\}^3} \int_{\mathbb{T}^3} \mathrm{d}k\, \delta(k_0 + \sum_{i=1}^{3}\sigma_i k_i)\delta(\omega_0 + \sum_{i=1}^{3}\sigma_i\omega_i) \prod_{\ell=0}^{3} \frac{1}{2\omega_\ell} \\ \times (W_1 W_2 W_3 + \sigma_1 W_0 W_2 W_3 + \sigma_2 W_0 W_1 W_3 + \sigma_3 W_0 W_1 W_2)\,. \tag{4.96}$$

Detailed derivations of these onsite collision operators using standard perturbation expansions can also be found in the literature: the third order collision operator in [21] and the fourth order collision operator in [1].

4.2.2 Solution of the Collisional Constraints

The above derivation of the Boltzmann collision operator is not mathematically rigorous. In fact, the argument used for neglecting the higher order terms and replacing "$t - s$" by "∞" in the derivation of the energy conservation δ-function are at present mathematically uncontrollable approximations. Here we assume that we are working in a regime in which these terms indeed can be neglected. Hence, instead of trying to find out under which conditions this would work for our particle chains, we start from the homogeneous Boltzmann equation and proceed to study its solutions to obtain predictions which can later be compared with other results on the chains, such as computer simulations.

The weakest part of the argument lies in the rule that "any integral over one of the k-components of the rapidly oscillating factor $\mathrm{e}^{-\mathrm{i}t\Omega_n(k)}$ leads to integrable decay in t and hence can be neglected". This is particularly suspect if the integration is performed over a one-dimensional space: typically, even the best decay estimates over a d-dimensional torus yield $\int_{\mathbb{T}^d} \mathrm{d}k\, \mathrm{e}^{-\mathrm{i}t\Omega(k)} = O(t^{-d/2})$ which is not integrable if $d = 1$. The decay of such oscillatory integrals is closely linked to the behaviour of the frequency function Ω_n around its zero set; consider, for example, the explicitly integrable example $\int_0^t \mathrm{d}s \int \mathrm{d}k\, \mathrm{e}^{-\mathrm{i}s\Omega_n(k)} = \int \mathrm{d}k\, \frac{\mathrm{i}}{\Omega_n(k)}(\mathrm{e}^{-\mathrm{i}t\Omega_n(k)} - 1)$ which is uniformly bounded in t, if $|\Omega_n(k)|^{-1}$ is integrable, but which can blow up as fast as $O(t)$, for instance, if the set of k with $\Omega_n(k) = 0$ has a non-zero measure.

These subtleties are also reflected in the Boltzmann collision operator: the two δ-functions need to be carefully integrated over before one can use the equation. In the well-studied continuum setup of Boltzmann equations, the dispersion relation is given by $\frac{1}{2}k^2$, $k \in \mathbb{R}^d$, and the integral contains $\delta(k_0 + k_1 - k_2 - k_3)\delta((k_0^2 + k_1^2 - k_2^2 - k_3^2)/2)$ which can be explicitly integrated over, yielding the "standard" rarefied gas collision operator.

For a lattice system, the solutions to $\Omega_n(k) = 0$ are not so easy to handle. As we will show below, there are however explicit solutions for the one-dimensional nearest neighbour dispersion relation, relevant to the FPU chains. After the solution manifold has been found, one still needs to choose some parametrization of the manifold to integrate over the energy conservation δ-function, since this will yield an additional factor to the remaining integrand.

In fact, even at this part of the procedure, some care is needed since the *trivial solutions* to the collisional constraints, to be discussed later, do not contribute to the total collision operator but would result in infinite factors if integrated over using the above rule. Hence, the trivial solutions need to be removed from the solution manifold before integrating out the collisional constraint δ-functions.

We begin with a result which shows that nearest neighbour dispersion relations are, in fact, somewhat pathological: they suppress all collisions involving three phonons.

4.2.2.1 $\mathscr{C}_3 = 0$ for Nearest Neighbour Dispersion Relations

Consider the nearest neighbour dispersion relation, $\omega(k) = (1 - 2\delta\cos(2\pi k))^{1/2}$ with $0 < \delta \le \frac{1}{2}$ (we ignore the overall prefactor here, since it does not affect the solution manifold). To estimate the collision energy $\Omega_3 = \omega_0 + \sigma_1\omega_1 + \sigma_2\omega_2$, consider the following parametrization of ω as the magnitude of a complex number: Since $0 < 2\delta \le 1$, we can define $r = \frac{1}{2}\text{arcosh}(2\delta)^{-1} \ge 0$. Then by explicit computation

$$\omega(k) = |z(k)|, \quad \text{for} \quad z(k;\delta) = \sqrt{\delta}(\mathrm{e}^{r} - \mathrm{e}^{-r-\mathrm{i}2\pi k}). \tag{4.97}$$

The above parametrization and the triangle inequality yield an upper bound for differences of dispersion relations. Namely, for $k, q \in \mathbb{T}$,

$$\begin{aligned} |\omega(k) - \omega(k+q)| &= \big|\,|z(k)| - |z(k+q)|\,\big| \\ &\le |z(k) - z(k+q)| = \mathrm{e}^{-r}\sqrt{\delta}|1 - \mathrm{e}^{-\mathrm{i}2\pi q}|. \end{aligned} \tag{4.98}$$

Since $|1 - \mathrm{e}^{-\mathrm{i}2\pi q}|^2 = 2 - 2\cos(2\pi q) = \delta^{-1}(2\delta - 2\delta\cos(2\pi q)) \le \delta^{-1}\omega(q)^2$, then

$$|\omega(k) - \omega(k+q)| \le \mathrm{e}^{-r}\omega(q). \tag{4.99}$$

If $\delta < \frac{1}{2}$, we have $r > 0$, and hence $\mathrm{e}^{-r} < 1$. In addition, then $\omega(k) \ge c_\delta$, with $c_\delta = (1-2\delta)^{1/2} > 0$. Thus if the momentum constraint $k_0 + \sigma_1 k_1 + \sigma_2 k_2 = 0$ holds, we have

$$\Omega_3 = \omega_0 + \omega_1 + \omega_2 \ge 3c_\delta, \qquad \text{for} \quad \sigma_1 = 1 = \sigma_2, \tag{4.100}$$

$$\Omega_3 = \omega_0 - \omega_1 + \omega_2 \ge (1 - \mathrm{e}^{-r})c_\delta, \qquad \text{for} \quad -\sigma_1 = 1 = \sigma_2, \tag{4.101}$$

$$\Omega_3 = \omega_0 + \omega_1 - \omega_2 \ge (1 - \mathrm{e}^{-r})c_\delta, \qquad \text{for} \quad \sigma_1 = 1 = -\sigma_2, \tag{4.102}$$

$$\Omega_3 = \omega_0 - \omega_1 - \omega_2 \le -(1 - \mathrm{e}^{-r})c_\delta, \qquad \text{for} \quad \sigma_1 = -1 = \sigma_2. \tag{4.103}$$

Therefore, with a nearest neighbour interactions and pinning, the momentum constraint keeps the three-phonon energy well separated from zero, at least a distance $(1 - \mathrm{e}^{-r})c_\delta$ apart. In such a case, one should set $\mathscr{C}_3 = 0$ in the phonon Boltzmann equation.

If $\delta = \frac{1}{2}$, such as in the FPU models, we have $r = 0$ and $c_\delta = 0$, and Eq. (4.99) does not rule out the existence of solutions to both constraints. However, it is possible to improve the bound (4.98), by recalling that an equality in the triangle inequality holds if and only if one of the numbers is a *non-negative* multiple of the other one. This requires that for the given k, q there should be $R \ge 0$ such that $(1 - \mathrm{e}^{-\mathrm{i}2\pi k_1}) = R(1 - \mathrm{e}^{-\mathrm{i}2\pi k_2})$, where $k_1 = k$, $k_2 = k + q$, or vice versa. For instance by a geometric argument, it is straightforward to see that this happens if and only if $k_1 = 0$ or $k_2 = 0$.

As above, by inspecting the four sign combinations, we thus find that $|\Omega_3| > 0$ unless $k_0 = 0$, $k_1 = 0$, or $k_2 = 0$. If $k_0 = 0$, we have $\Omega_3 = 0$ identically for $\sigma_1 = -\sigma_2$, and only at the point $k_1 = 0 = k_2$, if $\sigma_1 = \sigma_2$. If $k_0 \neq 0$, we have $\Omega_3 > 0$ for $\sigma_1 = 1 = \sigma_2$, and $\Omega_3 = 0$, only if one of the following conditions is satisfied:

1. $-\sigma_1 = 1 = \sigma_2$, and $k_2 = 0$,
2. $\sigma_1 = 1 = -\sigma_2$, and $k_1 = 0$,
3. $\sigma_1 = -1 = \sigma_2$, and $k_1 = 0$ or $k_2 = 0$.

However, these solutions do not contribute to the collision operator in the FPU models as the collision operator $\mathscr{C}_3$ defined in (4.93) contains a prefactor $\prod_{\ell=0}^{2} |\sin(\pi k_\ell)|$ which is zero on any of the above solution manifolds. Therefore, for the FPU-chains, the lowest order kinetic theory would imply using $\mathscr{C}_3 = 0$.

The above argument straightforwardly generalizes to higher dimensions. As shown in Appendix 18.1 of [21], also higher dimensional nearest neighbour dispersion relations with pinning have no solutions to the constraints in the three-phonon collision operator and, hence, $\mathscr{C}_3 = 0$. However, this property depends on the dispersion relation. For instance, consider the next-to-nearest neighbour interaction with $\hat{\alpha}(k) = (1 - \cos(2\pi k))^2$. Then the dispersion relation is $\omega(k) = 1 - \cos(2\pi k)$ and for $\sigma_1 = 1 = -\sigma_2$ we thus have

$$\Omega_3 = \omega_0 + \omega_1 - \omega_2 = 1 - \cos(2\pi k_0) - \cos(2\pi k_1) + \cos(2\pi(k_0 + k_1)), \tag{4.104}$$

which is zero if $k_1 = \frac{1}{2} - k_0$. This introduces solutions to the three-phonon collisional constraints in the next-to-nearest neighbour case.

4.2.2.2 $\mathscr{C}_4$ for Nearest Neighbour Dispersion Relations

From now on, we only consider the nearest neighbour dispersion relations. We do not claim that this case is representative of the general case: as the previous examples indicate, other types of behaviour might appear for other dispersion relations.

Consider thus the collisional constraints in the four-phonon collision operator, assuming a nearest neighbour dispersion relation. If $\sigma_i = 1$ for all i, by the above estimates, we have $\Omega_4 > 0$, unless $\delta = \frac{1}{2}$ and $k_i = 0$ for every $i = 0, 1, 2, 3$.

If $\sigma_i = -1$ for all i, we resolve the momentum constraint by integrating out k_3 which yields $k_3 = k_0 - k_1 - k_2$. Then

$$\begin{aligned}\Omega_4 &= \omega_0 - \omega_1 - \omega_2 - \omega_3 \\ &= \Omega_3(k_0, k_1, k_0 - k_1; 1, -1, -1) + \Omega_3(k_0 - k_1, k_2, k_0 - k_1 - k_2; 1, -1, -1).\end{aligned} \tag{4.105}$$

By the estimates derived in Sect. 4.2.2.1, we then have $\Omega_4 < 0$ uniformly in the case with pinning. In addition, if $\delta = \frac{1}{2}$, the only way to get $\Omega_4 = 0$ is that both of the terms above are zero, which happens only if $k_1 = 0$ or $k_1 = k_0$, *and* $k_2 = 0$ or $k_2 = k_0 - k_1$, i.e., $k_3 = 0$. This implies that the only solutions to $\Omega_4 = 0$ are the *trivial solutions* where two of the wave numbers k_1, k_2, k_3 are zero and the remaining one is equal to k_0.

The final degenerate case is found if $\sum_{i=1}^{3} \sigma_i = 1$. Then only one of σ_i is negative; for notational simplicity let us suppose it is σ_1 (the other cases can then be obtained by permutation of the indices). The momentum conservation implies then that $k_1 - k_0 = k_2 + k_3$, and hence

$$\begin{aligned} \Omega_4 &= \omega_0 - \omega_1 + \omega_2 + \omega_3 \\ &= \Omega_3(k_0, k_1, k_0 - k_1; 1, -1, 1) + \Omega_3(k_2, k_3, k_2 + k_3; 1, 1, -1) . \end{aligned} \tag{4.106}$$

Thus $\Omega_4 = 0$ only for the trivial solution in the unpinned case, namely, if $k_1 = k_0$ and $k_2 = 0 = k_3$.

The remaining sign-combinations have $1 + \sum_{i=1}^{3} \sigma_i = 0$, i.e., the polarization of phonons is preserved in the collision. These are also called collisions which *conserve the phonon number*. From the above results we can already conclude that every term in the collision operator which is not phonon number conserving has only trivial solutions to the energy constraint, and these solutions all require that two of the k_i, $i = 1, 2, 3$, are equal to zero and the last one is equal to k_0.

In particular, there are no trivial solutions in the pinned case, so these can again be safely neglected. The same holds for the FPU chains, due to the prefactor $\prod_{\ell=0}^{3} |\sin(\pi k_\ell)|$, and since the solution involves a two-dimensional integral and an integrand which has only a point-singularity. To have more convincing an argument, one should choose a regularization of the δ-function and to show that the contribution from the resulting ordinary integrals vanishes as the regularization is removed. An example of such a procedure is given in [15, Sect. 3] where it is used for showing that the trivial solutions in the number conserving case do not contribute to the linearized collision operator.

Here we are interested in the collision operators defined in (4.94) and (4.96). Using the permutation properties of the integrands multiplying the constraint δ-functions, we thus arrive at the following simplified forms for the collision operators: in the FPU chains, the standard kinetic argument yields

$$\begin{aligned} \mathscr{C}[W](k_0) = 9\pi\, (2\lambda_4)^2 \overline{\omega}^{-8} \int_{\mathbb{T}^3} \mathrm{d}k\, \delta(k_0 + k_1 - k_2 - k_3) \delta(\omega_0 + \omega_1 - \omega_2 - \omega_3) \prod_{\ell=0}^{3} \omega_\ell \\ \times (W_1 W_2 W_3 + W_0 W_2 W_3 - W_0 W_1 W_3 - W_0 W_1 W_2) , \end{aligned} \tag{4.107}$$

and for onsite nonlinear interactions

$$\mathscr{C}[W](k_0) = \frac{9\pi}{4}\lambda_4^2 \int_{\mathbb{T}^3} \mathrm{d}k\, \delta(k_0 + k_1 - k_2 - k_3)\delta(\omega_0 + \omega_1 - \omega_2 - \omega_3) \prod_{\ell=0}^{3} \omega_\ell^{-1}$$
$$\times (W_1 W_2 W_3 + W_0 W_2 W_3 - W_0 W_1 W_3 - W_0 W_1 W_2) \,. \tag{4.108}$$

The two operators are nearly identical, differing only by the powers of the $\omega(k_\ell)$ factors. This difference however has an important consequence to the properties of the solutions to these two equations: the onsite equation (4.108) predicts finite thermal conductivity from the Green–Kubo formula, whereas the FPU chain equation predicts an infinite result. Qualitatively, one can understand this by noticing that in the unpinned case ω has zeroes which leads to enhancement of collisions for the onsite anharmonicity, but which suppresses them in the FPU case. However, the qualitative argument is not sufficient to determine the magnitude of the effect, and we need to do a more careful study of the properties of the solutions to obtain predictions about the thermal conductivities.

4.2.2.3 Integration of the Collisional Constraints in FPU Chains

Let us begin with the unpinned nearest neighbour models relevant to the FPU chains. Since then $\omega(k) = \sqrt{2\overline{\omega}}|\sin(\pi k)|$, it is analytically simpler to reparametrize the wave-number integrals by changing variables from $k \in \mathbb{T}$ to $p = 2\pi k$ which belongs to the half-open interval $I = [0, 2\pi)$. Then $\omega(k)$ is replaced by $\tilde{\omega}(p) = \sqrt{2}\sin(p/2)$ where the absolute value is not needed since it was possible to restrict the values of p to an interval where $\sin(p/2)$ is positive.

Although very useful for derivation of explicit parametrizations of the solution manifold, it should be stressed that some care is needed with this choice: *it is crucial below that all arithmetic involving $p \in I$ is performed modulo I.* For instance, if $p_1 = \pi/2$ and $p_2 = 3\pi/2$, we have to use $p_1 - p_2 = +\pi$ in $\tilde{\omega}(p_1 - p_2)$ to get its value correctly. This rule of "modulo I arithmetic" will be applied without further mention below.

We make the change of variables also in the spatially homogeneous Wigner function and consider $\tilde{W}_t(p) = W_t(p/(2\pi))$. Let us for simplicity drop the tilde from here and from the reparametrized dispersion relation, and denote them simply by $W_t(p)$ and $\omega(p)$. With these conventions, the phonon Boltzmann equation of the FPU chain becomes $\partial_t W_t(p) = \mathscr{C}[W_t](p)$ where

$$\mathscr{C}[W](p_0) = \frac{9}{\pi}\lambda_4^2\overline{\omega}^{-8} \int_{I^2} \mathrm{d}p_1\mathrm{d}p_2\, \delta(\Omega(p)) \prod_{\ell=0}^{3} \omega_\ell$$
$$\times (W_1 W_2 W_3 + W_0 W_2 W_3 - W_0 W_1 W_3 - W_0 W_1 W_2) \,, \tag{4.109}$$

with $\omega_\ell = \omega(p_\ell)$, $W_\ell = W(p_\ell)$, and

$$\Omega(p) = \omega_0 + \omega_1 - \omega_2 - \omega_3 \,, \quad p_3 = (p_0 + p_1 - p_2) \bmod I \,. \tag{4.110}$$

The detailed solution of the collisional constraints, $\Omega(p) = 0$ in (4.110), can be found in [15]. By Corollary 3.3. there, we can conclude that both constraints are satisfied exactly for those p_ℓ, $\ell = 0, 1, 2, 3$, which satisfy one of the following three relations

1. $p_2 = p_0$,
2. $p_1 = p_2$, or
3. $p_1 = h(p_0, p_2) \bmod I$, where h is defined using the standard, non-periodic, arithmetic in

$$h(x, y) = \frac{y - x}{2} + 2 \arcsin\left(\tan \frac{|y - x|}{4} \cos \frac{y + x}{4} \right) \tag{4.111}$$

with arcsin denoting the principal branch with values in $[-\pi/2, \pi/2]$.

The solutions satisfying item 1 or 2 are called *perturbative* or *trivial*, while the solution satisfying 3 is called *non-perturbative*. The nomenclature can be understood by expanding the constraint Ω around small values of $p_\ell \in \mathbb{R}$ using $\omega(p) \approx 2^{-\frac{1}{2}}|p|(1 - p^2/24)$. The resulting equation then has only 1 and 2 as its solutions.

The remaining energy conservation δ-function can then be formally resolved by integrating over a suitably chosen direction in the p_1, p_2-variables. For instance, choosing the p_1-integral for this purpose would yield for any $p_2 \neq p_0$ and for any continuous periodic function G,

$$\begin{aligned} \int_I \mathrm{d}p_1 \, \delta(\Omega(p)) G(p_0, p_1, p_2) = {} & \frac{1}{|\partial_2 \Omega(p_0, p_2, p_2)|} G(p_0, p_2, p_2) \\ & + \frac{1}{|\partial_2 \Omega(p_0, h(p_0, p_2), p_2)|} G(p_0, h(p_0, p_2), p_2). \end{aligned} \tag{4.112}$$

However, this procedure needs to be used with some care: for instance, we have $\int_I \mathrm{d}p_2 \, |\partial_2 \Omega(p_0, p_2, p_2)|^{-1} = \infty$, and thus the first term on the right hand side of (4.112) would typically diverge when integrated over p_2.

This problem is resolved in the collision operator by the alternating signs in its integrand which guarantee that the integrand vanishes at the trivial solutions. This can even be proven rigorously if $1/W$ is sufficiently regular, say twice continuously differentiable, by using the following equivalent form for the collision operator:

$$\frac{9}{\pi} \lambda_4^2 \overline{\omega}^{-8} \int_{I^2} \mathrm{d}p_1 \mathrm{d}p_2 \, \delta(\Omega(p)) \prod_{\ell=0}^{3} (\omega_\ell W_\ell) \left(W_0^{-1} + W_1^{-1} - W_2^{-1} - W_3^{-1} \right) . \tag{4.113}$$

However, we do not wish to go into more detail here, but just add the contribution from the trivial solutions to the class of terms which are assumed to be negligible in the kinetic theory of FPU lattices.

If the trivial solutions are neglected, we can use results from [15], which rely on the explicit form of the non-trivial solution h, and obtain the fully integrated form

$$\mathscr{C}[W](p_0) = \frac{9\sqrt{2}}{\pi}\lambda_4^2\overline{\omega}^{-8}\int_0^{2\pi} \mathrm{d}p_2 \frac{1}{\sqrt{F_+(p_0,p_2)}}\prod_{\ell=0}^{3}\omega_\ell \\ \times (W_1W_2W_3 + W_0W_2W_3 - W_0W_1W_3 - W_0W_1W_2) \, , \tag{4.114}$$

where $p_1 = h(p_0, p_2)$, $p_3 = p_0 + p_1 - p_2$, and

$$F_\pm(x,y) = \left(\cos\frac{x}{2} + \cos\frac{y}{2}\right)^2 \pm 4\sin\frac{x}{2}\sin\frac{y}{2} \, . \tag{4.115}$$

(See for instance, Lemma 3.4 in [15] for more details. From the Lemma a weak convergence of the integrals with a regularized δ-function $\delta_\varepsilon(\Omega) = \varepsilon\pi^{-1}(\varepsilon^2 + \Omega^2)^{-1}$, $\varepsilon > 0$, and assuming continuity of W, can be established.)

The function F_- defined in (4.115) is related to the change of variables which corresponds to using p_2 instead of p_1 to integrate out the energy δ-function. Namely, as proven in Lemma 3.5 in [15],

$$\int_{I^2} \mathrm{d}p_0\mathrm{d}p_2 \frac{1}{\sqrt{F_+(p_0,p_2)}} G(p_0, h(p_0,p_2)) \\ = 2\int_{I^2} \mathrm{d}p_0\mathrm{d}p_1 \frac{\mathbb{1}(F_-(p_0,p_1) > 0)}{\sqrt{F_-(p_0,p_1)}} G(p_0,p_1) \, , \tag{4.116}$$

for any G for which either of the two integrals converges. The characteristic function restricts the integral to the subset of I^2 in which the argument of the square root is positive. Let us also mention that the change of variables becomes more involved if G also depends on p_2 directly since each pair (p_0, p_1), for which $F_-(p_0, p_1) > 0$, has two distinct values p_2 solving the energy constraint. Further details about these solutions can be found from the proof of the above mentioned Lemma 3.5.

4.2.2.4 Integration of the Collisional Constraints for Onsite Nonlinearity

To allow a comparison, let us also consider the case with an onsite nonlinearity. If the harmonic part has no pinning, then the above FPU discussion applies immediately, since these models have the same dispersion relation and differ only by the factors in the integrand. We thus find the following integrated form of the collision operator, if $\delta = \frac{1}{2}$ in the general nearest neighbour dispersion relation:

$$\mathscr{C}[W](p_0) = \frac{9}{2^{7/2}\pi}\lambda_4^2 \int_0^{2\pi} \mathrm{d}p_2 \frac{1}{\sqrt{F_+(p_0, p_2)}} \prod_{\ell=0}^{3} \omega_\ell^{-1}$$
$$\times (W_1W_2W_3 + W_0W_2W_3 - W_0W_1W_3 - W_0W_1W_2) \,, \qquad (4.117)$$

where $p_1 = h(p_0, p_2)$, $p_3 = p_0 + p_1 - p_2$, and F_+ has been defined in (4.115). (We continue to neglect contributions from the trivial solutions.)

The analytic structure of the solution manifold gets more complicated once the dispersion relation has pinning, i.e., when $\delta < \frac{1}{2}$. Nevertheless, it is still possible to find a function $h(p_0, p_2)$ such that the non-perturbative solution is parametrized by a condition

$$p_1 = h(p_0, p_2; \delta) \,. \qquad (4.118)$$

In other words, the enumeration by the three conditions in the previous section continues to hold, only with a function h which depends on δ. Naturally, $h(p_0, p_2; \frac{1}{2})$ is then given by (4.111).

Since to our knowledge the explicit form of the solution function h is not available in the literature, let us present some details for its derivation. For the moment, let us return to standard arithmetic for p_ℓ and consider $p_i \in \mathbb{R}$, $i = 0, 1, 2$, and take $p_3 = p_0 + p_1 - p_2$. Next change variables from p_ℓ to

$$u = \frac{p_2 - p_0}{2} = \frac{p_1 - p_3}{2} \,, \quad v = \frac{p_2 + p_0}{2} \,, \quad w = \frac{p_1 + p_3}{2} \,. \qquad (4.119)$$

Then

$$p_1 = w + u \,, \quad p_3 = w - u \,, \quad p_2 = v + u \,, \quad p_0 = v - u \,. \qquad (4.120)$$

Therefore, the energy constraint $\Omega_4 = 0$ is equivalent to

$$g(w, u) = g(v, u) \,, \qquad (4.121)$$

where, for simplicity, we set $\overline{\omega} = 1$ and then define

$$g(v, u) = \sqrt{1 - 2\delta \cos(v + u)} - \sqrt{1 - 2\delta \cos(v - u)} = \omega_2 - \omega_0 \,, \qquad (4.122)$$

and thus $g(w, u) = \omega_1 - \omega_3$.

Since $g(v, \pi n) = 0$ for all v and $n \in \mathbb{Z}$, (4.121) is solved by any $v, w \in \mathbb{R}$ if $u \in \pi\mathbb{Z}$. This corresponds to the trivial solution $p_2 = p_0 \bmod 2\pi$.

Assume thus $u \notin \pi\mathbb{Z}$. To fix a sign convention, let us next suppose that $w, v \in J = (-\pi, \pi]$ which can always be achieved without changing u by shifting p_ℓ by a suitably chosen integer multiple of 2π. Since

$$g(v,u) = 2\delta\frac{\cos(v-u) - \cos(v+u)}{\omega_2 + \omega_0} = 4\delta\frac{\sin u \sin v}{\omega_2 + \omega_0}, \tag{4.123}$$

where $\sin u \neq 0$ and $\omega_\ell \geq 0$, (4.121) can only hold if $\sin v$ and $\sin w$ have the same sign. For $w, v \in J$ this is equivalent to requiring $\text{sign}(w) = \text{sign}(v)$.

To solve (4.121), we first note that

$$g(v,u)^2 = 2(1 - 2\delta\cos u \cos v - \omega_0\omega_2), \tag{4.124}$$

$$g(w,u)^2 = 2(1 - 2\delta\cos u \cos w - \omega_1\omega_3). \tag{4.125}$$

Hence, if (4.121) is true, we need to have

$$\omega_1\omega_3 = -2\delta\cos u(\cos w - \cos v) + \omega_0\omega_2. \tag{4.126}$$

We square both sides one more time, and use the identities

$$\omega_1^2\omega_3^2 = 1 - 4\delta\cos u\cos w + 4\delta^2\cos^2 w - 4\delta^2\sin^2 u, \tag{4.127}$$

$$\omega_0^2\omega_2^2 = 1 - 4\delta\cos u\cos v + 4\delta^2\cos^2 v - 4\delta^2\sin^2 u, \tag{4.128}$$

which follow from the trigonometric relation $\cos(a+b)\cos(a-b) = \cos^2 a - \sin^2 b$.

The result can be written in terms of the variable $y = \cos w - \cos v$, and we find that (4.121) implies the equation

$$y^2 4\delta^2\sin^2 u - y4\delta(\cos u(1 - \omega_0\omega_2) - 2\delta\cos v) = 0. \tag{4.129}$$

Since $\sin u \neq 0$, this equation has exactly two solutions for y. The solution $y = 0$ has $\cos w = \cos v$ which yields only the solution $w = v$ for (4.121) with $w, v \in J$ since then w and v must have the same sign. But then $p_2 = p_1$, so $y = 0$ corresponds to the second trivial solution.

This leaves only the following candidate for a non-perturbative solution:

$$y = \frac{\cos u(1 - \omega_0\omega_2) - 2\delta\cos v}{\delta\sin^2 u}. \tag{4.130}$$

The denominator appears to lead to singularities at $u = 0$ and at $\delta = 0$. However, both singularities are removable. Namely, (4.130) implies that

$$y + 2\cos v = \cos u\frac{1 - 2\delta\cos u\cos v - \omega_0\omega_2}{\delta\sin^2 u} = \frac{4\delta\cos u\sin^2 v}{1 - 2\delta\cos u\cos v + \omega_0\omega_2}. \tag{4.131}$$

Since the left hand side is equal to $\cos w + \cos v$ and w, v have the same sign, we again get only one solution which expresses w as a function of u, v.

In summary, the above computation yields the following expression for the non-perturbative solution. Choose $p_0, p_2 \in J = (-\pi, \pi]$. Then also $v \in J$ and we have $p_1 = h(p_0, p_2; \delta) \bmod 2\pi$, where the non-perturbative solution is given by

$$h(p_0, p_2; \delta) = \frac{p_2 - p_0}{2} + \operatorname{sign}\left(\frac{p_2 + p_0}{2}\right) \arccos\Bigg[-\cos\frac{p_2 + p_0}{2} + 2\delta \frac{\sin p_0 + \sin p_2}{1 - \delta(\cos p_0 + \cos p_2) + \sqrt{(1 - 2\delta\cos p_0)(1 - 2\delta\cos p_2)}} \times \sin\frac{p_2 + p_0}{2} \Bigg] \tag{4.132}$$

with $\arccos \in [0, \pi]$ denoting the principal branch. In addition, to have the correct value of h at the apparent discontinuity $p_2 = -p_0$, we also define $\operatorname{sign}(0) = 1$: this yields $h(p_0, p_2) \to \pi - p_0 = h(p_0, -p_0)$, as $p_2 \searrow -p_0$, and $h(p_0, p_2) \to h(p_0, -p_0) - 2\pi$, as $p_2 \nearrow -p_0$. We skip the rest of the details, namely proving that the above arccos is always well-defined (i.e., its argument lies in $[-1, 1]$) and that the result always provides a solution to the original constraint problem.

In general, for a fixed p_2 the function h is continuous, one-to-one, and satisfies

$$\omega(p_0) + \omega(h(p_0, p_2)) - \omega(p_2) - \omega(p_0 + h(p_0, p_2) - p_2) = 0\,. \tag{4.133}$$

In Fig. 4.1 we display a few non-perturbative solutions, at $\delta = 0.4$ and at $\delta = 0.5$, for three different values of p_2. For small δ one finds

$$h(p_0, p_2; \delta) = \pi - p_0 - \delta\big(\sin p_0 + \sin p_2\big) + \mathscr{O}(\delta^2)\,, \tag{4.134}$$

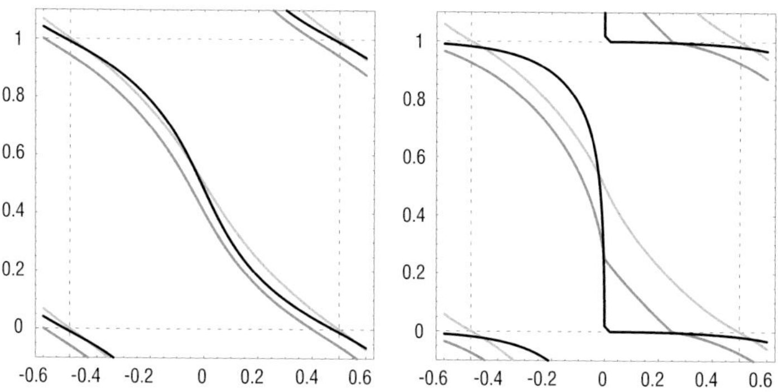

Fig. 4.1 The non-perturbative solution $k_1 = \frac{1}{2\pi} h(2\pi k_0, 2\pi k_2; \delta)$ as a function of k_0 for fixed k_2 and δ. On the *left*, $\delta = 0.4$, and on the *right*, $\delta = 0.5$. For each δ, three solutions are plotted, corresponding to $k_2 = 0.02$ (*black*), 0.25 (*dark grey*), and 0.5 (*light grey*). We plot $\mathbb{T} \times \mathbb{T}$ in the extended zone scheme, the *dashed lines* are the boundaries of a unit cell. For $\delta < 0.5$ the non-perturbative solution is smooth, while for $\delta = 0.5$ there are cusp singularities (reprinted from [1])

which reasonably well approximates the left hand side of Fig. 4.1. The existence of the above limiting solution was first noted in [11] and it served as an important motivation for looking closer at the predictions from one-dimensional kinetic theory of phonons.

To resolve the collisional constraints, we integrate over p_1, as in the FPU case. Since $\partial_{p_1}\Omega(p_0,p_1,p_2) = \omega'(p_1) - \omega'(p_0+p_1-p_2)$ and $\omega'(p) = \delta \sin p/\omega(p)$, this yields the collision operator

$$\mathscr{C}[W](p_0) = \frac{9}{16\pi\delta}\lambda_4^2 \int_{-\pi}^{\pi} \mathrm{d}p_2 \, \frac{1}{\omega_0\omega_2|\omega_3 \sin p_1 - \omega_1 \sin p_3|} \times (W_1W_2W_3 + W_0W_2W_3 - W_0W_1W_3 - W_0W_1W_2) \,, \tag{4.135}$$

where $p_1 = h(p_0,p_2;\delta)$ and $p_3 = p_1 + p_0 - p_2$.

4.3 Energy Transport in the Kinetic Theory of Phonons

4.3.1 Entropy and H-Theorem of the Phonon Boltzmann Equations

For ergodic systems, time averages will converge to ensemble averages. We have been working under the assumption that energy is the only ergodic variable for our particle chains. In addition, for infinite volume systems the canonical and microcanonical ensembles should agree, so to study the convergence towards equilibrium, one could inspect the evolution of the standard entropy density. Since it is maximized for the given energy exactly by the canonical ensemble, this should yield an increasing function, at least asymptotically for large times.

Consider thus a probability measure $\rho(q,p)\mathrm{d}^N q\, \mathrm{d}^N p$ for the state of the finite periodic system with N particles. The corresponding entropy is defined by

$$S[\rho] = -\int \mathrm{d}^N q\, \mathrm{d}^N p\, \rho(q,p) \ln \rho(q,p) \,. \tag{4.136}$$

For small couplings, the canonical Gibbs measure is approximately Gaussian. In general, for measures which are nearly Gaussian in the random vector X, and assuming that X has a mean μ and a covariance matrix C, we can use an approximation $\ln \rho(q,p) \approx -\frac{1}{2}\left(\ln[\det(2\pi C)] + (X-\mu)^T C^{-1}(X-\mu)\right)$. This implies that the entropy of such nearly Gaussian measures satisfies

$$S[\rho] \approx \frac{1}{2}\ln[\det(2\pi C)] + \frac{1}{2}\operatorname{Tr} 1 \,. \tag{4.137}$$

Thus the entropy per particle, $N^{-1}S[\rho]$, is then approximately equal to a constant plus $\frac{1}{2N}\ln(\det C)$. For a spatially homogeneous particle chain, one has

$C = \mathscr{F}^{-1}\widehat{C}\mathscr{F}$ where $\mathscr{F}$ denotes the Fourier-transform. Therefore, for spatially homogeneous nearly Gaussian states and large enough N

$$N^{-1}S[\rho] \approx \text{constant} + \frac{1}{2}\int \mathrm{d}k \ln \det F(k)\,, \tag{4.138}$$

where $F(k)$ is a 2×2 matrix defined by

$$F(k) = \int \mathrm{d}k' \begin{pmatrix} \kappa[\hat{q}(k')^*, \hat{q}(k)] & \kappa[\hat{q}(k')^*, \hat{p}(k)] \\ \kappa[\hat{q}(k')^*, \hat{p}(k)] & \kappa[\hat{p}(k')^*, \hat{p}(k)] \end{pmatrix}. \tag{4.139}$$

This can be expressed in terms of the phonon fields, and hence their Wigner function $W(k)$, using the relations given in (4.6):

$$F(k) = \frac{1}{2}\begin{pmatrix} \omega(k)^{-1}(W(k) + W(-k)) & -\mathrm{i}(W(k) - W(-k)) \\ \mathrm{i}(W(k) - W(-k)) & \omega(k)(W(k) + W(-k)) \end{pmatrix}. \tag{4.140}$$

Therefore, $\det F(k) = W(k)W(-k)$, and thus we obtain the following definition of entropy density for phonon systems in a kinetic scaling limit

$$S[W] = \int \mathrm{d}k \ln W(k)\,. \tag{4.141}$$

After the above preliminaries, it is satisfying to find out that the above reasoning leads to an entropy functional which satisfies an H-theorem for phonon Boltzmann equations. This holds quite generally but let us check it explicitly only for the homogeneous Boltzmann equations $\partial_t W(k) = \mathscr{C}[W(k)]$ where the collision operator $\mathscr{C}$ is given either by (4.107) or by (4.108). In both cases, there is a positive function G such that

$$\begin{aligned} \mathscr{C}[W](k_0) = &\int_{\mathbb{T}^3} \mathrm{d}k\, \delta(k_0 + k_1 - k_2 - k_3)\delta(\omega_0 + \omega_1 - \omega_2 - \omega_3) \prod_{\ell=0}^{3} [G(\omega_\ell)W_\ell] \\ &\times \left(W_0^{-1} + W_1^{-1} - W_2^{-1} - W_3^{-1}\right). \end{aligned} \tag{4.142}$$

Since the integrand, including both constraints, is symmetric under $0 \leftrightarrow 1$ and $2 \leftrightarrow 3$, and antisymmetric under $(0, 1) \leftrightarrow (2, 3)$, we find that

$$\begin{aligned} \partial_t S[W_t] &= \int \mathrm{d}k_0 W_0^{-1} \partial_t W_t(k_0) \\ &= \frac{1}{4}\int_{\mathbb{T}^{\{0,1,2,3\}}} \mathrm{d}k\, \delta(k_0 + k_1 - k_2 - k_3)\delta(\omega_0 + \omega_1 - \omega_2 - \omega_3) \\ &\quad \times \prod_{\ell=0}^{3} [G(\omega_\ell)W_\ell] \left(W_0^{-1} + W_1^{-1} - W_2^{-1} - W_3^{-1}\right)^2 \geq 0\,. \end{aligned} \tag{4.143}$$

Hence, the entropy $S[W_t]$ is increasing along solutions of the Boltzmann equation, i.e., it satisfies an H-theorem. The entropy production functional $D[W]$, satisfying $\partial_t S[W_t] = D[W_t]$, also has an explicit form which can be read off from the right hand side of (4.143).

4.3.2 *Steady States*

The H-theorem (4.143) allows to classify all steady states of the phonon Boltzmann equation. Namely, suppose that $\bar{W}$ is a steady state, i.e., $W_t = \bar{W}$ is a solution to the phonon Boltzmann equation. Since then the left hand side of (4.143) is zero, the integrand defining the entropy production has to vanish almost everywhere on the solution manifold.

We are only interested in nondegenerate steady states which correspond to functions W for which $W > 0$ almost everywhere. Then it is possible to apply the previous explicit solutions $p_1 = h$ of the collisional constraints to simplify the above problem. Namely, for any steady state W its inverse $f(p) = W(p)^{-1}$ then necessarily satisfies

$$f(p_0) + f(h(p_0, p_2; \delta)) - f(p_2) - f(p_0 - p_2 + h(p_0, p_2; \delta)) = 0\,, \qquad (4.144)$$

for almost every p_0 and p_2.

Due to the collisional constraints, any linear combination $f(k) = \beta\omega(k) + \alpha$ is obviously a solution to the functional equation (4.144). As shown in [20, 21], these are quite generally the only solutions for two- and higher-dimensional crystals.

The one-dimensional case is more intricate. For instance, $h(p_0, p_2; \delta) \to \pi - p_0$ when $\delta \to 0$. Hence, in the limit, any function f satisfying the symmetry requirement $f(p) = -f(\pi - p)$ is a solution to (4.144). However, we do not expect that any new solutions would appear for the nearest neighbour dispersion relations, i.e., if $\delta > 0$. This has even been rigorously proven to be the case for the unpinned dispersion relation with $\delta = \frac{1}{2}$; see Sect. 5 in [15] for details.

In summary, we expect that the only steady states for which W is integrable, are given by

$$W^{(\mathrm{eq})}(k) = \frac{1}{\beta(\omega(k) - \mu)}\,, \qquad (4.145)$$

where $\beta > 0$ and $\mu < \min \omega$. The last condition is necessary, since if there were a point k_0 for which $\omega(k_0) = \mu$, this would lead to a nonintegrable singularity at k_0 for W (the singularity $|x|^{-1}$ is not integrable around $x = 0$ in one dimension).

Using (4.142) it is straightforward to check that indeed $\mathscr{C}[W^{(\mathrm{eq})}] = 0$ for the functions defined in (4.145). Hence, all of them are true steady states of the phonon Boltzmann equation. The conservation of phonon number is reflected in the appearance of the second parameter μ for the steady states. It is expected to be a spurious chemical potential, and one expects that $\mu \to 0$ eventually as $t \to \infty$. However, this process is not captured by the standard kinetic approximation which we have applied here, and resolving the issue would require a different approach.

Let us also mention in passing that there are additional solutions to $\mathscr{C}[W] = 0$ if one allows W to be a distribution. For instance, $W(k) = \delta(k)$ formally provides a solution to the FPU collision operator. This is not surprising since the additional constraints given by such distributions can restrict the solution manifold further. In general, one should treat such distributional solutions with great care since it is far from obvious that the oscillatory terms, which are neglected in the derivation of the phonon Boltzmann equation, remain lower order contributions for such non-chaotic initial data. However, there are known examples where distributional solutions seem to play a physical role, such as in the kinetic theory of Bose-Einstein condensation: see [12] and the references therein for more details.

4.3.3 *Green–Kubo Formula and the Linearized Boltzmann Equation*

Let us now return to the Green–Kubo formula for thermal conductivity. We consider here the kinetic prediction for the leading harmonic part of the current-current correlation function, as derived in Sect. 4.1.5. We choose the basic one-parameter canonical Gibbs measure at temperature $T > 0$ as the initial steady state. Its limiting covariance as $\lambda \to 0$ then has a Wigner function $W^{(\mathrm{eq})}$ as given in (4.145) with $\beta = 1/T$ and $\mu = 0$. (As discussed earlier, we expect the true steady states to have $\mu = 0$ even though the steady states of the kinetic model can have $\mu \neq 0$.)

In fact, the results in Sect. 4.1.5 imply that the evolution of the harmonic part of the Green–Kubo correlator is determined by the phonon Boltzmann equation linearized around its stationary solution $W^{(\mathrm{eq})}(k;\beta) = 1/(\beta\omega(k))$. We begin with the following approximation for the Green–Kubo correlator obtained from (4.69) and (4.71):

$$C(t;\beta) \approx \overline{\omega}^4 \delta^2 \int_{\mathbb{T}} \mathrm{d}k\, \phi_0(k) \partial_\varepsilon W_t^{(\varepsilon)}(k)|_{\varepsilon=0}\,, \tag{4.146}$$

where we have written the covariance in terms of the Wigner function $W_t^{(\varepsilon)}(k)$ and defined

$$\phi_0(k) = \sin(2\pi k)\,. \tag{4.147}$$

The Wigner function is computed using expectation $\langle\cdot\rangle^{(\varepsilon)}$ over the stochastic process which starts from a perturbation of the stationary state whose initial Wigner function converges to $W_0^{(\varepsilon)}(k) = 1/(\beta\omega(k) - \varepsilon\phi_0(k))$ as $\lambda \to 0$.

Since $\varepsilon|\phi_0(k)| \leq 2\pi\varepsilon|k|$, the Wigner function is positive for sufficiently small $\varepsilon > 0$, even in the FPU-models for which $\omega(0) = 0$: this follows from the observation that it is always possible to find some $c_0 > 0$ such that $\omega(k) \geq c_0|k|$ in a neighbourhood of 0. Hence, $h_t(k) = \partial_\varepsilon W_t^{(\varepsilon)}(k)|_{\varepsilon=0}$ is well defined and its initial value is given by $h_0 = (W^{(\mathrm{eq})})^2\phi_0$. Its time-evolution can be determined using the equality $\partial_t h_t(k) = \partial_\varepsilon\partial_t W_t^{(\varepsilon)}(k)|_{\varepsilon=0}$ and the kinetic theory approximation $\partial_t W_t^{(\varepsilon)} \approx \mathscr{C}[W_t^{(\varepsilon)}]$ where $\mathscr{C}$ denotes the appropriate phonon Boltzmann collision operator. The explicit form is straightforward to compute using (4.142) and the fact that $\delta(\omega_0+\omega_1-\omega_2-\omega_3)\left(W_0^{-1} + W_1^{-1} - W_2^{-1} - W_3^{-1}\right) = 0$, for any of the functions $W = W^{(\mathrm{eq})}$ in (4.145). We obtain the linearized Boltzmann equation $\partial_t h_t = -\mathscr{L}h_t$ where $\mathscr{L}$ is the operator for which $\mathscr{L}h = LW^{-2}h$ with W^{-1} denoting multiplication by the function $\beta\omega(k)$ and

$$(Lf)(k_0) = \int_{\mathbb{T}^3} \mathrm{d}k\, \delta(k_0 + k_1 - k_2 - k_3)\delta(\omega_0 + \omega_1 - \omega_2 - \omega_3) \times \prod_{\ell=0}^{3} [G(\omega_\ell)W_\ell]\,(f_0 + f_1 - f_2 - f_3)\,, \tag{4.148}$$

with $W_\ell = 1/(\beta\omega(k_\ell))$ and $f_\ell = f(k_\ell)$.

The operator L is self-adjoint, even positive, on the Hilbert space $L^2(\mathbb{T})$: if $f \in L^2(\mathbb{T})$ is any sufficiently regular function, then by the symmetry properties of the integrand, the scalar product between f and Lf satisfies

$$\langle f, Lf\rangle = \frac{1}{4}\int_{\mathbb{T}^{\{0,1,2,3\}}} \mathrm{d}k\, \delta(k_0 + k_1 - k_2 - k_3)\delta(\omega_0 + \omega_1 - \omega_2 - \omega_3) \times \prod_{\ell=0}^{3} [G(\omega_\ell)W_\ell] \times |f_0 + f_1 - f_2 - f_3|^2 \geq 0\,. \tag{4.149}$$

In fact, the positivity of the operator L is closely related to the maximization of entropy at the steady state. Namely, by differentiating the entropy production (4.143) twice, we find

$$\partial_t\partial_\varepsilon^2 S[W_t^{(\varepsilon)}]|_{\varepsilon=0} = 2\langle f, Lf\rangle\,, \quad \text{with } f = W^{-2}h_t\,, \tag{4.150}$$

where we have used the above mentioned property that the integrand vanishes whenever either of the two factors involving differences of W^{-1} is left undifferentiated.

Since L is positive operator, any operator of the form $A^\dagger LA$, where A is a bounded operator and $A^\dagger$ denotes its adjoint, is also positive. Therefore, the operator $\tilde{L} = W^{-1}LW^{-1}$ is positive on $L^2(\mathbb{T})$, and thus the solution of the evolution equation $\partial_t\tilde{h}_t = -\tilde{L}\tilde{h}_t$ can be written as $\tilde{h}_t = \mathrm{e}^{-t\tilde{L}}\tilde{h}_0$ where each $\mathrm{e}^{-t\tilde{L}}$ is a contraction operator on $L^2(\mathbb{T})$.

These properties indicate that it is more natural to study the perturbations in terms of $\tilde{h}_t$ instead of h_t: any solution to $\partial_t h_t = -\mathscr{L} h_t$ provides a solution to $\partial_t \tilde{h}_t = -\tilde{L}\tilde{h}_t$ by setting $\tilde{h}_t = W^{-1} h_t$ and vice versa. Therefore, the Boltzmann equation linearized around a steady state W has a solution

$$h_t = W e^{-t\tilde{L}} W^{-1} h_0 \,, \tag{4.151}$$

for every initial perturbation h_0 for which $W^{-1}h_0 \in L^2(\mathbb{T})$.

As mentioned above, the Green–Kubo correlation function concerns the case with $h_0(k) = W(k)^2 \sin(2\pi k)$ and $W(k) = \beta^{-1}\omega(k)^{-1}$. Thus $W^{-1}h_0$ is equal to

$$\tilde{h}_0(k) = \beta^{-1} \frac{\sin(2\pi k)}{\omega(k)} \,, \tag{4.152}$$

which is a bounded function on the torus; for FPU-like models with $\omega(k) = O(|k|)$, the function h_0 is not continuous at $k = 0$ but its left and right limits exist and are finite. Hence, we can use the solution in (4.151) and conclude that the Green–Kubo correlation function at the steady state with temperature β^{-1} has a kinetic theory approximation

$$C(t;\beta) \approx \overline{\omega}^4 \delta^2 \langle \phi_0, h_t \rangle = \overline{\omega}^4 \delta^2 \int_{\mathbb{T}} \mathrm{d}k\, \phi_0(k) W(k) \left(e^{-t\tilde{L}} \tilde{h}_0 \right)(k) = \overline{\omega}^4 \delta^2 \langle \tilde{h}_0, e^{-t\tilde{L}} \tilde{h}_0 \rangle \,. \tag{4.153}$$

The right hand side is equal to the L^2-norm $\|\overline{\omega}^2 \delta e^{-\frac{1}{2}t\tilde{L}} \tilde{h}_0\|^2$, in particular, it is always positive. The question about the prediction of kinetic theory for the thermal conductivity at a certain equilibrium state hence boils down to the decay of the above norm under the semigroup $e^{-t\tilde{L}}$. As we will show next, the two Boltzmann equations discussed above prove that both integrable and non-integrable decay can occur in the kinetic theory of phonons.

4.3.4 *Kinetic Theory Prediction for Thermal Conductivity in Chains with Anharmonic Pinning*

Suppose the positive operator $\tilde{L}$ has a spectral gap of size $\delta_0 > 0$ above its zero eigenvalue and $\tilde{h}_0$ is orthogonal to its eigenspace of zero. Then spectral theory implies a bound $\|e^{-\frac{1}{2}t\tilde{L}} \tilde{h}_0\|^2 \le \|\tilde{h}_0\|^2 e^{-t\delta_0}$ which is integrable over t. Hence, in this case the kinetic theory prediction is always a finite conductivity and its leading behaviour in λ_4 can be computed using the kinetic approximation in (4.59). This yields

$$\kappa(\beta^{-1}) \approx \beta^2\overline{\omega}^4\delta^2 \int_0^\infty \mathrm{d}r\, \langle \tilde{h}_0, \mathrm{e}^{-r\tilde{L}}\tilde{h}_0\rangle = \beta^2\overline{\omega}^4\delta^2 \lim_{\varepsilon\to 0+} \int_0^\infty \mathrm{d}r\, \langle \tilde{h}_0, \mathrm{e}^{-r(\varepsilon+\tilde{L})}\tilde{h}_0\rangle$$
$$= \lim_{\varepsilon\to 0+} \beta^2\overline{\omega}^4\delta^2 \langle \tilde{h}_0, (\varepsilon+\tilde{L})^{-1}\tilde{h}_0\rangle\,, \tag{4.154}$$

where the middle equality is a consequence of dominated convergence theorem.

We have added the regulator $\varepsilon > 0$ here to make the operator $\tilde{L}$ invertible on the whole Hilbert space. Then the scalar product $\langle \tilde{h}_0, (\varepsilon + \tilde{L})^{-1}\tilde{h}_0\rangle$ is finite even if $\tilde{h}_0$ is not orthogonal to the zero subspace of $\tilde{L}$. This has the benefit that one can use the full operator $\tilde{L}$, instead of its restriction to the orthocomplement of the zero eigenspace, and study the expectation of the resolvent,

$$R(\varepsilon) = \langle \tilde{h}_0, (\varepsilon+\tilde{L})^{-1}\tilde{h}_0\rangle\,, \quad \varepsilon > 0\,, \tag{4.155}$$

instead of the asymptotic behaviour of the semigroup.

The above result implies that, should $\lim_{\varepsilon\to 0+} R(\varepsilon) = \infty$, then the kinetic prediction for the thermal conductivity at the corresponding steady state is also infinite (note that both $\tilde{L}$ and $\tilde{h}_0$ depend on the choice of the steady state). In particular, this happens if $\tilde{h}_0$ is not orthogonal to the zero eigenspace of $\tilde{L}$: otherwise, we have $R(\varepsilon) \geq \|P_0\tilde{h}_0\|^2/\varepsilon \to \infty$, as $\varepsilon \to 0$, where P_0 denotes the orthogonal projection to the zero eigenspace. Hence, the orthogonality condition mentioned in the beginning of this subsection is necessary for a finite prediction from kinetic theory.

Let $\tilde{\mu}(\mathrm{d}\alpha)$ denote the spectral measure obtained from the spectral decomposition of $\tilde{L}$ with respect to the vector $\tilde{h}_0$. Then $\tilde{\mu}$ is a positive Borel measure on the real axis whose support lies on the spectrum of $\tilde{L}$, on $\sigma(\tilde{L}) \subset \mathbb{R}_+$, and it is characterized by the condition

$$\langle \tilde{h}_0, f(\tilde{L})\tilde{h}_0\rangle = \int \tilde{\mu}(\mathrm{d}\alpha) f(\alpha)\,, \tag{4.156}$$

which holds, in particular, for every continuous real function f. Thus $\|\tilde{h}_0\|^2 = \int \tilde{\mu}(\mathrm{d}\alpha)$, and

$$R(\varepsilon) = \int_{\sigma(\tilde{L})} \tilde{\mu}(\mathrm{d}\alpha)\, \frac{1}{\varepsilon+\alpha}\,. \tag{4.157}$$

If $\tilde{h}_0 = 0$, we have $R(\varepsilon) = 0$ for all ε, and thus in this case the kinetic theory prediction would be zero conductivity. If $\tilde{h}_0 \neq 0$, such as for the above discussed particle chains, we can normalize the measure $\tilde{\mu}$ into a probability measure by dividing it by $\|\tilde{h}_0\|^2$. In addition, the second derivative of the function $\alpha \mapsto (\varepsilon+\alpha)^{-1}$ is positive, and thus it is convex. Therefore, we can apply Jensen's inequality in (4.157) to conclude that

$$R(\varepsilon) = \|\tilde{h}_0\|^2 \int_{\sigma(\tilde{L})} \frac{\tilde{\mu}(\mathrm{d}\alpha)}{\|\tilde{h}_0\|^2} \frac{1}{\varepsilon + \alpha} \geq \frac{\langle \tilde{h}_0, \tilde{h}_0 \rangle^2}{\varepsilon \langle \tilde{h}_0, \tilde{h}_0 \rangle + \langle \tilde{h}_0, \tilde{L}\tilde{h}_0 \rangle} . \tag{4.158}$$

The right hand side converges to $\langle \tilde{h}_0, \tilde{h}_0 \rangle^2 / \langle \tilde{h}_0, \tilde{L}\tilde{h}_0 \rangle$ as $\varepsilon \to 0$. Since $\tilde{L} = W^{-1}LW^{-1}$ and $\tilde{h}_0 = W\phi_0$, we find that

$$\beta^2 \overline{\omega}^4 \delta^2 \frac{\langle W\phi_0, W\phi_0 \rangle^2}{\langle \phi_0, L\phi_0 \rangle} \tag{4.159}$$

forms a *lower bound* for the kinetic theory prediction of thermal conductivity in the particle chains.

By (4.149), ψ can be a zero eigenvector of $\tilde{L}$ if and only if it belongs to $L^2(\mathbb{T})$ and $f = W^{-1}\psi$ satisfies (4.144), i.e., it is a collisional invariant. We have argued (and even proven for $\delta = 1/2$) that all collisional invariants are linear combinations of ω and 1. This implies that the zero subspace of $\tilde{L}$ is spanned by 1 and W. For the particle chains, $\tilde{h}_0(k) = W(k)\phi_0(k) = \omega'(k)/(2\pi\delta\overline{\omega}^2\beta)$ where ω' denotes the derivative of ω. Therefore, by periodicity, $\int_{\mathbb{T}} \mathrm{d}k\, \tilde{h}_0(k)(aW(k) + b) = 0$ for all $a, b \in \mathbb{C}$, and thus $\tilde{h}_0$ is indeed orthogonal to the proposed zero subspace of $\tilde{L}$.

Let us also sketch a possible proof for the spectral gap of $\tilde{L}$. Consider an arbitrary ψ and denote $f = W^{-1}\psi$. We begin from the integral representation (4.149) for $\langle f, Lf \rangle = \langle \psi, \tilde{L}\psi \rangle$. Let us mollify the singularities in the integrand by choosing a suitable regularizing function $\Phi : \mathbb{T}^4 \to [0, 1]$ and then using the bound $1 \geq \Phi(k)$ inside the integrand. The purpose of Φ is to regularize the collision cross section of the model, so that in the remaining integral one can use the explicit solution and define an operator L' such that $\langle f, Lf \rangle \geq \langle f, L'f \rangle$ and $L' = V - K$ where V is a positive multiplication operator and K is a self-adjoint integral operator. If we can then tune Φ so that V and $1/V$ are both bounded and $B = V^{-1/2}KV^{-1/2}$ is a compact operator, it follows that $1 - B$ has a spectral gap and its zero subspace consists of those functions ϕ for which $V^{-1/2}\phi$ is a collisional invariant. Then for every f which is orthogonal to the collisional invariants, we have $\langle f, Lf \rangle \geq \langle f, L'f \rangle \geq \delta' \|f\|^2$ for some $\delta' > 0$ independent of f. This estimate would then prove a gap also for $\tilde{L}$ in the pinned case: then W and $1/W$ are both bounded functions and there is $\delta_0 > 0$ such that $\langle \psi, \tilde{L}\psi \rangle \geq \delta_0 \|\psi\|^2$ for every ψ for which $W^{-1}\psi$ is orthogonal to the collisional invariants.

The above procedure is a variant of the standard argument used to prove a gap in kinetic theory when the "relaxation time" determined by the multiplication operator V is bounded from above and below (see the next section for discussion about the relaxation time approximation). The problem about using the standard argument directly for the linearized operator of the pinned chains lies in the non-integrable singularity related to the Jacobian of the change of variables which resolves the energy constraint. This singularity is the same one which makes the total collision cross section—obtained by neglecting all terms which contain W_0 in (4.135)—

infinite: since in the integrand $p_3 = p_1 - y$ when $p_2 = p_0 + y$, the integrand has a nonintegrable singularity of strength at least $|y|^{-1}$ at $p_2 = p_0$. Thus both the total collision cross section and the relaxation time function V are formally infinite for all k_0. The sign however is positive, and thus by reducing the strength of the collisions one can aim at approximating the linearized operator from below by an operator of the "standard form" with a finite relaxation time function. However, let us not try to complete the argument here but instead focus on its implications on the thermal conductivity.

To summarize, the above argument strongly indicates that the kinetic theory prediction for the thermal conductivity at temperature β^{-1} for chains with anharmonic pinning is a finite non-zero value, and for small λ_4 we should have

$$\kappa(\beta^{-1}) \approx \beta^2 \overline{\omega}^4 \delta^2 \langle \tilde{h}_0, \tilde{L}^{-1} \tilde{h}_0 \rangle \,, \tag{4.160}$$

where "$\tilde{L}^{-1}\tilde{h}_0$" denotes the unique $\phi \in L^2(\mathbb{T}^d)$ for which $\tilde{L}\phi = \tilde{h}_0$ (such ϕ can be found if $\tilde{L}$ has a spectral gap and $\tilde{h}_0$ is orthogonal to the zero eigenspace of $\tilde{L}$, as we have argued above). Inserting $W(k) = (\beta\omega(k))^{-1}$ and using the definitions of $\tilde{L}$ and $\tilde{h}_0$ this can be simplified to

$$\kappa(\beta^{-1}) \approx \beta^{-2} \overline{\omega}^4 \delta^2 \langle \omega^{-2}\phi_0, L^{-1}(\omega^{-2}\phi_0) \rangle \,, \tag{4.161}$$

where the operator L is explicitly given by

$$(L\psi)(k_0) = \beta^{-4}\lambda_4^2 \frac{9\pi}{4} \int_{\mathbb{T}^3} \mathrm{d}k \, \delta(k_0 + k_1 - k_2 - k_3)\delta(\omega_0 + \omega_1 - \omega_2 - \omega_3)$$
$$\times \prod_{\ell=0}^{3} \omega_\ell^{-2} \, (\psi_0 + \psi_1 - \psi_2 - \psi_3) \,. \tag{4.162}$$

Thus the dependence on the temperature and coupling factorizes:

$$\kappa(\beta^{-1}) \approx \beta^2 \lambda_4^{-2} \overline{\omega}^9 C(\delta) \,, \tag{4.163}$$

where the constant $C(\delta)$ is a function of the harmonic pinning parameter δ only,

$$C(\delta) = \delta^3 \frac{8}{9} \langle \nu^{-2}\phi_0, L_0^{-1}(\nu^{-2}\phi_0) \rangle \,, \quad \text{with} \tag{4.164}$$

$$(L_0\psi)(k_0) = \int_{-\pi}^{\pi} \frac{\mathrm{d}p_2}{2\pi} \left| \frac{\sin p_1}{\nu_1} - \frac{\sin p_3}{\nu_3} \right|^{-1} \prod_{\ell=0}^{3} \frac{1}{\nu_\ell^2} \, (\psi_0 + \psi_1 - \psi_2 - \psi_3) \,, \tag{4.165}$$

where $\psi_\ell = \psi(p_\ell/(2\pi))$, $p_0 = 2\pi k_0$, $p_3 = p_1 + p_0 - p_2$, and $p_1 = h(p_0, p_2; \delta)$, as defined in (4.132). In addition, we have used here ν to denote the normalized dispersion relation with $\overline{\omega} = 1$, that is, in the above formulae $\nu(k) = \sqrt{1 - 2\delta\cos(2\pi k)}$ and $\nu_\ell = \nu(p_\ell/(2\pi))$. The form is amenable for numerical inversion of the operator L_0, by choosing a suitable orthonormal basis for the subspace of $L^2(\mathbb{T})$ which consists of vectors orthogonal to 1 and ω.

The Jensen inequality lower bound given in (4.159) implies that

$$C(\delta) \geq \delta^3 \frac{8}{9} \frac{\langle \nu^{-1}\phi_0, \nu^{-1}\phi_0\rangle^2}{\langle \phi_0, L_0\phi_0\rangle}. \tag{4.166}$$

Using the symmetrized form in (4.149) for $\langle \phi_0, L_0\phi_0\rangle$ and changing the integration variable from k_0 to $p_0 = 2\pi k_0$ then yields

$$\begin{aligned} \delta^{-3}C(\delta) \geq{} & \frac{32}{9}\left(\int_{-\pi}^{\pi} \mathrm{d}p_0 \, \frac{\sin^2 p_0}{\nu_0^2}\right)^2 \\ & \times \left(\int_{-\pi}^{\pi} \mathrm{d}p_0 \int_{-\pi}^{\pi} \mathrm{d}p_2 \left|\frac{\sin p_1}{\nu_1} - \frac{\sin p_3}{\nu_3}\right|^{-1} \prod_{\ell=0}^{3} \frac{1}{\nu_\ell^2} |\psi_0 + \psi_1 - \psi_2 - \psi_3|^2\right)^{-1}, \end{aligned} \tag{4.167}$$

where $\psi_\ell = \sin p_\ell$. When $\delta \to 0$, we have $\nu_\ell \to 1$, $p_1 \to \pi - p_0$, and $p_1 \to \pi - p_2$ in the above. The remaining integrals can be computed explicitly, and the limit of the right hand side is found to be equal to $\pi^2/36 \approx 0.274$. The numerical inversion of the full operator in the $\delta \to 0$ limit in [1] resulted in the value 0.2756 which is very close to the above Jensen bound. In addition, evaluation of the right hand side of (4.167) by numerical integration shows that it depends only weakly on δ, decreasing to 0.2 at $\delta = 0.3$. However, the bound becomes ineffective for larger values of δ, going to zero as $\delta \to \frac{1}{2}$.

Therefore, the formula $0.274\,\delta^3\overline{\omega}^9\beta^2\lambda_4^{-2}$ provides a fairly good approximation for the lower bound in (4.159) for the kinetic theory prediction of the thermal conductivity, at least for small enough δ. This approximation was compared in [1] to the thermal conductivity measured in numerical simulations of large finite chains with boundary thermostats. The numerical simulations where performed with $\overline{\omega} = \delta^{-1/2}$ and $\lambda_3 = 0$, and instead of $\lambda_4 \to 0$ with fixed β one considers $\beta \to \infty$ with fixed λ_4. The two limits can be connected by a straightforward scaling argument which allows to compare the kinetic prediction with the conductivity observed in the simulations (see [1, Sect. 2] for details). As mentioned above, the Jensen inequality lower bound given in (4.159) is a good approximation of the numerically inverted value for small δ. In Fig. 4.2 we have given the comparison between the values from the numerical particle simulations and the Jensen bound. The agreement is surprisingly good and seems to indicate that for this model the kinetic theory gives good description of the dominant effects affecting thermal conduction in the pinned anharmonic chains.

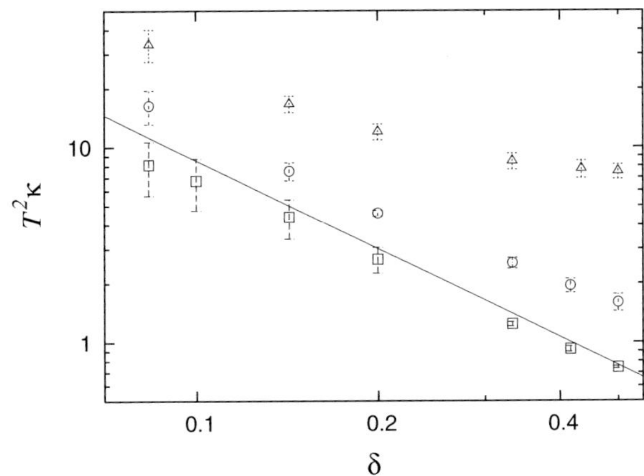

Fig. 4.2 Measured $T^2\kappa(T)$ compared against the kinetic theory computation for small δ and T (*straight line*). The data points are for $T = 0.1$ (*open square*), 0.4 (*open circle*), and 4 (*open triangle*) (reprinted from [1])

4.3.5 *Anomalous Energy Conduction in the Kinetic Theory of FPU Chains*

The kinetic approximation to the Green–Kubo correlation function in the FPU chains has been studied rigorously in [15]. As mentioned above, the FPU chains have $\delta = \frac{1}{2}$, and in [15] the frequency normalization was chosen as $1/\sqrt{2}$ since then $\omega(k) = |\sin(\pi k)|$ has no prefactors. For easy comparison, let us make this choice also in this subsection for the FPU chains: if needed, the scale $\overline{\omega}$ can be reintroduced in the results just as was done in the previous subsection.

With this choice and using (4.148), the Boltzmann collision operator linearized around a steady state $W(k) = \beta^{-1}\omega(k)^{-1}$ becomes

$$(Lf)(k_0) = 9\pi\lambda_4^2(2/\beta)^4 \int_{\mathbb{T}^3} \mathrm{d}k\, \delta(k_0 + k_1 - k_2 - k_3)\delta(\omega_0 + \omega_1 - \omega_2 - \omega_3) \times (f_0 + f_1 - f_2 - f_3)\,, \tag{4.168}$$

and it is related to the kinetic theory prediction of the Green–Kubo correlation function by

$$C(t;\beta) \approx \frac{1}{2^4}\langle \tilde{h}_0, \mathrm{e}^{-t\tilde{L}}\tilde{h}_0\rangle\,, \tag{4.169}$$

where $\tilde{L} = \beta^2\omega L\omega$ and $\tilde{h}_0(k) = 2\cos(\pi k)/\beta$ for $k \in [0,1)$. Choosing a normalization somewhat different from the previous subsection, we may rewrite this in a dimensionless form by defining $\psi(k) = \frac{1}{2}\cos(\pi k)$, $0 \le k < 1$, and

$L_0 = \frac{1}{9}\lambda_4^{-2}(\beta/2)^4\omega L\omega$. This yields

$$C(t;\beta) \approx \frac{1}{\beta^2}\langle\psi, e^{-tc\tilde{L}_0}\psi\rangle\,, \tag{4.170}$$

where the constant $c = 12^2\lambda_4^2\beta^{-2}\pi^{-1}$ and $\tilde{L}_0 = \omega L_0\omega$. In [15], this was called the kinetic conjecture and the above form coincides with the one given in [15, Eq. (1.18)] after a change of variables from $k \in [0,1)$ to $p \in [0,2\pi)$.

Explicitly, the operator L_0 can be written for $p \in I = [0,2\pi)$ as

$$\begin{aligned}(L_0 f)(p_0/(2\pi)) = \int_{I^3} \mathrm{d}p\,\delta(p_0+p_1-k_2-k_3)\delta(\omega_0+\omega_1-\omega_2-\omega_3)\\ \times (f_0+f_1-f_2-f_3)\,.\end{aligned} \tag{4.171}$$

Now the constraints can be explicitly integrated using the results mentioned in Sect. 4.2.2.3, yielding

$$L_0 = V - A\,, \quad \text{with} \tag{4.172}$$

$$(V\psi)(k) = V(2\pi k)\psi(k) \quad \text{and} \quad (A\psi)(k) = \int_0^1 \mathrm{d}k' 2\pi K(2\pi k, 2\pi k')\psi(k')\,, \tag{4.173}$$

where

$$V(p) = \int_0^{2\pi} \mathrm{d}p' K_2(p,p')\,, \quad K(p,p') = 2K_2(p,p') - K_1(p,p')\,, \quad \text{with} \tag{4.174}$$

$$K_1(p,p') = 4\frac{\mathbb{1}(F_-(p,p')>0)}{\sqrt{F_-(p,p')}} \quad \text{and} \quad K_2(p,p') = \frac{2}{\sqrt{F_+(p,p')}}\,. \tag{4.175}$$

In this formula, the multiplication operator arises from the "f_0"-term and the K_1 term in the integral kernel K from the "f_1"-term. By symmetry, the contributions from the "f_2"- and "f_3"-terms are equal, each contributing one K_2-term to the integral kernel. It probably comes as no surprise that the precise analysis of the semigroup generated by $\tilde{L}_0$ gets rather technical. However, this has been done in [15], and let us only repeat the main conclusions from the analysis here.

Unlike for the onsite anharmonic perturbation, the relaxation time for the FPU models is finite, and as the above formula shows, the operator $\tilde{L}_0$ can be written in the standard form $\tilde{V} - \tilde{A}$ where $\tilde{V} = \omega^2 V$ is a multiplication operator and $\tilde{A} = \omega A\omega$ is an integral operator. It is proven in [15, Lemma 4.1] that the function $\tilde{V}(k)$ is continuous and can be bounded from above and below by $|\sin(\pi k)|^{5/3}$. In particular $\tilde{V}(0) = 0$ and, consequently, the operator $\tilde{L}_0$ has no spectral gap.

In kinetic theory, it is a common practice to use the *relaxation time approximation* to approximate the linearized Boltzmann evolution. In our case, this amounts to dropping the operator $\tilde{A}$, i.e., approximating

$$\langle \psi, \mathrm{e}^{-t\tilde{L}_0}\psi \rangle \approx \langle \psi, \mathrm{e}^{-t\tilde{V}}\psi \rangle \,. \tag{4.176}$$

Since $\psi(0) = 1$, the decay of the relaxation time approximation is now entirely determined by the values of the potential near zero. The above bounds imply that $\tilde{V}(k) = O(|k|^{5/3})$, and thus the relaxation time approximation predicts $C(t;\beta) = \mathcal{O}(t^{-3/5})$ for large t. (To our knowledge, this decay of the relaxation time approximation was first derived in [19].)

However, the contribution arising from adding the integral operator $\tilde{A}$ is also singular, and more careful study is required to conclude that the relaxation time prediction continues to hold for the full semigroup. Fortunately, the above straightforward estimate gives the correct decay speed: it was shown in [15] that the resolvent of the semigroup satisfies

$$\begin{aligned}\left\langle \psi, \frac{1}{\varepsilon + \tilde{L}_0}\psi \right\rangle &= \left\langle \psi, \frac{1}{\varepsilon + \tilde{V}}\psi \right\rangle + \left\langle \psi, \frac{1}{\varepsilon + \tilde{V}}\tilde{A}\frac{1}{\varepsilon + \tilde{V}}\psi \right\rangle \\ &\quad + \left\langle \psi, \frac{1}{\varepsilon + \tilde{V}}\tilde{A}\frac{1}{\varepsilon + \tilde{L}}\tilde{A}\frac{1}{\varepsilon + \tilde{V}}\psi \right\rangle,\end{aligned} \tag{4.177}$$

where the first term is identical to the relaxation time approximation, and behaves as $\varepsilon^{-2/5}$ for small $\varepsilon > 0$. The second and third term are $\mathcal{O}(\varepsilon^{-1/5-\varepsilon})$ for any $\varepsilon > 0$. Although also this second contribution is divergent, the first term is dominant, and thus we confirm the prediction of the relaxation time approximation in this particular case.

Even the exact asymptotics can be found for this particular choice. Applying [15, Corollary 2.6] we find that the kinetic theory predicts for times with $t(\lambda_4/\beta)^2 \gg 1$, and for sufficiently small couplings λ_3 and λ_4,

$$C(t;\beta) \approx C_0 \beta^{-\frac{4}{5}} (\lambda_4^2 t)^{-\frac{3}{5}} \,. \tag{4.178}$$

Here $C_0 = c_0(\pi 12^{-2})^{3/5}/(2\pi\Gamma(2/5))$ is an explicit numerical constant. Evaluation of the Gamma-function and the integrals defining the constant "c_0" in (6.14) and (4.7) of [15] yields a numerical approximation $C_0 \approx 0.00386$.

Of course, for much longer than kinetic times, the terms neglected in the derivation of the Boltzmann equation might become important and alter the asymptotic decay. The above results also imply that, on the kinetic time scale, the energy spread is superdiffusive: the quadratic energy spread observable discussed in Sect. 4.1.5 should then be increasing as $S(t) = O(t^{7/5})$.

The energy spread has been analysed in more detail in [17]. The authors study the time evolution of *inhomogeneous* perturbations around a given thermal equilibrium state. This results in an evolution equation which corresponds to the phonon

Boltzmann evolution where the nonlinear collision operator has been replaced by the above linearized operator. Explicitly, the perturbation $f(t,x,k)$, defined via $W = (1+f)/(\beta\omega)$, evolves then by

$$\partial_t f(t,x,k) + \frac{1}{2\pi}\omega'(k)\partial_x f(t,x,k) = (\tilde{L}f(t,x,\cdot))(k)\,. \tag{4.179}$$

The result concerns L^2-integrable initial data varying at a scale ε^{-1}, with ε small. It is shown that for sufficiently long times, the solution then first thermalizes in the k-variable, which by the definition of f implies that it becomes independent of k. Diffusive relaxation in the spatial variable would then mean that at a time scale ε^{-2} the perturbation follows the heat equation $\partial_t f + \kappa(-\Delta)f = 0$, with $\kappa > 0$. However, this does not occur here: instead, it is shown that at the time scale $\varepsilon^{-8/5}$ the perturbation satisfies a fractional diffusion equation $\partial_t f + \kappa(-\Delta)^{4/5}f = 0$, with $\kappa > 0$.

This corresponds to a superdiffusive relaxation of the initial perturbation. Moreover, the fractional diffusion spreads local perturbations at time t only up to distances $t^{5/8}$. This is in apparent contradiction with the earlier claim that $S(t) = O(t^{7/5})$ which would indicate that the spatial spread occurs at a speed $O(t^{7/10})$, i.e., faster than predicted by the fractional diffusion equation.

The resolution lies in the tail behaviour of solutions to the fractional diffusion equation. Let us conclude the section with a somewhat heuristic argument which would explain the above results. By using Fourier-transform and a simple scaling argument, one finds that the solution to $\partial_t f + \kappa(-\Delta)^{4/5}f = 0$, with an initial data f_0, is given by a convolution of f_0 with an integral kernel K_t which satisfies a scaling relation $K_t(x) = t^{-p}F(x/t^p)$ where $p = 5/8$. Unlike the Gaussian heat kernel in (4.46), the function F is merely polynomially decreasing, with $F(y)$ decaying as $y^{-13/5}$ for large $|y|$ (the power $13/5$ is obtained by evaluating $1/p + 1$). Thus $\int \mathrm{d}y\, y^2 F(y) = \infty$, and the quadratic spread from fractional diffusion becomes immediately infinite, even if it is initially finite.

However, on the microscopic scale, the velocities of the ballistic harmonic evolution are bounded from above, and this will eventually cut off the above decay, and change it from the above powerlaw decay to an exponentially fast decay at spatial distances $O(t)$ from the source. Therefore, we would expect that the true microscopic distribution of an initially local perturbation at a large time t is $O(t^{-p})$ for distances $|x| = O(t^p)$, it is $O(t|x|^{-2-3/5})$ for $|x|$ between $O(t^p)$ and $O(t)$, and it becomes exponentially decreasing for distances larger than $O(t)$. For such functions, the value of $S(t)$ is entirely dominated by the midscale powerlaw tail which yields a term $O(t^{7/5})$, just as we obtained from the kinetic prediction for the Green–Kubo correlation function using the linearized Boltzmann equation.

4.4 Concluding Remarks

As the last two explicit examples show, kinetic theory of phonons is capable of uncovering detailed information about the decay of time-correlations and, via the Green–Kubo formula, about the thermal conductivities of classical particle chains. This is somewhat surprising, considering the various mathematical problems and uncertainties discovered along the way to the phonon Boltzmann equation and even in its analysis. However, the agreement between the kinetic prediction and numerical simulations for thermal conductivities in chains with anharmonic pinning, and the discovery of anomalous energy transport by fractional Brownian motion from the linearized Boltzmann equation in the FPU-β chains, present a strong case in favour of looking for further applications of the phonon Boltzmann equations, even in the somewhat degenerate one-dimensional case. After all, it is at present one of the very few general tools which allow computing the dependence of the thermal conduction properties on the parameters of the microscopic evolution directly, without introduction of additional fitting parameters.

We have also seen that some care is needed in the application of the phonon Boltzmann equation. Most importantly, the equation is closely tied to the scales on which the free streaming of phonons *begins* to alter its character due to the collisions. Hence, it describes the evolution up to the kinetic time-scale only, and it is possible that further changes are found at larger time-scales. Nevertheless, since the H-theorem implies that solutions to the kinetic equation push the system towards thermal equilibrium states, drastic changes in the character of the evolution should be the exception, not the rule.

The evaluation of the decay of Green–Kubo correlation functions is one of the robust applications of kinetic theory, and the thermal conductivity obtained by the perturbation procedure recalled here should in general yield its leading behaviour in the limit of weak perturbations. However, as a warning about the standard procedure, let us stress that the commonly used relaxation time approximation of the linearized collision operator is *only* an order-of-magnitude estimate of the real kinetic prediction which involves inverting the full linearized operator. In the pinned case, we found that the relaxation time approximation predicts zero thermal conductivity. This turned out to be misleading since already the straightforward Jensen bound for the full inverse proves that the kinetic prediction for the conductivity is non-zero. In contrast, for the anomalous conduction in the FPU-β chain the relaxation time approximation does capture the correct asymptotic decay of the kinetic prediction.

One major open question in the kinetic theory of phonons, and in fact of nonlinearly perturbed systems in general, is the precise manner of handling spatially inhomogeneous perturbations. The standard Boltzmann transport term, appearing on the left hand side of (4.179), might require adjusting to capture all effects relevant to the transport. One such example often found in the literature is an addition of a Vlasov-type term. For which systems, under which time-scales, and for which initial data, such corrections are necessary, remains unresolved at the moment.

One benefit from a better understanding of the behaviour of inhomogeneous perturbations could be a first-principles derivation of fluctuating hydrodynamics for these systems, including a precise dependence of its parameters on the microscopic evolution, in the limit of weak couplings. The application of fluctuating hydrodynamics to the transport in one-dimensional particle chains has been discussed in [22] and reviewed elsewhere in this volume. It appears to be the first model which is able to describe the anomalous transport in one-dimensional particle chains fully in agreement with computer simulations of the spread of localized perturbations. Connecting it directly to the microscopic dynamics would be a breakthrough in understanding the microscopic origin and precise nature of transport in crystalline structures, such as the present particle chains. Kinetic theory, the phonon Boltzmann equation in particular, could well provide some of the missing steps into this direction.

Acknowledgements I am most grateful to Herbert Spohn for his comments and suggestions for improvements. Most of the discussion here is based on his works and on our joint collaborations. The related research has been made possible by support from the Academy of Finland. I am also grateful to Matteo Marcozzi and Alessia Nota for their comments on the manuscript.

References

1. Aoki, K., Lukkarinen, J., Spohn, H.: Energy transport in weakly anharmonic chains. J. Stat. Phys. **124**, 1105–1129 (2006). doi:10.1007/s10955-006-9171-2
2. Benedetto, D., Castella, F., Esposito, R., Pulvirenti, M.: From the N-body Schrödinger equation to the quantum Boltzmann equation: a term-by-term convergence result in the weak coupling regime. Commun. Math. Phys. **277**(1), 1–44 (2008)
3. Buttà, P., Caglioti, E., Di Ruzza, S., Marchioro, C.: On the propagation of a perturbation in an anharmonic system. J. Stat. Phys. **127**(2), 313–325 (2007)
4. Butz, M.: Kinetic limit for wave propagation in a continuous, weakly random medium: self-averaging and convergence to a linear Boltzmann equation. Ph.D. thesis, Technische Universität München (2015)
5. Chen, T.: Localization lengths and Boltzmann limit for the Anderson model at small disorders in dimension 3. J. Stat. Phys. **120**, 279–337 (2005)
6. Erdős, L., Yau, H.T.: Linear Boltzmann equation as the weak coupling limit of a random Schrödinger equation. Commun. Pure Appl. Math. **53**(6), 667–735 (2000)
7. Erdős, L., Salmhofer, M., Yau, H.T.: Quantum diffusion of the random Schrödinger evolution in the scaling limit I. The non-recollision diagrams. Acta Math. **200**(2), 211–277 (2008)
8. Gérard, P., Markowich, P.A., Mauser, N.J., Paupaud, F.: Homogenization limits and Wigner transforms. Commun. Pure Appl. Math. **50**, 323–379 (1997)
9. Harris, L., Lukkarinen, J., Teufel, S., Theil, F.: Energy transport by acoustic modes of harmonic lattices. SIAM J. Math. Anal. **40**(4), 1392–1418 (2008)
10. Lanford, O.E., Lebowitz, J.L., Lieb, E.H.: Time evolution of infinite anharmonic systems. J. Stat. Phys. **16**(6), 453–461 (1977)
11. Lefevere, R., Schenkel, A.: Normal heat conductivity in a strongly pinned chain of anharmonic oscillators. J. Stat. Mech. **2006**(02), L02001 (2006). doi:10.1088/1742-5468/2006/02/L02001
12. Lu, X.: The Boltzmann equation for Bose–Einstein particles: regularity and condensation. J. Stat. Phys. **156**(3), 493–545 (2014). doi:10.1007/s10955-014-1026-7

13. Lukkarinen, J., Marcozzi, M.: Wick polynomials and time-evolution of cumulants. arXiv e-print (2015). arXiv.org:1503.05851
14. Lukkarinen, J., Spohn, H.: Kinetic limit for wave propagation in a random medium. Arch. Ration. Mech. Anal. **183**(1), 93–162 (2007). doi:10.1007/s00205-006-0005-9
15. Lukkarinen, J., Spohn, H.: Anomalous energy transport in the FPU-β chain. Commun. Pure Appl. Math. **61**(12), 1753–1786 (2008). doi:10.1007/s00205-006-0005-9
16. Lukkarinen, J., Spohn, H.: Weakly nonlinear Schrödinger equation with random initial data. Invent. Math. **183**(1), 79–188 (2011)
17. Mellet, A., Merino-Aceituno, S.: Anomalous energy transport in FPU-β chain. J. Stat. Phys. **160**(3), 583–621 (2015). doi:10.1007/s10955-015-1273-2
18. Mielke, A.: Macroscopic behavior of microscopic oscillations in harmonic lattices via Wigner-Husimi transforms. Arch. Ration. Mech. Anal. **181**, 401–448 (2006)
19. Pereverzev, A.: Fermi-Pasta-Ulam β lattice: Peierls equation and anomalous heat conductivity. Phys. Rev. E **68**(5), 056124 (2003)
20. Spohn, H.: Collisional invariants for the phonon Boltzmann equation. J. Stat. Phys. **124**, 1131–1135 (2006)
21. Spohn, H.: The phonon Boltzmann equation, properties and link to weakly anharmonic lattice dynamics. J. Stat. Phys. **124**(2–4), 1041–1104 (2006). doi:10.1007/s10955-005-8088-5
22. Spohn, H.: Nonlinear fluctuating hydrodynamics for anharmonic chains. J. Stat. Phys. **154**(5), 1191–1227 (2014). doi:10.1007/s10955-014-0933-y
23. Ziman, J.M.: Electrons and Phonons: The Theory of Transport Phenomena in Solids. Oxford University Press, London (1967)

Chapter 5
Thermal Conductivity in Harmonic Lattices with Random Collisions

Giada Basile, Cédric Bernardin, Milton Jara, Tomasz Komorowski, and Stefano Olla

Abstract We review recent rigorous mathematical results about the macroscopic behaviour of harmonic chains with the dynamics perturbed by a random exchange of velocities between nearest neighbor particles. The random exchange models the effects of nonlinearities of anharmonic chains and the resulting dynamics have similar macroscopic behaviour. In particular there is a superdiffusion of energy for unpinned acoustic chains. The corresponding evolution of the temperature profile is governed by a fractional heat equation. In non-acoustic chains we have normal diffusivity, even if momentum is conserved.

5.1 Introduction

Lattice systems of coupled anharmonic oscillators have been widely used in order to understand the macroscopic transport of the energy, in particular the superdiffusive behavior in one and two dimensional unpinned chains. While a lot of numerical experiments and heuristic considerations have been made (cf. [30, 31, 41] and many

G. Basile
Dipartimento di Matematica, Università di Roma La Sapienza, Roma, Italy
e-mail: basile@mat.uniroma.it

C. Bernardin
Laboratoire J.A. Dieudonné UMR CNRS 7351, Université de Nice Sophia-Antipolis, Parc Valrose, 06108 Nice Cedex 02, France
e-mail: cbernard@unice.fr

M. Jara
IMPA, Rio de Janeiro, Brazil
e-mail: mjara@impa.br

T. Komorowski
Institute of Mathematics, Polish Academy of Sciences, Warsaw, Poland
e-mail: komorow@hektor.umcs.lublin.pl

S. Olla (✉)
Ceremade, UMR CNRS 7534, Université Paris Dauphine, 75775 Paris Cedex 16, France
e-mail: olla@ceremade.dauphine.fr

S. Lepri (ed.), *Thermal Transport in Low Dimensions*, Lecture Notes in Physics 921, DOI 10.1007/978-3-319-29261-8_5

contributions in the present volume), very few mathematical rigourous scaling limits have been obtained until now.

For harmonic chains it is possible to perform explicit computations, even in the stationary state driven by thermal boundaries (cf. [38]). But since these dynamics are completely integrable, the energy transport is purely ballistic and they do not provide help in understanding the diffusive or superdiffusive behavior of anharmonic chains.

The *scattering* effect of the non-linearities can be modeled by stochastic perturbations of the dynamics such that they conserve total momentum and total energy, like a random exchange of the velocities between the nearest neighbor particles. We will describe the results for the one-dimensional chains, and we will mention the results in the higher dimensions in Sect. 5.9. In particular we will prove how the transport through the fractional Laplacian, either asymmetric or symmetric, emerges from microscopic models.

The *infinite dynamics* is described by the velocities and positions $\{(p_x, q_x) \in \mathbb{R}^2\}_{x\in\mathbb{Z}}$ of the particles. The formal Hamiltonian is given by

$$\mathscr{H}(p,q) := \frac{1}{2m}\sum_x p_x^2 + \frac{1}{2}\sum_{x,x'} \alpha_{x-x'} q_x q_{x'}, \tag{5.1}$$

where we assume that the masses are equal to 1 and that α is symmetric with a finite range or at most exponential decay $|\alpha_x| \leq Ce^{-c|x|}$. We define the Fourier transform of a function $f : \mathbb{Z} \to \mathbb{R}$ as $\hat{f}(k) = \sum_x e^{-2\pi i x k} f(x)$ for $k \in \mathbb{T}$ the unitary length torus. We assume that $\hat{\alpha}(k) > 0$ for $k \neq 0$. The function $\omega(k) = \sqrt{\hat{\alpha}(k)}$ is called the *dispersion relation* of the chain.

5.1.1 Unpinned Chains

We are particularly interested in the unpinned chain, i.e. $\hat{\alpha}(0) = 0$, when the total momentum is conserved even under the stochastic dynamics described below. Then, the infinite system is translation invariant under shift in q, and the correct coordinates are the interparticle distances (also called stretches, or strains):

$$r_x = q_x - q_{x-1}, \qquad x \in \mathbb{Z}. \tag{5.2}$$

When $\hat{\alpha}''(0) > 0$ we say that the chain is *acoustic* (i.e. there is a non-vanishing sound speed). We will see that this is a crucial condition for the superdiffusivity of the energy in one dimension. For unpinned acoustic chains we have that $\omega(k) \sim |k|$ as $k \to 0$.

5.1.2 Pinned Chains

When $\hat{\alpha}(0) > 0$, the system is pinned and translation invariance is broken. In this case $\omega'(k) \sim k$ as $k \to 0$. This is also the case for unpinned non-acoustic chains, where $\hat{\alpha}(0) = \hat{\alpha}''(0) = 0$. This fact is responsible for the diffusive behavior of the energy, cf. Sect. 5.7.

5.1.3 Dynamics with Stochastic Collisions

To the Hamiltonian dynamics we add random elastic collisions, where momenta of the nearest-neighbor particles are exchanged. This happens at independent random exponential times: each couple of particles labeled $x, x+1$ exchange their velocities p_x and p_{x+1} at exponential independent random times of intensity γ. Equivalently there are independent Poisson processes $\{N_{x,x+1}(t),\ \ x \in \mathbb{Z}\}$ of intensity γ, independent from the positions and velocities of all particles. The evolution of the system is described by the stochastic differential equations

$$
\begin{aligned}
\dot{q}_x &= p_x \\
\dot{p}_x &= -(\alpha * q(t))_x + (p_{x+1}(t^-) - p_x(t^-))\,\dot{N}_{x,x+1}(t) + (p_{x-1}(t^-) - p_x(t^-))\,\dot{N}_{x-1,x}(t)
\end{aligned}
\tag{5.3}
$$

where $\dot{N}_{x,x+1}(t) = \sum_j \delta(t - T_{x,x+1}(j))$, with $\{T_{x,x+1}(j)\}$ the random times when $N_{x,x+1}$ jumps, and $p_x(t^-)$ is the velocity of the particle x just before time t, i.e. $\lim_{s\downarrow 0} p_x(t-s)$.

The evolution of the probability density on the configurations then follows the Fokker-Planck equations:

$$
\partial_t f(t,p,q) = (A+S)f(t,p,q) \tag{5.4}
$$

where A is the Hamiltonian operator

$$
A = \sum_x \left(p_x \partial_{q_x} - (\partial_{q_x}\mathscr{H})\partial_{p_x} \right), \tag{5.5}
$$

while S is the generator of the random exchanges

$$
Sf(p,q) = \gamma \sum_x \left(f(p^{x,x+1},q) - f(p,q) \right), \tag{5.6}
$$

where $p^{x,x+1}$ is the configuration obtained exchanging p_x and p_{x+1}.

This stochastic perturbation of the Hamiltonian dynamics has the property to conserve the total energy, and in the unpinned case the resulting dynamics conserves also the total momentum ($\sum_x p_x$), and the *volume* or *strain* of the chain ($\sum_x r_x$). It has also the property that these are the only conserved quantities. In this sense it gives the necessary ergodicity to the dynamics [13, 23].

We have also considered other type of conservative random dynamics, like a continuous random exchange of the momenta of each triplets $\{p_{x-1}, p_x, p_{x+1}\}$. The intersection of the kinetic energy sphere $p_{x-1}^2 + p_x^2 + p_{x+1}^2 = C$ with the plane $p_{x-1} + p_x + p_{x+1} = C'$ gives a one dimensional circle. Then we define a dynamics on this circle by a standard Wiener process on the corresponding angle. This perturbation is locally more mixing, but it gives the same results for the macroscopic transport.

5.1.4 Equilibrium Stationary Measures

Due to the harmonicity of the interactions, the Gibbs equilibrium stationary measures are Gaussians. Positions and momenta are independent and the distribution is parametrized, according to the rules of statistical mechanics, by the temperature $T = \beta^{-1} > 0$. In the pinned case, they are formally given by

$$\nu_\beta(dq, dp) \sim \frac{e^{-\beta \mathscr{H}(p,q)}}{\mathscr{Z}} \, dq \, dp.$$

In the unpinned case, the correct definition should involve the r_x variables. Then the distribution of the r_x's is Gaussian and becomes uncorrelated in the case of the nearest neighbor interaction. For *acoustic unpinned chains* the Gibbs measures are parameterized by

$$\boldsymbol{\lambda} = (\beta^{-1}(\text{temperature}), \bar{p}(\text{velocity}), \tau(\text{tension})),$$

and are given formally by

$$\nu_{\boldsymbol{\lambda}}(dr, dp) \sim \frac{e^{-\beta [\mathscr{H}(p,q) - \bar{p} \sum_x p_x - \tau \sum_x r_x]}}{\mathscr{Z}} \, dr \, dp. \tag{5.7}$$

Non-acoustic chains are tensionless and the equilibrium measures have a different parametrization, see Sect. 5.7.

5.1.5 Macroscopic Space-Time Scales

We will mostly concentrate on the acoustic unpinned case (except Sect. 5.7). In this case the total Hamiltonian can be written as

$$\mathscr{H}(p,q) := \sum_x \frac{p_x^2}{2} - \frac{1}{4} \sum_{x,x'} \alpha_{x-x'} (q_x - q_{x'})^2, \tag{5.8}$$

where $q_x - q_{x'} = \sum_{y=x'+1}^{x} r_y$ for $x > x'$. We define the energy per atom as:

$$e_x(r, p) := \frac{p_x^2}{2} - \frac{1}{4}\sum_{x'} \alpha_{x-x'}(q_x - q_{x'})^2. \tag{5.9}$$

There are three conserved (also called *balanced*) fields: the energy $\sum_x e_x$, the momentum $\sum_x p_x$ and the strain $\sum_x r_x$. We want to study the macroscopic evolution of the spatial distribution of these fields in a large space-time scale. We introduce a scale parameter $\varepsilon > 0$ and, for any smooth test function $J : \mathbb{R} \to \mathbb{R}$, define the empirical distribution

$$\varepsilon \sum_x J(\varepsilon x)\mathbf{U}_x(\varepsilon^{-a}t), \qquad \mathbf{U}_x = (r_x, p_x, e_x). \tag{5.10}$$

We are interested in the limit as $\varepsilon \to 0$. The parameter $a \in [1, 2]$ corresponds to different possible scalings. The value $a = 1$ corresponds to the hyperbolic scaling, while $a = 2$ corresponds to the diffusive scaling. The intermediate values $1 < a < 2$ pertain to the superdiffusive scales.

The interest of the unpinned model is that there are three different macroscopic space-scales where we observe non-trivial behaviour of the chain: $a = 1, \frac{3}{2}, 2$.

5.2 Hyperbolic Scaling: The Linear Wave Equation

Let us assume that the dynamics starts with a random initial distribution $\mu_\varepsilon = \langle \cdot \rangle_\varepsilon$ of finite energy of size ε^{-1}, i.e. for some positive constant E_0:

$$\varepsilon \sum_x \langle e_x(0) \rangle_\varepsilon \le E_0. \tag{5.11}$$

where we denote $e_x(t) = e_x(p(t), q(t))$. We will also assume that some smooth macroscopic initial profiles

$$\mathfrak{u}_0(y) = (\mathfrak{r}_0(y), \mathfrak{p}_0(y), \mathfrak{e}_0(y))$$

are associated with the initial distribution, in the sense that:

$$\lim_{\varepsilon \to 0} \mu_\varepsilon \left(\left| \varepsilon \sum_x J(\varepsilon x)\mathbf{U}_x(0) - \int J(y)\mathfrak{u}_0(y)dy \right| > \delta \right) = 0, \qquad \forall \delta > 0, \tag{5.12}$$

for any test function J.

Then it can be proven [27] that these initial profiles are governed by the linear wave equation in the following sense

$$\lim_{\varepsilon\to 0} \varepsilon \sum_x J(\varepsilon x)\mathbf{U}_x(\varepsilon^{-1}t) = \int J(y)\mathfrak{u}(y,t)dy \tag{5.13}$$

where $\mathfrak{u}(y,t) = (\mathfrak{r}(y,t), \mathfrak{p}(y,t), \mathfrak{e}(y,t))$ is the solution of:

$$\partial_t \mathfrak{r} = \partial_y \mathfrak{p}, \qquad \partial_t \mathfrak{p} = \tau_1 \partial_y \mathfrak{r}, \qquad \partial_t \mathfrak{e} = \tau_1 \partial_y (\mathfrak{p}\mathfrak{r}) \tag{5.14}$$

with $\tau_1 = \frac{\hat{\alpha}''(0)}{8\pi^2}$ (the square of the speed of sound of the chain). Notice that in the non-acoustic case $\hat{\alpha}''(0) = 0$, there is no evolution at the hyperbolic scale.

Observe that the evolution of the fields of strain $\mathfrak{r}$ and momentum $\mathfrak{p}$ is autonomous of the energy field. Furthermore we can define the macroscopic *mechanical* energy as

$$\mathfrak{e}_{mech}(y,t) = \frac{1}{2}\left(\tau_1 \mathfrak{r}(y,t)^2 + \mathfrak{p}(y,t)^2\right) \tag{5.15}$$

and the temperature profile or *thermal* energy as

$$T(y,t) = \mathfrak{e}(y,t) - \mathfrak{e}_{mech}(y,t). \tag{5.16}$$

It follows immediately from (5.14) that $T(y,t) = T(y,0)$, i.e. the temperature profile does not change on the hyperbolic space-time scale.

There is a corresponding decomposition of the energy of the random initial configurations: long wavelengths (invisible for the exchange noise in the dynamics) contribute to the mechanical energy and they will evolve in this hyperbolic scale following the linear equations (5.14). This energy will eventually disperse to infinity at large time (in this scale). Because of the noise dynamics, short wavelength will contribute to the variance (temperature) of the distribution, and the corresponding profile does not evolve in this hyperbolic scale. See [27] for the details of this decomposition.

5.3 Superdiffusive Evolution of the Temperature Profile

As we have seen in the previous section, for acoustic chains the mechanical part of the energy evolves ballistically in the hyperbolic scale and eventually it will disperse to infinity. Consequently when we look at the larger superdiffusive time scale $\varepsilon^{-a}t$, $a > 1$, we start only with the thermal profile of energy, while the strain and momentum profiles are equal to 0. It turns out that, for acoustic chains, the temperature profile evolves at the time scale corresponding to $a = 3/2$:

$$\lim_{\varepsilon\to 0} \varepsilon \sum_x J(\varepsilon x)\ \langle e_x(\varepsilon^{-3/2}t)\rangle_\varepsilon = \int J(y)\mathfrak{T}(y,t)dy, \qquad t > 0, \tag{5.17}$$

where $\mathfrak{T}(y,t)$ solves the fractional heat equation

$$\partial_t \mathfrak{T} = -c|\Delta_y|^{3/4}\mathfrak{T}, \qquad \mathfrak{T}(y,0) = T(y,0) \tag{5.18}$$

where $c = \hat{\alpha}''(0)^{3/4}2^{-9/4}(3\gamma)^{-1/2}$. This is proven in [26]. We also have that the profiles for the other conserved quantities remain flat:

$$\begin{aligned} &\lim_{\varepsilon\to 0} \varepsilon \sum_x J(\varepsilon x)\ \langle r_x(\varepsilon^{-3/2}t)\rangle_\varepsilon = 0, \\ &\lim_{\varepsilon\to 0} \varepsilon \sum_x J(\varepsilon x)\ \langle p_x(\varepsilon^{-3/2}t)\rangle_\varepsilon = 0, \quad t > 0. \end{aligned} \tag{5.19}$$

Clearly the null value is due to the finite energy assumption, otherwise it will be the corresponding constants, i.e. $\lim_{\varepsilon\to 0} \varepsilon \sum_x \langle r_x(0)\rangle_\varepsilon$ and $\lim_{\varepsilon\to 0} \varepsilon \sum_x \langle p_x(0)\rangle_\varepsilon$.

In finite volume with given boundary conditions (periodic or else), the mechanical energy will persist, oscillating in linear waves. At the larger superdiffusive time scale, waves will oscillate fast giving a weak convergence for the initial profiles of strain and momentum to constant values.

5.4 The Diffusive Behavior of the Phonon-Modes

We have seen that (5.19) holds at the superdiffusive time scale and consequently at any larger time scale. But if we recenter the evolution of the strain and momentum around the propagation of the wave equation we see Gaussian fluctuations at the diffusive space time scale. For this purpose it is useful to introduce a microscopic approximation of the Riemann invariants (normal modes) of the wave equation:

$$f_x^{\pm}(t) = p_x(t) \pm \tau_1^{1/2}\left(r_x(t) \pm \frac{3\gamma - 1}{2}(r_{x+1}(t) - r_x(t))\right). \tag{5.20}$$

Once $f_x^{\pm}(t)$ are recentred on the Riemann invariants of the wave equation, they diffuse on the proper space-time scale, more precisely:

$$\varepsilon \sum_x J(\varepsilon x \mp \tau_1^{1/2}\varepsilon^{-1}t)\, f_x^{\pm}(\varepsilon^{-2}t) \underset{\varepsilon\to 0}{\longrightarrow} \int_{\mathbb{R}} J(y)\, \bar{\mathfrak{f}}^{\pm,d}(y,t)dy,$$

$$\partial_t\, \bar{\mathfrak{f}}^{\pm,d} = \frac{3\gamma}{2}\, \partial_y^2\, \bar{\mathfrak{f}}^{\pm,d}.$$

For a proof see [27].

5.5 Equilibrium Fluctuations

If we start with an equilibrium stationary measure corresponding to a certain temperature T, momentum $\bar{p}$ and strain $\bar{r}$, then of course there will be no evolution of the empirical fields defined by (5.10). But we shall observe the evolution of the equilibrium time correlations defined by

$$S^{\ell,\ell'}(x,t) =< u_x^{\ell}(t)\, u_0^{\ell'}(0) > - < u_0^{\ell}(0) >< u_0^{\ell'}(0) >, \quad u_x^1 = r_x,\ u_x^2 = p_x,\ u_x^3 = e_x, \tag{5.21}$$

where $\langle\cdot\rangle$ denotes the expectation with respect to the dynamics in the corresponding equilibrium. Let us assume for simplicity of notation that $\bar{r} = 0 = \bar{p}$, otherwise we have to shift x along the characteristics of the linear wave equation. At time $t = 0$, it is easy to compute the limit (in a distributional sense):

$$\lim_{\varepsilon\to 0} \varepsilon^{-1/2} S([\varepsilon^{-1}y], 0) = \delta(y) \begin{pmatrix} T & 0 & \tilde{\alpha}T \\ 0 & T & 0 \\ \tilde{\alpha}T & 0 & (\frac{1}{2} + \alpha^2)T^2 \end{pmatrix} \tag{5.22}$$

where $\tilde{\alpha}$ is a constant depending only on the interaction.

In the hyperbolic time scale $a = 1$ this correlation matrix evolves deterministically, i.e.

$$\lim_{\varepsilon\to 0} \varepsilon^{-1/2} S([\varepsilon^{-1}y], \varepsilon^{-1}t) = \bar{S}(y,t) \tag{5.23}$$

where

$$\partial_t \bar{S}^{11}(y,t) = \partial_y \bar{S}^{22}(y,t), \qquad \partial_t \bar{S}^{22}(y,t) = \tau_1 \partial_y \bar{S}^{11}(y,t), \tag{5.24}$$

while for the energy correlations

$$\partial_t \bar{S}^{33}(y,t) = 0. \tag{5.25}$$

In particular if $\bar{p} = \bar{r} = 0$, energy fluctuations do not evolve at the hyperbolic time scale. As for the energy profile out of equilibrium, the evolution is at a further time scale. By a duality argument (cf. [26]), the evolution of the energy correlations occurs at the superdiffusive time scale with $a = 3/2$, namely

$$\lim_{\varepsilon\to 0} \varepsilon^{-1/2} S^{33}([\varepsilon^{-1}y], \varepsilon^{-3/2}t) = \tilde{S}^{33}(y,t), \tag{5.26}$$

where $\tilde{S}^{33}(y,t)$ is the solution of:

$$\partial_t \tilde{S}^{33} = -c|\Delta_y|^{3/4} \tilde{S}^{33}. \tag{5.27}$$

5.6 The Phonon Boltzmann Equation

A way to understand the energy superdiffusion in the one-dimensional system is to analyze its kinetic limit, i.e. a limit for weak noise where the number of stochastic collisions per unit time remains bounded. In the non-linear case it corresponds to a weak non-linearity limit as proposed first in his seminal paper [37] in 1929 by Peierls. He intended to compute thermal conductivity for insulators in analogy with the kinetic theory of gases. The main idea is that at low temperatures the lattice vibrations responsible for the energy transport can be described as a gas of interacting particles (phonons) characterized by a wave number k. The time-dependent distribution function of phonons solves a Boltzmann type equation. Over the last years, starting from the work of Spohn [40], several papers are devoted to achieve phononic Boltzmann type equations from microscopic dynamics of oscillators.

A rigorous derivation can be achieved for the chain of harmonic oscillators perturbed by a stochastic exchange of velocities [7]. The main tool is the introduction of a Wigner function, which describes the energy density of the phonons. Let $\hat{\psi}$ be the complex field

$$\hat{\psi}(k) = \frac{1}{\sqrt{2}}\big(\omega(k)\hat{q}(k) + i\hat{p}(k)\big), \qquad k \in \mathbb{T},$$

where $\hat{p}$, $\hat{q}$ are the Fourier transform of the variables p, q and $\omega(k)$ is the dispersion relation of chain. The energy of the chain can be expressed in terms of the fields $\hat{\psi}$, namely $\mathscr{H} = \int_{\mathbb{T}} dk\, |\hat{\psi}(k)|^2$. The evolution of the field $\hat{\psi}$ due to the pure harmonic Hamiltonian without noise ($\gamma = 0$) reads

$$\partial_t \hat{\psi}(k,t) = -i\omega(k)\hat{\psi}(k,t),$$

therefore the quantities $|\hat{\psi}(k)|^2$ are preserved by the harmonic dynamics.

The Wigner distribution is defined in analogy to the usual one in quantum mechanics

$$\begin{aligned} W^{\varepsilon}(y,k,t) &= (\varepsilon/2) \int_{\mathbb{R}} e^{2\pi i \xi y} \widehat{W}^{\varepsilon}(\xi,k,t) d\xi, \\ \widehat{W}^{\varepsilon}(\xi,k,t) &:= \langle \hat{\psi}(k - \varepsilon\xi/2, t)^* \; \hat{\psi}(k + \varepsilon\xi/2, t) \rangle_{\varepsilon} \end{aligned} \tag{5.28}$$

where $\langle \cdot \rangle_{\varepsilon}$ denotes the expectation value with respect to the initial measure, chosen in such a way that the average of the total energy, is of order ε^{-1}, i.e. $\varepsilon \langle \mathscr{H} \rangle_{\varepsilon} \le E_0$, see (5.11). We also require that the averages of all p_x and q_x are zero.

The Wigner distribution $W^{\varepsilon}(y,k,t)$ defined by (5.28) gives a different energy distribution from the one considered in the previous sections, i.e. $< e_{[\varepsilon^{-1}y]}(t) >$. But in the macroscopic limit, as $\varepsilon \to 0$, they are equivalent [26].

We look at the evolution of the Wigner function on a time scale $\varepsilon^{-1}t$, with the strength of the noise of order ε, i.e. we consider the dynamics defined by (5.3) with

γ replaced by $\varepsilon\gamma$, like in a Boltzmann-Grad limit. This evolution is not autonomous and is given by

$$\begin{aligned}\partial_t \hat{W}^{\varepsilon}(\xi,k,t) =& - i\xi\,\omega'(k)\hat{W}^{\varepsilon}(\xi,k,t) + \gamma\,\mathscr{C}\hat{W}^{\varepsilon}(\xi,k,t)\\ &-\frac{\gamma}{2}\mathscr{C}\hat{Y}^{\varepsilon}(\xi,k,t) - \frac{\gamma}{2}\mathscr{C}\hat{Y}^{\varepsilon}(\xi,k,t)^* + \mathscr{O}(\varepsilon),\end{aligned} \tag{5.29}$$

where $\hat{Y}^{\varepsilon}$ is the field $\hat{Y}^{\varepsilon}(\xi,k,t) = \langle\hat{\psi}(k+\varepsilon\xi/2,t)\hat{\psi}(k-\varepsilon\xi/2,t)\rangle_{\varepsilon}$ and $\mathscr{C}$ is a linear operator. The transport term is due to the harmonic Hamiltonian, while the "collision" operator $\mathscr{C}$ is related to the stochastic noise.

It turns out that the field $\hat{Y}^{\varepsilon}$ is the Wigner distribution associated to the *difference* between the kinetic energy and the potential energy. This is fast oscillating on the time scale ε^{-1} and in the limit $\varepsilon \to 0$ it disappears after time integration. Therefore in the limit $\varepsilon \to 0$ the Wigner function W^{ε} weakly converges to the solution of the following linear Boltzmann equation

$$\partial_t W(y,k,t) + \frac{1}{2\pi}\omega'(k)\,\partial_y W(y,k,t) = \mathscr{C}W(y,k,t). \tag{5.30}$$

It describes the evolution of the energy density distribution, over the physical space $\mathbb{R}$, of the phonons, characterized by a wave number k and traveling with velocity $\omega'(k)$. We remark that for the unpinned acoustic chains $\omega'(k)$ remains strictly positive for small k. The collision term has the following expression

$$\mathscr{C}f(k) = \int_{\mathbb{T}} dk' R(k,k')[f(k') - f(k)],$$

where the kernel R is positive and symmetric. One can write an exact expression on R, nevertheless its crucial feature is that R behaves like k^2 for small k, due to the fact that the noise preserves the total momentum. Naïvely, it means that phonons with small wave numbers travel with a finite velocity, but they have low probability to be scattered, thus their mean free paths have a macroscopic length (ballistic transport). This intuitive picture has an exact statement in the probabilistic interpretation of (5.30). The equation describes the evolution of the probability density of a Markov process $(K(t), Y(t))$ on $\mathbb{T}\times\mathbb{R}$, where $K(t)$ is a reversible jump process and $Y(t)$ is an additive functional of K, namely $Y(t) = \int_0^t \omega'(K_s)ds$. A phonon with the wave number k waits in its state an exponentially distributed random time $\tau(k)$ with mean value $\sim k^{-2}$ for small k. Then it jumps to another state k' with probability $\sim k'^2 dk'$. The additive functional $Y(t)$ describes the position of the phonon and can be expressed as

$$Y(t) = \sum_{i=1}^{\mathscr{N}_t} \tau(X_i)\omega'(X_i),$$

where $\{X_i\}_{i\geq 1}$ is the Markov chain given by the sequence of the states visited by the process $K(t)$. Here $\mathcal{N}_t$ denotes the number of jumps up to the time t. The tail distribution of the random variables $\{\tau(X_i)\omega'(X_i)\}$ with respect to the stationary measure π of the chain behaves like

$$\pi\left[|\tau(X_i)\omega'(X_i)| > \lambda\right] \sim \frac{1}{\lambda^{3/2}}.$$

Therefore the variables $\tau(X_i)\omega'(X_i)$ have an infinite variance with respect to the stationary measure. We remark that the variance is exactly the expression of the thermal conductivity obtained in [6]. The rescaled process $N^{-2/3}Y(Nt)$ converges in distribution to a stable symmetric Lévy process with index $3/2$ [3, 25]. As a corollary, the rescaled solution of the Boltzmann equation $W(N^{2/3}y, k, Nt)$ converges, as $N \to +\infty$, to the solution of the fractional diffusion equation

$$\partial_t u(y,t) = -|\Delta|^{3/4} u(y,t).$$

A different, more analytic approach can be found in [34].

In the pinned or the non-acoustic cases, $\omega' \sim k$ for small k, $\tau(X_i)\omega'(X_i)$ has finite variance with respect to the stationary measure. In particular this variance coincides with the thermal diffusivity computed by the Green-Kubo formula [6]. Then one can prove that the rescaled solution $W(N^{1/2}y, k, Nt)$ converges to the heat equation

$$\partial_t u(y,t) = D\Delta u(y,t),$$

with D given by the thermal diffusivity.

The results described above give a two step approach to the diffusion or superdiffusion of the energy: first a kinetic limit where the Boltzmann phonon equation is obtained in the weak noise limit, then a superdiffusive or diffusive rescaling of the solution of this equation. The results described in Sect. 5.3 concern a simple space-time rescaling, without any weak noise approximation.[1] Still Boltzmann equation helps to understand the proof of (5.17), that goes under the following lines.

Let us just consider the superdiffusive case and consider the evolution of the Wigner distribution at time $\varepsilon^{-3/2}t$ and for the noise of intensity γ. Equation (5.29) becomes

$$\begin{aligned}\partial_t \hat{W}^\varepsilon(\xi,k,t) =& -i\xi\,\omega'(k)\varepsilon^{-1/2}\hat{W}^\varepsilon(\xi,k,t) + \gamma\varepsilon^{-3/2}\mathcal{C}\hat{W}^\varepsilon(\xi,k,t)\\ &-\frac{\gamma}{2}\varepsilon^{-3/2}\mathcal{C}\left(\hat{Y}^\varepsilon(\xi,k,t) + \hat{Y}^\varepsilon(\xi,k,t)^*\right) + \mathcal{O}(\varepsilon),\end{aligned} \tag{5.31}$$

[1] In the non-linear cases we cannot expect that the two step approach would give the same result of the direct rescaling of the dynamics. In the β-FPU the kinetic limit seems to give a different superdiffusion scaling than the direct limit [33, 41].

This looks like a very singular limit. Still due to fast oscillations the $\hat{Y}^\varepsilon$ terms disappears after time integration. Furthermore, because the number of collisions per unit time tends to infinity, the limit of the Wigner distribution *homogenizes* in the variable k, i.e. its limit becomes a function $W(y,t)$ independent of variable k. Assume that the above facts have been proven, and consider the case of the simple random exchange of the velocities, that give a scattering rate of the form: $R(k,k') = R(k)R(k')$, with $\int R(k)dk = 1$. The argument below, that follows the line of Mellet et al. [34], can be generalized to various rate functions [26]. The Laplace transform w_ε in time of the Wigner distribution satisfies the equation:

$$\begin{aligned}&\left(\lambda + 2\gamma\varepsilon^{-3/2}R(k) + i\omega'(k)\xi\varepsilon^{-1/2}\right) w_\varepsilon(\lambda,\xi,k)\\ &\quad = \widehat{W}_\varepsilon(\xi,k,0) + \varepsilon^{-3/2}2\gamma R(k)\int R(k')dk' + \mathcal{O}(\varepsilon). \qquad (5.32)\end{aligned}$$

By dividing the above expression by $D_\varepsilon(\lambda,\xi,k) = \varepsilon^{3/2}\lambda + 2\gamma R(k) + i\omega'(k)\xi\varepsilon$, multiplying by $2\gamma R(k)$ and integrating in k one obtains

$$a_\varepsilon \int w_\varepsilon(\lambda,\xi,k)R(k)dk - \int \frac{2\gamma R(k)\widehat{W}_\varepsilon(\xi,k,0)}{D_\varepsilon(\lambda,\xi,k)}dk = \mathcal{O}(\varepsilon) \qquad (5.33)$$

where, since $R(k) \sim k^2$ and due to the assumptions made on $\omega'(k)$:

$$a_\varepsilon = 2\gamma\varepsilon^{-3/2}\left(1 - \int \frac{2\gamma R^2(k)}{D_\varepsilon(\lambda,\xi,k)}dk\right) \underset{\varepsilon\to 0}{\longrightarrow} \lambda + c|\xi|^{3/2}.$$

Thanks to the homogenization property of $\widehat{W}_\varepsilon$

$$\int w_\varepsilon(\lambda,\xi,k)R(k)dk \underset{\varepsilon\to 0}{\longrightarrow} w(\lambda,\xi).$$

Furthermore

$$\frac{2\gamma R(k)}{D_\varepsilon(\lambda,\xi,k)} \underset{\varepsilon\to 0}{\longrightarrow} 1,$$

and we conclude that the limit function $w(\lambda,\xi)$ satisfies the equation

$$\left(\lambda + c|\xi|^{3/2}\right) w(\lambda,\xi) = \int \widehat{W}(\xi,k,0)dk \qquad (5.34)$$

where $\widehat{W}(\xi,k,0)$ is the limit of the initial condition. Equation (5.34) is the Laplace-Fourier transform of the fractional heat equation. Assuming instead that $\omega'(k) \sim k$, a similar argument gives the normal heat equation [26].

5.7 Non-acoustic Chains: Beam Dynamics

We have seen in the previous sections that the ballistic behaviour at the hyperbolic scale and the superdiffusive behavior of the energy strictly depends on the positivity of the sound velocity τ_1. In the case $\hat{\alpha}(0) = \hat{\alpha}''(0) = 0$ all the coefficients in these evolutions are null and in fact the limit of the energy follows a regular diffusion. Notice that the dynamics is still momentum conserving.

A typical example is given by the following choice of the interaction:

$$\alpha_0 = 6\alpha, \quad \alpha_1 = -4\alpha, \quad \alpha_2 = \alpha, \quad \alpha_x = 0, \quad x > 2 \tag{5.35}$$

(then $\hat{\alpha}(k) = \alpha 4 \sin^4(\pi k)$) that corresponds to the Hamiltonian

$$\mathscr{H} = \sum_x \left[\frac{p_x^2}{2} + \frac{\alpha}{2} (q_{x+1} - 2q_x + q_{x-1})^2 \right]. \tag{5.36}$$

Notice that the expression (5.7) is not defined for any value of the tension τ, this is why we also call these chains tensionless. Basically, the equilibrium energy does not change by *pulling* the chain. It does change by *bending* it, this is why the relevant quantities are defined by the local *curvatures* or *deflections*:

$$\kappa_x = -\Delta q_x = 2q_x - q_{x+1} - q_{x-1}. \tag{5.37}$$

The relevant balanced quantities are now (κ_x, p_x, e_x).

The invariant equilibrium measures are formally given by

$$\frac{e^{-\beta\left[\mathscr{H} - \bar{p}\sum_x p_x - \mathfrak{L}\sum_x \kappa_x\right]}}{\mathscr{Z}} \, d\kappa \, dp \tag{5.38}$$

where the parameter $\mathfrak{L}$ is called *load*. Notice that these measures would be non-translation invariant in the coordinates r_x's.

Under these conditions the sound velocity τ_1 is always null, and there is no ballistic evolution of the chain. In fact it turns out that the three quantities $(\mathfrak{k}(t,y), \mathfrak{p}(t,y), \mathfrak{e}(t,y))$ evolve on the diffusive space-time scale. By defining

$$\mathfrak{e}_{mech}(t,y) = \frac{1}{2}\left(\mathfrak{p}(t,y)^2 + \alpha \mathfrak{k}^2\right) \quad \text{and} \quad \mathfrak{T}(t,y) = \mathfrak{e}(t,y) - \mathfrak{e}_{mech}(t,y),$$

after the corresponding space-time scaling we obtain [28]:

$$\begin{cases} \partial_t \mathfrak{k} &= -\,\partial_y^2 \mathfrak{p}, \\ \partial_t \mathfrak{p} &= \alpha \partial_y^2 \mathfrak{k} + \gamma \partial_y^2 \mathfrak{p}, \\ \partial_t \mathfrak{T} &= D_\gamma \partial_y^2 \mathfrak{T} + \gamma \left(\partial_y \mathfrak{p}\right)^2. \end{cases} \tag{5.39}$$

The first two equations are the damped Euler-Bernoulli beam equations. The third one describes the diffusive behavior of the energy. The thermal diffusivity $D_\gamma = \frac{C\alpha}{\gamma} + \gamma$, where C is an explicit constant independent from γ and α. In particular for constant initial values of $\mathfrak{k}$ and $\mathfrak{p}$, the thermal energy (i.e. temperature) profile follows a normal heat equation with the thermal diffusivity that can be computed explicitly (cf. [28]). Notice also that a gradient of the macroscopic velocity induce a local increase of the temperature.

These models provide rigorous counter-examples to the usual conjecture that the momentum conservation in one dimension always implies superdiffusivity of the energy (cf. [19, 31]). The presence of a non-vanishing sound velocity seems a necessary condition.

5.8 A Simpler Model with Two Conserved Quantities

In this section we consider the nearest neighbors unpinned harmonic chain (with mass 1 and coupling forces $\alpha_1 = \alpha_{-1} = -\frac{\alpha_0}{2}$) but we add different stochastic collisions with the properties that they conserve the total energy and some extra quantity (that we call "volume") but no longer momentum and stretch. By defining $a = \sqrt{\alpha_0}$ and the field $\{\eta_x \in \mathbb{R}\ ;\ x \in \mathbb{Z}\}$ by $\eta_{2x} = ar_x$ and $\eta_{2x+1} = p_x$, the Hamiltonian equations are reduced to

$$\dot{\eta_x} = a(\eta_{x+1} - \eta_{x-1}).$$

The stochastic collisions are such that at random times given by independent Poisson clocks $N_{x,x+1}(t)$ of intensity γ the kinetic energy at site x is exchanged with the corresponding potential energy. The simplest way to do it is to exchange the variable η_x with η_{x+1}. Because of the form of the noise the total energy $\sum_x \frac{\eta_x^2}{2m}$ and the "volume" $\sum_x \eta_x = \sum_x (p_x + ar_x)$ are the only conserved quantities of the dynamics [14]. By reducing the number of conserved quantities from 3 (energy, momentum, stretch) to 2 (energy, volume) we expect to see easily the influence of the other conserved quantity on the superdiffusion of energy. The nature of the superdiffusion for models with two conserved quantities are studied in the nonlinear fluctuating hydrodynamics framework by Spohn and Stoltz in [43].

In the hyperbolic time scaling, starting from an initial distribution associated to some smooth macroscopic initial volume-energy profiles $(v_0(y), \mathfrak{e}_0(y))$, we can prove that these initial profiles evolve following the linear wave equation $(v(t,y), e(t,y))$ which is solution of

$$\partial_t v = 2a\partial_y v, \quad \partial_t \mathfrak{e} = a\partial_y (v^2).$$

As in Sect. 5.2 we can introduce the *mechanical energy* $\mathfrak{e}_{mech}(y,t) = \frac{v^2(y,t)}{2}$ and the thermal energy $T(y,t) = \mathfrak{e}(y,t) - \mathfrak{e}_{mech}(y,t)$. The later remains constant in time.

Mutatis mutandis the discussion of Sects. 5.3–5.5 can be applied to this model with two conserved quantities with very similar conclusions[2] [15, 16]. The interesting difference is that in (5.18) and (5.27) the fractional Laplacian has to be replaced by the *skew fractional Laplacian*:

$$\partial_t \mathfrak{T} = -c\{\,|\Delta_y|^{3/4} - \nabla_y|\Delta_y|^{1/4}\}\,\mathfrak{T}, \qquad \mathfrak{T}(y,0) = T(y,0) \tag{5.40}$$

for a suitable explicit constant $c > 0$. The skewness is produced here by the interaction of the (unique) sound mode with the heat mode. In the models of Sect. 5.1 which conserve three quantities, there are two sound modes with opposite velocities. The skewness produced by each of them is exactly counterbalanced by the other one so that it is not seen in the final equations (5.18) and (5.27).

5.8.1 The Extension Problem for the Skew-Fractional Laplacian

For the model with two conserved quantities introduced in this section, the derivation of the skew fractional heat equation, at least at the level of the fluctuations in equilibrium as defined in Sect. 5.5, can be implemented by means of the so-called *extension problem* for the fractional Laplacian [17, 44]. As we will see, this extension problem does not only provides a different derivation, but it also clarifies the role of the other (fast) conservation law (i.e. the volume). It can be checked that for any $\beta > 0$ and any $\rho \in \mathbb{R}$, the product measure with Gaussian marginals of mean ρ and variance (temperature) β^{-1} are stationary under the dynamics of $\{\eta_x(t); x \in \mathbb{Z}\}$. Let us assume that the dynamics starts from a stationary state. For simplicity we assume $\rho = 0$ and $a = 1$. The space-time energy correlation function $S_\varepsilon(x,t)$ is defined here by

$$S_\varepsilon(x,t) = \big\langle (\eta_x(\varepsilon^{-3/2}t)^2 - \beta^{-1})\,(\eta_0(0)^2 - \beta^{-1})\big\rangle.$$

It turns out that the energy fluctuations are driven by volume correlations. Therefore it makes sense to define the *volume* correlation function as

$$G_\varepsilon(x,y,t) = \big\langle \eta_x(\varepsilon^{-3/2}t)\eta_y(\varepsilon^{-3/2}t)\;(\eta_0(0)^2 - \beta^{-1})\big\rangle.$$

Let $f : [0,T] \times \mathbb{R} \to \mathbb{R}$ be a smooth, regular function and for each $t \in [0,T]$, let $u_t : \mathbb{R} \times \mathbb{R}_+ \to \mathbb{R}$ be the solution of the boundary-value problem

$$\begin{cases} -\partial_x u + \gamma \partial_y^2 u = 0 \\ \qquad \partial_y u(x,0) = \partial_x f(x,t). \end{cases}$$

---

[2] In [15] only the equilibrium fluctuations are considered but the methods developed in [26, 27] can be applied also to the models considered in this section.

It turns out that $\partial_x u_t(x,0) = \mathscr{L} f_t(x)$, where $\mathscr{L}$ is the skew fractional Laplacian defined in (5.40).

For test functions $f : \mathbb{R} \to \mathbb{R}$ and $u : \mathbb{R} \times \mathbb{R}_+ \to \mathbb{R}$ define

$$\langle S_\varepsilon(t), f\rangle_\varepsilon = \sum_{x\in\mathbb{Z}} S_\varepsilon(x,t) f(\varepsilon x),$$

$$\langle G_\varepsilon(t), u\rangle_\varepsilon = \sum_{\substack{x,y\in\mathbb{Z}\\ x<y}} G_\varepsilon(x,y,t)\; u\big(\tfrac{\varepsilon}{2}(x+y), \sqrt{\varepsilon}(y-x)\big).$$

After an explicit calculation we have:

$$\tfrac{d}{dt}\Big\{\langle S_\varepsilon(t) f_t\rangle_\varepsilon - \tfrac{\varepsilon^{1/2}}{2\gamma}\langle G_\varepsilon(t), u_t\rangle_\varepsilon + \varepsilon\langle S_\varepsilon(t), u_t(\cdot,0)\rangle_\varepsilon\Big\} = \langle S_\varepsilon(t), \partial_x f_t\rangle_\varepsilon$$

plus error terms that vanish as $\varepsilon \to 0$. From this observation it is not very difficult to obtain that, for any smooth function f of compact support,

$$\lim_{\varepsilon\to 0}\langle S_\varepsilon(t), f\rangle_\varepsilon = \int P(t,x) f(x)dx,$$

where $P(t,x)$ is the fundamental solution of the skew fractional heat equation (5.40).

Let us explain in more details why the introduction of the test function u_t solves the equation for the energy correlation function $S_\varepsilon(x,t)$. It is reasonable to parametrize volume correlations by its distance to the diagonal $x = y$. The microscopic current associated to the energy $\eta_x(t)^2$ is equal to $\eta_x(t)\eta_{x+1}(t)$, which can be understood as the volume correlations around the diagonal $x = y$. Volume evolves in the hyperbolic scale with speed 2. Fluctuations around this transport evolution appear in the diffusive scale and are governed by a diffusion equation. This means that at the hyperbolic scale ε^{-1}, fluctuations are of order $\varepsilon^{-1/2}$, explaining the non-isotropic space scaling introduced in the definition of $\langle G_\varepsilon(t), u\rangle_\varepsilon$. It turns out that the couple $(S_\varepsilon, G_\varepsilon)$ satisfies a closed system of equations, which can be checked to be a semidiscrete approximation of the system

$$\begin{cases} \sqrt{\varepsilon}\partial_t u_t = -\partial_x u_t + \gamma\partial_y^2 u_t\\ \partial_y u_t(x,0) = \partial_x f_t(x)\\ \partial_t f_t = \partial_x u_t(x,0). \end{cases}$$

Therefore, the volume serves as a fast variable for the evolution of the energy, which corresponds to a slow variable. The extension problem plays the role of the cell problem for the homogenization of this fast-slow system of evolutions.

5.9 The Dynamics in Higher Dimension

One of the interesting features of the harmonic dynamics with energy and momentum conservative noise is that they reproduce, at least qualitatively, the expected behavior of the non-linear dynamics, also in higher dimensions. In particular in the three or higher dimension the thermal conductivity, computed by the Green-Kubo formula, is finite, while it diverges logarithmically in two dimension (always for non-acoustic systems), cf. [5, 6].

In dimension $d \geq 3$ it can be also proven that equilibrium fluctuations evolve diffusively, i.e. that the asymptotic correlation $\tilde{S}^{33}(y, t)$, defined in Sect. 5.5 but with a diffusive scaling, satisfies [4]:

$$\partial_t \tilde{S}^{33} = D \partial_y^2 \tilde{S}^{33} \tag{5.41}$$

where D can be computed explicitly by Green-Kubo formula in terms of ω and scattering rate R [5, 6]. Similar finite diffusivity and diffusive evolution of the fluctuations are proven for pinned models ($\hat{\alpha}(0) > 0$), see [4]. In two dimensions, while the logarithmic divergence of the Green-Kubo expression of the thermal conductivity is proven in [6], the corresponding diffusive behaviour at the logarithmic corrected time scale is still an open problem. For the result obtained from the kinetic equation see [2]. The two-dimensional model is particularly interesting in light of the large thermal conductivity measured experimentally on graphene [46], an essentially a two-dimensional material.

5.10 Thermal Boundary Conditions and the Non-equilibrium Stationary States

Traditionally the problem of thermal conductivity has been approached by considering the stationary non-equilibrium state for a finite system in contact with heat bath at different temperatures (cf.[31, 38]). This set-up is particularly suitable for numerical simulations and convenient because it gives an straight operational definition of the thermal conductivity in terms of the stationary flux of energy, avoiding to specify the macroscopic evolution equation. But for the theoretical understanding and corresponding mathematical proofs of the thermal conductivity phenomena, this is much harder than the non-stationary approach described in the previous section. This because the stationary state conceals the space-time scale.

The finite system consists of $2N+1$ atoms, labelled by $x = -N, \dots, N$, with end points connected to two heat baths at temperature T_l and T_r. These baths are modeled by Langevin stochastic dynamics, so that the evolution equations are given by

$$\begin{aligned}
\dot{q}_x &= p_x, \qquad x = -N, \dots, N, \\
\dot{p}_x &= -(\alpha_N * q(t))_x + (p_{x+1}(t^-) - p_x(t^-))\, \dot{N}_{x,x+1}(t) + (p_{x-1}(t^-) - p_x(t^-))\, \dot{N}_{x-1,x}(t) \\
&\quad + \delta_{x,-N}\left(-p_{-N} + \sqrt{2T_l}\, dw_{-N}(t)\right) + \delta_{x,N}\left(-p_N + \sqrt{2T_r}\, dw_N(t)\right),
\end{aligned} \tag{5.42}$$

where $w_{-N}(t), w_N(t)$ are two independent standard Brownian motions, and the coupling α_N is properly defined in order to take into account the boundary conditions. For this finite dynamics there is a (non-equilibrium) unique stationary state, where the energy flow from the hot to the cold side. Observe that because of the exchange noise between the atoms, the stationary state is not Gaussian, unlike in the case studied in [38].

Denoting the stationary energy flux by J_N, the thermal conductivity of the finite chain is defined as

$$\kappa_N = \lim_{|T_l - T_r| \to 0} \frac{(2N+1)J_N}{T_l - T_r}. \tag{5.43}$$

For a finite N it is not hard to prove that κ_N can be expressed in terms of the corresponding Green-Kubo formula. For the periodic boundary unpinned acoustic case this identification gives $\kappa_N \sim N^{1/2}$ (cf. [5]). For the noise that conserves only the energy, but not momentum (like independent random flips of the signs of the momenta), the system has a finite thermal diffusivity and the limit $\kappa_N \to \kappa$, as $N \to +\infty$, can be computed explicitly, as proven in [11].

The natural question is about the macroscopic evolution of the temperature profile in a non-stationary situation, and the corresponding stationary profile. It turns out [7] that this macroscopic equation is given by a fractional heat equation similar to (5.17) with a proper definition of the fractional Laplacian $|\Delta|^{3/4}$ on the interval $[-1, 1]$ subject to the boundary conditions $\mathfrak{T}(-1) = T_l, \mathfrak{T}(1) = T_r$. This is defined by using the following orthonormal basis of functions on the interval $[-1, 1]$:

$$u_n(y) = \cos\left(n\pi(y+1)/2\right), \qquad n = 0, 1, \dots \tag{5.44}$$

Any continuous function $f(y)$ on $[-1, 1]$ can be expressed in terms of a series expansion in u_n. Then we define $|\Delta|^s u_n(y) = (n\pi/2)^{2s} u_n(y)$. Observe that for $s = 1$ we recover the usual definition of the Laplacian. For $s \neq 1$ this is *not* equivalent to other definitions of the fractional Laplacian in a bounded interval, e.g. [47].

Correspondingly the stationary profile is given by

$$\begin{aligned} &|\Delta|^s \mathfrak{T} = 0, \text{ in } (-1, 1) \\ &\mathfrak{T}(-1) = T_l, \ \mathfrak{T}(1) = T_r, \end{aligned} \tag{5.45}$$

namely $\mathfrak{T}(y) = \frac{1}{2}(T_l + T_r) + \frac{1}{2}(T_l - T_r)\theta(y)$, with

$$\theta(y) = c_s \sum_{m \text{ odd}} \frac{1}{m^{2s}} \cos\big(m\pi(y+1)/2\big),$$

c_s such that $\theta(-1) = 1$. This expression corresponds with the formula computed directly in [32] using a continuous approximation of the covariance matrix of the stationary state for the dynamics with fixed boundaries.

5.11 The Non-linear Chain

From the above rigorous results on the harmonic chain with the random collision dynamics, and the arguments of Spohn from fluctuating hydrodynamics and mode couplings (cf. [41, 42]), we can conjecture the corresponding behaviour in the anharmonic case.

Let us consider just nearest neighbor interaction given by the potential energy $V(q_x - q_{x-1})$ of an anharmonic spring. We assume $V : \mathbb{R} \to (0, +\infty)$ is smooth and that it grows quadratically at infinity. Define the energy of the oscillator x as

$$e_x(r,p) := \frac{p_x^2}{2} + V(r_x) \tag{5.46}$$

The dynamics is defined as the solution of the Newton equations

$$\dot{q}_x(t) = p_x, \quad \dot{p}_x(t) = -\left(V'(r_x) - V'(r_{x+1})\right), \quad x \in \mathbb{Z} \tag{5.47}$$

plus a random exchange of velocities as in the previous sections, regulated by an intensity γ. The equilibrium Gibbs measures are parametrized by

$$\boldsymbol{\lambda} = (\beta^{-1} \text{ (temperature)}, \bar{p} \text{ (velocity)}, \tau \text{ (tension)}),$$

and are given explicitly by

$$d\nu_{\boldsymbol{\lambda}} = \prod_x e^{-\beta(V(r_x) - \tau r_x + \frac{(p_x - \bar{p})^2}{2}) - \mathcal{G}(\boldsymbol{\lambda})} \, dr_x \, dp_x \tag{5.48}$$

When a random exchange of velocity is present ($\gamma > 0$) it can be proven that these are the only *regular* translation invariant stationary measures [13, 23]. We have

$$\nu_\lambda(p_x) = \bar{p}, \quad \nu_\lambda(r_x) = -\frac{1}{\beta}\partial_\tau \mathcal{G} := \bar{r}, \quad \nu_\lambda(e_x) = \frac{1}{\beta} - \partial_\beta \mathcal{G} - \frac{\tau}{\beta}\partial_\tau \mathcal{G} := \bar{e}.$$

These thermodynamical relations can be inverted to express the parameters $(\beta^{-1}, \bar{p}, \tau)$ in terms of $(\bar{p}, \bar{r}, \bar{e})$. It turns out that the tension is then equal to a nonlinear function $\bar{\tau}(\bar{r}, \bar{u})$ of the average stretch $\bar{r}$ and the average internal energy $\bar{u} = \bar{e} - \bar{p}^2/2$.

After the hyperbolic rescaling of the dynamics, we expect that the empirical distribution of the balanced quantities converge to the system of hyperbolic equations:

$$\begin{aligned}
\partial_t \bar{r} &= \partial_y \bar{p} \\
\partial_t \bar{p} &= \partial_y \bar{\tau} \\
\partial_t \bar{e} &= \partial_y \left(\bar{p}\bar{\tau}\right).
\end{aligned} \tag{5.49}$$

This limit can be proven, under certain condition on the boundaries, in the smooth regime, if the microscopic dynamics is perturbed by a random exchange of velocities between nearest neighbors particles by using relative entropy methods [13, 22, 36].

The limit should be still valid after shocks develop, with the limit profile given by an entropic weak solution. This is a main open problem in hydrodynamic limits.

After a long time, the (entropy) solution of (5.49) should converge (maybe in a weak sense) to some mechanical equilibrium characterized by:

$$\bar{p}(y) = p_0, \qquad \bar{\tau}(\bar{r}(y), \bar{u}(y)) = \tau_0, \quad \text{for some constants } p_0, \tau_0. \tag{5.50}$$

It is very hard to characterize all possible stationary solutions that satisfy (5.50). Probably they are generically very irregular. But certainly if we start with a smooth initial condition that satisfies (5.50), they do not move. Also by the relative entropy methods, it is possible to prove that starting with such initial profiles, the empirical distribution of the balanced quantities will converge at the hyperbolic space-time scale to such a stationary solution at any time.

Still we do know that the microscopic dynamics will converge to a global equilibrium, so this implies that there exists a larger time scale such that these profiles will evolve and eventually reach also thermal equilibrium.

There is a numerical evidence and heuristic arguments about the divergence of the Green-Kubo formula defining the thermal diffusivity for such one dimensional systems, so we expect that the larger time scale at which these profiles evolve is superdiffusive.

From the nonlinear fluctuation hydrodynamics [41], one can conjecture the following: the space-time scale is $(\varepsilon^{-1}x, \varepsilon^{-2a}t)$, and the temperature $T(x,t) = \beta^{-1}(x,t)$ evolves following some fractional heat equation, possibly non-linear. If V is symmetric and $\tau_0 = 0$, then $a = 3/4$, and in all other cases $a = 5/6$.

5.12 The Disordered Chain

The effect of disorder on transport and phonons localization properties in chains of oscillators has attracted a lot of interest [19, 31]. Randomness can appear at the level of the masses of particles or at the level of the potentials. We consider only the case of random masses with non random potential V or the case of non random masses and non random interaction potential V with random harmonic pinnings. In the first case, the Hamiltonian is then given by

$$\mathcal{H}(p,q) := \sum_x \frac{p_x^2}{2m_x} + \sum_{x,x'} V(q_x, q_x')$$

where $\{m_x\}$ are positive random variables, while in the second case, the Hamiltonian is given by

$$\mathcal{H}(p,q) := \sum_x \frac{p_x^2}{2m} + \sum_{x,x'} V(q_x, q_x') + \sum_x \nu_x q_x^2$$

where $\{\nu_x\}$ are positive random variables, $m > 0$ being the mass of the particles. The presence of randomness is relevant for the thermal properties of the system but the fact that randomness affect potentials or masses is not.

For one dimensional unpinned disordered harmonic chains it is known that the behavior of the conductivity is very sensitive to the boundary conditions since it can diverge as $\sqrt{N}$ or vanish as $1/\sqrt{N}$ with the systems length N [1, 18, 39, 45]. If harmonic pinning is added localization of normal modes leads an exponential decay of the heat current and a zero conductivity. The situation in higher dimensions, even in the case of harmonic interactions, is still under debate but it is expected that conductivity is finite in dimension $d \geq 3$ if disorder is sufficiently weak [29]. About the effect of nonlinearities, numerical evidences suggest that a very small amount of anharmonicity in pinned chains is sufficient to restore a diffusive regime with a positive finite value of the conductivity [20]. However it is a challenging open question to decide if the transition from an insulator to a conductor occurs at zero or some finite small value of anharmonicity [8, 20, 35].

It is suggestive to think that a stochastic noise could affect transport properties of harmonic chains in some rough sense similar to the addition of nonlinearities. This question has been first address in [9] in the Green-Kubo formula setting, revisited in [21] from the non equilibrium stationary state point of view (see Sect. 5.10) and extended in [10, 24] to incorporate weakly nonlinear chains. In these papers, the authors consider a disordered harmonic chain, or weakly nonlinear in [10, 24], with a stochastic noise which consists to flip, independently for each particle, at independent random exponential times of mean $1/\lambda$, $\lambda > 0$, the velocity of the particle. Notice that this energy conserving noise is very different from the noise considered in the rest of the paper since it does not conserve momentum. In particular, for ordered pinned and unpinned nonlinear chains, this noise is sufficient to provide a finite conductivity κ [12]. However, it turns out that for an ordered harmonic chain, $\kappa \sim \lambda^{-1}$ and, as expected, increases to infinity as the strength of the noise $\lambda \downarrow 0$. In [10, 24] it is proved that localization effects persist: $\kappa = \mathcal{O}(\lambda)$ for a pinned disordered chain with a small anharmonic potential, and $\kappa \sim \lambda$ for a pinned harmonic chain. As far as we know disordered chains with energy-momentum conserving noise have never been investigated.

Acknowledgements We thank Herbert Spohn for many inspiring discussions on this subject.

The research of Cédric Bernardin was supported in part by the French Ministry of Education through the grant ANR-EDNHS. The work of Stefano Olla has been partially supported by the European Advanced Grant *Macroscopic Laws and Dynamical Systems* (MALADY) (ERC AdG 246953) and by a CNPq grant *Sciences Without Frontiers*. Tomasz Komorowski acknowledges the support of the Polish National Science Center grant UMO-2012/07/B/ST1/03320.

References

1. Ajanki, O., Huveneers, F.: Rigorous scaling law for the heat current in disordered harmonic chain. Commun. Math. Phys. **301**, 841–883 (2011)
2. Basile, G.: From a kinetic equation to a diffusion under an anomalous scaling. Ann. Inst. Henri Poincare Prob. Stat. **50**(4), 1301–1322 (2014). doi:10.1214/13-AIHP554
3. Basile, G., Bovier, A.: Convergence of a kinetic equation to a fractional diffusion equation. Markov Proc. Rel. Fields **16**, 15–44 (2010)
4. Basile, G., Olla, S.: Energy diffusion in harmonic system with conservative noise. J. Stat. Phys. **155**(6), 1126–1142 (2014). doi:10.1007/s10955-013-0908-4
5. Basile, G., Bernardin, C., Olla, S.: A momentum conserving model with anomalous thermal conductivity in low dimension. Phys. Rev. Lett. **96**, 204–303 (2006). doi:10.1103/PhysRevLett.96.204303
6. Basile, G., Bernardin, C., Olla, S.: Thermal conductivity for a momentum conservative model. Commun. Math. Phys. **287**, 67–98 (2009)
7. Basile, G., Olla, S., Spohn, H.: Energy transport in stochastically perturbed lattice dynamics. Arch. Ration. Mech. **195**(1), 171–203 (2009)
8. Basko, D.M.: Weak chaos in the disordered nonlinear Schrödinger chain: destruction of Anderson localization by Arnold diffusion. Ann. Phys. **326**, 1577–1655 (2011)
9. Bernardin, C.: Thermal conductivity for a noisy disordered harmonic chain. J. Stat. Phys. **133**(3), 417–433 (2008)
10. Bernardin, C., Huveneers, F.: Small perturbation of a disordered harmonic chain by a noise and an anharmonic potential. Probab. Theory Relat. Fields **157**(1–2), 301–331 (2013)
11. Bernardin, C., Olla, S.: Fourier law and fluctuations for a microscopic model of heat conduction. J. Stat. Phys. **118**(3/4), 271–289 (2005)
12. Bernardin, C., Olla, S.: Transport properties of a chain of anharmonic oscillators with random flip of velocities. J. Stat. Phys. **145**, 1224–1255 (2011)
13. Bernardin, C., Olla, S.: Thermodynamics and non-equilibrium macroscopic dynamics of chains of anharmonic oscillators. Lecture Notes (2014). Available at https://www.ceremade.dauphine.fr/~olla/
14. Bernardin, C., Stoltz, G.: Anomalous diffusion for a class of systems with two conserved quantities. Nonlinearity **25**(4), 1099–1133 (2012)
15. Bernardin, C., Goncalves, P., Jara, M.: 3/4 fractional superdiffusion of energy in a system of harmonic oscillators perturbed by a conservative noise. Arch. Ration. Mech. Anal. 1–38 (2015). doi:10.1007/s00205-015-0936-0. Issn 1432-0673 [online first]
16. Bernardin, C., Goncalves, P., Jara, M., Sasada, M., Simon, M.: From normal diffusion to superdiffusion of energy in the evanescent flip noise limit. J. Stat. Phys. **159**(6), 1327–1368 (2015)
17. Caffarelli, L., Silvestre, L.: An extension problem related to the fractional Laplacian. Commun. Partial Differ. Equ. **32**(8), 1245–1260 (2007)
18. Casher, A., Lebowitz, J.L.: Heat flow in regular and disordered harmonic chains. J. Math. Phys.**12**, 1701–1711 (1971)
19. Dhar, A.: Heat transport in low dimensional systems. Adv. Phys. **57**(5), 457–537 (2008)
20. Dhar, A., Lebowitz, J.L.: Effect of phonon–phonon interactions on localization. Phys. Rev. Lett. **100**, 134301 (2008)
21. Dhar, A., Venkateshan, K., Lebowitz, J.L.: Heat conduction in disordered harmonic lattices with energy-conserving noise. Phys. Rev. E **83**(2), 021108 (2011)
22. Even, N., Olla, S.: Hydrodynamic limit for an Hamiltonian system with boundary conditions and conservative noise. Arch. Ration. Mech. Appl. **213**, 561–585 (2014). doi:10.1007/s00205-014-0741-1
23. Fritz, J., Funaki, T., Lebowitz, J.L.: Stationary states of random Hamiltonian systems. Probab. Theory Relat. Fields **99**, 211–236 (1994)

24. Huveneers, F.: Drastic fall-off of the thermal conductivity for disordered lattices in the limit of weak anharmonic interactions. Nonlinearity **26**(3), 837–854 (2013)
25. Jara, M., Komorowski, T., Olla, S.: A limit theorem for an additive functionals of Markov chains. Ann. Appl. Probab. **19**(6), 2270–2300 (2009)
26. Jara, M., Komorowski, T., Olla, S.: Superdiffusion of energy in a chain of harmonic oscillators with noise. Commun. Math. Phys. **339**, 407–453 (2015). doi:10.1007/s00220-015-2417-6
27. Komorowski, T., Olla, S.: Ballistic and superdiffusive scales in macroscopic evolution of a chain of oscillators. Nonlinearity (2016). arXiv:1506.06465
28. Komorowski, T., Olla, S.: Diffusive propagation of energy in a non-acoustic chain (2016, preprint)
29. Kundu, A., Chaudhuri, A., Roy, D., Dhar, A., Lebowitz, J.L., Spohn, H.: Heat transport and phonon localization in mass-disordered harmonic crystals. Phys. Rev. B **81**, 064301 (2010)
30. Lepri, S., Livi, R., Politi, A.: Heat conduction in chains of nonlinear oscillators. Phys. Rev. Lett. **78**, 1896 (1997)
31. Lepri, S., Livi, R., Politi, A.: Thermal conduction in classical low-dimensional lattices. Phys. Rep. **377**, 1–80 (2003)
32. Lepri, S., Meija-Monasterio, C., Politi, A.: A stochastic model of anomalous heat transport: analytical solution of the steady state. J. Phys. A: Math. Gen. **42**, 025001 (2009)
33. Lukkarinen, J., Spohn, H.: Anomalous energy transport in the FPU-β chain. Commun. Pure Appl. Math. **61**, 1753–1786 (2008). doi:10.1002/cpa.20243
34. Mellet, A., Mischler, S., Mouhot, C.: Fractional diffusion limit for collisional kinetic equations. Arch. Ration. Mech. Anal. **199**(2), 493–525 (2011)
35. Oganesyan, V., Pal, A., Huse, D.: Energy transport in disordered classical spin chains. Phys. Rev. B **80**, 115104 (2009)
36. Olla, S., Varadhan, S.R.S., Yau, H.T.: Hydrodynamic limit for a Hamiltonian system with weak noise. Commun. Math. Phys. **155**, 523–560 (1993)
37. Peierls, R.E.: Zur kinetischen Theorie der Waermeleitung in Kristallen. Ann. Phys. Lpz. **3**, 1055–1101 (1929)
38. Rieder, Z., Lebowitz, J.L., Lieb, E.: Properties of harmonic crystal in a stationary non-equilibrium state. J. Math. Phys. **8**, 1073–1078 (1967)
39. Rubin, R.J., Greer, W.L.: Abnormal lattice thermal conductivity of a one-dimensional, harmonic, isotopically disordered crystal. J. Math. Phys. **12**, 1686–1701 (1971)
40. Spohn, H.: The phonon Boltzmann equation, properties and link to weakly anharmonic lattice dynamics. J. Stat. Phys. **124**(2–4), 1041–1104 (2006)
41. Spohn, H.: Nonlinear fluctuating hydrodynamics for anharmonic chains. J. Stat. Phys. **154**(5), 1191–1227 (2014)
42. Spohn, H.: Fluctuating hydrodynamics approach to equilibrium time correlations for anharmonic chains. In: Lepri, S. (ed.) Thermal Transport in Low Dimensions: From Statistical Physics to Nanoscale Heat Transfer. Springer, Heidelberg (2016)
43. Spohn, H., Stoltz, G.: Nonlinear fluctuating hydrodynamics in one dimension: the case of two conserved fields. J. Stat. Phys. **160**, 861–884 (2015). doi:10.1007/s10955-015-1214-0
44. Stroock, D.W., Varadhan, S.R.S.: Diffusion processes with boundary conditions. Commun. Pure Appl. Math. **24**, 147–225 (1971)
45. Verheggen, T.: Transmission coefficient and heat conduction of a harmonic chain with random masses: asymptotic estimates on products of random matrices. Commun. Math. Phys. **68**, 69–82 (1979)
46. Xu, Y., Li, Z., Duan, W.: Thermal and thermoelectric properties of graphene. Small **10**(11), 2182–2199 (2014). doi:10.1002/smll.201303701
47. Zola, A., Rosso, A., Kardar, M.: Fractional Laplacian in a bounded interval. Phys. Rev. E **76**, 21116 (2007)

Chapter 6
Simulation of Heat Transport in Low-Dimensional Oscillator Lattices

Lei Wang, Nianbei Li, and Peter Hänggi

Abstract The study of heat transport in low-dimensional oscillator lattices presents a formidable challenge. Theoretical efforts have been made trying to reveal the underlying mechanism of diversified heat transport behaviors. In lack of a unified rigorous treatment, approximate theories often may embody controversial predictions. It is therefore of ultimate importance that one can rely on numerical simulations in the investigation of heat transfer processes in low-dimensional lattices. The simulation of heat transport using the non-equilibrium heat bath method and the Green-Kubo method will be introduced. It is found that one-dimensional (1D), two-dimensional (2D) and three-dimensional (3D) momentum-conserving nonlinear lattices display power-law divergent, logarithmic divergent and constant thermal conductivities, respectively. Next, a novel diffusion method is also introduced. The heat diffusion theory connects the energy diffusion and heat conduction in a straightforward manner. This enables one to use the diffusion method to investigate the objective of heat transport. In addition, it contains fundamental information about the heat transport process which cannot readily be gathered otherwise.

L. Wang (✉)
Department of Physics and Beijing Key Laboratory of Opto-electronic Functional Materials and Micro-nano Devices, Renmin University of China, Beijing 100872, People's Republic of China
e-mail: phywanglei@ruc.edu.cn

N. Li
Center for Phononics and Thermal Energy Science, School of Physics Science and Engineering, Tongji University, Shanghai 200092, People's Republic of China
e-mail: nbli@tongji.edu.cn

P. Hänggi
Institute of Physics, University of Augsburg, Augsburg 86135, Germany
e-mail: Peter.Hanggi@physik.uni-augsburg.de

S. Lepri (ed.), *Thermal Transport in Low Dimensions*, Lecture Notes in Physics 921, DOI 10.1007/978-3-319-29261-8_6

6.1 Simulation of Heat Transport with Non-equilibrium Heat Bath Method and Equilibrium Green-Kubo Method

We start out by considering numerically heat transport in typical momentum-conserving nonlinear lattices, from 1D to 3D. The numerical simulations are performed with two different methods: the non-equilibrium heat bath method and the celebrated equilibrium Green-Kubo method. Our major focus will be on the length dependence of thermal conductivities where the asymptotic behavior towards the thermodynamic limit is of prime interest. As a result, numerical simulations are usually taken on lattices employing very large up to even huge system sizes. Therefore, in order to get a compromise between better numerical accuracy and acceptable computational cost, a 5-th order Runge-Kutta algorithm [1] is applied for the simulations of the dissipative systems in the former case, while an embedded Runge-Kutta-Nystrom algorithm of orders 8(6) [2, 3] is applied for the simulations of the conservative Hamiltonian systems in the latter case.

6.1.1 Power-Law Divergent Thermal Conductivity in 1D Momentum-Conserving Nonlinear Lattices

Heat conduction induced by a small temperature gradient is expected to satisfy the Fourier's law in the stationary regime:

$$j = -\kappa \nabla T, \tag{6.1}$$

where j denotes the steady state heat flux, ∇T denotes the small temperature gradient, and κ denotes the thermal conductivity. In practical numerical simulations, the temperature difference ΔT is usually fixed for convenience. In this setup, for a system with length L, the steady state heat flux j should be inversely proportional to L: $j = -\kappa \Delta T/L$, if Fourier's law is obeyed and κ is a constant. However, for many 1D momentum-conserving lattices [4, 5], it is numerically found that j decays as $L^{-1+\alpha}$ with a positive α. This finding indicates that the thermal conductivity κ is length dependent and diverges with L as $\kappa \propto L^{\alpha}$ in the thermodynamical limit $L \to \infty$. The Fourier's law is broken and the heat conduction is called anomalous.

For this anomalous heat conduction, transport theories from different approaches make different predictions for the divergency exponent α [5]. The renormalization group theory [6] for 1D fluids predicts a universal value of $\alpha = 1/3$ and it is claimed that the thermal conductivity of oscillator chains including the Fermi-Pasta-Ulam (FPU) lattices should diverge in this universal way [7]. Early Mode-Coupling Theories (MCT) predict one universal value of $\alpha = 2/5$ for all 1D FPU lattices [4], while another MCT taking the transverse motion into account predicts $\alpha = 1/3$ [8, 9]. Later, a self-consistent MCT proposes that there should be two universality classes instead of one. It states that models with asymmetric interaction potentials are characterized by a divergency exponent $\alpha = 1/3$, while models with symmetric potentials are characterized by a larger value of $\alpha = 1/2$ [10–12]. The value of $\alpha = 1/3$ is also predicted by calculating the relaxation rates of phonons [13].—For an intriguing discussion of the physical existence of anharmonic phonons and its interrelation between a phonon mean free path and its associated mean phonon relaxation time we refer the interested readers to recent work [14].

A theory based on Peierls-Boltzmann equation is applied for the FPU-β lattice, and $\alpha = 2/5$ is predicted [15]. In [16], energy current correlation function is studied for the FPU-β lattice and $\alpha = 2/5$ is found with small nonlinearity approximation. Similar to the discrepancy among theoretical predictions, numerical results are also not convergent. For example, an early numerical study suggests $\alpha = 2/5$ [17], while some recent studies support $\alpha = 1/3$ [18–20].

In this section, we numerically study heat conduction in typical 1D nonlinear lattices with the following Hamiltonian

$$H = \sum_i [\frac{p_i^2}{2} + V(u_i - u_{i-1})], \tag{6.2}$$

where p_i and u_i denotes the momentum and displacement from equilibrium position for i-th particle. For convenience, dimensionless units are applied and the mass of all particles can be set as unity. The interaction potential energy between particles i and $i-1$ is $V_i \equiv V(u_i - u_{i-1})$. The interaction force is correspondingly obtained as $f_i = -\partial V_i/\partial u_i$. The local energy belongs to the particle i is defined here as $E_i = \frac{p_i^2}{2} + \frac{1}{2}(V_i + V_{i+1})$, i.e., neighboring particles share their interaction potential energy equally. The instantaneous local heat flux is then defined as $j_i \equiv \frac{1}{2}(\dot{u}_i + \dot{u}_{i+1})f_{i+1}$ and the total heat flux is defined as $J(t) \equiv \sum_i j_i(t)$.

The interaction potential takes the general FPU form as $V(u) = \frac{1}{2}k_2u^2 + \frac{1}{3}k_3u^3 + \frac{1}{4}k_4u^4$. The following three types of lattices will be studied, i.e., (1) the FPU-$\alpha\beta$ lattices with $k_2 = k_4 = 1, k_3 = 1$ (in short as FPU-$\alpha1\beta$ lattice); and $k_2 = k_4 = 1$, $k_3 = 2$ (in short as FPU-$\alpha2\beta$ lattice); (2) the FPU-β lattice with $k_2 = k_4 = 1$, $k_3 = 0$; and (3) the purely quartic or the qFPU-β lattice with $k_2 = k_3 = 0$, $k_4 = 1$. The interactions in the FPU-$\alpha\beta$ lattices are asymmetric, i.e., $V(u) \neq V(-u)$, while the interactions in other lattices are symmetric. In the former case, the temperature pressure is nonvanishing, finite in the thermodynamic limit [21].

6.1.1.1 Non-equilibrium Heat Bath Method

Firstly, we calculate the thermal conductivity κ_{NE} according to the definition Eq. (6.1) with the non-equilibrium heat bath method. The subscript 'NE' indicates that the calculation is in non-equilibrium steady states. To this end, fixed boundary conditions are applied, i.e., $u_0 = u_N = 0$, with N being the total number of particles. Since the lattice constant a has been set as unity in the dimensionless units, the lattice length L is simply equivalent to the particle number N as $L = Na = N$. Two

Langevin heat baths with temperatures $T = 0.5$ and 1.5 are coupled to the two ends of the lattice, respectively. The equation of motion of the particle coupled to the heat bath is described by the following Langevin dynamics

$$\ddot{u} = f - \lambda \dot{u} + \xi, \tag{6.3}$$

where f denotes the interaction force generated from other particles, ξ denotes a Wiener process with zero mean and variance $2\lambda k_B T$, and λ denotes the relaxation coefficient of the Langevin heat bath. Generally, the resulting heat flux approaches zero in both limits $\lambda \to 0$ and $\lambda \to \infty$ [4]. In practice, the λ has been optimized to be 0.2 so as to maximize the heat currents. In order to achieve better performance, we usually put more than one particles into the heat bath in each end [22, 23].

To avoid the effect of possible slow convergence [24], the simulations have been performed long enough time until the temperature profiles are well established and the heat currents along the lattice become constant, see Fig. 6.1. The temperature gradient $\nabla T \equiv \frac{dT}{dL}$ is calculated by linear least squares fitting of the temperature profile in the central region, aiming to greatly reduce the boundary effects.

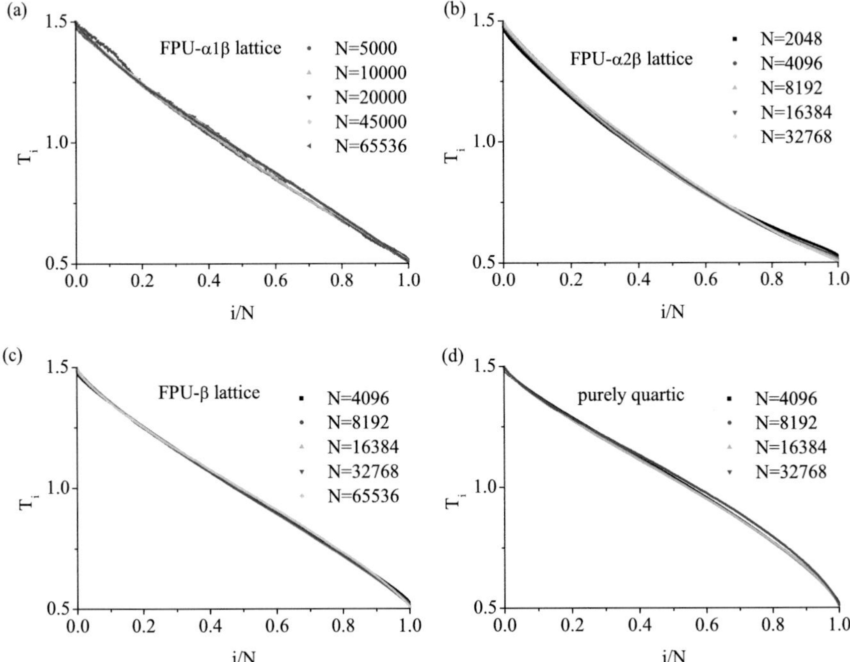

Fig. 6.1 Temperature profiles for (**a**) the FPU-$\alpha 1\beta$, (**b**) the FPU-$\alpha 2\beta$, (**c**) the FPU-β, and (**d**) the purely quartic lattices with various length L. Only the temperature profiles in the central region are taken into account in calculating the temperature gradient $\frac{dT}{dL}$, i.e., the left and right 1/4 of the lattices are excluded in order to remove boundary effects

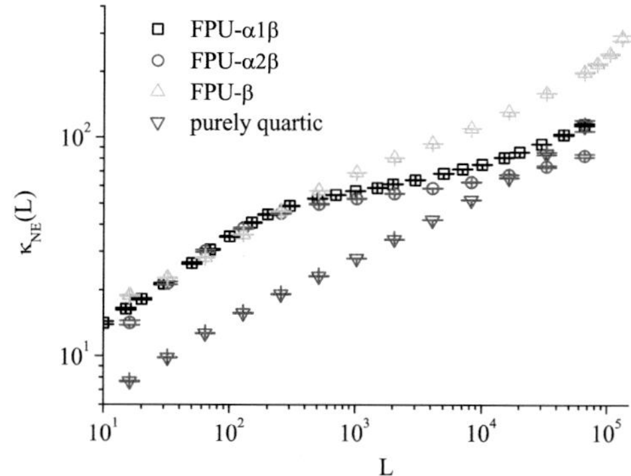

Fig. 6.2 Thermal conductivity $\kappa_{\rm NE}$ versus lattice length L. The reversed tendency can be roughly seen in the *rightmost part* of the figure

This so evaluated thermal conductivity $\kappa_{\rm NE}(L)$ are plotted for different 1D lattices in Fig. 6.2. For the lattices with asymmetric interactions, a very flat length-dependence of thermal conductivity is observed with L ranging from several hundreds to ten thousands sites. However, for even longer lengths, say, $L > 1 \times 10^4$, the running slope of the thermal conductivity $\kappa_{\rm NE}(L)$ as the function of L starts to grow again. By comparing the results in the two FPU-$\alpha\beta$ lattices, we see that the asymptotic tendency of curving up of $\kappa_{\rm NE}(L)$ is not affected even for the case of strong asymmetry ($k_3 = 2$ is the maximum value that keeps the potential single well, at the given $k_2 = k_4 = 1$). With the increase of asymmetry, the tendency of curving up of $\kappa_{\rm NE}(L)$ can only be postponed as shown in Fig. 6.2. Such a phenomenon can also be observed in the FPU-β lattice, although the effect is much more slight. Because of this, the thermal conductivity in this lattice displays a little bit slower divergence as $L^{1/3}$ in a certain length regime [18]. This finite-size Fourier-like behavior has been repeatedly observed recently in many 1D lattices [23, 25–29], its physical reason is, however, still not clear.

6.1.1.2 Green-Kubo Scheme

The major drawback of the non-equilibrium heat bath method is the boundary effects due to coupling with heat baths which are difficult to be sufficiently removed. In addition, the temperature difference cannot be set as small values in this method, otherwise the net heat currents can hardly be distinguished from the statistical fluctuations. Therefore, the systems prepared in the non-equilibrium heat bath method are far from ideal close-to-equilibrium states. Numerical difficulties also prevent us from simulating even longer lattices. We thus turn to calculate the equilibrium heat current autocorrelation functions with the Green-Kubo method, which provides an alternative way of determining the divergency exponent α [30]. No heat bath enters into the lattice dynamics.

In a finite lattice with N particles, the heat current correlation function $c_N(\tau)$ is defined as

$$c_N(\tau) \equiv \frac{1}{N}\langle J(t)J(t+\tau)\rangle_t. \tag{6.4}$$

where $\langle\cdot\rangle$ denotes the ensemble average, which is equivalent to the time average for chaotic and ergodic systems considered here. Compared with $c_N(\tau)$ for finite lattice, its value in thermodynamic limit is much more meaningful, i.e.,

$$c(\tau) \equiv \lim_{N\to\infty} c_N(\tau). \tag{6.5}$$

According to the Green-Kubo formula [30], the thermal conductivity κ_{GK} is integrated as

$$\kappa_{\mathrm{GK}} \equiv \frac{1}{k_B T^2}\int_0^\infty c(\tau)d\tau. \tag{6.6}$$

The Boltzmann constant k_B is set to unity in the dimensionless units. But k_B is kept in formulas for the completeness of understanding. For anomalous heat conduction, the above integral does not converge due to the slow time decay of $c(\tau)$ in asymptotic limit. In common practice, the length-dependent thermal conductivity is calculated by introducing a cutoff time $t_s = L/v_s$, instead of infinity, as the upper limit of the integral, i.e.,

$$\kappa_{\mathrm{GK}}(L) \equiv \frac{1}{k_B T^2}\int_0^{L/v_s} c(\tau)d\tau, \tag{6.7}$$

where the constant v_s is the speed of sound and the subscript 'GK' denotes that the calculation is based on the Green-Kubo formula. Since we are only interested in the divergency exponent α of $\kappa_{\mathrm{GK}}(L)$, its exact value is not relevant to any conclusion we made.

However, in numerical calculations, only lattices with finite N can possibly be simulated. The $c_N(\tau)$ generally depends on the lattice length N, and the finite-size effects must be taken into consideration very carefully. We next present the simulation with a very long lattice length of $N = 20{,}000$ followed by the discussion of finite-size effects.

The simulations are carried out in lattices with periodic boundary conditions, which is known to provide the best convergence to thermodynamic limit properties. Microcanonical simulations are performed with zero total momentum and identical energy density ϵ which corresponds to the same temperature $T = 1$ for all lattices. The energy density ϵ equals to $0.864, 0.846, 0.867$ and 0.75, for the 1D FPU-$\alpha 1\beta$ lattice, the FPU-$\alpha 2\beta$ lattice, the FPU-β lattice, and the purely quartic lattice, respectively.

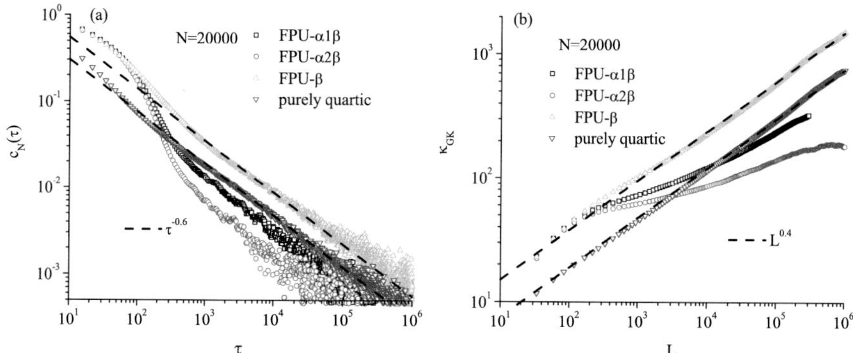

Fig. 6.3 (**a**) The heat current correlation $c_N(\tau)$ versus time lag τ. Lines with slope $-3/5$ are drawn for reference. Data for the FPU-β and the purely quartic lattices fit them quite well in the long τ regime. (**b**) $\kappa_{\text{GK}}(L)$. Lines with slope $2/5$ are drawn for reference. For the FPU-β lattice, the slope $2/5$ fits very well in the regime of L from 10^3 to 10^6. It fits for the purely quartic lattice in even wider regime from 10^2 to 10^6

The time τ dependence of $c_N(\tau)$ are shown in Fig. 6.3a. In a very large window of τ, the $c_N(\tau)$ for the FPU-β lattice and the purely quartic lattice follows a power-law decay of $c_N(\tau) \propto \tau^{\gamma}$ with $\gamma = -3/5$ very well. It should be pointed out that the decay exponent γ is related to the divergency exponent α as $\gamma = \alpha - 1$ resulted from Eq. (6.7). While for the FPU-$\alpha\beta$ lattices, the $c_N(\tau)$ decays very fast before it approaches an asymptotic power-law decay behavior.

The corresponding length dependence of the thermal conductivity $\kappa_{\text{GK}}(L)$ from Eq. (6.7) is plotted in Fig. 6.3b. For the FPU-β lattice, the best fit of the data from $L = 10^4$ upward gives rise to a divergency exponent $\alpha = 0.42$, which strongly prefers the theoretical prediction of $\alpha = 2/5$ to $\alpha = 1/3$. As for the purely quartic lattice, the best fit of the data for L from 10^2 upward, covering four orders of magnitude, yields $\alpha = 0.41$, which is even closer to the prediction of $\alpha = 2/5$.

Heat current correlation loss. To quantitatively evaluate the finite-size effects, we plot the heat current correlation $c_N(\tau)$ for the 1D purely quartic lattice with sizes from $N = 128$ to $N = 131{,}072$ in Fig. 6.4a. We define a heat-current correlation loss (CCL) $\Delta c_N(\tau)$ induced by the finite-size effects as

$$\Delta c_N(\tau) \equiv c(\tau) - c_N(\tau). \tag{6.8}$$

The relative loss $\lambda_N(\tau)$ is defined as

$$\lambda_N(\tau) \equiv \frac{\Delta c_N(\tau)}{c(\tau)} = 1 - \frac{c_N(\tau)}{c(\tau)}. \tag{6.9}$$

By setting a certain critical value η for the relative loss $\lambda_N(\tau)$, a characteristic time lag, the cutoff time lag $\tau_c(N)$ can be obtained for each length N by solving,

$$\lambda_N(\tau_c(N)) = \eta. \tag{6.10}$$

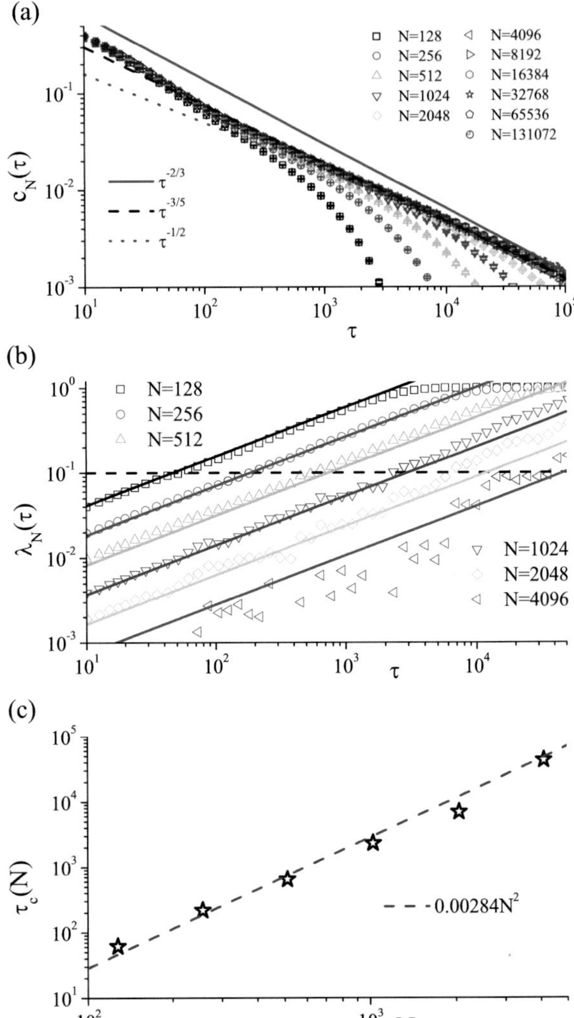

Fig. 6.4 (**a**) The heat current correlation $c_N(\tau)$ for the purely quartic lattice for various lattice length N. Lines with different slopes $-2/3$, $-3/5$ and $-1/2$, respectively, are drawn for reference. (**b**) $\lambda_N(\tau)$ as a function of τ for various lattice length N. *Oblique solid lines* from the top down stand for the fittings of $\lambda_N(\tau)$, $3N^{-1.16}\tau^{0.58}$, for N ranging from 128 to 4096. The *horizontal dashed line* refers to $\lambda_N(\tau) = 0.1$. $\lambda_N(\tau)$ cross this line at the cut off time lag $\tau_c(N)$. (**c**) The cutoff time lag $\tau_c(N)$ as a function of the lattice length N. The *blue dashed line* stands for the expectation in Eq. (6.13), $2.84 \times 10^{-3}N^2$

Since the asymptotic $c(\tau)$ can never be actually calculated, we thus need to approximately replace $c(\tau)$ with $c_N(\tau)$ for a finite long enough lattice instead. Under any existing criterion [31], the length of $N = 131{,}072$ is long enough for correlation times $\tau \leq 5 \times 10^4$. Therefore, the asymptotic $c(\tau)$ refers to $c_{131072}(\tau)$ in the descriptions of our numerical simulations hereafter.

In Fig. 6.4b, the relative loss $\lambda_N(\tau)$ for the purely quartic lattice for various length N is plotted . The $\lambda_N(\tau)$ is larger in shorter lattices as one should expect. Interesting enough, all the data of $\lambda_N(\tau)$ as the correlation time τ for various N fit the following universal relation quite well:

$$\lambda_N(\tau) \approx 3N^{-1.16}\tau^{0.58}. \tag{6.11}$$

This relation implies that the cutoff time lag $\tau_c(N)$ should follow a square-law dependence on the lattice length N:

$$\tau_c(N) \approx (\frac{\eta}{3})^{\frac{1}{0.58}} N^2. \tag{6.12}$$

For the critical value of $\eta = 0.1$, namely, the $c_N(\tau)$ decreases to 90 % of the value of $c(\tau)$ at $\tau = \tau_c(N)$. The cutoff time lag $\tau_c(N)$ as the function of length N can be obtained from Eq. (6.10) as

$$\tau_c(N) \approx 2.84 \times 10^{-3} N^2. \tag{6.13}$$

This is shown in Fig. 6.4c where good agreement with numerical data can be observed.

It is reasonable to expect that Eq. (6.11) should also remain valid in larger N regime. We are thus able to estimate the value of relative loss $\lambda_N(\tau)$ in Fig. 6.3a, which is no more than 10 %. Given the fact that $c_N(\tau) \propto \tau^{\gamma}$ was fitted over four orders of magnitude of τ, the underestimate of δ induced therefrom must not be higher than $|\log_{10} 0.9|/4 \approx 0.01$. It is noticed that such an error is much smaller than the difference between the three theoretical expectations $\gamma = -2/3, \gamma = -3/5$ and $\gamma = -1/2$. The conclusion that $c(\tau)$ is best fitted as $c(\tau) \propto \tau^{-3/5}$ should not be affected by this finite-size effect. We also expect that the cases in 2D [32] and 3D [33] purely quartic lattices are also similar.

In summary, we have numerically calculated the length-dependent thermal conductivities κ in a few typical 1D lattices by using both non-equilibrium heat bath and equilibrium Green-Kubo methods. Consistent results are obtained for thermal conductivity divergency exponent α. For the FPU-β and the purely quartic lattices, the thermal conductivities κ follows a power-law length-dependence of $\kappa \propto L^{0.4}$ very well, for a wide regime of L. While for the asymmetric FPU-$\alpha\beta$ lattices, large finite-size effects are observed. As a result, κ increases with lattice length very slowly in a wide range of L. Our numerical simulations indicate that κ regains its increase in yet longer length L [23, 26]. This is also consistent with some recent studies [27, 28].

The studies of the heat current correlation loss in the purely quartic lattice indicate that for a not-very-large N, $c_N(\tau)$ is close enough to the asymptotic $c(\tau)$ within a very long correlation time τ window. Therefore, we are able to extract $\kappa_{\text{GK}}(L)$ from Eq. (6.7) with an effective very long L by performing simulations in a lattice with relatively much smaller size N, i.e. $L = v_s\tau_c >> N$. The research area of the investigation of heat transport applicable for the Green-Kubo method is thus greatly broadened [34].

6.1.2 *Logarithmic Divergent Thermal Conductivity in 2D Momentum-Conserving Nonlinear Lattices*

For 2D and 3D momentum-conserving systems with higher dimensionality, consistent predictions are achieved from different theoretical approaches. The linear response approaches based on the renormalization group [6] and mode-coupling theory [4, 35–37] both predict that the heat current autocorrelation function $c(\tau)$ decays with the correlation time τ as $c(\tau) \propto \tau^{\gamma}$, where $\gamma = -1$ and $-3/2$ for 2D and 3D, respectively. These predictions indicate that the resulted thermal conductivity is logarithmically divergent as $\kappa \propto \ln L$ in 2D systems and a finite value in 3D systems.

In 2D momentum-conserving systems, a logarithmic divergence of thermal conductivity of $\kappa \propto \ln L$ is reported by numerical simulations in the FPU-β lattice with rectangle [38, 39] and disk [40] geometries where vector displacements are considered. However, a power-law divergent thermal conductivity of $\kappa \propto L^{\alpha}$ is also observed in the 2D FPU-β lattices with scalar displacements [41].

In this subsection we systematically study the heat conduction in a few 2D square lattices with a scalar displacement field $u_{i,j}$, where the schematic 2D setup is plotted in Fig. 6.5a. The scalar 2D Hamiltonian reads

$$H = \sum_{i=1}^{N_X}\sum_{j=1}^{N_Y}[\frac{p_{i,j}^2}{2} + V(u_{i,j} - u_{i-1,j}) + V(u_{i,j} - u_{i,j-1})], \tag{6.14}$$

where N_X and N_Y denotes the number of layers in X and Y directions. The inter-particle potential takes the FPU form of $V(u) = \frac{1}{2}k_2u^2 + \frac{1}{3}k_3u^3 + \frac{1}{4}k_4u^4$. Dimensionless units is applied and all the particle masses has been set to unity. In order to justify the logarithmic divergence and also study the influence of inter-particle coupling, we choose three types of lattices, i.e., the FPU-$\alpha 2\beta$ lattice: $k_2 = k_4 = 1$, $k_3 = 2$; the FPU-β lattice: $k_2 = k_4 = 1$, $k_3 = 0$; and the purely quartic lattice: $k_2 = k_3 = 0$, $k_4 = 1$.

The interaction forces between a particle labeled (i, j) and its nearest right and up neighbors are $f_{i,j}^X = -dV(u_{i,j} - u_{i-1,j})/du_{i,j}$ and $f_{i,j}^Y = -dV(u_{i,j} - u_{i,j-1})/du_{i,j}$. The local heat currents in the two directions are defined as $j_{i,j}^X = \frac{1}{2}(\dot{u}_{i,j} + \dot{u}_{i+1,j})f_{i+1,j}^X$ and $j_{i,j}^Y = \frac{1}{2}(\dot{u}_{i,j} + \dot{u}_{i,j+1})f_{i,j+1}^Y$, respectively.

6.1.2.1 Non-equilibrium Heat Bath Method

We first calculate the thermal conductivity κ_{NE} in non-equilibrium stationary states. The fixed boundary conditions are applied in the X-direction, while periodic boundary conditions are applied in the Y-direction. The left- and right-most columns are coupled to Langevin heat baths with temperatures $T_L = 1.5$ and $T_R = 0.5$, respectively, see Fig. 6.5a. Heat currents through the X-direction along with the

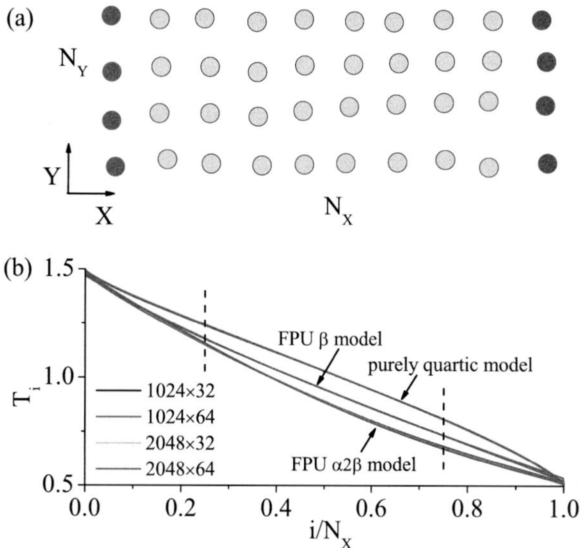

Fig. 6.5 (**a**) Scheme of a 2D square lattice. Heat current along the X axis is calculated. (**b**) Temperature profiles of different lattices. Curve groups from down top correspond to the FPU-$\alpha 2\beta$, FPU-β, and purely quartic lattices, respectively. Only the data in the central region between the *two vertical dashed lines* are taken into account in calculating the temperature gradient ∇T

direction of temperature gradient are measured. For each lattice with the largest size (2048×64), the average heat current is performed over time period of $2 \sim 4 \times 10^7$ in dimensionless units after long enough transient time. The temperature of each column is defined as the average temperature of all the particles in that column, i.e.,

$$T(i) \equiv \frac{1}{N_Y} \sum_{j=1}^{N_Y} T(i,j) = \frac{1}{N_Y} \sum_{j=1}^{N_Y} \langle \dot{u}_{i,j}^2 \rangle,$$

The temperature profiles of different lattices for various $N_X \times N_Y$ are plotted in Fig. 6.5b. Those profiles with different sizes for the same lattice are all overlapped with each other, which indicates that the temperature gradient ∇T along the X-direction can be well established. It is also confirmed that the temperature profiles and the heat currents along the lattices approach constant values which are independent of the overall time used here. The thermal conductivity κ for 2D systems is defined as:

$$\kappa_{\mathrm{NE}} = -\frac{\langle J \rangle}{N_Y \nabla T},$$

where J stands for the total heat current, and the temperature gradient ∇T is along the X direction. Since the lattice constant a is set to unity, the lattice length L is simply equivalent to the number of layers N_X in X-direction. As shown in Fig. 6.5b, the shapes of temperature profiles are obviously nonlinear. Such a nonlinearity is caused by boundary effects rather than the intrinsic temperature dependence of the thermal conductivity, as can be concluded by examining the temperature dependence of κ_{NE} for different lattices in Fig. 6.6d. To reduce this boundary effect, the temperature gradient ∇T is calculated by a linear least-squares fitting of the temperature profiles in the central region where the left- and right- most 1/4 of the lattices are excluded.

In Fig. 6.6a–c, the thermal conductivities κ_{NE} versus L with different widths of N_Y are plotted in linear-log scales. In the large lattice size region, the narrow lattices with smaller N_Y posses higher values of thermal conductivities κ_{NE}. This is not a surprise since the narrow 2D lattices are much closer to a 1D lattice where high thermal conductivities are expected. The length dependence of κ_{NE} for the 2D FPU-$\alpha 2\beta$ lattice (Fig. 6.6a) with $N_Y = 64$ becomes flat, indicating that κ_{NE} increases much more slowly than a logarithmic growing. In contrast, the thermal conductivities κ_{NE} for the 2D FPU-β lattice diverges with length L more rapidly than a logarithmic divergence, as can seen from Fig. 6.6b and its inset. This is actually a power-law divergence of $\kappa \propto L^{\alpha}$ and the divergency exponent α can be estimated from the best fit of the last four points as $\alpha = 0.27 \pm 0.02$. However, for 2D purely

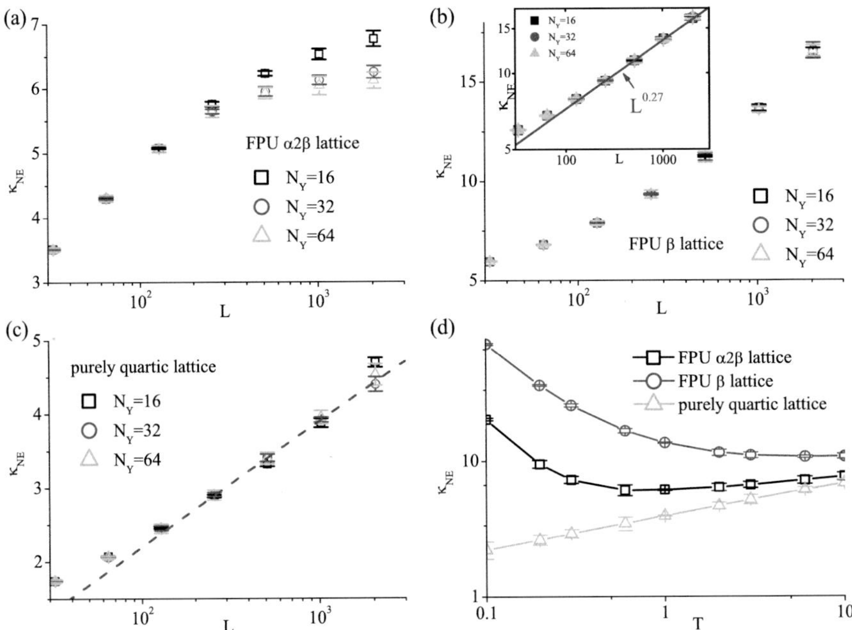

Fig. 6.6 Thermal conductivity κ_{NE} in 2D (**a**) the FPU-$\alpha\beta$, (**b**) the FPU-β and (**c**) the purely quartic lattices versus lattice length L for various N_Y. The *dashed line* that indicate logarithmic growth is drown for reference. *Inset* of (**b**): data for the FPU-β lattice in double logarithmic scale. *Solid line* corresponds to the power-law divergence $L^{0.27}$. (**d**) κ_{NE} versus temperature T in various lattices

quartic lattices, the $\kappa_{\rm NE}$ displays a logarithmic growth as $\kappa \propto \ln L$ over at least one order of magnitude in length scale, as in Fig. 6.6c.

6.1.2.2 Green-Kubo Method

Similar to the situation in 1D lattices [23], finite-size and finite-temperature-gradient effects from the non-equilibrium heat bath method are considerable and not easy to be removed. We shall turn to the equilibrium Green-Kubo method [30] to seek higher accuracy of numerical results.

In the Green-Kubo simulation, periodic boundary conditions are applied in both the X- and Y-directions. The autocorrelation function of total heat current $c^X_{N_X,N_Y}(\tau)$ in the X direction is defined as

$$c^X_{N_X,N_Y}(\tau) \equiv \frac{1}{N_X N_Y}\langle J^X(t)J^X(t+\tau)\rangle_t, \tag{6.15}$$

where $J^X(t) \equiv \sum_{i,j} j^X_{i,j}(t)$ is the instantaneous total heat current in the X-direction. For simplicity, the subscripts N_X and N_Y of $c^X_{N_X,N_Y}(\tau)$ are omitted as $c^X(\tau)$ hereafter except in case of necessity. The length-dependent thermal conductivity $\kappa_{\rm GK}(L)$ from the Green-Kubo method can be defined as

$$\kappa_{\rm GK}(L) \equiv \frac{1}{k_B T^2} \lim_{N_X\to\infty} \lim_{N_Y\to\infty} \int_0^{L/v_s} c^X(\tau)d\tau, \tag{6.16}$$

where v_s is again the speed of sound. Microcanonical simulations are performed with zero total momentum [4] and specified energy density ϵ which corresponds to the same temperature $T = 1$ for different lattices. The energy density ϵ equals 0.887, 0.892 and 0.75 for the 2D FPU-$\alpha 2\beta$, FPU-β, and purely quartic lattices, respectively. A number of independent runs (64 for 1024×1024 and fewer for smaller lattices) are carried out. Simulations of the largest lattices (1024×1024) are performed for about total time of 10^7 in the dimensionless units.

The decays of $c^X(\tau)$ with the correlation time τ for different 2D lattices are plotted in Fig. 6.7a–c. To eliminate the finite-size effects, we have performed simulations by varying N_X and N_Y and only consider the asymptotic behavior which is the part of curves overlapping with each other. Within the range of standard error, the satisfactory overlap with each other is clearly observed. In the specific cases with $N_X = N_Y$, the average of the autocorrelation function of $[c^X(\tau) + c^Y(\tau)]/2$ is plotted instead. This is equivalent to double the simulation time steps to achieve higher accuracy without actually performing any more computation. And due to the symmetry of the square lattice, it is obvious to find that $c^X_{512,1024}(\tau) = c^Y_{1024,512}(\tau)$, where the simulations for the two lattices can be carried out in the same run.

It is observed that in Fig. 6.7a, the $c^X(\tau)$ in the 2D FPU-$\alpha 2\beta$ lattices decays much faster than the theoretical prediction of $c^X(\tau) \propto \tau^{-1}$ in a wide regime of correlation time (τ). As a result, the integrated $\kappa_{\rm GK}$ displays a saturation behavior

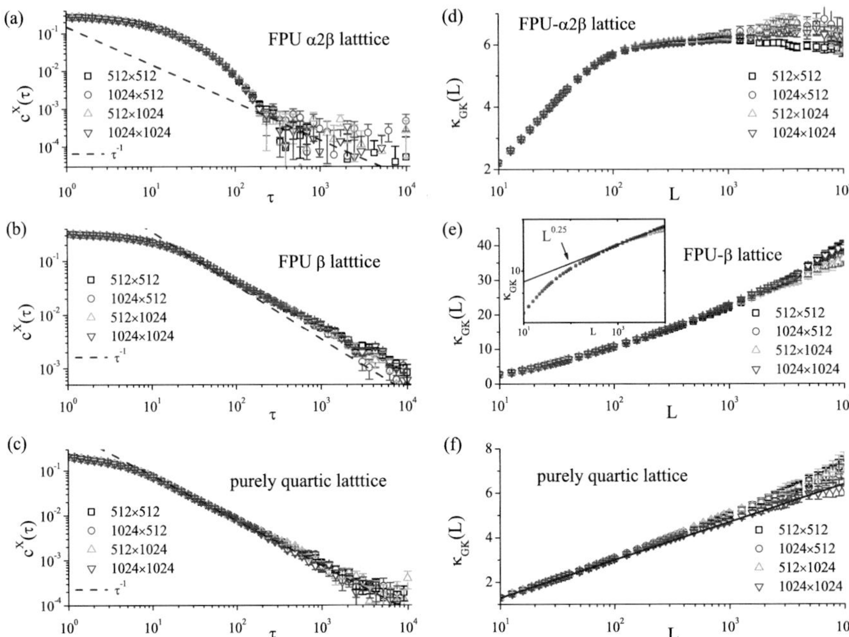

Fig. 6.7 (**a**)–(**c**), $c^X(\tau)$ for in $N_X \times N_Y$ lattices. Lines correspond to τ^{-1} are drawn for reference. (**a**) The FPU-$\alpha 2\beta$ lattice. $c^X(\tau)$ decays much faster than τ^{-1} in short τ regime. The decay tend to slow down for $\tau > 300$. (**b**) The FPU-β lattice. $c^X(\tau)$ decays obviously slower than τ^{-1} in a quite wide regime of τ . (**c**) The purely quartic lattice. $c^X(\tau)$ follows τ^{-1} very well in nearly three orders of magnitude of τ. (**d**)–(**f**), $\kappa_{\rm GK}(L)$ in the X-direction in $N_X \times N_Y$ lattices. (**d**) The FPU-$\alpha 2\beta$ lattice. A flat $\kappa_{\rm GK}$ is again observed for $L < 2000$. Thereafter κ tend to rise up. It is easy to understand that slow down of the decay of $c(\tau)$ cannot instantly induce a visible rise up of $\kappa_{\rm GK}$, since $c(\tau)$ has already decayed to a too low value. (**e**) The FPU-β lattice. In a wide regime of τ, c_τ decays obviously slower than τ^{-1}. *Inset*: data plotted in double logarithmic scale. *Solid line* corresponds to $L^{0.25}$. (**f**) The purely quartic lattice. $\kappa_{\rm GK}$ for 1024×1024 follows the straight line very well in nearly three orders of magnitude of τ, This strongly supports a logarithmically divergent thermal conductivity. The slight rise for smaller lattices is due to the finite-size effect

with the length for large L in Fig. 6.7d. The rapid decay of $c^X(\tau)$ tends to slow down for yet longer time τ. However, the asymptotic behavior cannot be numerically confirmed due to large fluctuations. In Fig. 6.7b, the $c^X(\tau)$ in the 2D FPU-β lattice decays evidently more slowly than τ^{-1}, which gives rise to a power-law divergence of thermal conductivity of $\kappa_{\rm GK} \propto L^\alpha$ in Fig. 6.7e. The best fit of this divergency exponent α in the regime of $L > 10^3$ is obtained as $\alpha = 0.25 \pm 0.01$. For the 2D purely quartic lattice as seen in Fig. 6.7c, the $c^X(\tau)$ decays as the predicted behavior of $c^X(\tau) \propto \tau^{-1}$ for nearly three orders of magnitudes of correlation time τ. This finding strongly supports a logarithmic diverging thermal conductivity of $\kappa \propto \ln L$, which can be clearly observed in Fig. 6.7f. In all cases, the tendency of $\kappa_{\rm GK}$ from Green-Kubo method (shown in the right column of Fig. 6.7) is in good agreement with that of $\kappa_{\rm NE}$ from non-equilibrium heat bath method (shown in Fig. 6.6).

In summary, we have extensively studied heat conduction in three 2D nonlinear lattices with both non-equilibrium heat bath method and Green-Kubo method. The roles of harmonic and asymmetric terms of the inter-particle coupling are clearly observed by comparing the results for the purely quartic lattice and the other two latices. In the 2D purely quartic lattice, the heat current autocorrelation function $c(\tau)$ is found to decay as τ^{-1} in three orders of magnitude from 10^1 to 10^4. This strongly supports a logarithmically divergent thermal conductivity of $\kappa \propto \ln L$ consistent with the theoretical predictions. For the 2D FPU-β lattice, our non-equilibrium and equilibrium calculations suggest a power-law divergence with a divergency exponent $\alpha = 0.27 \pm 0.02$ and 0.25 ± 0.01, respectively. A very significant finite-size effect which results a flat length dependence of $\kappa(L)$ is observed in the 2D FPU-$\alpha\beta$ lattice with asymmetric potential. Most existing numerical studies on 2D lattices with asymmetric interaction terms suggest a logarithmically divergent behavior as $\kappa \propto \ln L$, e.g., the Fermi-Pasta-Ulam (FPU)-β lattice with rectangle [38, 39] and disk [40] geometries. It might be due to that the effect of the harmonic term is largely offset by that of the asymmetric term, thus yielding a logarithmic-like divergence of thermal conductivity.

Similar to the findings of 1D lattices where κ tends to diverge with length in the same way in the thermodynamic limit for all kinds of lattices [23], it should also be expected that κ will diverge as $\log L$ in long enough 2D FPU-$\alpha\beta$ and FPU-β lattices as already observed for 2D purely quartic lattice. However, in order to see such an asymptotic divergence, 2D lattices with much larger sizes have to be simulated which is beyond the scope of our current studies.

We should emphasize that such numerical studies are not only of theoretical importance. Progresses in nano-technology have made it possible to experimentally measure the size dependence of thermal conductivities in some 1D [42] and 2D [43–46] nano-scale materials.

6.1.3 Normal Heat Conduction in a 3D Momentum-Conserving Nonlinear Lattice

For 3D momentum-conserving systems, all the above-mentioned theories predict that the heat current autocorrelation function decays with correlation time τ as τ^{γ} with $\gamma = -3/2$, which gives rise to a normal heat conduction. However, numerical simulations are not so conclusive [47, 48]. It is only reported in 2008, by non-equilibrium simulations, that the running divergency exponent $\alpha_L \equiv d \ln \kappa / d \ln L$ of the 3D FPU-β lattice shows a power law decay in L, thus vanishes in the thermodynamic limit [49]. Normal heat conduction in 3D systems is therefore verified. However, according to the Green-Kubo formula, any power-law decay of $c(\tau) \propto t^{\gamma}$ with $\gamma < -1$ will yield a finite value of thermal conductivity signaturing a normal heat conduction behavior [30]. In order to confirm the theoretical prediction of the specific value of $\gamma = -3/2$, the heat current autocorrelation function $c(\tau)$ must be directly calculated by using the equilibrium Green-Kubo method.

We investigate the decay of the heat current autocorrelation function in a 3D cubic lattice with a scalar displacement field $u_{i,j,k}$. The 3D Hamiltonian reads

$$H = \sum_{i=1}^{N_X} \sum_{j=1}^{N_Y} \sum_{k=1}^{N_Z} \left[\frac{p_{i,j,k}^2}{2} + V(u_{i,j,k} - u_{i-1,j,k}) \right.$$
$$\left. + V(u_{i,j,k} - u_{i,j-1,k}) + V(u_{i,j,k} - u_{i,j,k-1}) \right], \quad (6.17)$$

where $V(u) = \frac{1}{4}u^4$ takes the purely quartic form. We choose this purely quartic potential due to its simplicity and high nonlinearity, where close-to-asymptotic behaviors can be achieved in shorter time and space scales. This model can also be regarded as the high temperature limit of the FPU-β model. Periodic boundary conditions, which provide the best convergence to thermodynamic limits, are applied in all three directions, i.e., $u_{N_X,j,k} = u_{0,j,k}$, $u_{i,N_Y,k} = u_{i,0,k}$ and $u_{i,j,N_Z} = u_{i,j,0}$. The interactions between a particle (i,j,k) and its nearest neighbors are: $f_{i,j,k}^X = -dV(u_{i,j,k} - u_{i-1,j,k})/du_{i,j,k}$, $f_{i,j,k}^Y = -dV(u_{i,j,k} - u_{i,j-1,k})/du_{i,j,k}$ and $f_{i,j,k}^Z = -dV(u_{i,j,k} - u_{i,j,k-1})/du_{i,j,k}$. The local heat current in three directions are defined as $j_{i,j,k}^X = \frac{1}{2}(\dot{u}_{i,j,k} + \dot{u}_{i+1,j,k})f_{i+1,j,k}^X$, $j_{i,j,k}^Y = \frac{1}{2}(\dot{u}_{i,j,k} + \dot{u}_{i,j+1,k})f_{i,j+1,k}^Y$, and $j_{i,j,k}^Z = \frac{1}{2}(\dot{u}_{i,j,k}+\dot{u}_{i,j,k+1})f_{i,j,k+1}^Z$, respectively. For convenience and simplicity, $N_X = N_Y = W$ is always chosen and the focus is on the heat conduction in the Z direction with different cross section area of W^2 and length of N_Z.

The heat current autocorrelation function $c^Z(\tau)$ in the Z direction for a given W is defined as [30]

$$c^Z(\tau) \equiv \lim_{N_Z \to \infty} \frac{1}{W^2 N_Z} \langle J^Z(t) J^Z(t+\tau) \rangle_t, \quad (6.18)$$

where $J^Z(t) \equiv \sum_{i,j,k} j_{i,j,k}^Z(t)$ is the instantaneous total heat current in Z direction. The length-dependent thermal conductivity $\kappa_{\rm GK}^Z(L)$ is defined as

$$\kappa_{\rm GK}^Z(L) \equiv \frac{1}{k_B T^2} \int_0^{L/v_s} c^Z(\tau) d\tau, \quad (6.19)$$

where the constant v_s is the speed of sound [4, 5]. v_s is of order 1 for the present lattice. Microcanonical simulations are performed with zero total momentum[4, 50, 51] and fixed energy density $\epsilon = 0.75$ which corresponds to the temperature $T = 1$. Due to the presence of statistical fluctuations, the simulation must be carried out long enough, otherwise the real decay exponents of the autocorrelation function cannot be determined with good enough accuracy. We perform the calculations by 64-thread parallel computing. The simulation of the largest lattice ($W = 64$ and $N_Z = 128$) is performed for the total time of 5×10^6 in dimensionless units.

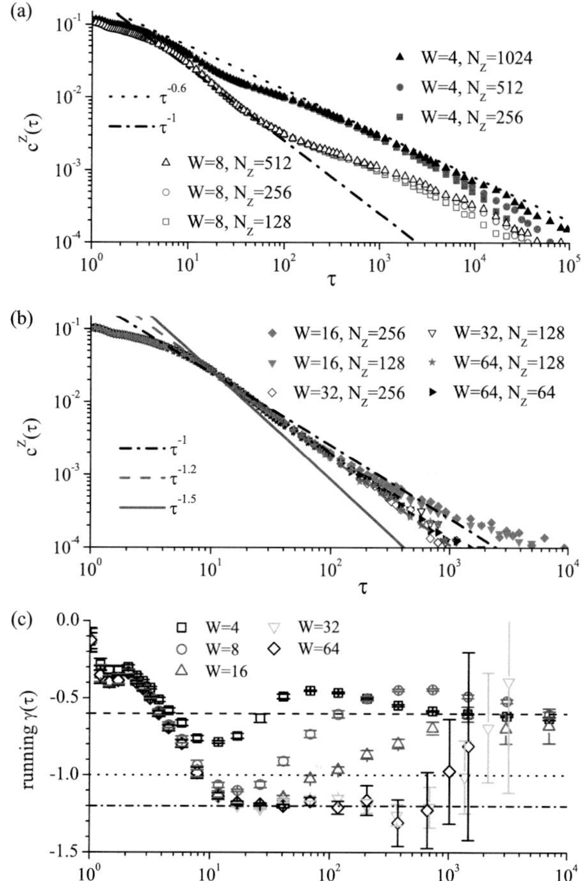

Fig. 6.8 (**a**) $c^Z(\tau)$ versus τ for various N_Z and $W = 4$ and 8. Lines with slopes -0.6 and -1 are plotted for reference. (**b**) $c^Z(\tau)$ versus τ for various N_Z and $W = 16$, 32, and 64. Lines with slopes -1, -1.2, and -1.5 are plotted for reference. The larger the W, the longer the curve follows the power law decay $\tau^{-1.2}$, which suggests that $c(\tau)$ follows $\tau^{-1.2}$ as W is large enough.
(**c**) Dependence of the running exponent $\gamma(\tau) \equiv d \ln c^Z(\tau)/d \ln \tau$ on the time lag τ. Lines for $\gamma = -0.6$, -1, and -1.2 are depicted for reference. One can see that for $W \geq 16$, the bottoms of the curves stop decreasing and tend to saturate at a W-independent value -1.2. The curves stay here for a longer time as W increases

The decay of the autocorrelation function $c^Z(\tau)$ with the correlation time τ is plotted in Fig. 6.8a, b. For a given width W, we perform simulations by varying the lattice length N_Z and consider only the asymptotic behavior shown in the part of curves overlapping with each other to avoid finite-size effects. In the short-time region, typically, for $\tau < 10^1$, all the curves of $c^Z(\tau)$ are relatively flat. This corresponds to the ballistic transport regime when τ is shorter than or comparable to the phonon mean lifetime. Except in this ballistic regime, the $c^Z(\tau)$ for $W = 4$ always decays more slowly than τ^{-1}, implying that the cross-section is too small to display a genuine 3D behavior. The $c^Z(\tau)$ for $W = 8$ decays faster than τ^{-1} in the regime $\tau \in (10, 30)$, showing a weak 3D behavior. In longer time, the decay exponent γ becomes less negative and reverts to the exponent similar to that for the case $W = 4$. For $W = 16$, the curves show a 3D behavior longer and finally again revert to the 1D-like behavior. For width $W = 32$ and 64 where the similar reversal is expected, we fail to observe the 3D behavior in the asymptotic time limit due to the presence of large statistical fluctuations. This picture indicates a crossover from

3D to 1D behavior appearing at a W-dependent threshold for the correlation time as $\tau_c(W)$. Below this critical time τ_c, the system displays a 3D behavior, while a 1D behavior is recovered above τ_c. In a macroscopic system, where the width and the length are comparable, only 3D behaviors can be observed. This might be a consequence of the universality of Fourier's law in the macroscopic world in nature.

Furthermore, one can see that, for $W \geq 8$, the larger the width, the longer the $c^Z(\tau)$ shows a power-law decay of $\tau^{-1.2}$. This suggests that the asymptotic behavior of the autocorrelation function should be $c^Z(\tau) \propto \tau^{-1.2}$ when W is large enough. It should be emphasized that the decay exponent is different from the traditional theoretical prediction of $\gamma = -3/2$. Interestingly, the numerically observed $\gamma = -1.2$ agrees with the formula $\gamma = -2d/(2+d)$ (for $d = 3$) based on the hydrodynamic equations for a normal fluid with an added thermal noise [6]. However, in that paper the authors limit the validity of this formula to $d \leq 2$. Our result $\gamma = -1.2$ is also compatible with the value $\gamma = -0.98 \pm 0.25$ for the 3D FPU model reported in [47].

In order to illustrate the 3D-1D crossover more clearly, we plot the τ-dependence of the running decay exponent $\gamma(\tau)$ defined as

$$\gamma(\tau) \equiv \frac{d \ln c^Z(\tau)}{d \ln \tau} \tag{6.20}$$

in Fig. 6.8c. For $W = 4$, the bottom of the running decay exponent $\gamma(\tau)$ is at -0.8, showing the absence of a 3D behavior. For $W = 8$, the bottom drops to about -1.1, showing a weak 3D behavior. For $W \geq 16$, the bottoms of the $\gamma(\tau)$ tend to saturate at a W-independent value -1.2. As W increases, the $\gamma(\tau)$ stay at this value for a longer time. This indicates that $\gamma = -1.2$ is the asymptotic decay exponent for a "real" 3D system. Since the decay exponent $\gamma = -1$ in 2D lattices, the threshold time $\tau_c(W)$ of the 3D-1D crossover can thus be reasonably defined as $\gamma(\tau_c(W)) = -1$. It can be estimated that $\tau_c(8) \approx 35$ and $\tau_c(16) \approx 90$. For $W \geq 32$, it is hard to estimate the threshold time due to large statistical errors.

The length-dependent thermal conductivity $\kappa^Z_{\mathrm{GK}}(L)$ for various W and N_Z is plotted in Fig. 6.9. similar to the 1D purely quartic lattice as shown in Fig. 6.8a, b, the 3D quartic lattice $c(\tau)$ can be correctly calculated for quite large correlation time τ by simulating a not-very-long lattice N_Z, i.e. $L = v_s\tau >> N_Z$. We are thus able to evaluate $\kappa^Z_{\mathrm{GK}}(L)$ for an effective length L, which is much longer than N_Z. For $W = 4$, the 3D behavior is nearly absent, similar to the picture shown in Fig. 6.8. As a result, the κ^Z_{GK} approaches to a 1D power-law divergence as $L^{0.4}$ directly. For $W = 8$, beyond trivial ballistic regime, the κ^Z_{GK} increases slowly at first, indicating the tendency to 3D behavior, and then inflects up to the 1D-like power-law behavior. For $W = 16$, the inflection occurs at a larger length L. Finally for $W = 32$ and $W = 64$, although the similar inflection is expected to occur at even larger length L, we are not able to see it due to numerical difficulties.

One can conclude that the 3D system should display normal heat conduction behavior if the cross-section area W^2 is large enough. Based on the threshold correlation time $\tau_c(W)$ defined earlier, a threshold length $N^Z_c(W)$ can be defined

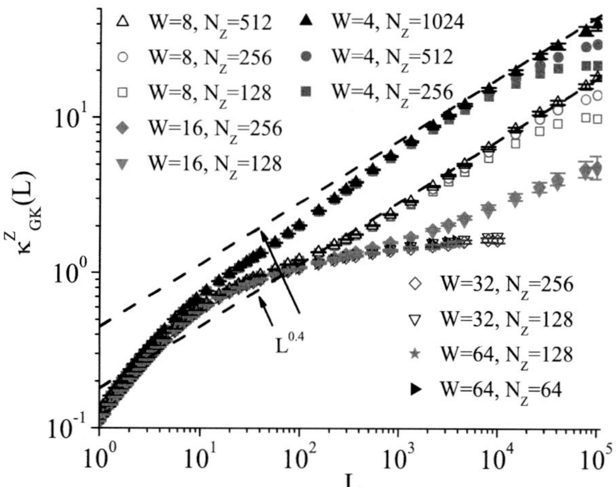

Fig. 6.9 $\kappa_{\rm GK}^{Z}(L)$ for various W and N_Z. For a given W, results for different values of N_Z are plotted in order to distinguish finite N_Z effects. For each W we plot error bars only for the data for the longest N_Z. One can see the tendency of the curves to become flat as W increases, indicating the presence of normal heat conduction. *Dashed lines* with slope 0.4 are drawn for reference

accordingly by requiring $N_c^Z(W) \equiv v_s\tau_c(W)$. The threshold length $N_c^Z(W)$ determined here is shorter than the estimation made by Saito and Dhar [49], in which a lattice width $W = 16$ shows a 3D behavior up to $L = 16384$. In a recent experimental study, an apparent 1D-like anomalous heat conduction behavior appears in multiwall nanotubes with diameters around 10 nm and lengths of a few μm [42]. It seems that our estimation agrees with the experimental result. However, more detailed experimental measurements of heat conduction in shorter samples or samples with larger cross-section of silicon nanowire [52, 53] or graphene [43], are necessary to give an accurate estimation of the threshold length $N_c^Z(W)$.

A running exponent $\alpha(L)$ is defined as the local slope of $\kappa_{\rm GK}^{Z}(L)$ as

$$\alpha(L) \equiv \frac{d\ln\kappa_{\rm GK}^{Z}(L)}{d\ln L} = \frac{L}{\kappa_{\rm GK}^{Z}(L)}c^Z(\tau). \tag{6.21}$$

In the 3D regime, the $c^Z(\tau)$ behaves as $c^Z(\tau) \sim \tau^\gamma$ as shown in Fig. 6.8. As $L \to \infty$, the $\kappa_{\rm GK}^{Z}(L)$ approaches a constant κ since $\gamma < -1$. Then one can obtain

$$\alpha(L) \sim \frac{L}{\kappa}L^\gamma = \frac{1}{\kappa}L^{\gamma+1}. \tag{6.22}$$

where $\alpha(L)$ decays asymptotically as $L^{-0.2}$ for $\gamma = -1.2$. This power-law decay of the exponent $\alpha(L)$ with the length L as $\alpha(L) \propto L^{-0.2}$ quantitatively explains the result previously found in [49].

In summary, we have numerically studied heat conduction in 3D momentum-conserving nonlinear lattices by the Green-Kubo method. The main findings are: (1) For a fixed width $W \geq 8$, a 3D-1D crossover was found to occur at a W-dependent threshold of a lattice size $N_c^Z(W)$. Below N_c^Z the system displays a 3D behavior while it displays a 1D behavior above N_c^Z. (2) In the 3D regime, the heat current autocorrelation function $c^Z(\tau)$ decays asymptotically as τ^γ with $\gamma = -1.2$. This value being more negative than -1 indicates normal heat conduction, which is consistent with the theoretical expectation. (3) The exponent $\gamma = -1.2$ implies that the running exponent $\alpha(L)$ follows a power-law decay, $\alpha \propto L^{-0.2}$, which also agrees very well with that reported in [49]. (4) The detailed value $\gamma = -1.2$ however deviates significantly from the conventional theoretical expectation of $\gamma = -1.5$.

6.2 Simulation of Heat Transport with the Diffusion Method

In the numerical studies of heat transport in nonlinear lattices, the most frequently used methods are the direct non-equilibrium heat bath method [4] and the equilibrium Green-Kubo method [30]. For the non-equilibrium heat bath method, the system is connected with two heat baths in both ends and driven into a stationary state. The averaged heat flux j is recorded which gives rise to the thermal conductivity κ through the relation of $j = -\kappa \nabla T$. For the equilibrium Green-Kubo method, the system is prepared from microcanonical dynamics without heat bath. The autocorrelation function $C_{JJ}(t)$ of the total heat flux is recorded and the thermal conductivity κ can be obtained by integrating $C_{JJ}(t)$ via the Green-Kubo formula.

Besides the non-equilibrium heat bath method and Green-Kubo method, a novel diffusion method is recently proposed by Zhao in studying the anomalous heat transport and diffusion processes of 1D nonlinear lattices [54]. This is also an equilibrium method, while the statistics can be drawn from microcanonical or canonical dynamics. In contrast to the Green-Kubo method, this diffusion method relies on the information of the autocorrelation function of the local energy. In Hamiltonian dynamics, the total energy is always a conserved quantity. Due to this very property of energy conservation, it is then rigorously proved in a heat diffusion theory recently developed by Liu et al, stating that the energy diffusion method is equivalent to the Green-Kubo method in the sense of determining the system's thermal conductivity [55]. In particular, the energy diffusion method is able to provide more information than that from the Green-Kubo method. The real-time spatiotemporal excess energy density distribution $\rho_E(x,t)$ plays the role of a generating function which is essential for the analysis of underlying heat conduction behavior.

In principle, the thermal conductivity κ can be generally expressed in a length-dependent form as $\kappa \propto L^\alpha$ with L the system length. For system with normal heat conduction, the heat divergency exponent $\alpha = 0$ implies that κ is length-independent obeying the Fourier's heat conduction law. The heat divergency exponent $\alpha = 1$ represents a ballistic heat conduction behavior. For $0 < \alpha < 1$, the system displays the so-called anomalous heat conduction behavior. On the other hand, the Mean Square Displacement (MSD) $\langle \Delta x^2(t) \rangle_{\mathrm{E}}$ of the excess energy generally grows with time asymptotically as $\langle \Delta x^2(t) \rangle_{\mathrm{E}} \propto t^\beta$ where the energy diffusion exponent β classifies the diffusion behaviors. The $\beta = 1$ and 2 represent the normal and ballistic energy diffusion behaviors, respectively. For $1 < \beta < 2$, the diffusion process is called superdiffusion.

In 1D nonlinear lattice systems, the heat conduction and energy diffusion originating from the same energy transport process are closely related. What is the relation between the heat conduction and energy diffusion processes? The relation formula between heat divergency and energy diffusion exponents $\alpha = \beta - 1$ is proposed by Cipriani et al. by investigating single particle Levy walk diffusion process [56]. This same formula is then formally derived as a natural result from the heat diffusion theory [55]. This relation formula tells that: (1) Normal energy diffusion with $\beta = 1$ corresponds to normal heat conduction with $\alpha = 0$ and vice versa. This is the case for 1D ϕ^4 lattice and Frenkel-Kontorova lattice [54]. (2) Ballistic energy diffusion with $\beta = 2$ implies ballistic heat conduction with $\alpha = 1$ and vice versa. The 1D Harmonic lattice and Toda lattice fall into this class [54]. (3) Superdiffusive energy diffusion with $1 < \beta < 2$ yields anomalous heat conduction with $0 < \alpha < 1$ and vice versa. The 1D FPU-β lattice is verified to posses energy superdiffusion with $\beta = 1.40$ and anomalous heat conduction with $\alpha = 0.40$ [54, 57] belonging to this class.

Besides total energy, total momentum is another conserved quantity for 1D nonlinear lattices without on-site potential, such as the FPU-β lattice. It is commonly believed that the conservation of momentum is essential for the actual heat conduction behavior. Predictions from mode coupling theory [4] and renormalization group theory [6] claim that momentum conservation should give rise to anomalous heat conduction in one dimensional systems. However, there is one exception to these predictions: the 1D coupled rotator lattice, which displays normal heat conduction behavior despite its momentum conserving nature [22, 58]. This unusual phenomenon stimulates the efforts to explore the interplay between energy transport and momentum transport. The transport coefficient corresponding to the momentum transport is the bulk viscosity. For momentum conserving system, in principle, there should also be a formal connection between the momentum transport and the momentum diffusion. This very momentum diffusion theory has also been developed for 1D momentum-conserving lattices [59]. Due to the complexity of bulk viscosity, the momentum diffusion theory is more complicated than the heat diffusion theory. Nevertheless, there seems to be a relation between the actual behaviors of energy and momentum transport implied from extensive numerical studies.

In the following, the energy diffusion method will be first introduced. The heat diffusion theory will be derived in the framework of linear response theory. Numerical simulations for two typical 1D nonlinear lattices will be used to verify the validity of this heat diffusion theory. Some results from the energy diffusion method will be shown for typical 1D lattices. The momentum diffusion method will then be discussed. The momentum diffusion theory will be derived in the same sense as heat diffusion theory for the 1D lattice systems. Some numerical results from the momentum diffusion method will be displayed and potential connection between momentum and heat transports will be discussed in the final part.

6.2.1 Energy Diffusion

In this part, we first introduce the energy diffusion method in the investigation for energy diffusion process of 1D lattices. The heat diffusion theory will be derived in the linear response regime and verified by numerical simulations. Some numerical results from this energy diffusion method will be presented to demonstrate the advantages of this novel method.

6.2.1.1 Heat Diffusion Theory

The heat diffusion theory [55, 59] unifies energy diffusion and heat conduction in a rigorous way. The central result reads

$$\frac{d^2\langle \Delta x^2(t)\rangle_{\mathrm{E}}}{dt^2} = \frac{2C_{JJ}(t)}{k_B T^2 c_{\mathrm{V}}}\,, \tag{6.23}$$

where k_B is the Boltzmann constant and c_{v} is the volumetric specific heat. The autocorrelation of total heat flux $C_{JJ}(t)$ on the right hand side is the term entering the Green-Kubo formula from which thermal conductivity can be calculated. The MSD $\langle \Delta x^2(t)\rangle_{\mathrm{E}}$ of energy diffusion describes the relaxation process in which an initially nonequilibrium energy distribution evolves towards equilibrium:

$$\langle \Delta x^2(t)\rangle_{\mathrm{E}} \equiv \int (x - \langle x\rangle_E)^2 \rho_E(x,t)dx = \langle x^2(t)\rangle_E - \langle x\rangle_E^2\,. \tag{6.24}$$

This normalized fraction of excess energy $\rho_E(x,t)$ at a certain position x at time t reads

$$\rho_E(x,t) = \frac{\delta\langle h(x,t)\rangle_{\mathrm{neq}}}{\delta E} = \frac{\delta\langle h(x,t)\rangle_{\mathrm{neq}}}{\int \delta\langle h(x,0)\rangle_{\mathrm{neq}}dx}. \tag{6.25}$$

Here the excess energy distribution is proportional to the deviation $\delta\langle h(x,t)\rangle_{\mathrm{neq}} \equiv \langle h(x,t)\rangle_{\mathrm{neq}} - \langle h(x)\rangle_{\mathrm{eq}}$, where $\langle\cdot\rangle_{\mathrm{neq}}$ denotes the expectation value in the nonequilibrium diffusion process, $\langle\cdot\rangle_{\mathrm{eq}}$ denotes the equilibrium average, and $h(x,t)$ denotes the local Hamiltonian density. For isolated energy-conserving systems, this total excess energy, $\delta E = \int \delta\langle h(x,t)\rangle_{\mathrm{neq}}dx$ remains conserved. Therefore, the normalized condition $\int \rho_E(x,t)dx = 1$ is fulfilled during the time evolution as a result of energy conservation.

In the linear response regime, the deviation of local excess energy can be explicitly derived in terms of equilibrium spatiotemporal correlation $C_{hh}(x,t;x',0)$ of local Hamiltonian density $h(x,t)$ as

$$\delta\langle h(x,t)\rangle_{\mathrm{neq}} = \frac{1}{k_B T}\int C_{hh}(x,t;x',0)\eta(x')dx', \tag{6.26}$$

where $C_{hh}(x,t;x',t') \equiv \langle \Delta h(x,t)\Delta h(x',t')\rangle_{\text{eq}}$, with $\Delta h(x,t) = h(x,t) - \langle h(x)\rangle_{\text{eq}}$, and the $-\eta(x)h(x)$ represents a small perturbation switched off suddenly at time $t=0$, with $\eta(x) \ll 1$.

Therefore, the normalized excess energy distribution can be derived from Eqs. (6.25) and (6.26) as

$$\rho_E(x,t) = \frac{1}{\mathcal{N}} \int C_{hh}(x-x',t)\eta(x')dx', \tag{6.27}$$

where $\mathcal{N} = k_B T^2 c_{\text{V}} \int \eta(x)dx$ is the normalization constant.

The key point which connects energy diffusion and heat conduction is the local energy continuity equation due to energy conservation

$$\frac{\partial h(x,t)}{\partial t} + \frac{\partial j(x,t)}{\partial x} = 0, \tag{6.28}$$

where $j(x,t)$ is the local heat flux density. One can then obtain

$$\frac{\partial^2 C_{hh}(x,t)}{\partial t^2} = \frac{\partial^2 C_{jj}(x,t)}{\partial x^2}, \tag{6.29}$$

By defining the total heat flux $J_L = \int_{-L/2}^{L/2} j(x,t)dx$ and the autocorrelation function of total heat flux $C_{JJ}(t) \equiv \lim_{L\to\infty} \langle J_L(t)J_L(0)\rangle_{\text{eq}}/L = \int_{-\infty}^{\infty} C_{jj}(x,t)dx$, the central result (6.23) of the heat diffusion theory can be derived.

As a result of energy conservation, the heat diffusion theory of Eq. (6.23) gives the general relation between energy diffusion and heat conduction. The actual behavior of energy diffusion or heat conduction can be normal or anomalous while the relation (6.23) remains to be the same:

1. For normal energy diffusion, the MSD increases asymptotically linearly with time, i.e.

   $$\langle \Delta x^2(t)\rangle_{\text{E}} \cong 2D_E t, \tag{6.30}$$

   in the infinite time limit $t \to \infty$. Here D_E is the so-called thermal diffusivity. According to Eq. (6.23), the corresponding thermal conductivity κ can be obtained as

   $$\kappa = \int_0^\infty \frac{C_{JJ}(t)}{k_B T^2} dt = \frac{c_{\text{V}}}{2} \lim_{t\to\infty} \frac{d\langle \Delta x^2(t)\rangle_{\text{E}}}{dt} = c_{\text{V}} D_E. \tag{6.31}$$

   This is nothing but the Green-Kubo expression for normal heat conduction.
2. For ballistic energy diffusion, the MSD is asymptotically proportional to the square of time as

   $$\langle \Delta x^2(t)\rangle_{\text{E}} \propto t^2. \tag{6.32}$$

Substituting this expression into Eq. (6.23), one can deduce that $C_{JJ}(t)$ is a non-decaying constant, reflecting the ballistic nature of heat conduction as well.

3. For superdiffusive energy diffusion, the MSD obeys

$$\langle \Delta x^2(t) \rangle_{\mathrm{E}} \propto t^{\beta}, \ \ 1 < \beta < 2. \tag{6.33}$$

From Eq. (6.23), the decay of $C_{JJ}(t)$ is a slow process as $C_{JJ}(t) \propto t^{\beta-2}$ and the integral of $C_{JJ}(t)$ diverges. In this situation, no finite superdiffusive thermal conductivity exists. The typical way in practice is to introduce an upper cutoff time $t_s \sim L/v_s$ with v_s the speed of sound due to renormalized phonons [60]. A length-dependent superdiffusive thermal conductivity can be obtained through Eq. (6.23):

$$\kappa \sim \frac{1}{k_B T^2} \int_0^{L/v_s} C_{JJ}(t) dt = \frac{c_{\mathrm{V}}}{2} \frac{d\langle \Delta x^2(t) \rangle_{\mathrm{E}}}{dt} \bigg|_{t \sim L/v_s} \propto L^{\beta-1}. \tag{6.34}$$

The length-dependent anomalous thermal conductivity is usually expressed as $\kappa \propto L^{\alpha}$. One can immediately obtain the scaling relation between energy diffusion and heat conduction

$$\alpha = \beta - 1, \tag{6.35}$$

which is a general relation and not limited to superdiffusive energy diffusion only.

4. For subdiffusive energy diffusion, the MSD follows asymptotically

$$\langle \Delta x^2(t) \rangle_{\mathrm{E}} \propto t^{\beta}, \ \ 0 < \beta < 1. \tag{6.36}$$

From the relation in (6.23), the autocorrelation function of total heat flux reads asymptotically

$$C_{JJ}(t) \propto \beta(\beta - 1) t^{\beta-2}. \tag{6.37}$$

The $C_{JJ}(t)$ remains integrable and the thermal conductivity can be derived as

$$\kappa = \int_0^{\infty} \frac{C_{JJ}(t)}{k_B T^2} dt = \frac{c_{\mathrm{V}}}{2} \lim_{t \to \infty} \frac{d\langle \Delta x^2(t) \rangle_{\mathrm{E}}}{dt} \sim \lim_{t \to \infty} t^{\beta-1} = 0. \tag{6.38}$$

This vanishing integral of $C_{JJ}(t)$ is not surprising, if one notices that the asymptotic prefactor of $C_{JJ}(t)$ in (6.37) is a negative value due to $\beta - 1 < 0$.

6.2.1.2 Numerical Verification of the Heat Diffusion Theory

The heat diffusion theory of Eq. (6.23) developed for continuous system is general and applicable also for discrete system. We choose two typical 1D nonlinear lattices to demonstrate the validity of this heat diffusion theory. One is the purely quartic FPU-β (qFPU-β) lattice which is the high temperature limit of FPU-β lattice, where energy diffusion is superdiffusive and heat conduction is anomalous or length-dependent [18, 23]. The dimensionless Hamiltonian with finite $N = 2M + 1$ atoms reads

$$H = \sum_i H_i = \sum_i \left[\frac{1}{2}p_i^2 + \frac{1}{4}(u_{i+1} - u_i)^4\right], \tag{6.39}$$

where p_i and u_i denote momentum and displacement from equilibrium position for i-th atom, respectively. The index i is numerated from $-M$ to M.

The other one is the ϕ^4 lattice which shows normal energy diffusion as well as normal heat conduction [54, 61, 62]. The dimensionless Hamiltonian reads

$$H = \sum_i H_i = \sum_i \left[\frac{1}{2}p_i^2 + \frac{1}{2}(u_{i+1} - u_i)^2 + \frac{1}{4}u_i^4\right]. \tag{6.40}$$

In the numerical simulations, we adopt the microcanonical dynamics where energy density E per atom is set as the input parameter. Algorithm with higher accuracy of fourth-order symplectic method [63, 64] can be used to integrate the equations of motions. Periodic boundary conditions $u_i = u_{i+N}$ and $p_i = p_{i+N}$ are applied and the equilibrium temperature T can be calculated from the definition $T = T_i = \langle p_i^2 \rangle$, where $\langle \cdot \rangle$ denotes time average which equals to the ensemble average due to the chaotic and ergodic nature of these two nonlinear lattices. The volumetric specific heat need to be calculated via the relation $c_v = (\langle H_i^2 \rangle - \langle H_i \rangle^2)/T^2$ which is independent of the choice of index i. For qFPU-β lattice, the volumetric specific heat $c_v = 0.75$ is a temperature-independent constant.

In order to define the discrete expression of excess energy density distribution $\rho_E(i,t)$, we first introduce the energy-energy correlation function, reading:

$$C_E(i,t;j,t=0) \equiv \frac{\langle \Delta H_i(t) \Delta H_j(0) \rangle}{k_B T^2 c_v}, \tag{6.41}$$

where $\Delta H_i(t) \equiv H_i(t) - \langle H_i(t) \rangle$. Applying a localized, small initial excess energy perturbation at the central site, $\eta(i) = \varepsilon \delta_{i,0}$ in Eq. (6.27), the discrete excess energy distribution can be obtained:

$$\rho_E(i,t) = \sum_j C_E(i,t;j,0)\eta(j)/\varepsilon = C_E(i,t:j=0,t=0), \quad -M \le i \le M. \tag{6.42}$$

The MSD $\langle \Delta x^2(t)\rangle_{\rm E}$ of energy diffusion of Eq. (6.24) for discrete lattice can be defined as

$$\langle \Delta x^2(t)\rangle_{\rm E} \equiv \sum_i i^2 \rho_E(i,t) = \sum_i i^2 C_E(i,t;j=0,t=0), \ -M \le i \le M, \quad (6.43)$$

by noticing that $\langle x(t)\rangle_{\rm E} = 0$.

The second derivative of $\langle \Delta x^2(t)\rangle_{\rm E}$ can be numerically obtained as

$$\frac{d^2\langle \Delta x^2(t)\rangle_{\rm E}}{dt^2} \approx \frac{\langle \Delta x^2(t+\Delta t)\rangle_{\rm E} - 2\langle \Delta x^2(t)\rangle_{\rm E} + \langle \Delta x^2(t-\Delta t)\rangle_{\rm E}}{(\Delta t)^2} \quad (6.44)$$

where Δt is the time difference between two consecutive recorded $\langle \Delta x^2(t)\rangle_{\rm E}$. The autocorrelation function of total heat flux $C_{JJ}(t)$ for discrete lattice system is defined as $C_{JJ}(t) = \langle \Delta J(t)\Delta J(0)\rangle/N$, with $J(t) = \sum_i j_i(t)$. The local heat flux $j_i(t) = -\dot{u}_i \partial V(u_i - u_{i-1})/\partial u_i$ is derived from local energy continuity equation where $V(x)$ denotes the form of potential energy in Hamiltonian.

In Fig. 6.10, we verify the main relation (6.23) for 1D qFPU-β lattice. The MSD of energy diffusion $\langle \Delta x^2(t)\rangle_{\rm E}$ as the function of time is plotted in Fig. 6.10a. Its

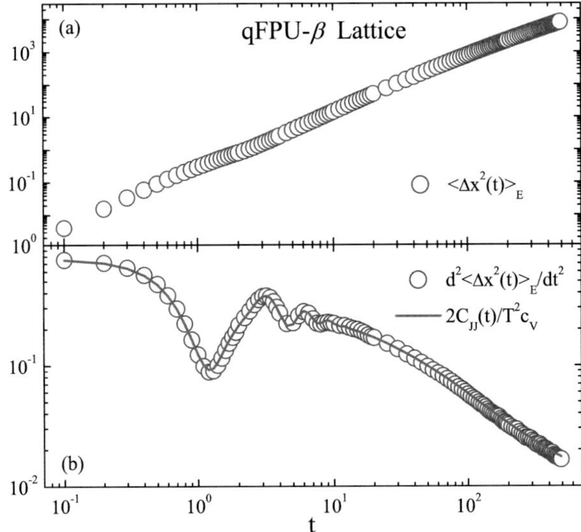

Fig. 6.10 Numerical verification of the main relation of Eq. (6.23) of heat diffusion theory for 1D qFPU-β lattice. (**a**) The MSD $\langle \Delta x^2(t)\rangle_{\rm E}$ of energy diffusion as the function of time t in log-log scale. (**b**) The second derivative of MSD $d^2\langle \Delta x^2(t)\rangle_{\rm E}/dt^2$ (*hollow circles*) and the rescaled autocorrelation function of total heat flux $2C_{JJ}(t)/(T^2 c_{\rm v})$ (*solid line*) as the function of time t in log-log scale. The perfect agreement between them demonstrates the validity of the main relation of Eq. (6.23). The Boltzmann constant k_B has been set as unity when applying dimensionless units in numerical simulations. The simulations are performed for a qFPU-β lattice with the average energy density per atom $E = 0.015$ and the number of atoms $N = 601$

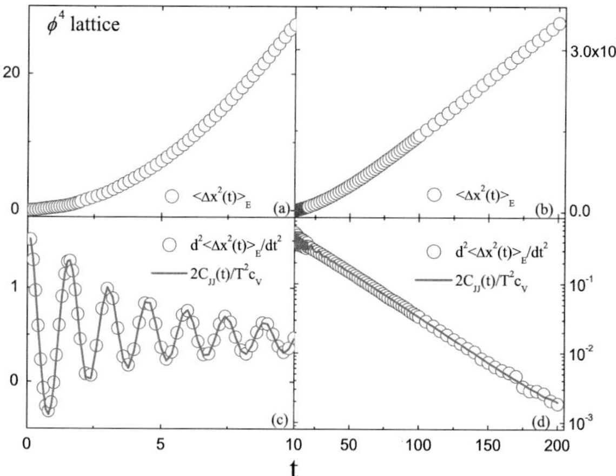

Fig. 6.11 Numerical verification of the main relation of Eq. (6.23) of heat diffusion theory for 1D ϕ^4 lattice. (**a**) and (**b**): the MSD $\langle \Delta x^2(t) \rangle_{\text{E}}$ of energy diffusion as the function of time t for $0 < t < 10$ and $10 < t < 200$, respectively. (**c**) and (**d**): the second derivative of MSD $d^2\langle \Delta x^2(t) \rangle_{\text{E}}/dt^2$ (*hollow circles*) and the rescaled autocorrelation function of total heat flux $2C_{JJ}(t)/(T^2 c_{\text{v}})$ (*solid line*) as the function of time t in linear-linear scale for $0 < t < 10$ and log-linear scale for $10 < t < 200$, respectively. The perfect agreement between them demonstrates the validity of the main relation of Eq. (6.23). The simulations are performed for a ϕ^4 lattice with the average energy density per atom $E = 0.4$ and the number of atoms $N = 501$

second derivative $d^2\langle \Delta x^2(t) \rangle_{\text{E}}/dt^2$ is extracted out and directly compared with the rescaled autocorrelation function of total heat flux $2C_{JJ}(t)/(k_B T^2 c_{\text{v}})$ in Fig. 6.10b. The perfect agreement between them justifies the validity of the main relation (6.23) predicted from heat diffusion theory. The numerical results for 1D ϕ^4 lattice are also plotted in Fig. 6.11 and same conclusion can be obtained. It should be pointed out that the 1D qFPU-β lattice displays superdiffusive energy diffusion and anomalous heat conduction, while 1D ϕ^4 lattice shows normal energy diffusion and heat conduction. These facts can be observed by noticing that the autocorrelation function $C_{JJ}(t)$ eventually follows a power law decay as $C_{JJ}(t) \propto t^{-0.60}$ for qFPU-β lattice and an exponential decay as $C_{JJ}(t) \propto e^{-t/\tau}$ for ϕ^4 lattice where τ represents a characteristic relaxation time.

6.2.1.3 Energy Diffusion Properties for Typical 1D Lattices

The heat diffusion theory formally connects the energy diffusion and heat conduction as described in Eq. (6.23). It enables us to use energy diffusion method to investigate the heat conduction process. More importantly, the energy diffusion method is able to provide more information about the heat transport process than the Green-Kubo method or the direct non-equilibrium heat bath method.

The key information from the energy diffusion method is the spatiotemporal distribution of excess energy $\rho_E(i,t)$, from which the MSD of energy diffusion $\langle \Delta x^2(t)\rangle_{\mathrm{E}}$ can be generated. The second derivative of $\langle \Delta x^2(t)\rangle_{\mathrm{E}}$ gives rise to the autocorrelation function of total heat flux $C_{JJ}(t)$ which finally yields the thermal conductivity via Green-Kubo formula. With the knowledge of $\rho_E(i,t)$, one can resolve the expression of $\langle \Delta x^2(t)\rangle_{\mathrm{E}}$ or $C_{JJ}(t)$, but not vice versa. Therefore, in determining the actual heat conduction behavior, the excess energy distribution $\rho_E(i,t)$ plays the essential role of a generating function.

To illustrate the importance [59]: the coupled rotator lattice

$$H = \sum_i \left[\frac{1}{2}p_i^2 + (1 - \cos(u_{i+1} - u_i))\right], \tag{6.45}$$

and the amended coupled rotator lattice

$$H = \sum_i \left[\frac{1}{2}p_i^2 + (1 - \cos(u_{i+1} - u_i)) + \frac{K}{2}(u_{i+1} - u_i)^2\right], \tag{6.46}$$

where an additional quadratic interaction potential term is added.

The excess energy distributions $\rho_E(i,t)$ for coupled rotator lattice are plotted in Fig. 6.12a. For sufficiently large times, the excess energy distribution $\rho_E(i,t)$ evolves

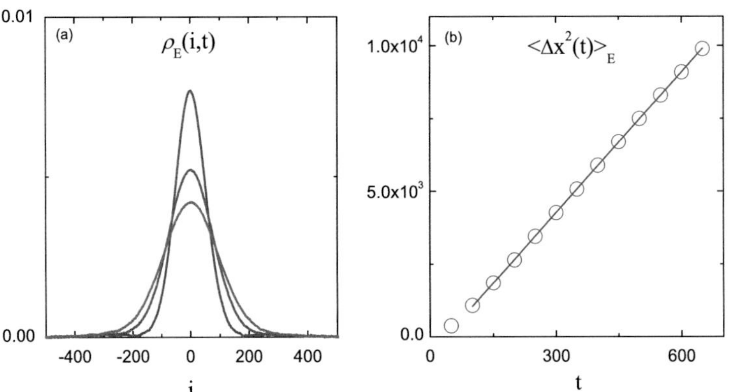

Fig. 6.12 Energy diffusion processes in the 1D coupled rotator lattice. (**a**) Spatial distribution of the energy autocorrelation $\rho_E(i,t) = C_E(i,t;j=0,t=0)$. The correlation times are $t = 200, 400$ and 600 from top to the bottom in the central part, respectively. The distribution of $\rho_E(i,t)$ follows the Gaussian normal distribution as $\rho_E(i,t) \sim \frac{1}{\sqrt{4\pi D_E t}} e^{-\frac{i^2}{4D_E t}}$. (**b**) The MSD of the energy diffusion $\langle \Delta x^2(t)\rangle_{\mathrm{E}}$ as the function of time. The *solid straight line* is the best fit for the MSD $\langle \Delta x^2(t)\rangle_{\mathrm{E}}$ implying a normal diffusion process. The simulations are performed for a coupled rotator lattice with the average energy density per atom $E = 0.45$ and the number of atoms $N = 1501$

very closely into a Gaussian distribution function with its profile perfectly given by

$$\rho_E(i,t) \sim \frac{1}{\sqrt{4\pi D_E t}} e^{-\frac{i^2}{4D_E t}}, \tag{6.47}$$

where D_E denotes the diffusion constant for energy diffusion. As a result, the MSD of energy diffusion $\langle \Delta x^2(t)\rangle_{\rm E}$ depicts at a linear time dependence

$$\langle \Delta x^2(t)\rangle_{\rm E} \sim \sum_i i^2 \rho_E(i,t) = \sum_i i^2 \frac{1}{\sqrt{4\pi D_E t}} e^{-\frac{i^2}{4D_E t}} = 2D_E t, \tag{6.48}$$

for sufficiently long time as can be seen in Fig. 6.12b, being the hall mark for normal diffusion. Accordingly, heat diffusion theory for normal energy diffusion implies that the heat conduction behavior is normal as well, with the thermal conductivity given by $\kappa = c_{\rm v} D_E$.

This normal energy diffusion behavior also occurs for other 1D lattice systems with normal heat conduction, such as ϕ^4 lattice and Frenkel-Kontorova lattice [54].

For the 1D amended coupled rotator lattice described in Eq. (6.46), the excess energy distributions $\rho_E(i,t)$ are plotted in Fig. 6.13a. Besides the central peak, there are also two side peaks moving outside with a constant sound velocity v_s. It is amazing that this excess energy distribution $\rho_E(i,t)$ of 1D nonlinear lattice closely

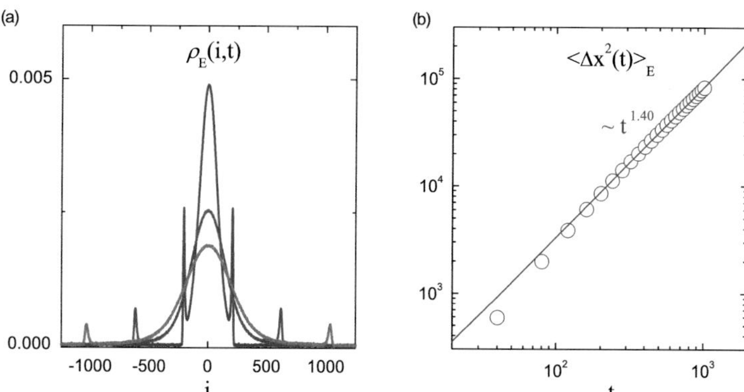

Fig. 6.13 Energy diffusion processes in the 1D amended coupled rotator lattice. (**a**) Spatial distribution of the energy autocorrelation $\rho_E(i,t) = C_E(i,t;j=0,t=0)$. The correlation times are $t = 200, 600$ and 1000 from top to the bottom in the central part, respectively. Besides the central peak, there are two side peaks moving outside with the constant sound velocity. This is the Levy walk distribution giving rise to a superdiffusive energy diffusion. (**b**) The MSD of the energy diffusion $\langle \Delta x^2(t)\rangle_{\rm E}$ as the function of time. The *solid straight line* is the best fit for the superdiffusive MSD as $\langle \Delta x^2(t)\rangle_{\rm E} \propto t^{1.40}$. The simulations are performed for an amended coupled rotator lattice with the average energy density per atom $E = 1$. The number of atoms $N = 2501$ and $K = 0.5$

resembles to the single particle Levy walk distribution $\rho_{LW}(x,t)$ [56]

$$\rho_{LW}(x,t) \propto \begin{cases} t^{-1/\mu}\exp\left(\frac{-ax^2}{t^{2/\mu}}\right), & |x| \lesssim t^{1/\mu} \\ tx^{-\mu-1}, & t^{1/\mu} < |x| < vt \\ t^{1-\mu}, & |x| = vt \\ 0, & |x| > vt \end{cases} \tag{6.49}$$

where a is an unknown constant and v is the particle velocity. This Levy walk distribution $\rho_{LW}(x,t)$ is a result of a particle moving ballistically between consecutive collisions with a waiting time distribution $\psi(t) \propto t^{-\mu-1}$ and a velocity distribution $f(u) = [\delta(u-v) + \delta(u+v)]/2$.

The MSD for the Levy walk distribution $\rho_{LW}(x,t)$ follows a time dependence of $\langle \Delta x^2(t) \rangle_{LW} \propto t^{\beta}$ with $\beta = 3 - \mu$. In Fig. 6.13b, the time dependence of MSD $\langle \Delta x^2(t) \rangle_{\mathrm{E}}$ of energy diffusion for the 1D amended coupled rotator lattice is plotted where the best fit indicates that $\beta = 1.40$. This will in turn correspond to a $\mu = 1.60$ in the Levy walk scenario. According to the relation formula of Eq. (6.35) of heat diffusion theory, the corresponding heat conduction should be anomalous with a divergent length dependent thermal conductivity of $\kappa \propto L^{\alpha}$ with $\alpha = 0.40$ [23].

It is very interesting to notice that there is a characteristic μ for the Levy walk distribution. By requiring that the heights of the central peak and side peaks decay with a same rate in the Levy walk distribution (6.49), one can obtain $-1/\mu = 1-\mu$ which gives rise to the golden ratio $\mu = (\sqrt{5}+1)/2 \approx 1.618$. As a result, the corresponding characteristic energy superdiffusion exponent $\beta = (5-\sqrt{5})/2 \approx 1.382$ and the anomalous heat conduction exponent $\alpha = (3-\sqrt{5})/2 \approx 0.382$ can be derived. Interesting enough, Lee-Dadswell et al. derived the same exponent $\alpha = (3-\sqrt{5})/2$ as the converging value of a Fibonacci sequence in a toy model within the framework of hydrodynamical theory in 2005 [65]. Actually, this exponent is not far from the existing numerical results [18, 23, 66].

6.2.2 *Momentum Diffusion*

In the following part, the momentum diffusion method will be introduced for momentum conserving systems. A momentum diffusion theory will be derived in the same sense as heat diffusion theory. The numerical results reflecting the momentum diffusion properties will then be presented for several 1D nonlinear lattices. Based on the numerical results, the possible connection between momentum and energy transports will be discussed in the final part.

6.2.2.1 Momentum Diffusion Theory

In analogy to the heat diffusion, one can also construct a momentum diffusion theory [59] for lattice systems which reads:

$$\frac{d^2\langle\Delta x^2(t)\rangle_{\mathrm{P}}}{dt^2} = \frac{2}{k_B T} C_{J^P}(t), \tag{6.50}$$

where $\langle\Delta x^2(t)\rangle_{\mathrm{P}}$ denotes the momentum diffusion and $C_{J^P}(t)$ is the centered autocorrelation function of total momentum flux.

The MSD of excess momentum $\langle\Delta x^2(t)\rangle_{\mathrm{P}}$ can be defined as

$$\langle\Delta x^2(t)\rangle_{\mathrm{P}} = \sum_i i^2 \rho_P(i,t), \ -M \le i \le M. \tag{6.51}$$

The excess momentum distribution $\rho_P(i,t)$ describes the nonequilibrium relaxation process of momentum due to a small kick of short duration to the j-th atom. The kick occurs with a constant impulse I, yielding a force kick at site j as $f_j(t) = I\delta(t)$. The normalized $\rho_P(i,t)$ is given by

$$\rho_P(i,t) = \frac{\langle p_i(t)\rangle_{\mathrm{re}}}{\sum_i \langle p_i(t)\rangle_{\mathrm{re}}}, \tag{6.52}$$

where $\langle p_i(t)\rangle_{\mathrm{re}}$ represents the response of momentum of i-th atom to the small perturbation of $-f_j(t)u_j$. In the linear response regime, it can be obtained that $\langle p_i(t)\rangle_{\mathrm{re}} = IC_P(i,t;j,0)$, where

$$C_P(i,t;j,0) = \frac{\langle\Delta p_i(t)\Delta p_j(0)\rangle}{k_B T} \tag{6.53}$$

is the autocorrelation function for the excess momentum fluctuation. The sum $\sum_i C_P(i,t;j,0) = 1$ at time $t = 0$ and remains normalized due to the conservation of momentum. As a result, the excess momentum distribution $\rho_P(i,t)$ assumes the form

$$\rho_P(i,t) = C_P(i,t;j=0,t=0), \tag{6.54}$$

if the kick is put at the atom with index $j = 0$.

The centered autocorrelation function of momentum flux $C_{J^P}(t)$ in Eq. (6.50) is given by

$$C_{J^P}(t) = \frac{1}{N}\langle\Delta J^P(t)\Delta J^P(0)\rangle, \ J^P = \sum_i j_i^P, \tag{6.55}$$

where the local momentum flux $j_i^P = -\partial V(u_i - u_{i-1})/\partial u_i$ with $V(x)$ the form of interaction potential is obtained from the discrete momentum continuity relation

$$\frac{dp_i}{dt} - j_i^P + j_{i+1}^P = 0. \tag{6.56}$$

It should be emphasized that here the momentum flux $\Delta J^P(t)$, unlike for energy flux, cannot be replaced with $J^P(t)$ itself. This is so because the equilibrium average is typically non-vanishing with $\langle J^P(t)\rangle = N\Lambda$, where Λ denotes a possibly non-vanishing internal equilibrium pressure in cases where the interaction potential is not symmetric.

The transport coefficient of momentum conduction related to the momentum diffusion is the bulk viscosity η. However, the presence of a finite, isothermal sound speed v_s implies that here the momentum spread contains a ballistic component which should be subtracted [67, 68] to yield the effective bulk viscosity η:

$$\eta \equiv \frac{1}{k_B T}\int_0^\infty C_{J^P}(t)dt - \frac{1}{2}v_s^2 t. \tag{6.57}$$

In case that the momentum diffusion occurs normal, one can invoke the concept of a finite momentum diffusivity

$$D_P \equiv \frac{1}{2}\lim_{t\to\infty}\left(\frac{d\langle\Delta x^2(t)\rangle_{\mathrm{P}}}{dt} - v_s^2 t\right). \tag{6.58}$$

Therefore, for the discrete lattices, this effective bulk viscosity η precisely equals the momentum diffusivity times the atom mass m (set to unity in the dimensionless units), namely

$$\eta = D_P. \tag{6.59}$$

If the excess momentum density spreads not normally, the limit in Eq. (6.58) no longer exits. The integration in Eq. (6.57) formally diverges, thus leading to an infinite effective bulk viscosity.

In the practice, it is found that the finite effective bulk viscosity and normal heat conduction always emerge in pair, so does the infinite effective bulk viscosity and anomalous heat conduction. This constitutes an alternative implementation of the investigation of heat conduction behavior in lattice systems.

6.2.2.2 Momentum Diffusion Properties for Typical 1D Lattices

The momentum diffusion theory connects the momentum diffusion and momentum transport via Eqs. (6.50) and (6.57). Similar to what we have discussed for the

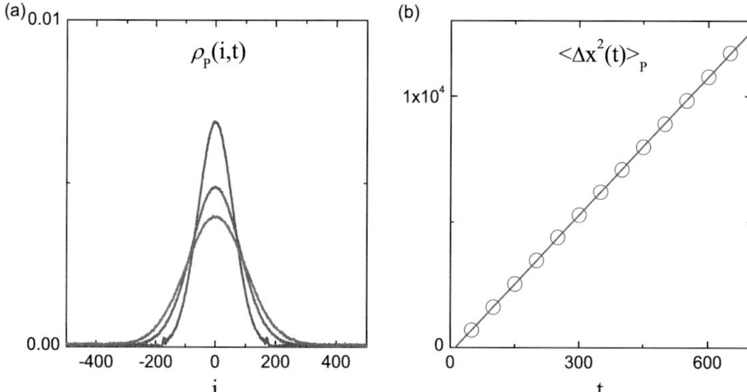

Fig. 6.14 Momentum diffusion processes in the 1D coupled rotator lattice. (**a**) Spatial distribution of the energy autocorrelation $\rho_P(i,t) = C_P(i,t;j=0,t=0)$. The correlation times are $t = 200, 400$ and 600 from top to the bottom in the central part, respectively. The distribution of $\rho_P(i,t)$ follows the Gaussian normal distribution as $\rho_P(i,t) \sim \frac{1}{\sqrt{4\pi D_P t}} e^{-\frac{i^2}{4D_P t}}$. (**b**) The MSD of the momentum diffusion $\langle \Delta x^2(t) \rangle_\text{P}$ as the function of time. The *solid straight line* is the best fit for the MSD $\langle \Delta x^2(t) \rangle_\text{P}$ implying a normal diffusion process. The simulations are performed for a coupled rotator lattice with the average energy density per atom $E = 0.45$ and the number of atoms $N = 1501$

energy diffusion, the excess momentum distribution $\rho_P(i,t)$ is the most important information we need to gather for momentum diffusion method.

We still consider the 1D coupled rotator lattice of Eq. (6.45) and amended coupled rotator lattice of Eq. (6.46) [59]. In Fig. 6.14a, the excess momentum distributions $\rho_P(i,t)$ for different correlation times $t = 200, 400$ and 600 are plotted. At sufficiently large times, the distributions $\rho_P(i,t)$ follow the Gaussian distribution as

$$\rho_P(i,t) \sim \frac{1}{\sqrt{4\pi D_P t}} e^{-\frac{i^2}{4D_P t}}, \tag{6.60}$$

where D_P represents the momentum diffusion constant. The MSD $\langle \Delta x^2(t) \rangle_\text{P}$ of momentum diffusion thus grows linearly with time at large times, i.e. $\langle \Delta x^2(t) \rangle_\text{P} \sim 2D_P t$, as can be seen from Fig. 6.14. There is no ballistic component for the distributions $\rho_P(i,t)$ in 1D coupled rotator lattice. Accordingly, the finite effective bulk viscosity η can be simply obtained as $\eta = D_P$. This result is consistent with the fact that the 1D coupled rotator lattice displays normal heat conduction behavior.

For the 1D amended coupled rotator lattice, the energy diffusion is superdiffusive and its heat conduction is anomalous. The excess momentum distributions $\rho_P(i,t)$ for different correlation times at $t = 200, 600$ and 1000 are plotted in Fig. 6.15a. In contrast to the coupled rotator lattice, here only two side peaks moving outside with a constant sound velocity v_s exist. To evaluate the true behavior of momentum conduction, this ballistic component within the momentum diffusion should be

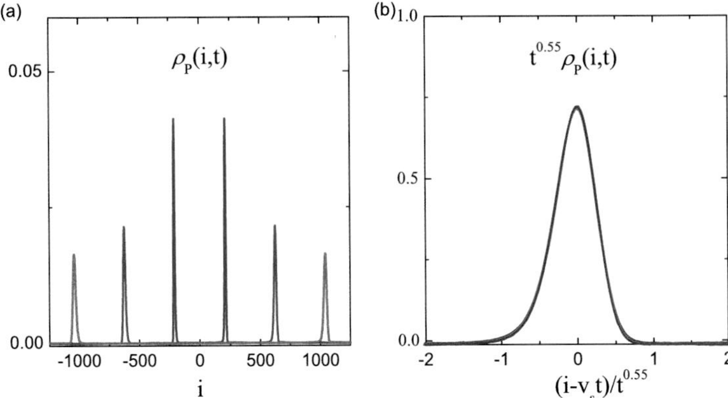

Fig. 6.15 Momentum diffusion processes in the 1D amended coupled rotator lattice. (**a**) Spatial distribution of the energy autocorrelation $\rho_P(i,t) = C_P(i,t;j=0,t=0)$. The correlation times are $t = 200, 600$ and 1000 from top to the bottom, respectively. There is no central peak and two side peaks moves outside with the constant sound velocity v_s. (**b**) The self-diffusion of the side peak of the momentum spreading. The rescaled side peaks $t^\delta \rho_P(i,t)$ of four different correlation times at $t = 400, 600, 800$ and 1000 all collapse into the same curve in the center of the rescaled moving frame $(i - v_s t)/t^\delta$ with a scaling exponent $\delta = 0.55$. The scaling exponent $\delta = 0.55 > 0.50$ implies the self-diffusion is superdiffusive and the effective bulk viscosity η is infinite. The simulations are performed for an amended coupled rotator lattice with the average energy density per atom $E = 1$. The number of atoms $N = 2501$ and $K = 0.5$

subtracted. One should instead analyze the self-diffusion behavior of the side peaks of the distributions $\rho_P(i,t)$. In Fig. 6.15b, the rescaled side peaks $t^\delta \rho_P(i,t)$ as the function of rescaled position of the peak center $(i - v_s t)/t^\delta$ are plotted for four different correlation times at $t = 400, 600, 800$ and 1000. With the choice of $\delta = 0.55$, the rescaled distributions collapse into a single curve all together. This rescaling behavior with $\delta = 0.55$ implies a superdiffusive self-diffusion for the side peaks, while normal self-diffusion would require for $\delta = 0.50$.

The integration of Eq. (6.57) is then divergent, giving rise to an infinite effective bulk viscosity η. This infinite η is consistent with the finding that the heat conduction is anomalous, since the energy diffusion is superdiffusive as can be observed in Fig. 6.13b. From our perspective and our own numerical results, the infinite bulk viscosity η and divergent length-dependent thermal conductivity always emerge in pair. However, there are some other numerical results and approximate theories indicating that finite bulk viscosity and anomalous heat conduction might coexist for symmetric lattices such as FPU-β lattice [65, 69]. This is still an open issue and deserves more investigation in the future.

In summary, a novel diffusion method is introduced to investigate the heat transport in 1D nonlinear lattices. The heat and momentum diffusion theories formally relate the diffusion processes to their corresponding conduction processes. The properties of energy and momentum diffusions for typical 1D lattices are presented and more fundamental information about transport processes can be provided from this novel method.

Acknowledgements This work is supported by the National Natural Science Foundation of China under Grant No. 11275267(L.W.), Nos. 11334007 and 11205114 (N.L.), the Fundamental Research Funds for the Central Universities, and the Research Funds of Renmin University of China 15XNLQ03 (L.W.), the Program for New Century Excellent Talents of the Ministry of Education of China with Grant No. NCET-12-0409 (N.L.), the Shanghai Rising-Star Program with grant No. 13QA1403600 (N.L.). Computational resources were provided by the Physical Laboratory of High Performance Computing at Renmin University of China(L.W.) and Shanghai Supercomputer Center (N.L.).

References

1. James, M.L., Smith, G.M., Wolford, J.C.: Applied Numerical Methods for Digital Computation. HarperCollins College Publishers, New York (1993)
2. Dormand, J.R., El-Mikkawy, M.E.A., Prince, P.J.: IMA J. Numer. Anal. **7**, 423 (1987)
3. Dormand, J.R., El-Mikkawy, M.E.A., Prince, P.J.: IMA J. Numer. Anal. **11**, 297 (1991)
4. Lepri, S., Livi, R., Politi, A.: Phys. Rep. **377**, 1 (2003)
5. Dhar, A.: Adv. Phys. **57**(5), 457 (2008)
6. Narayan, O., Ramaswamy, S.: Phys. Rev. Lett. **89**, 200601 (2002)
7. Mai, T., Narayan, O.: Phys. Rev. E **73**, 061202 (2006)
8. Wang, J.S., Li, B.: Phys. Rev. Lett. **92**, 074302 (2004)
9. Wang, J.S., Li, B.: Phys. Rev. E **70**, 021204 (2004)
10. Delfini, L., Lepri, S., Livi, R., Politi, A.: Phys. Rev. E **73**, 060201 (2006)
11. Delfini, L., Lepri, S., Livi, R., Politi, A.: J. Stat. Mech. Theory Exp. **2007**(02), P02007 (2007)
12. Delfini, L., Denisov, S., Lepri, S., Livi, R., Mohanty, P.K., Politi, A.: Eur. Phys. J. Special Topics **146**, 21 (2007)
13. Santhosh, G., Kumar, D.: Phys. Rev. E **77**, 011113 (2008)
14. Liu, S., Liu, J., Hänggi, P., Wu, C., Li, B.: Phys. Rev. B **90**, 174304 (2014)
15. Pereverzev, A.: Phys. Rev. E **68**, 056124 (2003)
16. Lukkarinen, J., Spohn, H.: Commun. Pure Appl. Math. **61**(12), 1753 (2008)
17. Lepri, S., Livi, R., Politi, A.: Phys. Rev. Lett. **78**(10), 1896 (1997)
18. Mai, T., Dhar, A., Narayan, O.: Phys. Rev. Lett. **98**(18), 184301 (2007)
19. Delfini, L., Lepri, S., Livi, R., Politi, A.: Phys. Rev. Lett. **100**, 199401 (2008)
20. Dhar, A., Narayan, O.: Phys. Rev. Lett. **100**, 199402 (2008)
21. Prosen, T., Campbell, D.K.: Phys. Rev. Lett. **84**, 2857 (2000)
22. Gendelman, O.V., Savin, A.V.: Phys. Rev. Lett. **84**(11), 2381 (2000)
23. Wang, L., Wang, T.: Europhys. Lett. **93**, 54002 (2011)
24. Delfini, L., Lepri, S., Livi, R., Mejia-Monasterio, C., Politi, A.: J. Phys. A Math. Theor. **43**(14), 145001 (2010)
25. Zhong, Y., Zhang, Y., Wang, J., Zhao, H.: Phys. Rev. E **85**, 060102 (2012)
26. Wang, L., Hu, B., Li, B.: Phys. Rev. E **88**, 052112 (2013)
27. Savin, A.V., Kosevich, Y.A.: Phys. Rev. E **89**, 032102 (2014)
28. Das, S.G., Dhar, A., Narayan, O.: J. Stat. Phys **154**, 204 (2014)
29. Chen, S., Wang, J., Casati, G., Benenti, G.: Phys. Rev. E **90**, 032134 (2014)
30. Kubo, R., Toda, M., Hashitsume, N.: Statistical Physics II. Springer Series in Solid State Sciences, vol. 31. Springer, Berlin (1991)
31. Chen, S., Zhang, Y., Wang, J., Zhao, H.: Phys. Rev. E **89**, 022111 (2014)
32. Wang, L., Hu, B., Li, B.: Phys. Rev. E **86**, 040101 (2012)

33. Wang, L., He, D., Hu, B.: Phys. Rev. Lett. **105**, 160601 (2010)
34. Wang, L., Xu, L., Zhao, H.: Phys. Rev. E **91**, 012110 (2015)
35. Ernst, M.H., Hauge, E.H., van Leeuwen, J.M.J.: Phys. Rev. A **4**, 2055 (1971)
36. Ernst, M., Hauge, E., van Leeuwen, J.: J. Stat. Phys. **15**(1), 7 (1976)
37. Ernst, M., Hauge, E., van Leeuwen, J.: J. Stat. Phys. **15**(1), 23 (1976)
38. Lippi, A., Livi, R.: J. Stat. Phys. **100**, 1147 (2000)
39. Yang, L., Grassberger, P., Hu, B.: Phys. Rev. E **74**, 062101 (2006)
40. Xiong, D., Wang, J., Zhang, Y., Zhao, H.: Phys. Rev. E **82**, 030101 (2010)
41. Grassberger, P., Yang, L.: eprint arXiv:cond-mat/0204247
42. Chang, C.W., Okawa, D., Garcia, H., Majumdar, A., Zettl, A.: Phys. Rev. Lett. **101**, 075903 (2008)
43. Ghosh, S., Bao, W., Nika, D.L., Subrina, S., Pokatilov, E.P., Lau, C.N., Balandin, A.A.: Nat. Mater. **9**, 555 (2010)
44. Xu, X., Wang, Y., Zhang, K., Zhao, X., Bae, S., Heinrich, M., Tinh Bui C., Xie, R., Thong, J.T.L., Hong, B.H., Loh, K.P., Li, B., Oezyilmaz, B.: eprint arXiv:1012.2937
45. Wang, Z., Xie, R., Bui, C.T., Liu, D., Ni, X., Li, B., Thong, J.T.L.: Nano Lett. **11**, 113 (2011)
46. Nika, D.L., Askerov, A.S., Balandin, A.A.: Nano Lett. **12**, 3238 (2012)
47. Shiba, H., Yukawa, S., Ito, N.: J. Phys. Soc. Jpn. **75**(10), 103001 (2006)
48. Shiba, H., Ito, N.: J. Phys. Soc. Jpn. **77**(5), 054006 (2008)
49. Saito, K., Dhar, A.: Phys. Rev. Lett. **104**, 040601 (2010)
50. Lepri, S., Livi, R., Politi, A.: Europhys. Lett. **43**, 271 (1998)
51. Lepri, S., Livi, R., Politi, A.: Phys. Rev. E **68**, 067102 (2003)
52. Morales, A.M., Lieber, C.M.: Science **279**(5348), 208 (1998)
53. Ma, D.D.D., Lee, C.S., Au, F.C.K., Tong, S.Y., Lee, S.T.: Science **299**(5614), 1874 (2003)
54. Zhao, H.: Phys. Rev. Lett. **96**, 140602 (2006)
55. Liu, S., Hänggi, P., Li, N., Ren, J., Li, B.: Phys. Rev. Lett. **112**, 040601 (2014)
56. Cipriani, P., Denisov, S., Politi, A.: Phys. Rev. Lett. **94**, 244301 (2005)
57. Wang, L., Wu, Z., Xu, L.: Phys. Rev. E **91**, 062130 (2015)
58. Giardinà, C., Livi, R., Politi, A., Vassalli, M.: Phys. Rev. Lett. **84**, 2144 (2000)
59. Li, Y., Liu, S., Li, N., Hänggi, P., Li, B.: New J. Phys. **17**, 043064 (2015)
60. Li, N., Li, B., Flach, S.: Phys. Rev. Lett. **105**, 054102 (2010)
61. Hu, B., Li, B., Zhao, H.: Phys. Rev. E **61**, 3828 (2000)
62. Aoki, K., Kusnezov, D.: Phys. Lett. A **265**(4), 250 (2000)
63. Skokos, C., Krimer, D.O., Komineas, S., Flach, S.: Phys. Rev. E **79**, 056211 (2009)
64. Laskar, J., Robutel, P.: Celest. Mech. Dyn. Astron. **80**(1), 39 (2001)
65. Lee-Dadswell, G.R., Nickel, B.G., Gray, C.G.: Phys. Rev. E **72**, 031202 (2005)
66. Zhao, H., Wen, Z., Zhang, Y., Zheng, D.: Phys. Rev. Lett. **94**, 025507 (2005)
67. Helfand, E.: Phys. Rev. **119**, 1 (1960)
68. Resibois, P., de Leener, M.: Classical Kinetic Theory of Fluids. Wiley, New York/London/Sidney/Toronto (1977)
69. Spohn, H.: J. Stat. Phys. **154**, 1191 (2014)

Chapter 7
Simulation of Dimensionality Effects in Thermal Transport

Davide Donadio

Abstract The discovery of nanostructures and the development of growth and fabrication techniques of one- and two-dimensional materials provide the possibility to probe experimentally heat transport in low-dimensional systems. Nevertheless measuring the thermal conductivity of these systems is extremely challenging and subject to large uncertainties, thus hindering the chance for a direct comparison between experiments and statistical physics models. Atomistic simulations of realistic nanostructures provide the ideal bridge between abstract models and experiments. After briefly introducing the state of the art of heat transport measurement in nanostructures, and numerical techniques to simulate realistic systems at atomistic level, we review the contribution of lattice dynamics and molecular dynamics simulation to understanding nanoscale thermal transport in systems with reduced dimensionality. We focus on the effect of dimensionality in determining the phononic properties of carbon and semiconducting nanostructures, specifically considering the cases of carbon nanotubes, graphene and of silicon nanowires and ultra-thin membranes, underlying analogies and differences with abstract lattice models.

7.1 Introduction

The rise of nanoscience, starting with the discovery of C_{60} and other carbon fullerenes [1] in the 1980s, and of carbon nanotubes [2] and semiconducting nanostructures [3] in the 1990s, has provided suitable platforms to probe experimentally the physical properties of systems with reduced dimensionality. As expected

D. Donadio (✉)
Department of Chemistry, University of California Davis, One Shields Avenue, Davis, CA 95616, USA

Donostia International Physics Center, Paseo Manuel de Lardizabal 4, 20018 Donostia-San Sebastian, Spain

IKERBASQUE, Basque Foundation for Science, Bilbao, Spain

Max Planck Institute for Polymer Research, Ackermannweg 10, 55128 Mainz, Germany
e-mail: ddonadio@ucdavis.edu

S. Lepri (ed.), *Thermal Transport in Low Dimensions*, Lecture Notes in Physics 921, DOI 10.1007/978-3-319-29261-8_7

from theoretical predictions, low-dimensional nanostructures display very different electronic properties from their three-dimensional bulk counterparts. One of the most striking examples was the direct observation of Dirac's cones in graphene by X-ray diffraction [4]. In general, one expects very different density of electronic states depending on the dimensionality of the systems, which can be exploited for specific application, for example to enhance the efficiency of thermoelectric energy conversion [5, 6]. However, such direct consequences of quantum confinement become appreciable only when the confined dimension of the nanostructure is reduced below a certain threshold, typically of the order of few nanometers, which may be difficult to attain [7].

In non-metallic systems the main heat carriers are phonons, i.e. quantized lattice vibrations. Analogously to the electronic structure, also the phononic properties of materials and, as a consequence, their thermal properties are deeply affected by dimensionality reduction [8]. The symmetry and dimensionality of nanostructures determine their phonon density of states, dispersion relations and the selection rules for scattering processes, thus impacting their heat capacity and their thermal conductivity. However, the way in which dimensionality affects thermal conductivity has not been fully clarified. On the one hand, following the predictions for non-linear models , e.g. the Fermi-Pasta-Ulam model, one would expect that the thermal conductivity of one- and two-dimensional nanostructure would be very large or even divergent [9–13]. On the other hand, from changes in phonon dispersion relations and from the growing impact of surface scattering one may argue that dimensionality reduction hampers thermal transport [14], by limiting phonon mean free paths below the characteristic size of the nanostructure, be it the diameter of a nanowire, or the thickness of a thin film or a membrane, which imposes the so called "Casimir limit" [15]. These two scenarios stem from different standpoints but are not incompatible: only few materials are truly one or two-dimensional and can compare to ideal statistical physics models, and at the same time the application of Casimir limit to phonon scattering in nanostructures is oversimplified.

Experiments on different systems show a complex reality, in which dimensionality reduction may either boost or limit thermal transport depending on the systems and the configuration of the measurements. Very high, even possibly diverging, thermal conductivity, and extremely long phonon mean free paths were measured in graphene [16–20], and carbon nanotubes [21, 22], whereas a considerable suppression of thermal conductivity was observed in silicon nanowires [23–25], thin films and membranes [26–30]. Nevertheless the measurements of thermal conductivity in nanostructures are very challenging, and yield large uncertainties, as they are very sensitive to the experimental conditions. For example the actual value of the thermal conductivity of suspended graphene is still debated and experimental estimates range from 2000 to $8000\,\mathrm{W\,m^{-1}\,K^{-1}}$. In addition, even if techniques for a spectroscopic characterization of thermal transport have been recently developed [31, 32], so far they could not directly ascertain the origin of the observed enhancement or suppression of thermal conductivity. In this scenario molecular simulations of nanostructures emerge as powerful tools to bridge the gap between simple models and experiments. The main advantage of molecular modeling is that complexity and details can be gradually introduced into models, thus realizing "gedanken experiments" that allow one to ascertain the origin of complex physical phenomena.

In the next section the simulation approaches to heat transport in bulk and low-dimensional systems are briefly outlined. The application of such simulation methods to elucidate the effects of dimensionality on thermal transport in carbon and silicon nanostructures are reported in Sects. 7.3 and 7.4, respectively. Section 7.5 summarizes the main findings and suggests future perspectives.

7.2 Simulation Tools

Several atomistic simulation tools are available to investigate thermal transport in nanostructures and compute the thermal conductivity (κ) of materials. Simulation methods can be sorted into two classes, namely Lattice Dynamics (LD) and Molecular Dynamics (MD). It is often useful to combine distinct complementary approaches, as they involve different approximations and limitations, which make them suitable to probe different transport regimes in systems of various size and complexity. Combining different methods sheds light on a broader variety of aspects of nanoscale heat transport. Hereafter the main-stream LD and MD approaches are briefly described.

7.2.1 *Anharmonic Lattice Dynamics*

In the harmonic approximation the normal modes of vibrations of a periodic system of N particles are determined by diagonalizing the *dynamical matrix* $\mathbf{D}$ for each momentum vector $\mathbf{q}$:

$$\mathbf{D}(\mathbf{q})\mathbf{e}_\lambda(\mathbf{q}) = \omega_\lambda^2(\mathbf{q})\mathbf{e}_\lambda(\mathbf{q}), \tag{7.1}$$

which provides the frequencies $\omega_\lambda(\mathbf{q})$ and the normalized displacement vectors $\mathbf{e}_\lambda(\mathbf{q})$. $\mathbf{D}(\mathbf{q})$ is defined as:

$$D_{ij}(\mathbf{q}) = \frac{1}{\sqrt{m_i m_j}} \frac{\partial^2 \Phi}{\partial x_{i,\alpha} \partial x_{j,\beta}} \exp(i\mathbf{r}_{ij} \cdot \mathbf{q}) \tag{7.2}$$

where Φ is the potential energy of the system at equilibrium, m_i is the mass of atom i, and $\mathbf{r_{ij}}$ the distances between pairs of atoms i and j, and α and β indicate the Cartesian components. In a system of N particles $\mathbf{D}$ is then a $3N \times 3N$ matrix. In a first principles framework, the matrix elements of $\mathbf{D}$ can be computed either by finite differences or by density functional perturbation theory [33].

Considering that quantized lattice vibrations can be treated as particles, their propagation in the diffusive regime can be described using the Boltzmann transport equation (BTE), in analogy with the diffusion of a gas:

$$\frac{\partial n_\lambda}{\partial t} + \mathbf{v}_\lambda \cdot \nabla n_\lambda = \left(\frac{dn_\lambda}{dt}\right)_{scattering}, \tag{7.3}$$

considering, however, that the phonons are spin-less quantum particles that obey to Bose-Einstein statistics. The righthand side of (7.3) accounts for the scattering processes that create or destroy phonons: namely, anharmonic scattering processes, isotopic, defect and boundary scattering. The non-equilibrium occupation function can be written as $n_\lambda = n_\lambda^0 + \delta n$, and, assuming small temperature gradients (∇T), one can linearize (7.3) by treating δn perturbatively. In stationary conditions the first term of (7.3) vanishes, and the linearized BTE is written as:

$$\mathbf{v}_\lambda \cdot \nabla T \frac{\partial n_\lambda^0}{\partial T} = \left(\frac{dn_\lambda}{dt}\right)_{scattering}. \tag{7.4}$$

The linearized BTE can be solved at different levels of accuracy and complexity. The simplest approach is to calculate the life time of each phonon, assuming that the population of all the other modes is the one at equilibrium (n_λ^0) [34]. The resulting expression for κ is the sum over the contribution of all the phonon modes, integrated to convergence over the first Brillouin Zone of the system, sampled with a grid of $N_\mathbf{q}$ q-points:

$$\kappa = \frac{1}{N_\mathbf{q}} \sum_{\lambda,\mathbf{q}} \kappa_\lambda(\mathbf{q}) = \frac{1}{N_\mathbf{q}} \sum_{\lambda,\mathbf{q}} C_\lambda(\mathbf{q}) \mathrm{v}_\lambda^2(\mathbf{q}) \tau_\lambda(\mathbf{q}) \tag{7.5}$$

where C_λ is the heat capacity per unit volume of each vibrational state, v_λ is the component of the group velocity in the direction of transport and τ_λ is the phonon lifetime. Even if it is approximated, this is a very useful expression that allows one to resolve the contribution to κ of each phonon branch at each frequency ω.

Heat capacity and group velocities are usually obtained from the dispersion relations in harmonic approximation, though anharmonic corrections are also possible. To compute the phonon lifetimes τ_λ, one has to consider all the scattering processes occurring in materials, namely phonon-phonon anharmonic scattering (normal and Umklapp), boundary and defects scattering (Fig. 7.1). In a perfectly crystalline material, neglecting electron-phonon interactions, the only possible phonon scattering mechanism is anharmonicity, and the main contribution comes from 3-phonon processes. Two viable processes exist: either two phonons (ω_1,ω_2) annihilate into a third one (ω_3), or one phonon (ω_1) decays into two phonons (ω_2,ω_3). Energy and momentum conservation determine the selection rules for three-phonon scattering:

$$\omega_1(\mathbf{q}) \pm \omega_2(\mathbf{q}') - \omega_3(\mathbf{q}'') = 0 \tag{7.6}$$

$$\mathbf{q} \pm \mathbf{q}' - \mathbf{q}'' = \mathbf{Q} \tag{7.7}$$

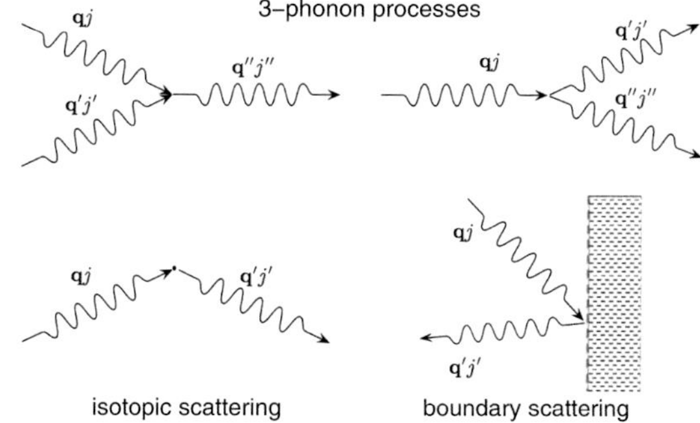

Fig. 7.1 Phonon scattering mechanisms in a crystalline thermal conductor (adapted from [39]). Copyright (2013) by The American Physical Society

where **Q** is a reciprocal lattice vector. Normal processes imply $\mathbf{Q} = 0$, while in Umklapp processes **Q** is finite. Only Umklapp scattering processes dissipate energy thus contributing to limit κ. Computing anharmonic scattering rates and phonon lifetimes (τ_{anh}) requires the knowledge of the third derivatives of the interatomic potential.

In the single mode relaxation time approximation (SMRTA) phonon lifetimes are expressed as the inverse of spectral line width [35, 36], treating on equal footing normal and Umklapp processes, which in fact contribute to scattering heat carriers with substantially different weights. In this approach the contribution from defect and boundary scattering, τ_{def} and τ_B, add up to τ_{anh} through Matthiessen's rule:

$$\frac{1}{\tau} = \frac{1}{\tau_{anh}} + \frac{1}{\tau_B} + \frac{1}{\tau_{def}}. \tag{7.8}$$

While SMRTA has been widely used and can be implemented efficiently to treat large nanostructured systems [37], it may result highly inaccurate at temperatures much lower than the Debye temperature, where normal scattering processes dominate, as it is the case for carbon based materials at room temperature, and especially for systems with reduced dimensionality [38, 39]. One can show that SMRTA provides a lower boundary to the thermal conductivity obtained by solving exactly the linearized BTE.

Methods to solve exactly the linearized BTE have been proposed, either using a self-consistent iterative numerical approach [40], or by minimizing a variational functional [39]. Exact approaches have demonstrated to be accurate and predictive for both bulk crystals [38, 41] and nanostructures with low dimensionality [42–45], especially when combined with *ab initio* calculations of the harmonic and anharmonic force constants. So far the exact solution of the BTE has been applied only to crystalline systems with relatively small unit cells, also because the *ab initio* calculations are computationally very demanding. In turn the variational BTE approach has proved extremely powerful to ascertain the effect of reduced dimensionality on heat transport in graphene and two-dimensional materials [44, 45].

7.2.2 Equilibrium Molecular Dynamics

Molecular Dynamics (MD) is a method designed to compute the properties of solids or liquids by taking thermodynamic averages over a trajectory obtained integrating the classical equations of motion of the particles in the simulation box [46]. In the simplest case MD simulations are performed at equilibrium conditions in the microcanonical ensemble, i.e. at constant energy, volume and number of particles. As the typical size scale of MD simulations goes from few hundreds to millions of atoms, to represent extended (bulk) systems periodic boundary conditions are applied to the simulation cell. Linear response and transport coefficients can be calculated from the fluctuations of the respective conjugate flux via Green-Kubo relations [47]. In the case of heat transport, κ is calculated from fluctuations of the heat flux via the heat flux autocorrelation function (HFACF). For a system in equilibrium, in the absence of a temperature gradient, the net heat current averages to zero over time, but the integral of its correlation function is finite and proportional to its thermal conductivity. The Green-Kubo expression for each component of the thermal conductivity tensor can be written as

$$\kappa_{\alpha\beta} = \frac{1}{k_B T^2} \lim_{t\to\infty} \lim_{V\to\infty} \frac{1}{V} \int_0^t \langle J_\alpha(t') J_\beta(0) \rangle dt', \tag{7.9}$$

where $\mathbf{J}$ is the heat flux, k_B is Boltzmann's constant, T is the system temperature and V its volume. In systems with anharmonic interactions the HFACF should decay to zero for large t, and its integral should saturate at a constant value. In practice, at long times the HFACF becomes noisy due to poor statistical sampling, and the integral drifts or presents large oscillations because of statistical noise. Therefore, $\kappa_{\alpha\beta}$ is taken as the stationary value of (7.9) before it starts drifting due to the accumulated numerical noise.

While κ is a second-order 3×3 tensor, one is normally interested in the diagonal components. In the case of 1D materials, such as nanotubes and nanowires, the system can be oriented in such a way that only one component is not zero. In 2D systems there are only two independent non-vanishing components κ_{xx}, κ_{yy}, which may be equivalent by symmetry, for example in the case of graphene. For 3D materials with cubic symmetry, like bulk silicon or diamond, there are three equivalent non-zero components $\kappa_{xx} = \kappa_{yy} = \kappa_{zz}$.

The advantage of equilibrium MD is that one does not have to make any assumption on the type of phonon scattering and all orders of anharmonicity are taken into account. In addition, with empirical potentials one can simulate relatively large systems, up to 10^7 particles, with no specific requirements of being crystalline: amorphous, polycrystalline, defective and liquid systems can be studied. On the other hand, convergence of the numerical integration of Eq. (7.9) in time and size needs to be checked thoroughly, and, depending on the material under investigation, it may occur only for very large samples and for fairly long time-scales. In addition, MD is based on Newtonian dynamics and quantum effects are not taken into account, thus it cannot provide quantitative predictions on κ at temperatures much lower than the Debye temperature (Θ_D).

7.2.3 Non-equilibrium Molecular Dynamics

Another, perhaps more intuitive way of using MD is to simulate stationary non-equilibrium conditions. One defines two regions of the simulation cell as heat source and heat sink, and generates an energy flux between them, through a part of the system where the atoms dynamics is unperturbed. Following Fourier's law, at stationary conditions a temperature gradient ∇T, proportional to the energy flux J, is established, and the thermal conductivity is given by the proportionality constant:

$$\kappa = -\frac{J}{\nabla T}. \tag{7.10}$$

A typical setup for NEMD simulations, is shown in Fig. 7.2. There are several ways of controlling the temperature in the thermal reservoirs. An option is to apply thermostats: stochastic local thermostats, e.g. Langevin, are preferable, as one would prefer to have fully phonon-absorbing reservoirs, to avoid that phonons travel through the heat sink uncontrollably if periodic boundary conditions are applied [48]. Alternatively one can apply the reverse-NEMD approach, in which the sta-

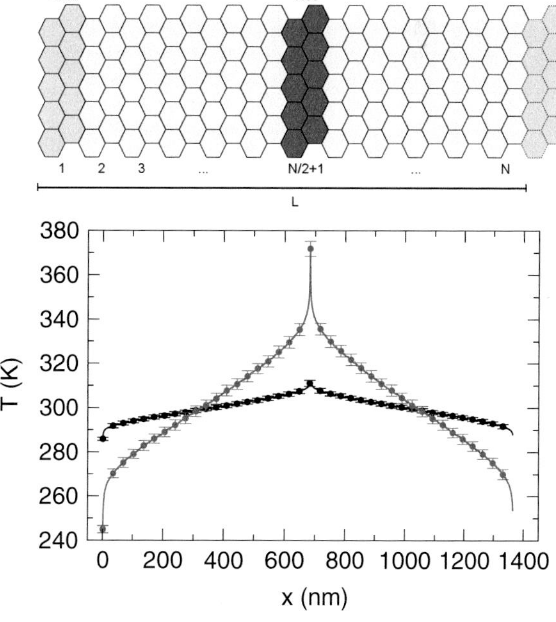

Fig. 7.2 Setup of a non-equilibrium molecular dynamics simulation (*top panel*) and temperature profile for a graphene patch 1.4 μm long with different imposed heat flux at stationary conditions (*bottom panel*)

tionary heat flux is induced by swapping the momenta of particles between the heat source and sink. This is a particularly elegant method, since, together with a stationary heat flux, one automatically achieves energy and momentum conservation [49].

NEMD simulations are closer in realization to experimental measurements, and are suitable to probe the onset of non-linear effects due to large temperature gradients. In turn, size scaling of NEMD to obtain the thermal conductivity of extended systems is tricky, as phonon mean free paths are truncated at the distance l between the heat source and sink. One needs to perform simulations for a set of values of l and extrapolate for l going to infinity, as convergence can rarely be achieved. The standard way of extrapolating assumes that the inverse of $\kappa(l)$ is linear with $1/l$:

$$\frac{1}{\kappa(l)} = \frac{1}{\kappa_\infty} + \frac{C}{l} \tag{7.11}$$

so that the intercept of the linear fit gives $1/\kappa_\infty$. In fact this extrapolation scheme assumes that the distribution of phonons that mostly contribute to heat transport have a narrow distribution of mean free paths, which is often not the case [50]. This feature of NEMD can also be exploited to resolve the relative contribution of phonons with a certain mean free path to the total κ of a material, in accordance with thermal conductivity spectroscopy experiments [31].

If size scaling is performed correctly NEMD and equilibrium MD should provide results in agreement within statistical uncertainty, as it was demonstrate for several bulk systems [51–53]. More rigorously, it was demonstrated by renormalization group analysis and mode coupling theory that for momentum conserving systems there is a strict correspondence between the decay of the heat flux autocorrelation function, $C_{jj}(t) \propto t^{-\beta}$, in Eq. (7.9) and the function $\kappa(l)$ in NEMD [11, 54]. Although consensus has not been reached on the exact values of β it is now well accepted that in three dimensions that $\beta > 1$, in two dimensions $\beta = 1$ and in one dimension $\beta < 1$. These exponents imply that in one-dimensional systems $\kappa(l)$ diverges like l^α with $\alpha = 1 - \beta$, in 2D $\kappa(l) \propto log(l)$, and in 3D κ is finite [10, 13, 55, 56]. This picture, resulting from the combination of analytical calculations and numerical simulations on lattice models, has stimulated further investigations on real systems but has not been confirmed either by experiments or by simulations of nanostructures.

7.2.4 Empirical Interatomic Potentials

On crystalline systems with few atoms in the unit cell it is possible to perform *ab initio* anharmonic lattice dynamics calculations using density functional perturbation theory and the $n+1$ theorem [33, 39, 41]. On the other hand, given the large size and time scales required for the convergence of MD simulations, it is usually impossible or impractical to combine MD and first-principles approaches to compute thermal transport in nanostructures. For this reasons one has to rely on empirical potentials, whose accuracy and transferability need to be probed.

For the materials considered in this book Chapter, mostly Carbon and Silicon, a few analytical empirical models have been shown to provide reliable estimates of the harmonic and anharmonic vibrational properties and of the total thermal conductivity of the most common bulk crystalline polymorphs. The most used many-body potentials for MD simulations of carbon nanostructures, such as graphene and carbon nanotubes, assume the analytical forms proposed by either Tersoff [57] or Brenner, suitably re-parametrized to reproduce correctly the phonon dispersion relations of graphene [58]. The Tersoff potential provides a sufficiently reliable estimate of the thermal conductivity of silicon and other semiconductors [53]. Alternatively the Stillinger-Weber potential has been extensively used to study thermal transport in silicon-based systems [51], and the environment dependent interatomic potential (EDIP) have been shown to provide an estimate of the thermal conductivity of bulk silicon very close to the experimental one [59].

7.3 Carbon Based Nanostructures

The rich variety of nanostructures that can be formed from graphitic carbon provides an ideal platform to probe the effect of dimensionality on lattice thermal transport, and hopefully compare directly to non-linear models. Thermal transport in graphene and carbon nanotubes has been studied extensively searching for cases of breakdown of Fourier's law and divergence of thermal conductivity with size [19–21]. While experimental measurements highlighted extremely high thermal conductivity and phonon mean free paths exceeding tens of nanometers, they could not provide a final assessment on the divergence of κ in graphene and CNTs, due to the limited size of the samples, the presence of unavoidable intrinsic defects, and technical issues, such as, for example, contact resistance. Atomistic simulations, using a combination of the techniques discussed in the previous section, are thus necessary to rationalize the experimental results. In the following subsection we discuss heat transport in graphene and in a (10,0) CNT, which is however representative of the general case of single wall CNTs. Atomistic models of these systems are shown in Fig. 7.3.

7.3.1 Graphene

Being a single sheet of atoms, atomically flat, graphene is the only truly two-dimensional material that may be compared directly to non-linear two-dimensional models. However, graphene atoms can vibrate perpendicularly to the plane and form ripples. At finite temperature ripples confer stability to graphene impacting its thermal properties: for example negative thermal expansion coefficients are a consequence of out-of-plane vibrations and ripples.

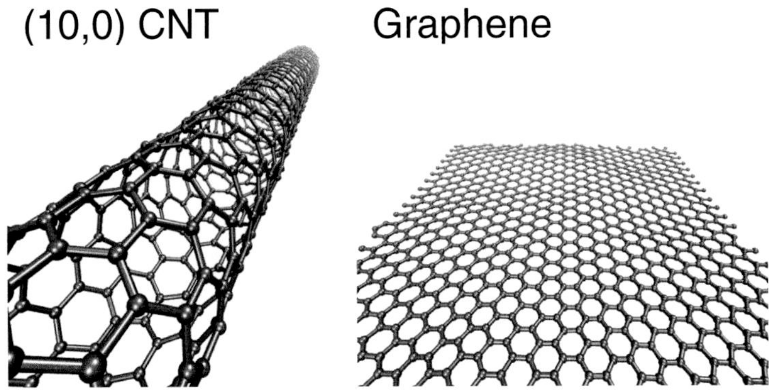

Fig. 7.3 Atomistic models of a (10,0) single wall carbon nanotube and of graphene

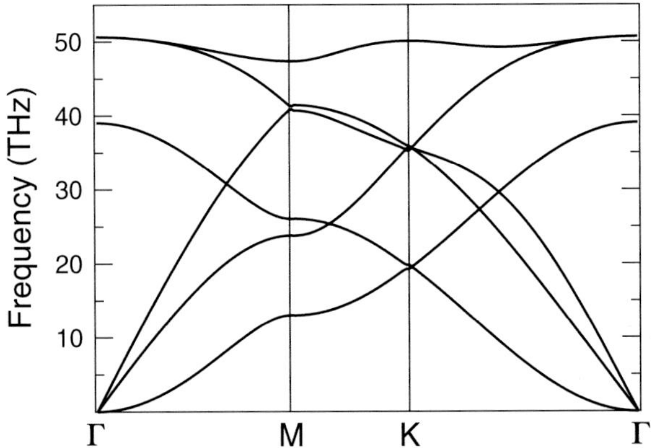

Fig. 7.4 Phonon bands of graphene computed with an optimized parameterization of the Tersoff potential [58]

The phonon spectrum of graphene (Fig. 7.4) resembles very closely the one of graphite [60]. The low frequency spectrum is characterized by a longitudinal in-plane mode with very high group velocity, a stiff in-plane transverse acoustic mode, and a softer flexural mode with quadratic dispersion near the Γ-point ($\omega \propto q^2$). The acoustic modes extend to frequencies as high as 42 THz, and are complemented at higher frequency by three optical modes that give rise to the extensively studied Raman peaks of sp^2 carbon.

The earliest successful model of heat transport in graphite [61] was based on a Debye approach for a two-dimensional gas of phonons with lower cutoff frequency of 4 THz. Since inter-planar interactions are neglected the model applies seamlessly also to graphene. Klemens and Pedraza neglected also out-of-plane vibration with the argument that these modes have very low group velocity and would not carry a significant amount of energy. This model provided reasonable estimates of κ in good, yet fortuitous agreement with experiments. More recent calculations based on a self-consistent solution of the BTE indeed demonstrated that flexural out-of-plane modes provide a large contribution to the thermal conductivity of graphene, due to their relatively high density at low frequency and their very long mean free path [62].

Several studies addressed thermal transport in graphene by MD simulations providing estimates of κ at room temperature from few hundreds to few thousands $\mathrm{W\,m^{-1}\,K^{-1}}$. Such variation of MD results stems from the sensitivity of κ to the functional form and the parameterization of the empirical potentials and from the difficulties in converging MD simulations both in terms of time and size. In addition any quantitative estimate of κ of carbon-based materials at room temperature would suffer badly from the lack of quantum effects. Even if, given the very high Debye temperature of carbon-based materials, MD results can only be considered qualitative, equilibrium and non-equilibrium simulations provide useful insight in the physics of thermal transport in graphene. Equilibrium MD simulations show that κ of graphene converges from above as a function of the size of the simulation cell, thus underlying the importance of sampling well low-frequency flexural modes, which assume a fundamental role in scattering heat carriers [63]. A calculation of the heat flux autocorrelation function, well converged in terms of sampling, shows that for all the sizes of models considered it decays faster than the inverse of time, which makes the argument in Eq. (7.9) integrable, indicating that the system has a finite thermal conductivity. Since the only conceptual difference between graphene models and two-dimensional non linear lattice models, for which κ diverges logarithmically, is that in graphene the atoms can move out-of-plane, we performed an equilibrium MD simulation, in which the carbon atoms were frozen in-plane. This fully two-dimensional model exhibits $1/t$ decay of the heat flux autocorrelation function, indicating that diverging κ is restored [63] (Fig. 7.5).

While equilibrium MD simulations suggest that the thermal conductivity of graphene in the thermodynamic limit is finite, they do not provide direct information on the dependence of κ on the size of the systems. The very large variation of

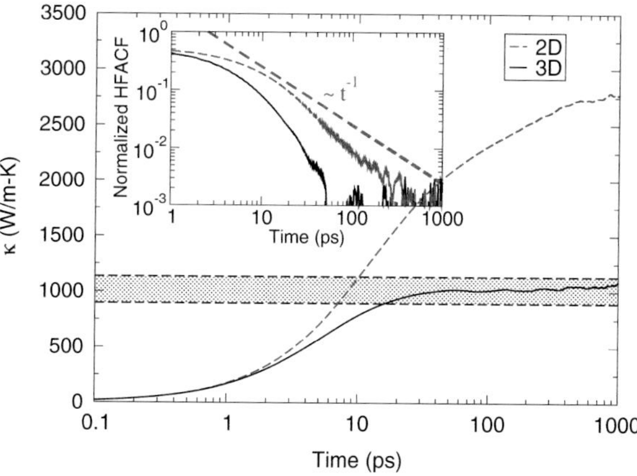

Fig. 7.5 Comparison of the heat flux autocorrelation functions (*inset*) and of their integrals (main graph) between a model of graphene with three-dimensional degrees of freedom (*solid black lines*) and one, in which motion is confined in-plane (*dashed blue lines*), from [63]. Copyright (2013) by The American Physical Society

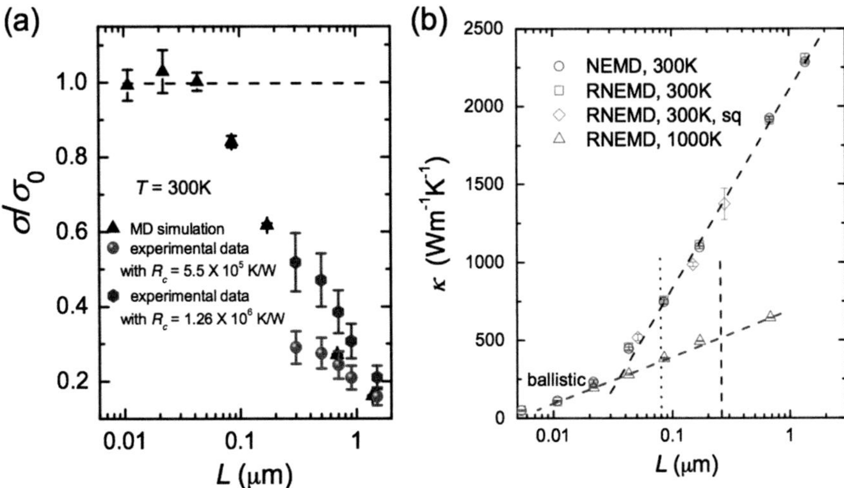

Fig. 7.6 (**a**) Normalized conductance of graphene as a function of length in non-equilibrium MD simulations and in experiments (**a**). Experimental data assuming two different contact resistances are reported. Transport is ballistic as long as $\sigma/\sigma_0 \sim 1$. (**b**) Thermal conductivity of graphene from non-equilibrium MD as a function of the size of the graphene model at T = 300 and 1000 K. At both temperature κ is not saturating for models 1.5 μm long. Different MD methods give consistent results. Adapted from [20]

the experimental estimate of κ, especially close to room temperature, may be partly justified by the different thermal conductivity as a function of the length of the graphene patches measured. Non-equilibrium MD simulations performed along with systematic experimental measurements showed that thermal transport in graphene at 300 K is ballistic up to ~100 nm (Fig. 7.6a). Simulations of larger models exhibit an apparently logarithmic divergence of the thermal conductivity both at room temperature and at 1000 K (Fig. 7.6b), in agreement with measurements that show similar diverging trend up to 10 μm at room temperature [20]. Non-equilibrium MD simulations of even larger systems, up to 100 μm, suggest that κ saturates for the largest models considered [64]. Similarly, calculations performed solving the BTE self-consistently confirm the size dependence of κ in the micrometer regime, but show that for larger lengths (L) κ is not proportional to $\log(L)$ [43]. This study also underscored the substantial contribution of out-of-plane modes with long mean free paths to the total thermal conductivity: for example such contribution would amount to 75 % for a 10 μm long graphene patch.

The application of a variational approach to BTE, which removes critical instabilities inherent in the self-consistent solution for two-dimensional systems [39], finally resolved the debate on anomalous heat diffusion and the divergence of κ in graphene. The exact solution of the BTE shows that the thermal conductivity of graphene converges in the thermodynamic limit [44], in substantial agreement with equilibrium MD results, but convergence occurs for samples of the order

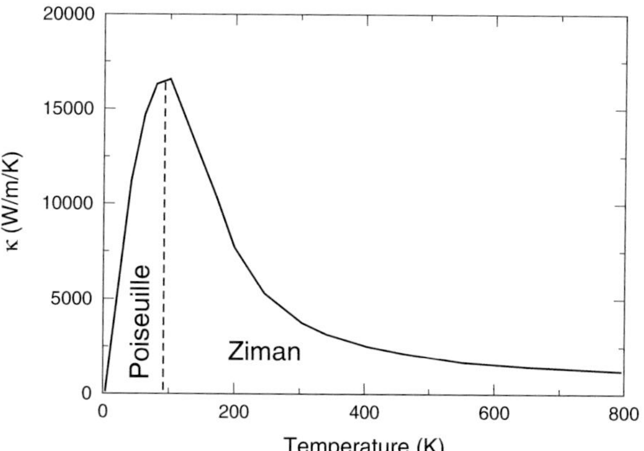

Fig. 7.7 Illustration of the different transport regimes in a graphene ribbon with a width of 100 μm. Data from [45]

of 1 mm. The reason is that the main heat carriers are not single phonons, but collective excitations with mean free paths of the order of 100 μm. This approach sheds new light on heat transport in two-dimensional systems, which is substantially different from three-dimensional materials. Due to the reduced dimensionality of the dual space, the selection rules for three-phonon scattering related to momentum conservation (Eq. (7.7)) lead to a predominance of Normal processes over Umklapp processes. While in standard solids it occurs only at cryogenic temperatures, e.g. $\sim$20 K for Silicon, in layered materials this condition is verified over a broad temperature range, up to 800 K, leading to hydrodynamic phonon transport [45, 65]. In graphene, as well as in other two-dimensional systems, such as graphane, hexagonal BN and MoS_2, the balance between non-dissipative processes and resistive ones dictates a Poiseuille regime at low temperature, before a peak of conductivity is reached, and a Ziman regime occurs at higher temperature (Fig. 7.7). In both cases normal processes dominate. In Poiseuille flow normal scattering shifts the phonon distribution, and heat flux is eventually dissipated by extrinsic events, for example boundary scattering, whereas in the Ziman regime the shifted phonon distribution dissipates heat relaxing to equilibrium via Umklapp and isotopic scattering. This scenario has to be compared to that occurring in standard three-dimensional solids, in which transport regimes evolve from ballistic to diffusive in a narrow temperature window of few tens of K.

7.3.2 *Carbon Nanotubes*

Momentum conserving one-dimensional models exhibit anomalous thermal transport with length-dependent thermal conductivity $\kappa \propto L^{\alpha}$ [11]. Carbon nanotubes, which can be grown up to tens of μm, were proposed as the most promising systems, for which anomalous heat conduction can be observed. Experiments agree that κ is of the order of thousands W m^{-1} K^{-1}, and is length dependent for lengths of the order of μm, thus supporting the hypothesis of a breakdown of Fourier's law [19, 21]. However, in the light of the recent theoretical studies on graphene reported in the previous section, it is likely that experimental observations of anomalous heat conduction may be explained in terms of collective excitations and hydrodynamic transport regime. While this issue has not been clarified yet, it is worth analyzing the available results of lattice dynamics and molecular dynamics simulations, mostly focusing on single wall carbon nanotubes (SWCNT).

Ideal SWCNTs have cylindric symmetry and extend periodically in one dimension. Such geometry yields four invariances, namely three translations and one free rotation along the tube axis, which result in four acoustic modes. Two of them have quadratic dispersion ($\omega \propto q^2$) near the Γ point and correspond to transverse sound waves. The longitudinal mode retains linear dispersion relation and has a group velocity higher than the torsional mode associated with the axial rotation, which has also $\omega \propto q$ [66] (Fig. 7.8). These features are independent on the chirality of the nanotubes, yet thinner wires have stiffer longitudinal modes and softer transverse flexure modes. The large number of atoms in the unit cell gives rise to a large number of optical "breathing-like" modes, which have nevertheless large group velocities at finite q-points.

The contributions from acoustic and higher order breathing modes make the thermal conductivity of CNTs very high, as it was observed both in experiments and in simulations [22, 67–69]. The reason for the extremely high thermal conductivity of CNTs is also related to their symmetry and one-dimensional periodicity. The selection rules for phonon-phonon interactions in a one-dimensional Brillouin zone limit the possibility of dissipative scattering (Umklapp), thus leading to very low scattering rates for flexural modes, and consequently to large κ [42] (Fig. 7.9).

The possibility that heat transport would be anomalous [21], implying divergence of κ due to the low dimensionality of nanotubes, was explicitly addressed by MD simulations, which, however, provided contradicting results. Equilibrium MD simulations suggest a scenario similar to the one drawn for graphene, exhibiting integrable heat flux autocorrelation functions [70, 71]. On the other hand non-equilibrium simulations suggest anomalous heat conduction [72]. Such discrepancy may possibly arise from non linear effects related to the large temperature gradients that are necessarily used in non-equilibrium simulations. So high temperature gradients may also correspond to those induced in experiments, in which 100 K differences over few micrometers are not uncommon. Simulations show that in these conditions thermal energy is transmitted efficiently by low-frequency mechanical waves that get coherently excited [73]. In the light of the recent discovery of hydrodynamic transport in graphene [45, 65], it is also likely that similar phenomena happen in carbon nanotubes, thus leading to anomalous transport over the typical length scales adopted in simulations. However, this hypothesis is yet to be probed.

A consequence of the low dimensionality of CNTs as well as of graphene is that thermal transport is extremely sensitive to any perturbation of the perfect crystallinity of these systems. MD simulations show that the thermal conductivity of carbon nanotubes is largely reduced by topological (Stone-Wales) and point defects. These defects affect the mean free paths of medium and high frequency phonons,

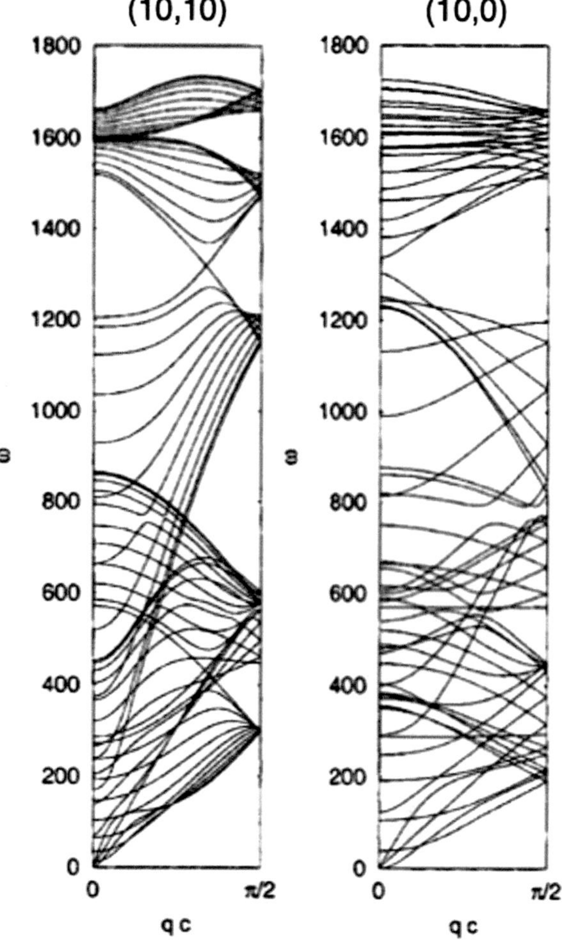

Fig. 7.8 Phonon dispersion relations of a (10,10) and of a (10,0) single wall carbon nanotube, adapted from [66]. Copyright (2004) by The American Physical Society

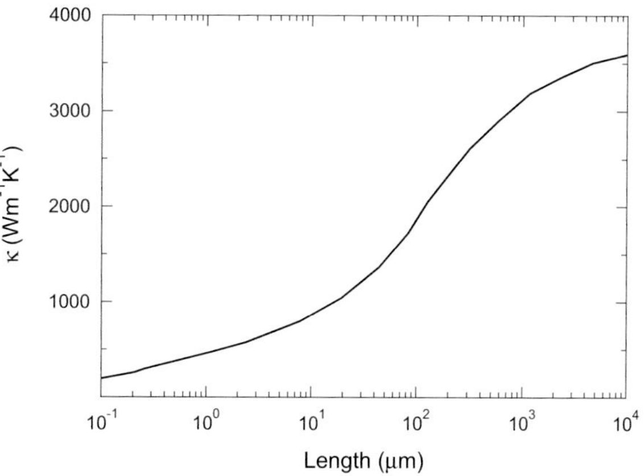

Fig. 7.9 Thermal conductivity of a (10,10) single wall carbon nanotube calculated by solving self-consistently the Boltzmann transport equation. Data from [42]

while low frequency phonons are still transmitted ballistically over micrometer lengths. As a result κ is much more sensitive to the concentration of defects rather than to their atomistic structure, and converges to similar values for different types of defects [74]. κ at room temperature can be reduced up to ten times by high concentrations of vacancies or di-vacancies. On the other hand the interaction either with other nanotubes, as in a CNT network, with a fluid medium or with a substrate dramatically affects the propagation of low frequency modes. Depending on the type of interaction with the substrate κ can be also reduced about ten-fold [70], with respect to that of suspended CNTs. Extremely low κ was predicted for CNT networks and pellets, however, in this case, heat transport is controlled by the very high contact resistance between two different CNTs, by the length of the CNTs, and eventually by the topology of the network [75].

7.4 Nanostructured Silicon

Nanoscale silicon has been investigated thoroughly under several aspects, including thermal transport, due to its capital importance for technology, especially for applications in electronics and energy conversion. The constant reduction of the size of transistors, which has rapidly followed Moore's law down to few atomic layers, has made thermal dissipation at the nanoscale a crucial issue in nano-electronics. At the same time, finding a reliable and reproducible way of reducing the thermal conductivity of silicon without hampering its bulk electronic properties would enable silicon-based thermoelectric devices. Huge reduction of the thermal conductivity was observed for silicon nanostructures with reduced dimensionality, both for nanowires (1D) and for ultra thin films and membranes (2D). The general understanding is that κ is reduced by diffusive surface scattering, which becomes more and more effective the smaller the diameter of nanowires or the thickness of two-dimensional structures. However, this rather simplistic picture does not reconcile with the large enhancement of κ predicted for non-linear models, confirmed to a certain extent by experiments on carbon nanostructure.

Atomistic simulations can solve the conundrum, as they make it possible to disentangle different effects connected to nanostructuring. For example it is possible to check the effect of dimensionality reduction by simulating heat transport in ideally crystalline systems with diameter or thickness still unattainable to experiments. In this respect it is worth noting that crystalline silicon nanowires with diameter as small as 1 nm were produced by oxide-assisted growth followed by etching [76] and membranes as thin as $\sim$8 nm were obtained via advanced lithographic processes [77]. However, their surface structure is subject to reconstruction, oxidation and roughness (see Fig. 7.10), which cannot be easily controlled during fabrication or simply upon exposure to standard environmental conditions, and measuring thermal transport in these systems is extremely challenging.

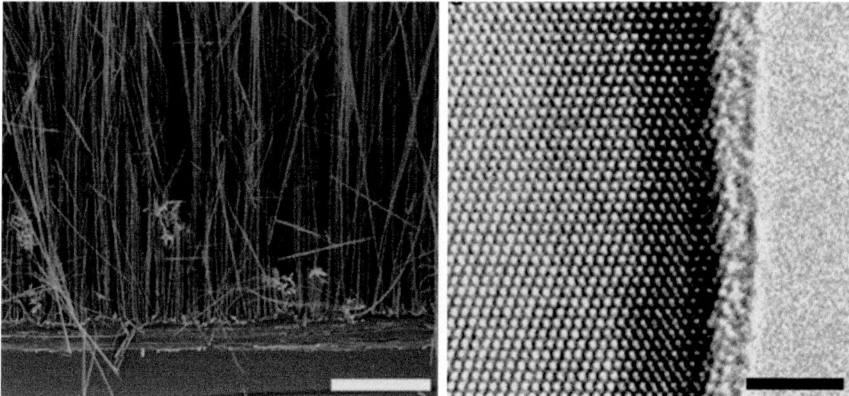

Fig. 7.10 SEM (*left panel*) and TEM (*right panel*) silicon nanowires produced by vapor-liquid-solid growth adapted from [24]. The scale bar in the SEM image (*left*) corresponds to 10 μm and the one in the TEM image (*right*) to 3 nm. The TEM image shows that the wire is crystalline but exhibits a disordered surface oxide layer about 4 atomic layers thick

7.4.1 Silicon Nanowires

Thermal transport in silicon nanowires was initially studied to probe whether silicon nanostructures could be employed as high-performance thermoelectrics. Assuming that phonons would be scattered diffusively by surfaces, a significant reduction of phonon mean free path was expected, leading to highly reduced thermal conductivity. Early predictions from simplified kinetic models, assuming that phonon mean free paths are chopped by diffuse surface scattering, were actually confirmed by the first equilibrium MD simulations, in which the interaction between silicon atoms was modeled using the Stillinger-Weber potential [78]. In this pioneering study the authors found that the thermal conductivity of model silicon wires with frozen boundary conditions, i.e. constrained surface atoms, is up to two orders of magnitude lower than the reference bulk value and is directly proportional to the diameter of the wires. The observation that κ is temperature independent over a broad range of T, and that a model based on the BTE matches well the MD data lead to the conclusion that surface scattering is the key player in determining thermal transport in thin silicon nanowires. The measurements that followed, even though on wires with much larger cross section, confirmed the trends predicted by simulations [23]. Nevertheless these early models do not feature realistic surface structuring, nor are the changes in phonon dispersion relations properly taken into account.

In examining the effect of dimensionality reduction in thin silicon nanowires, one should first consider how the phonon dispersion relations change with respect to the bulk reference for ideally crystalline wires, so that surface effects are ruled out completely. Let us consider for example three crystalline SiNWs with diameter of 1.1, 2 and 3 nm grown in (001) direction. The surface reconstruction of these wires, shown in Fig. 7.11, was formerly optimized using density functional theory

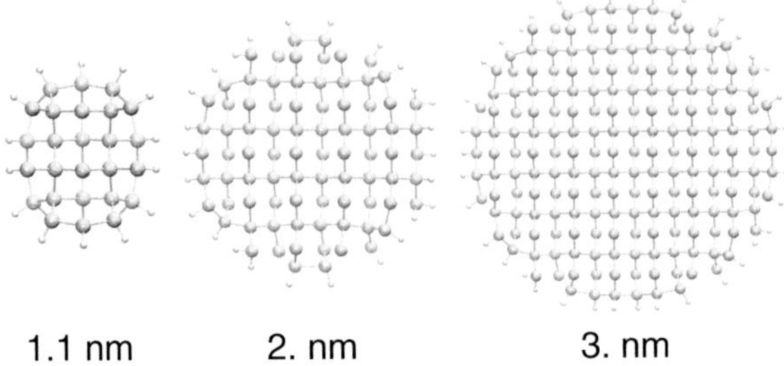

Fig. 7.11 Crystalline thin silicon nanowire models with hydrogenated and ideally reconstructed surfaces [79]

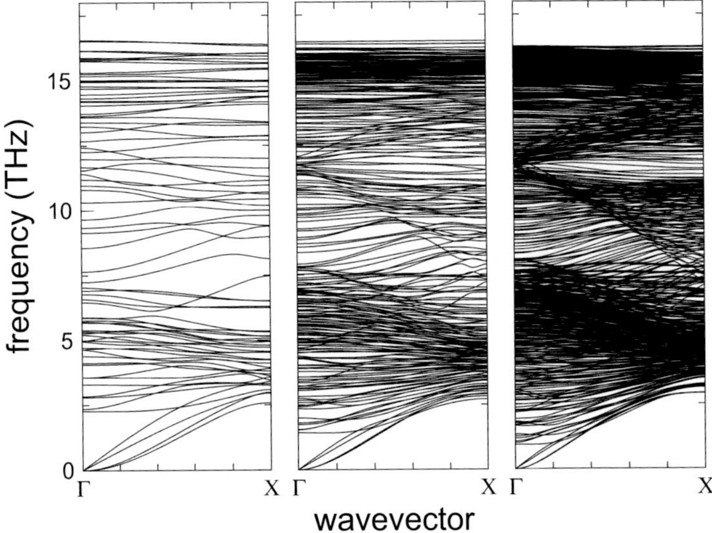

Fig. 7.12 Phonon dispersion relations of crystalline silicon nanowires with diameter of 1.1 nm (*left panel*), 2 nm (*center*) and 3 nm (*right*). High frequency Si-H bending and stretching modes are not shown

[79], and was proven to be stable also when the interaction between silicon atoms are modeled using the Tersoff bond-order interatomic potential.

The phonon dispersion relations of these systems computed by harmonic lattice dynamics using Eq. (7.1) are shown in Fig. 7.12. As in the case of carbon nanotubes, the most immediate consequence of dimensionality reduction is that nanowires exhibit four acoustic modes, corresponding to three translational and one rotational invariant transformations. The two transverse acoustic modes, also dubbed flexural modes, have quadratic dispersion in the vicinity of the Γ point. Calculations show that the thinner the wire, the softer these flexural modes are, and their dispersion relations remain quadratic for larger wavevectors. The other two acoustic modes have linear dispersion in q and correspond to torsional and longitudinal waves,

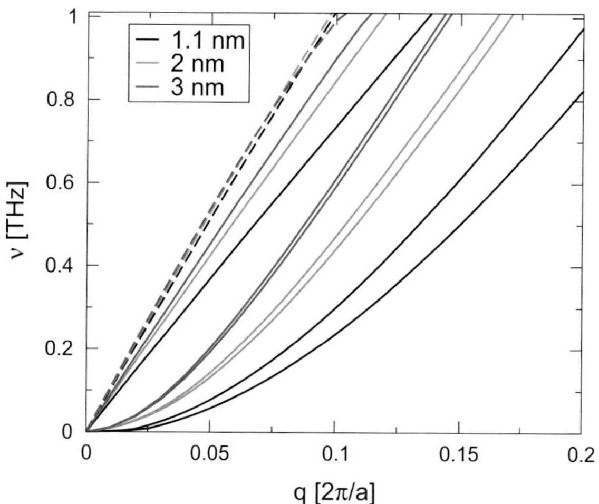

Fig. 7.13 Comparison of the dispersion relations of the acoustic phonons for thin silicon nanowires with different diameters. Flexural modes with $\omega \propto q^2$ and torsional modes with $\omega \propto q$ are plotted with *solid lines*, while longitudinal modes are plotted with *dashed lines*

respectively. Both the q^2 dispersion of the flexural modes and the presence of the torsional mode, originating from the cylindric symmetry, are a direct consequences of dimensional reduction and are independent on the material that constitutes the nanowires, the way interatomic forces are modeled or the surface structure. This scenario is very different from that of one-dimensional non linear models, which exhibit only a single longitudinal branch, since motion is restricted in one dimension.

The longitudinal acoustic (LA) modes exhibit higher group velocities, which are not affected by the diameter of the wires, whereas the torsional modes become softer the thinner the wires (Fig. 7.13). Due to the fairly large number of atoms in the unit cell of the nanowires considered here, higher frequency phonon branches appear at relatively low frequency. These modes provide viable channels to heat transport, however their group velocities remains limited, and significantly smaller than those of the acoustic phonons of bulk silicon. It is worth noting that the acoustic modes in silicon nanowires span a much more limited range of frequencies (<5 THz) than in bulk, where LA modes can extend up to 12 THz. From these observations it is difficult to draw precise conclusions on the thermal conductivity of crystalline wires, especially on its dependence on the diameter, but one would expect a reduction with respect to bulk, independently of surface scattering [14].

Both equilibrium and non-equilibrium MD simulations demonstrate a non-monotonic dependence of κ on the diameter of crystalline wires with similar structure [80, 81]. NEMD data from [80] in Fig. 7.14 show that κ decreases reducing the diameter, reaching a minimum for wires ~3.5 nm thick. Below this diameter κ grows for thinner wires, and may even approach the bulk reference (~200 W m^{-1} K^{-1}), as in the case of the EMD estimate for the 1.1 nm wire. The location of the minimum of κ would depend on the details of the model, including crystalline orientation of the wire, and surface reconstruction, as well as on the simulation setup, but it is a reproducible feature in nanowires. The increasing

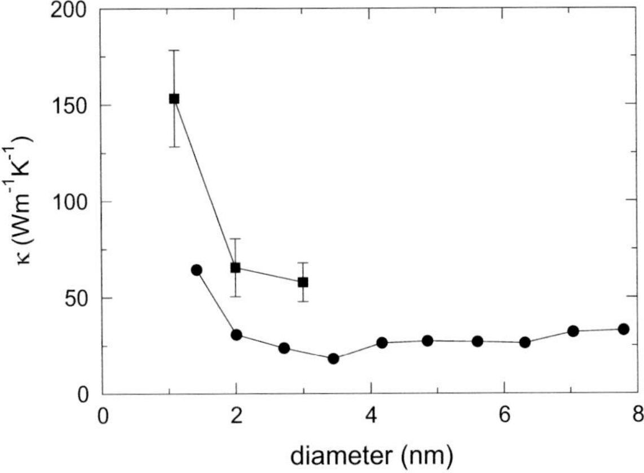

Fig. 7.14 Thermal conductivity of crystalline silicon nanowires at room temperature computed by non-equilibrium [80] and equilibrium molecular dynamics [81]

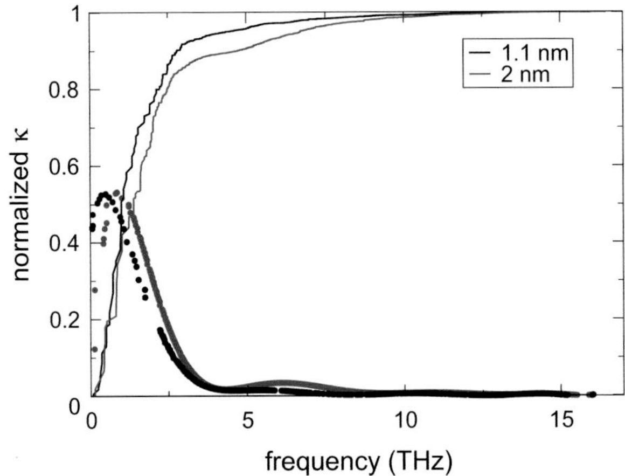

Fig. 7.15 Cumulative (*solid lines*) and differential (*dots*) contribution to the thermal conductivity of silicon nanowires with diameters of 1.1 and 2 nm as a function of the phonon frequency

κ of the thinnest silicon nanowires aligns to the general trend of enhanced thermal conductivity in systems with reduced dimensionality, observed also for carbon based materials. However, no simulation work has so far observed tendency to divergence, or enhancement of κ beyond the bulk value.

An analysis of the spectral contribution to κ as a function of phonon frequency in wires of 1.1 and 2 nm (Fig. 7.15), obtained from Eq. (7.5) in which lifetimes were computed by MD, shows that the distribution of heat carriers is shifted toward lower frequencies with respect to the bulk. This observation complies with the smaller frequency range of acoustic phonons in nanowires, and with the presence of flat bands with fairly small group velocities at higher frequency, which limits significantly phonon mean free paths. This type of analysis yields a detailed explanation of the origin of the minimum in thermal conductivity, which takes into account the combination of density of vibrational states, group velocities and the phonon lifetimes. Experimental verification of the prediction of a minimum of κ as a function of diameter is nevertheless still lacking, since surface scattering would completely bleach the direct effect of dimensionality reduction.

Thermal measurements on silicon nanowires with larger diameters have highlighted the role of surface scattering of phonons in dictating a very low thermal conductance [25]. MD simulations predict that even more dramatic reduction of κ, up to 100 times lower than the bulk value at room temperature, can be achieved in the thin silicon wires in Fig. 7.11 covered with a thin layer of amorphous material [81, 82]. Given the very large surface to volume ratio, these systems are extremely sensitive to surface modification. Figure 7.16 shows that the reduction of thermal conductivity (plotted as the ratio of $\kappa_{bulk}/\kappa_{wire}$) is rather abrupt with the thickness of the amorphous shell: a rough layer of $\sim$6 Å of amorphous silicon, corresponding to about five atomic layers, is sufficient to abate κ to 90 times less than bulk crystalline silicon at room temperature. Further surface amorphization does not induce further significant reduction of κ. Remarkably, the thermal conductivity of these crystalline-core/amorphous-shell ultra-thin wires, for which the most of the volume remains anyway crystalline, can be lower than that of amorphous silicon computed with the same modeling techniques and the same empirical potential [52]. These wires are among the few systems that have been predicted to break the *amorphous limit*

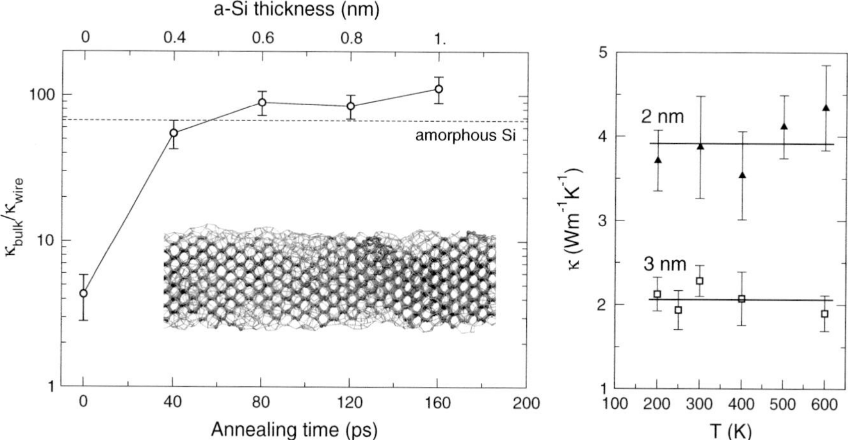

Fig. 7.16 (*Left panel*) Thermal conductivity of a core-shell silicon nanowire with a diameter of 3 nm as a function of the thickness of amorphous shell from equilibrium molecular dynamics. The amorphous shell is generated by thermal annealing close to the melting temperature of silicon, and its thickness is proportional to annealing time. The *dashed line* corresponds to the thermal conductivity of bulk amorphous silicon computed by molecular dynamics using the same empirical potential [52]. (*Right panel*) Thermal conductivity of core-shell nanowires with diameter of 2 and 3 nm as a function of temperature [82]

of κ at constant chemical composition [83]. However, while several experiments have demonstrate a drastic reduction of κ in nanowires, no measurements on ultra-thin wires with diameter of $\sim$3 nm have been performed so far, thus theoretical predictions of κ below the amorphous limit have not yet been confirmed.

The temperature dependence of κ (Fig. 7.16) and an analysis of the relative contributions of phonons to thermal transport demonstrates analogies to the characterization of heat carriers in disordered solids [84]. As in amorphous silicon, vibrational modes in core-shell wires can be classified into low-frequency propagating phonons, which can be described by BTE, and higher-frequency diffusive modes, which contribute to heat transport via an energy transfer mechanism that can be described in harmonic approximation using Fermi's golden rule [82, 84]. Both propagating and diffusive modes provide significant contribution to κ, which depend on the diameter of the wire and the thickness of the amorphous shell: for example for a wire with a diameter of 3 nm and an amorphous shell of 1 nm, propagating modes with frequencies up to 2 THz contribute about one half of the total κ at room temperature.

Given the delocalized character of both propagating and diffusive modes, which typically extend for several tens of nanometers, over the whole models considered in the simulation studies, one would argue that surface amorphization in thin wires deeply modifies the character of the vibrational modes, rather than just scattering phonons. It is interesting to point out that low frequency propagating phonons in rough wires can still exhibit remarkably long mean free paths of the order of 1 μm even in wires as thin as 2 nm (Fig. 7.17), against the common wisdom that the diameter of the wire marks the upper limit of the mean free path of the heat carriers (Casimir limit). Very long phonon mean free paths were indeed recently observed in SiGe nanowires [85].

Heat transport in thicker wires with diameters larger than 10 nm exhibit similar characteristics as those discussed for thin wires, however elastic scattering calculations on open systems and molecular dynamics simulations of periodic systems show that surface features become less effective in limiting κ [86, 87],

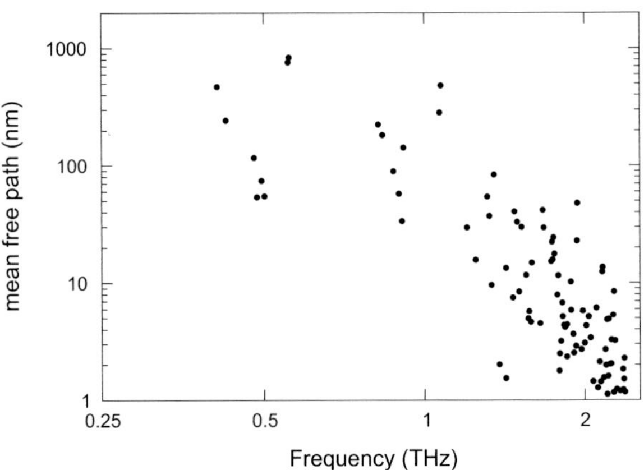

Fig. 7.17 Mean free paths of propagating phonons in a 2 nm thick silicon nanowire with few atoms thick amorphous shell

and the microscopic origin of the reduction of κ reported in some experiments remains debated. On the other hand, in thick wires mesoscopic models including surface scattering entail satisfactory results, provided that the correct phonon dispersion relations are used [88]. Besides surface amorphization, several other possible strategies to reduce κ in semiconducting wires were devised using atomistic simulations, such as introducing screw dislocations [89], surface faceting [90], alloying [91], and crystalline core-shell structuring [92].

To conclude, it is worth mentioning that atomistic modeling predicted that suspended semiconducting wires would exhibit quantized thermal conductance at low temperature in ballistic regime [93]. The quantum of conductance is universal, yet temperature-dependent, and corresponds to $\pi^2 k_B^2 T/3h$. The possibility to probe the quantum of conductance is made viable by the harmonic phonon features described at the beginning of this session, which are general for suspended one-dimensional nanostructures. This prediction was verified experimentally few years later and quantized thermal conductance was measured in silicon nitride phonon waveguides [94].

7.4.2 Ultra-Thin Silicon Membranes

Whereas suspended silicon nanowires display a high potential for tuning and optimization of the thermal conductivity that would be very attractive for thermo-electric energy conversion, their use in devices is complicated by their fragility and by the lack of scalable fabrication processes. In turn, quasi two-dimensional ultra thin membranes appear as more versatile systems for applications that rely heavily on phononic properties, including sensors, nanomechanical resonators and thermoelectrics. Recent experiments demonstrated intriguing phononic and thermal properties of silicon membranes, including flexural acoustic modes with quadratic dispersion relations [95], quasi-ballistic transport at the micro scale at room temperature [96], and yet a significant reduction of thermal conductivity compared to bulk[30]. A compelling microscopic interpretation of these results can be obtained by atomistic modeling, which shows to what extent they emerge from the interplay of dimensionality reduction and surface features.

The experimental measurements of κ in thin films and suspended membranes can be interpreted by mesoscopic kinetic models [34] based on bulk phonon dispersion relations [96], in which surface scattering is modeled empirically, in analogy with the Sondheimer model for electronic transport [97] (Fig. 7.18). However, comparing lattice dynamics results to molecular dynamics simulations, it was shown that for films or membranes of the order of, or thinner than, $\sim$10 nm large discrepancies may arise, thus indicating that models based on bulk properties cannot be used to predict the thermal conductivity of ultra-thin membranes [98].

As in the case of silicon nanowires discussed in the previous section, also for membranes it is necessary to assess the effect of dimensionality reduction on the harmonic and anharmonic phonon properties. If the interatomic interactions

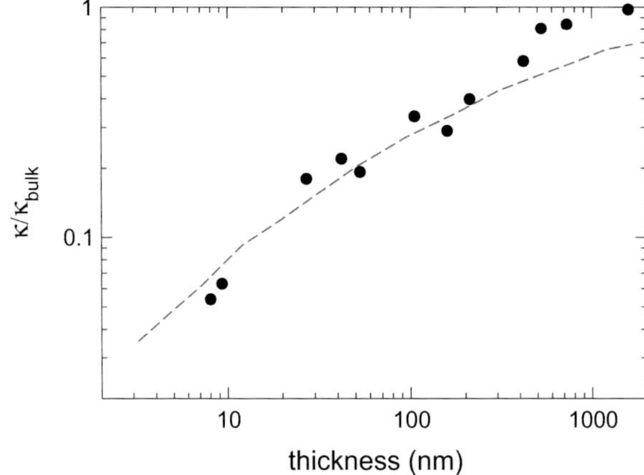

Fig. 7.18 Room temperature thermal conductivity of suspended silicon membranes normalized by the thermal conductivity of natural bulk silicon (148 W m^{-1} K^{-1} at 300 K) [30, 99]. The *dashed line* is the result of a kinetic model with diffusive surface scattering [96]

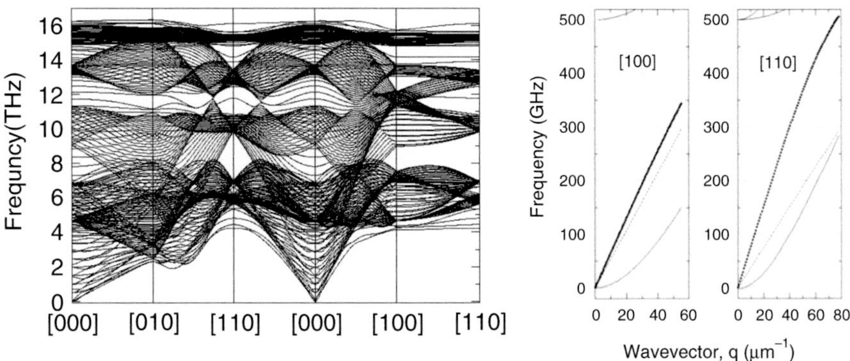

Fig. 7.19 Phonon dispersion relations of a 5 nm thick silicon membrane, and detail of the acoustic modes in proximity of the Γ point [100]

are modeled with inexpensive empirical potentials [57], the phonon dispersion relations of atomistic models of silicon membranes with thickness up to few tens of nanometers can be calculated by diagonalization of the dynamical matrix computed from Eq. (7.2). Lattice dynamics calculations of periodic crystalline models with ideal surface reconstruction allow us to asses the general effects of dimensionality reduction on phonons. Figure 7.19 reports the phonon dispersion relations of a 5 nm thick membrane. The large number of atoms in the unit cell gives rise to a large number of optical bands, which extend to low frequency and mix with the acoustic modes. The features of these bands, originated from removing periodicity in one direction, cannot be reproduced correctly by zone folding of bulk phonons, especially for the thinnest membranes, as they have derivative equal to zero at the

Γ point and $\omega \propto q^2$. The detail of the acoustic dispersions at the Γ point shows that one of the transverse acoustic modes converts into a flexural mode with $\omega \propto q^2$. The flexural modes, related to out-of-plane vibration, are very sensitive to the thickness of the membrane: thinner membranes exhibit softer flexural modes (Fig. 7.20a). The dispersion relations of these modes was measured by Brillouin light scattering experiments, which showed a remarkable agreement with modeling [95, 99]. Also the speed of sound of the longitudinal acoustic branch depends on the thickness of the membranes, yet weakly. In general it is lower than in the bulk, but in the sub-10 nm regime the speed of sound of the LA modes decreases with thickness. On the other hand the speed of sound of the in-plane transverse modes, which remain linear in q, is not affected by variations of thickness (Fig. 7.20b).

Whereas the characterization of the acoustic phonons from atomistic modeling exhibit a remarkable agreement with experiments, the thermal conductivity of pristine crystalline membranes, computed by equilibrium MD, is in sharp disagreement with the measurements reported in [30]. Crystalline membranes with ideal surfaces exhibit a reduction of κ with respect to bulk silicon, however not as large as the one probed by experiments (Fig. 7.21). Nevertheless, as opposed to the case of nanowires, κ decreases monotonically with decreasing thickness reaching $1/3$ of κ_{bulk} for $\sim$1 nm thick membranes. No evidence of possible divergence of κ was observed.

The discrepancy with experiments indicates that pristine crystalline models are not representative for real systems. Since silicon membranes are exposed to air during and after fabrication, a layer of native oxide forms at their surfaces. Such layer is about 1 nm thick and can exhibit nanoscale roughness. When models with rough and/or oxidized surfaces as those shown in Fig. 7.21b–d are considered, the simulations reproduce the experimental κ very well (Fig. 7.21e), thus indicating that also for ultra-thin silicon membranes surface properties dictate the major reduction of κ. Recent experiments, in which the native oxide layer is removed by wet etching

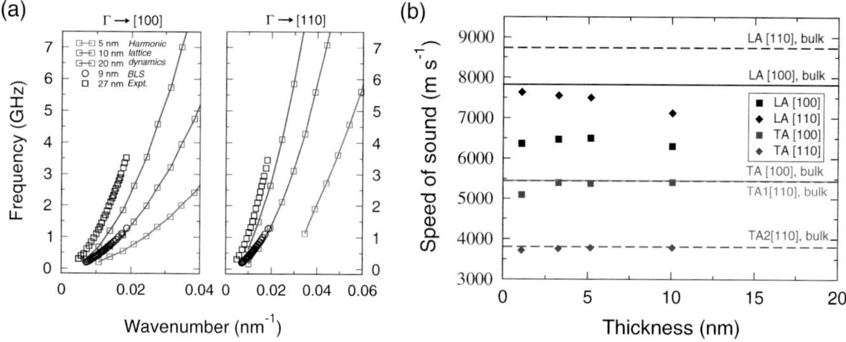

Fig. 7.20 Dispersion of the flexural modes of silicon membranes of different thickness (**a**), and dependence of the group velocities of in-plane longitudinal and transverse acoustic modes on membrane thickness (**b**). Data from [99, 100]

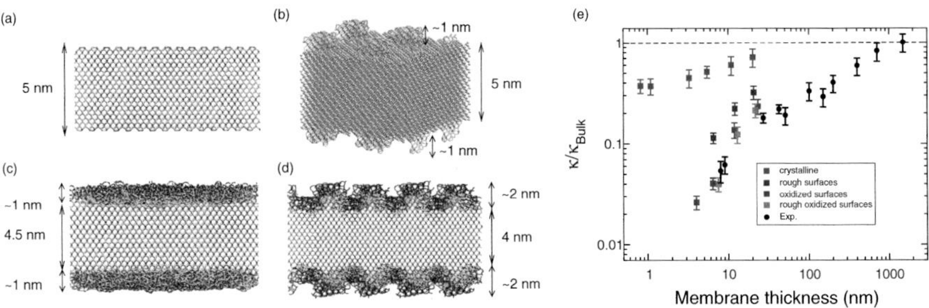

Fig. 7.21 Models of silicon membranes oriented in the (001) direction, with pristine crystalline surfaces (**a**), surface roughness (**b**), flat native oxide layers (**c**) and rough native oxide (**d**), and their thermal conductivity (**e**). Data from [99, 100]

and eventually let re-grow, confirm the prominent role of surface oxidation and roughness on thermal transport. This effect is more appreciable in the thinnest samples measured ($\sim$8 nm) for which an increase of κ of about 2.3 times (from 8 to 18 W m^{-1} K^{-1}) upon etching was observed. Further exposure of the same sample to air for several hours leads to a reduction of κ to 12 W m^{-1} K^{-1}. In thicker samples the same cycle of etching and re-oxidation lead to smaller variations of κ [99].

Non equilibrium MD simulations, which allow one to calculate the accumulation function of κ as a function of phonon mean free path, show that surface scattering shifts the major contribution to the total thermal conductivity at room temperature from phonons with a broad range of mean free paths up to 1 μm for pristine crystalline membranes, to a much narrower distribution of mean free paths smaller than 80 nm for membrane models with rough native oxide at the surfaces.

Dimensionality reduction in membranes enables other approaches to control heat transport via surface modifications. For example, it was suggested that drilling holes or depositing nanoscale pillars, of the same or of a different material would modify the band structure and enhance phonons scattering through local resonances, thus reducing the thermal conductivity [101–104].

7.5 Conclusions

In this chapter we have illustrated a limited, yet representative, set of problems in which atomistic simulations are exploited to shed light on nanoscale heat transport in materials with reduced dimensionality. Even though models for specific materials are employed, common features emerge, stemming for example from symmetries and invariances associated with dimensionality reduction. Phonon dispersion relations and vibrational density of states are deeply affected by changes in dimensionality, especially concerning acoustic phonons. Specifically, the emergence of flexural phonons with quadratic dispersion relations sizably influences both sound and heat propagation in one and two-dimensional nanostructures. Flexural modes in

2D membranes and 1D wires are characteristic of low dimensional systems in a three-dimensional space, in which each atom has three degrees of freedom. This aspect marks a major difference with non-linear low-dimensional systems, such as the Fermi-Pasta-Ulam model, in which motion is also confined in a space with reduced dimensionality. In addition, except for graphene, which is a truly atomically thick two-dimensional array of atoms, all the other systems known in nature have a complex three-dimensional structure, even though they may extend in one or two dimensions only.

Starting from this premises, it is then understandable that, at variance with non-linear models, no clear evidence of diverging thermal conductivity was found in atomistic simulations of nanostructures, even though both graphene and carbon nanotubes exhibit extremely high thermal conductivity, and diffusive (Fourier) heat transport has been predicted to occur only at macroscopic size scales. Phenomena like large ballistic phonon mean free path and the emergence of "second sound" at relatively high temperature, which lead to quasi-diverging κ as in non-linear models, actually stem from the reduced phase-space available to anharmonic phonon scattering and the presence of flexural modes.

A further general consequence of dimensionality reduction is the very large surface-to-volume ratio, which makes heat transport in low-dimensional nanostructures extremely sensitive to surface modifications, such as roughening, faceting, functionalization, oxidation and interactions with substrates. A possible analogy with statistical physics can be made considering the difference between momentum conserving and non-conserving models, such as the Frenkel-Kontorova model, where lack of momentum conservation, e.g from the interaction with an underlying fixed potential, suppresses divergence. The tremendous impact of surfaces on nanoscale heat transport was predicted in many simulation studies, and probed in several experiments. These works have demonstrated that accurate modeling needs to include chemical specificity and benefits from direct feedback with experiments.

References

1. Kroto, H.W., Heath, J.R., O'Brien, S.C., Curl, R.F.: Nature **318**, 162 (1985)
2. Iijima, S.: Nature **354**(6348), 56 (1991)
3. Alivisatos, A.P.: Science **271**, 933 (1996)
4. Sprinkle, M., Siegel, D., Hu, Y., Hicks, J., Tejeda, A., Taleb-Ibrahimi, A., Fèvre, P.L., Bertran, F., Vizzini, S., Enriquez, H., Chiang, S., Soukiassian, P., Berger, C., de Heer, W.A., Lanzara, A., Conrad, E.H.: Phys. Rev. Lett. **103**(22), 226803 (2009)
5. Hicks, L.D., Dresselhaus, M.S.: Phys. Rev. B **47**(24), 16631 (1993)
6. Dresselhaus, M.S., Chen, G., Tang, M.Y., Yang, R., Lee, H., Wang, D., Ren, Z., Fleurial, J.P., Gogna, P.: Adv. Mater. **19**(8), 1043 (2007)
7. Zhao, X., Wei, C.M., Yang, L., Chou, M.Y.: Appl. Phys. Lett. **92**(23), 236805 (2004)
8. Ziman, J.M.: Electrons and Phonons. Oxford University Press, Oxford (1960)
9. Aoki, K., Kusnezov, D.: Phys. Rev. Lett. **86**, 4029 (2001)
10. Narayan, O., Ramaswamy, S.: Phys. Rev. Lett. **89**(20), 200601 (2002)
11. Lepri, S., Livi, R., Politi, A.: Phys. Rep. **377**(1), 1 (2003)

12. Lepri, S., Livi, R., Politi, A.: Chaos **15**(1), 015118 (2005)
13. Wang, L., Hu, B., Li, B.: Phys. Rev. E **86**, 040101 (2012)
14. Balandin, A., Wang, K.L.: Phys. Rev. B **58**(3), 1544 (1998)
15. Casimir, H.: Physica **5**(6), 495 (1938)
16. Balandin, A.A., Ghosh, S., Bao, W., Calizo, I., Teweldebrhan, D., Miao, F., Lau, C.N.: Nano Lett. **8**(3), 902 (2008)
17. Ghosh, S., Bao, W., Nika, D.L., Subrina, S., Pokatilov, E.P., Lau, C.N., Balandin, A.A.: Nat. Mater. **9**(7), 555 (2010)
18. Chen, S., Wu, Q., Mishra, C., Kang, J., Zhang, H., Cho, K., Cai, W., Balandin, A.A., Ruoff, R.S.: Nat. Mater. **11**(1), 203 (2012)
19. Balandin, A.A.: Nat. Mater. **10**(8), 569 (2011)
20. Xu, X., Pereira, L.F.C., Wang, Y., Wu, J., Zhang, K., Zhao, X., Bae, S., Tinh Bui, C., Xie, R., Thong, J.T.L., Hong, B.H., Loh, K.P., Donadio, D., Li, B., Özyilmaz, B.: Nat. Commun. **5**, 3689 (2014)
21. Chang, C., Okawa, D., Garcia, H., Majumdar, A., Zettl, A.: Phys. Rev. Lett. **101**(7), 075903 (2008)
22. Yu, C., Shi, L., Yao, Z., Li, D., Majumdar, A.: Nano Lett. **5**(9), 1842 (2005)
23. Li, D., Wu, Y., Kim, P., Shi, L., Yang, P., Majumdar, A.: Appl. Phys. Lett. **83**(14), 2934 (2003)
24. Hochbaum, A.I., Chen, R., Delgado, R.D., Liang, W., Garnett, E.C., Najarian, M., Majumdar, A., Yang, P.: Nature **451**(7175), 163 (2008)
25. Chen, R., Hochbaum, A.I., Murphy, P., Moore, J., Yang, P., Majumdar, A.: Phys. Rev. Lett. **101**(10), 105501 (2008)
26. Asheghi, M., Leung, Y.K., Wong, S.S., Goodson, K.E.: Appl. Phys. Lett. **71**(13), 1798 (1997)
27. Ju, Y.S., Goodson, K.E.: Appl. Phys. Lett. **74**(20), 3005 (1999)
28. Liu, W., Asheghi, M.: J. Appl. Phys. **98**(12), 123523 (2005)
29. Liu, X., Wu, X., Ren, T.: Appl. Phys. Lett. **98**(17), 174104 (2011)
30. Chávez-Ángel, E., Reparaz, J.S., Gomis-Bresco, J., Wagner, M.R., Cuffe, J., Graczykowski, B., Shchepetov, A., Jiang, H., Prunnila, M., Ahopelto, J., Alzina, F., Sotomayor Torres, C.M.: APL Mater. **2**(1), 012113 (2014)
31. Minnich, A.J., Johnson, J.A., Schmidt, A.J., Esfarjani, K., Dresselhaus, M.S., Nelson, K.A., Chen, G.: Phys. Rev. Lett. **107**(9), 095901 (2011)
32. Regner, K.T., Sellan, D.P., Su, Z., Amon, C.H., Mcgaughey, A.J.H., Malen, J.A.: Nat. Commun. **4**, 1640 (2013)
33. Baroni, S., de Gironcoli, S., Dal Corso, A., Giannozzi, P.: Rev. Mod. Phys. **73**(2), 515 (2001)
34. Callaway, J.: Phys. Rev. **113**(4), 1046 (1959)
35. Fabian, J., Allen, P.: Phys. Rev. Lett. **77**(18), 3839 (1996)
36. Maradudin, A., Fein, A.: Phys. Rev. **128**, 2589 (1962)
37. Savic, I., Donadio, D., Gygi, F., Galli, G.: Appl. Phys. Lett. **102**(7), 073113 (2013)
38. Ward, A., Broido, D.A., Stewart, D.A., Deinzer, G.: Phys. Rev. B **80**(12), 125203 (2009)
39. Fugallo, G., Lazzeri, M., Paulatto, L., Mauri, F.: Phys. Rev. B **88**(4), 045430 (2013)
40. Broido, D., Ward, A., Mingo, N.: Phys. Rev. B **72**(1), 014308 (2005)
41. Broido, D.A., Malorny, M., Birner, G., Mingo, N., Stewart, D.A.: Appl. Phys. Lett. **91**(23), 231922 (2007)
42. Lindsay, L., Broido, D.A., Mingo, N.: Phys. Rev. B **80**(12), 125407 (2009)
43. Lindsay, L., Li, W., Carrete, J., Mingo, N., Broido, D.A., Reinecke, T.L.: Phys. Rev. B **89**(15), 155426 (2014)
44. Fugallo, G., Cepellotti, A., Paulatto, L., Lazzeri, M., Marzari, N., Mauri, F.: Nano Lett. **14**(11), 6109 (2014)
45. Cepellotti, A., Fugallo, G., Paulatto, L., Lazzeri, M., Mauri, F., Marzari, N.: Nat. Commun. **6**, 1 (2015)
46. Allen, M.P., Tildesley, D.J.: Computer Simulation of Liquids. Oxford University Press, Oxford (1989)
47. Kubo, R., Yokota, M., Nakajima, S.: J. Phys. Soc. Jpn. **12**(11), 1203 (1957)
48. Jund, P., Jullien, R.: Phys. Rev. B **59**(21), 13707 (1999)
49. Müller-Plathe, F.: J. Chem. Phys. **106**(14), 6082 (1997)

50. Sellan, D.P., Landry, E.S., Turney, J.E., McGaughey, A.J.H., Amon, C.H.: Phys. Rev. B **81**(21), 214305 (2010)
51. Schelling, P., Phillpot, S., Keblinski, P.: Phys. Rev. B **65**(14), 144306 (2002)
52. He, Y., Donadio, D., Galli, G.: Appl. Phys. Lett. **98**(14), 144101 (2011)
53. He, Y., Savic, I., Donadio, D., Galli, G.: Phys. Chem. Chem. Phys. **14**(47), 16209 (2012)
54. Dhar, A.: Adv. Phys. **57**(5), 457 (2008)
55. Lippi, A., Livi, R.: J. Stat. Phys. **100**(5), 1147 (2000)
56. Saito, K., Dhar, A.: Phys. Rev. Lett. **104**(4), 040601 (2010)
57. Tersoff, J.: Phys. Rev. B **39**(8), 5566 (1989)
58. Lindsay, L., Broido, D.A.: Phys. Rev. B **81**(20), 205441 (2010)
59. Sun, L., Murthy, J.Y.: Appl. Phys. Lett. **89**(17), 171919 (2006)
60. Nicklow, R., Wakabayashi, N., Smith, H.G.: Phys. Rev. **5**, 4951 (1972)
61. Klemens, P.G., Pedraza, D.F.: Carbon **32**(4), 735 (2002)
62. Lindsay, L., Broido, D.A., Mingo, N.: Phys. Rev. B **82**(11), 115427 (2010)
63. Pereira, L.F.C., Donadio, D.: Phys. Rev. B **87**(12), 125424 (2013)
64. Barbarino, G., Melis, C., Colombo, L.: Phys. Rev. B **91**(3), 035416 (2015)
65. Lee, S., Broido, D., Esfarjani, K., Chen, G.: Nat. Commun. **6**, 7290 (2015)
66. Mahan, G., Jeon, G.: Phys. Rev. B **70**(7), 075405 (2004)
67. Kim, P., Shi, L., Majumdar, A., McEuen, P.L.: Phys. Rev. Lett. **87**(21), 215502 (2001)
68. Berber, S., Kwon, Y.K., Tománek, D.: Phys. Rev. Lett. **84**(cond-mat/0002414. 20), 4613 (2000)
69. Lukes, J.R., Zhong, H.: J. Heat Transf. **129**(6), 705 (2007)
70. Donadio, D., Galli, G.: Phys. Rev. Lett. **99**(25), 255502 (2007)
71. Pereira, L.F.C., Savic, I., Donadio, D.: New J. Phys. **15**(10), 105019 (2013)
72. Savin, A.V., Hu, B., Kivshar, Y.S.: Phys. Rev. B **80**(19), 195423 (2009)
73. Zhang, X., Hu, M., Poulikakos, D.: Nano Lett. **12**(7), 3410 (2012)
74. Sevik, C., Sevincli, H., Cuniberti, G., Cagin, T.: Nano Lett. **11**(11), 4971 (2011)
75. Prasher, R.S., Hu, X.J., Chalopin, Y., Mingo, N., Lofgreen, K., Volz, S., Cleri, F., Keblinski, P.: Phys. Rev. Lett. **102**, 105901 (2009)
76. Ma, D., Lee, C.S., Au, F., Tong, S.Y., Lee, S.T.: Science **299**, 1874 (2003)
77. Shchepetov, A., Prunnila, M., Alzina, F., Schneider, L., Cuffe, J., Jiang, H., Kauppinen, E.I., Sotomayor Torres, C.M., Ahopelto, J.: Appl. Phys. Lett. **102**(19), 192108 (2013)
78. Volz, S., Chen, G.: Appl. Phys. Lett. **75**(14), 2056 (1999)
79. Vo, T., Williamson, A.J., Galli, G.: Phys. Rev. B **74**(4), 045116 (2006)
80. Ponomareva, I., Srivastava, D., Menon, M.: Nano Lett. **7**(5), 1155 (2007)
81. Donadio, D., Galli, G.: Phys. Rev. Lett. **102**(19), 195901 (2009)
82. Donadio, D., Galli, G.: Nano Lett. **10**(3), 847 (2010)
83. Chiritescu, C., Cahill, D.G., Nguyen, N., Johnson, D., Bodapati, A., Keblinski, P., Zschack, P.: Science **315**(5810), 351 (2007)
84. Allen, P., Feldman, J.: Phys. Rev. B **48**(17), 12581 (1993)
85. Hsiao, T.K., Chang, H.K., Liou, S.C., Chu, M.W., Lee, S.C., Chang, C.W.: Nat. Nanotechnol. **8**, 534–538 (2013)
86. He, Y., Galli, G.: Phys. Rev. Lett. **108**(21), 215901 (2012)
87. Duchemin, I., Donadio, D.: Appl. Phys. Lett. **100**(22), 223107 (2012)
88. Mingo, N., Yang, L., Li, D., Majumdar, A.: Nano Lett. **3**, 1713–1716 (2003)
89. Xiong, S., Ma, J., Volz, S., Dumitrică, T.: Small **10**(9), 1756 (2014)
90. Sansoz, F.: Nano Lett. **11**(12), 5378 (2011)
91. Chan, M.K.Y., Reed, J., Donadio, D., Mueller, T., Meng, Y.S., Galli, G., Ceder, G.: Phys. Rev. B **81**(17), 174303 (2010)
92. Hu, M., Giapis, K.P., Goicochea, J.V., Zhang, X., Poulikakos, D.: Nano Lett. **11**(2), 618 (2011)
93. Rego, L., Kirczenow, G.: Phys. Rev. Lett. **81**(1), 232 (1998)
94. Schwab, K., Henriksen, E., Worlock, J., Roukes, M.: Nature **404**(6781), 974 (2000)

95. Cuffe, J., Chávez, E., Shchepetov, A., Chapuis, P.O., El Boudouti, E.H., Alzina, F., Kehoe, T., Gomis-Bresco, J., Dudek, D., Pennec, Y., Djafari-Rouhani, B., Prunnila, M., Ahopelto, J., Sotomayor Torres, C.M.: Nano Lett. **12**(7), 3569 (2012)
96. Johnson, J.A., Maznev, A.A., Cuffe, J., Eliason, J.K., Minnich, A.J., Kehoe, T., Torres, C.M.S., Chen, G., Nelson, K.A.: Phys. Rev. Lett. **110**(2), 025901 (2013)
97. Sondheimer, E.H.: Adv. Phys. **1**(1), 1 (1952)
98. Turney, J.E., McGaughey, A.J.H., Amon, C.H.: J. Appl. Phys. **107**(2), 024317 (2010)
99. Neogi, S., Sebastian Reparaz, J., Pereira, L.F.C., Graczykowski, B., Wagner, M.R., Sledzinska, M., Shchepetov, A., Prunnila, M., Ahopelto, J., Sotomayor Torres, C.M., Donadio, D., ACS Nano **9**(4), 3820 (2015)
100. Neogi, S., Donadio, D.: Eur. Phys. J. B **88**, 73 (2015)
101. Yu, J.K., Mitrovic, S., Tham, D., Varghese, J., Heath, J.R.: Nat. Nanotechnol. **5**(10), 718 (2010)
102. He, Y., Donadio, D., Lee, J.H., Grossman, J.C., Galli, G., ACS Nano **5**(3), 1839 (2011)
103. Davis, B.L., Hussein, M.I.: Phys. Rev. Lett. **112**(5), 055505 (2014)
104. Graczykowski, B., Sledzinska, M., Alzina, F., Gomis-Bresco, J., Reparaz, J.S., Wagner, M.R., Sotomayor Torres, C.M.: Phys. Rev. B **91**(7), 075414 (2015)

Chapter 8
Experimental Probing of Non-Fourier Thermal Conductors

Chih-Wei Chang

Abstract Even though theoretical studies on non-Fourier thermal conduction have continued for many decades, experimental investigations on the topic just started in recent years. In this chapter, I will give my personal reflections on our past and recent experimental progress in elucidating the non-Fourier thermal conduction phenomena in SiGe nanowires, Si-Ge interfaces, multiwall nanotubes, and ultralong singlewall nanotubes. Non-Fourier thermal conduction includes the conventional ballistic thermal conduction in ordinary materials and the anomalous thermal transport in low dimensional systems. Due to finite-size effects, many of the non-Fourier thermal conduction behaviors observed previously can be attributed to conventional ballistic thermal conduction phenomena. On the other hand, significant experimental progress has been made to reveal the anomalous thermal conduction on low-dimensional materials of macroscopic sizes.

8.1 Introduction

In 1822, French mathematician and physicist Joseph Fourier published his work entitled Théorie Analytique de la Chaleur (The Analytic Theory of Heat), which established the phenomenological theory for thermal conduction in solids; or in general, the correct mathematical description of diffusion process. Fourier's seminal work later inspired Georg S. Ohm to discover Ohm's law for electrical conduction in 1827 and Adolf E. Fick to discover Fick's law for gas diffusion in 1855, which are the empirical foundations of many phenomena in macroscopic world. Under steady state, Fourier's law explicitly states that the heat flux density (J) is proportional to the temperature gradient, and the proportional constant is the thermal conductivity (κ), i.e.

$$J = -\kappa \nabla T \tag{8.1}$$

C.-W. Chang (✉)
Center for Condensed Matter Sciences, National Taiwan University, Taipei, Taiwan
e-mail: cwchang137@ntu.edu.tw

S. Lepri (ed.), *Thermal Transport in Low Dimensions*, Lecture Notes in Physics 921, DOI 10.1007/978-3-319-29261-8_8

On the other hand, similar to electrical conductivity in Ohm's law, Fourier's law also implicitly suggests that κ is an intrinsic property of a material and should be independent of the sample sizes.

Theoretical physicists started to notice anomalous violations of Fourier's law during 1940s [1]. Intensive investigations resurged in late 1990s when powerful computers became available [2]. It should be noted that when referring to "violations of Fourier's law" or "breakdown of Fourier's law" in the literatures, it often suggests violations of the implication of Fourier's law; that is, κ is no longer an intrinsic property of a material and could be size-dependent in low dimensional systems. In one-dimensional (1D) systems, κ would display a power-law divergence with respect to the length (L), i.e. $\kappa \sim L^{\alpha}$ ($\alpha = 0$ is Fourier's law). In general, $\alpha > 0$ can be regarded as non-Fourier thermal conduction. Thus, non-Fourier thermal conduction includes the following:

1. *Conventional ballistic thermal conduction* ($\alpha = 1$): in analogy to ballistic conduction of electrons, it represents a dissipationless thermal transport phenomenon. Experimentally, we will discuss our recent results of SiGe nanowires, Si-Ge interfaces, and multiwall nanotubes.
2. *Anomalous thermal conduction* ($0 < \alpha < 1$): although the underlying physics remains unclear, the anomalous thermal transport phenomena are known to occur in many 1D systems [3, 4]. Experimentally, we will discuss recent progress in measuring κ's of ultralong nanotubes.

Before we discuss the experiments on non-Fourier thermal conductors, we should ask the definition of a 1D system, i.e. what kinds of materials can be considered as a 1D, instead of a 3D system?

Unfortunately, I find the definition is poorly understood in literatures even though theorists have focused on studying anomalous thermal conduction of atomic chains for long time. Many theoretical models have focused on atomic chains that have only one degree of freedom for each atom. Such models are certainly 1D systems but they do not exist in real world. Other models have considered atomic chains with three dimensional degrees of freedom for each atom. However, although similar systems do exist in materials such as polyethylene, the associated thermal conduction measurements on individual, isolated molecular chain remain to be challenging. On the other hand, many kinds of nanotubes and nanowires have been synthesized. Can these nanotubes or nanowires be considered as 1D systems?

Although a rigorous definition of a 1D system needs more theoretical works, we note that there are three scales naturally emerging from the discussion of 1D systems. Empirically, there are three possible criteria for defining a 1D thermal conductor:

(A) *Aspect ratio*: when the length (L) of a sample is much larger than the diameter (d) of the sample, i.e. $L \gg d$ [5, 6].
(B) *Phonon mean free path*: when the dominant phonon mean free path (l) of a sample is much larger than the diameter of the sample, i.e. $l \gg d$.

(C) *Phonon wavelength*: when the dominant phonon wavelength (λ) for heat conduction of a sample is much larger than the diameter of the sample, i.e. $\lambda \gg d$ [6].

Experimentally, we can test which of the above statements is a correct definition for a 1D system. After a brief contemplation, we find that (A) is unlikely otherwise a man's long beard, a woman's long hair, a long optic fiber, and a long electric cable (any of them can easily have aspect ratio $L/d \gg 10^5$) would be common 1D materials. Violation of Fourier's law would be commonplace if (A) were true. In this chapter, we will experimentally investigate whether (B) or (C) is a correct definition for 1D thermal conductors.

8.2 Experimental Methods for Probing Non-Fourier Thermal Conduction

Fourier's law is a mathematical description for heat diffusing across a solid, which involves a lot of collisions during the process. In a particle picture of the process, it is similar to the energy transfer of gas molecules confined in a box with one end of the box being hotter than the other end. Different materials would involve different scattering processes that give rise to finite κ. Therefore, the averaged path of a gas molecule (in analogy to a phonon) undergoing two collisions can be viewed as the phonon mean free path (l). If the sample size is less than the l, conventional (i.e. $\alpha = 1$) ballistic thermal conduction would occur. Thus the operational definition of (conventional) ballistic thermal conduction means that there is no thermal resistance inside a sample and the thermal resistance exclusively happens at the contacts, i.e. the length-dependent thermal resistance ($1/K$, where K is thermal conductance) measurement should yield a constant $1/K$ for a ballistic thermal conductor shorter than l.

Following the operational definition, we now discuss the experimental procedures for determining l of a sample subjected to a temperature gradient, or equivalently, the distance of (conventional) ballistic thermal conduction. As shown in Fig. 8.1a, when the length (L) of the sample is less than l, heat transfer would be dissipationless and thermal resistance only occurs at the two contacts. Thus ballistic thermal conduction will give length-independent thermal resistance for $L < l$. Following the definition of $\kappa \equiv KL/A$ (where A is the cross sectional area of the sample), it will result in a length-dependent of κ for $L < l$ as well, as shown in Fig. 8.1b. When $L > l$, collision processes start to destroy the ballistic properties of phonons, one then have the ordinary result obeying Fourier's law, i.e. $1/K$ increases linearly with L, or equivalently, κ is length-independent, as shown in Fig. 8.1b.

Unfortunately, so far not many experiments have followed the operational definition to determine l of materials. In fact, many experiments studied dependences on lateral dimensions that are transverse to the temperature gradient, instead of varying the longitudinal dimensions that is parallel to the temperature gradient.

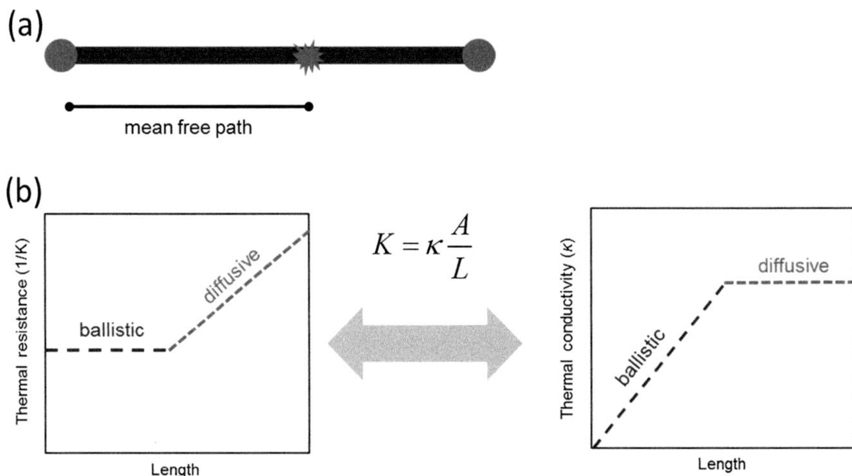

Fig. 8.1 (**a**) An illustration of a ballistic thermal conductor connected to a heat source and a heat sink. The phonon scatter inside the thermal conductor destroys the wave properties of phonons and leads to a finite phonon mean free path. (**b**) The corresponding data when probing the length dependent thermal transport of (**a**). Both *1/K* vs. *L* (*left*) and κ vs. *L* (*right*) relations are plotted here

Taking Si as an example, Goodson's group studied thickness-dependent, in-plane κ of silicon-on-insulator wafer by repeated thermal oxidation and wet etching [7, 8]. The l was estimated to be 300 nm instead of 43 nm predicted by a dispersionless theory [8]. Liu and Asheghi further studied κ of doped Si films with thinner thickness [9]. Li et al. studied diameter-dependent κ of Si nanowires and found that the κ decreases with decreasing diameter [10]. In these works, the estimation of l was based on arguments of Hendrik Casimir, namely the Casimir limit [11]. The Casimir limit states that the l of a thin film/nanowire would be limited by its thickness/diameter provided that the surface scatters phonons diffusively. However, how a rough surface scatter phonon diffusively remains highly unknown. Many theories of surface scatterings of phonons remain unjustified so far. Thus the above experiments based on the Casimir limit should be regarded as indirect estimation of l. In fact, the experimental verification of quantum thermal conduction provides a good example that the lateral confinement of the SiN_x (200 nm wide, 60 nm thick) beam does not necessarily affect the ballistic thermal conduction along the beam (~4 μm) [12]. Moreover, recent experimental observation of $l \sim 200$ nm in holey silicon with lateral confinements as small as 20 nm clearly demonstrates the invalidity of the Casimir limit [13]. Thus l is, in general, not limited by the lateral dimension of a sample.

Recently, there have been great interests in utilizing optical techniques to probe l's of materials, these include time-domain thermoreflectance (TDTR) [14, 15], frequency-domain thermoreflectance (FDTR) techniques [16]. In these optical techniques, heat is generated after the material absorbs the laser light. Then the temperature variation can be probed by another laser beam based on the temperature-dependence of the refractive index of the material. Experimentally, aluminum or gold thin film is commonly employed as a thermal transducer to efficiently convert light into heat. However, it inevitably introduces another problem of thermal contact resistances in these optical measurements, which is difficult to model and certainly introduces large systematic uncertainties in the experiments [17]. For example, the thermal transport at the interface has been suggested to play an important role in enhancing the anisotropy in an otherwise isotropic material [17]. Thus the interpretation of optical measurement results without considering the role of interface will lead to inconsistent conclusions. In TDTR or FDTR, changes of κ's were observed by changing spot sizes or the modulation frequency of the pump laser, which were used for estimating how much ballistic phonons contribute the heat conduction. However, because experimental complexities associated with laser spot sizes, laser modulation frequencies, temperature profiles, and interface phonon scatterings are quite involved, deducing the l's is not straightforward in these experiments [17]. Wilson and Cahill recently pointed out that parts of the previous discrepancies originated from anisotropic thermal transport in-plane and cross-plane, which are different for experiments varying spot sizes or changing modulation frequencies [17]. In addition, identifying the optical signals from phononic and electronic responses is sometimes nontrivial. Finally, theoretical interpretations of the optical measurement results remain controversial. For example, although it is commonly assumed that the observed deviation to be due to ballistic phonons, *ab-initio* calculations suggest that agreements on Si are better if simply considering harmonic and anharmonic phonon channels but without directly incorporating ballistic phonons [18]. In short, the interpretation of the experimental data based on TDTR or FDTR is model-dependent and nontrivial so far. Deducing l of Si from these works is not direct, either.

One should keep in mind that the operational definition of ballistic thermal conduction is based in Fig. 8.1 which demands the variation of the length to be parallel to the temperature gradient. Yet carrying out the desired experimental procedure shown in Fig. 8.1 is not a simple task in practice. Because sometimes the data were collected from different samples of various sizes, they may inevitably exhibit large uncertainties due to variations of sample qualities. Furthermore, many previous works have ignored the contribution of contact thermal resistance, which is likely to give erroneous interpretation of the experimental data, as will be discussed in the following.

Physically, thermal transport measurements usually require connecting leads to a sample, which would often result in finite contact thermal resistances. Thus the experimentally measured total thermal resistance ($1/K_{total}$) can be expressed as:

$$\frac{1}{K_{total}} = \frac{1}{K_{sample}} + \frac{1}{K_{contact}}, \tag{8.2}$$

where $1/K_{contact}$ is contact thermal resistance, which can be reasonably modeled as a constant, and $1/K_{sample}$ is sample resistance, which can be described as

$$1/K_{sample} = \begin{cases} 1/K_Q, & \text{for } L \leq l, \\ (1/K_Q)\frac{L}{l}, & \text{for } L > l, \end{cases} \tag{8.3}$$

by assuming that ballistic behavior takes place and $1/K_{sample}$ becomes constant when $L < l$, whereas diffusive behavior dominates and is proportional to L when $L > l$. If ballistic thermal conductance does not take place at all (i.e. $l = 0$), we simply have the diffusive behavior $1/K_{sample} = aL$ ($a > 0$) for all $L > 0$, and consequently the thermal conductivity $\kappa \equiv KL/A$ yields a constant regardless of L, obeying Fourier's law. For this reason, the observation that κ decreases with decreasing L is often regarded as a signal of non-Fourier thermal conduction phenomena, or the occurrence of ballistic thermal conduction. However, because the decrease of κ with decreasing L may be also explained by the presence of a nonzero contact resistance, reading the non-Fourier signal from the κ vs. L data alone could be fallacious. For example, when contact thermal resistance is finite, κ follows

$$\kappa \equiv \frac{K_{total}L}{A} = \frac{L/A}{aL + 1/K_{contact}} \tag{8.4}$$

which leads to $\kappa \to 0$ as $L \to 0$. To illustrate the fallacy, we plot κ vs. L and $1/K_{total}$ vs. L using Eqs. (8.3) and (8.4) in Fig. 8.2a, b, with added noises to simulate the

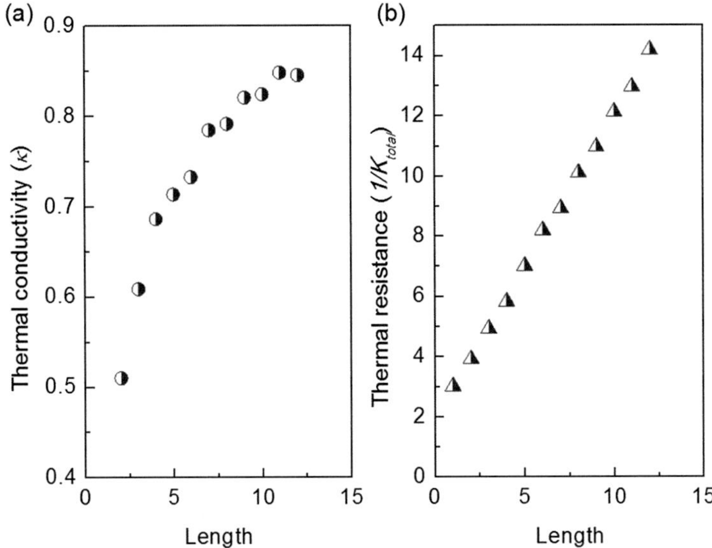

Fig. 8.2 Simulated (**a**) κ vs. L and (**b**) $1/K_{total}$ vs. L relations for $1/K_{contact} = 3$. It can be seen that finite contact thermal resistance will create a fallacious ballistic thermal conduction effect in the κ vs. L relation even if the sample is a diffusive conductor [19]

experimental data. If one merely focuses on the decrease of κ, the phonon mean free path l could be erroneously identified to be larger than 10, significantly deviating from the correct value ($l = 0$).

To determine the value of l rigorously, one has to perform regression analysis on the experimental data (of κ or $1/K_{total}$) in response to all range of L. A reliable result of regression analysis not only gives the best-fitted values of l, $1/K_Q$, and $1/K_{contact}$ but should also yield stringent standard errors for them. If regression analyses yields $l > 0$ with a small standard error, ballistic thermal conduction can be claimed.

To easily identify the ballistic thermal conduction from the experimental data, it is recommended to plot the $1/K_{total}$ vs. L relation. The $1/K_{total}$ vs. L relation would allow us to separate the ballistic thermal conduction from the unwanted effect of contact thermal resistance by reading the offset at $L \rightarrow 0$, as shown in Fig. 8.2b. The ballistic thermal conduction can be seen by observing the deviation from diffusive behavior, extrapolating from the data of large L's.

We now reexamine the previous experimental data of Ge and Si [20, 21], whose κ vs. L relation is shown in Fig. 8.3a, c), respectively. Because the effective lengths were collected from different samples of different dimensions, we employ the relation $A/K_{total} = L/\kappa$ and plot normalized thermal resistance A/K_{total} vs. L relations in Fig. 8.3b, d instead. Surprisingly, the A/K_{total} vs. L relation shown in Fig. 8.3b, d does not deviate from diffusive transport behavior down to $L = 42$ nm for Ge film and $L = 1$ μm for Si membranes even if the authors claim to find evidence for room temperature ballistic thermal conduction over $L = 1$ μm for Si. Apparently, the misinterpretation is due to the ignorance of contact thermal resistance in their data. On the other hand, the finite offset observed in Fig. 8.3d is an interesting effect, which indicates that optical heating methods do not necessarily yield zero interface thermal resistance as previously thought. Nevertheless, because the data in Fig. 8.3b, d do not deviate from the diffusive behavior, there is no observable ballistic thermal conduction of Si or Ge at room temperature. Recently,

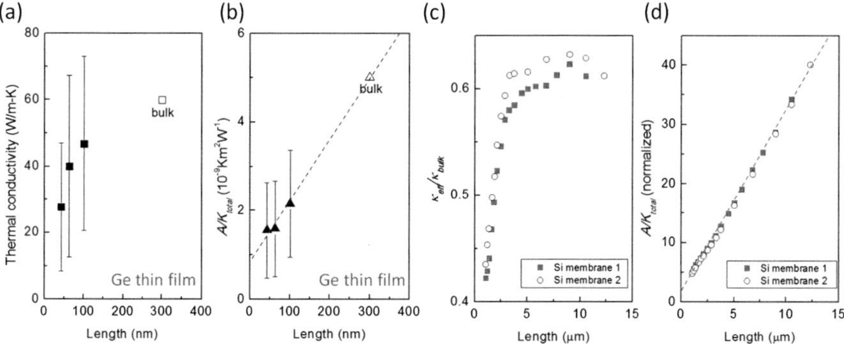

Fig. 8.3 (**a**) κ vs. L and (**b**) A/K_{total} vs. L relations for Ge film. (**c**) κ vs. L and (**d**) A/K_{total} vs. L relations for Si membranes. The diffusive behaviors are denoted by the *gray dashed line* in (**b**) and (**d**), respectively. The data are reproduced from Ref. [20, 21]

there is another experiment extending the investigated regimes to smaller length scales (~30 nm) using patterned Al as nanostructured transducers [22]. But if we employ the same analyses by plotting their κ vs. L data into A/K_{total} vs. L relation, we again find no evidence for the claimed ballistic thermal conduction. These examples cast strong concerns on the adequacies of the experimental methods for probing ballistic thermal conduction.

In this section, we discuss the operational definition of conventional ballistic thermal conduction. We emphasize that, to experimentally determining l of a given material, one must measure the $1/K_{total}$ vs. L relation (or A/K_{total} vs. L relation) and observe the departure from Fourier's law. Unfortunately, many previous experiments either disobeyed the operational definition or ignored the contribution from contact thermal resistances. We reanalyze the previous experiments on Si and find that the claim of $l > 1$ μm for Si at room temperature is incorrect.

8.3 Non-Fourier (Ballistic) Thermal Conduction in SiGe Nanowires

Recently, we have also recognized the issues mentioned in the previous section and directly measured the length-dependent κ and $1/K$ of homogeneously-alloyed SiGe nanowires [23]. SiGe nanowires are a model alloy system in which the role of alloy scatterings of phonons in nanoscale dimensions can be thoroughly investigated. Interestingly, theoretical predictions on l of SiGe vary five orders of magnitudes, ranging from 10 nm to 700 μm [24–27]. The large discrepancies thus motivate us to rigorously investigate the ballistic thermal conduction phenomena in SiGe nanowires.

Experimentally, $Si_{1-x}Ge_x$ ($x = 0.1$–0.6) nanowires with a wide range of structural variations and alloy concentrations were made by chemical vapor deposition methods. The scanning transmission electron microscope (STEM) images and the associated data reveal that $Si_{0.9}Ge_{0.1}$ nanowires are homogeneously alloyed with few defects or twin structures. On the other hand, $Si_{0.4}Ge_{0.6}$ nanowires exhibit apparent twin boundaries, stacking faults, striped compositional variations, surface roughness, and defects, as shown in Fig. 8.4.

To investigate the thermal conduction of individual SiGe nanowires, we fabricated microscale thermal conductivity test fixtures consisting of suspended heaters and sensors shown in Fig. 8.5. The thermal conductivity test fixture consisted of two suspended silicon nitride (SiN_x) membranes each supported by five 420 μm-long and 2 μm-wide SiN_x beams. Nanowires with chosen lengths or diameters were picked up and placed on the test fixture by a sharpened tungsten tip operated by a piezodriven manipulator inside a scanning electron microscope (SEM). In-situ deposition of Pt/C composites was then carried out to rigidly bond the nanowire to the test fixture. To measure the thermal conductance (K) of the nanowire, Joule heating was supplied to the heater and the temperature rises of the heater and sensor

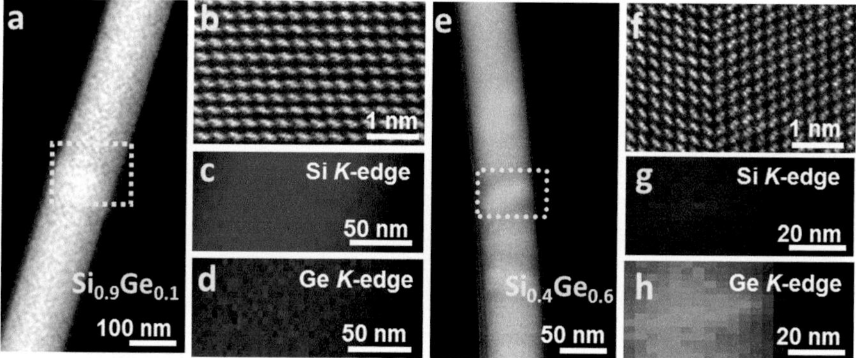

Fig. 8.4 (**a**) A representative STEM image of a homogeneously alloyed $Si_{0.9}Ge_{0.1}$ nanowire. (**b**) An atomic-resolution image of (**a**) showing the crystalline lattice. (**c** and **d**), Si and Ge K-edge elemental mappings of the *yellow-dotted area* of (**a**) showing homogeneously Si and Ge distributions. (**e**) A representative STEM image of a $Si_{0.4}Ge_{0.6}$ nanowire showing uniform intensity distributions radially and aperiodic striped distributions axially. (**f**) An atomic-resolution image of (**e**) displaying a twin boundary. (**g** and **h**), Si and Ge K-edge elemental mappings of the *yellow-dotted area* of (**e**) showing that Si and Ge elements are homogeneously alloyed with striped composition variations [23]

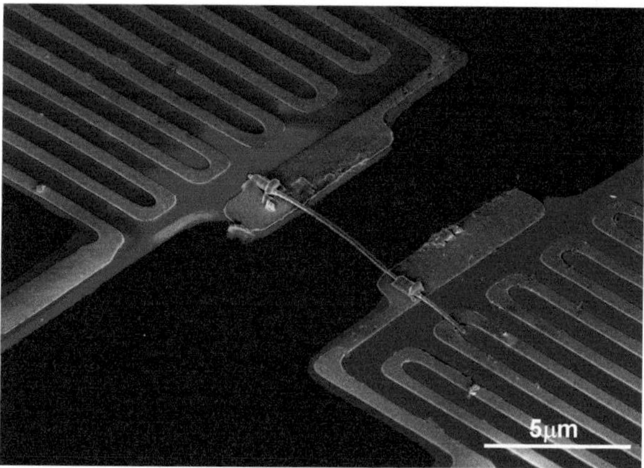

Fig. 8.5 A false colored SEM image of a thermal conductivity test fixture consisting of suspended heater and sensor pads with a SiGe nanowire anchored on it

were measured. Under steady state, K can be obtained using the relation

$$K = \frac{P}{\Delta T_H - \Delta T_S}\left(\frac{\Delta T_S}{\Delta T_H + \Delta T_S}\right) \tag{8.5}$$

where P is the Joule heating power, ΔT_H and ΔT_S is the temperature raise on the heater and the sensor, respectively. Due to the linear relation of resistance with respect to temperature of the Pt film resistors, the temperature variations of the heater and the sensor can be directly obtained by measuring their resistance. The thermal conductivity κ was evaluated by incorporating the length and the diameter of the nanowire determined by SEM. All the measurements were carried out at pressure $<10^{-5}$ mbar to eliminate unwanted heat convection.

We have systematically measured the length dependent thermal transport for more than twenty SiGe nanowires of different diameters, structures, and alloy concentrations. Due to the variations of the cross sectional area (A) between different samples, we plot the normalized thermal resistance (A/K) vs. L relation and κ vs. L in Fig. 8.6. Figure 8.6a shows that the κ increases linearly with L for $L < 8.3$ μm, followed by a slope change at $L = 8.3$ μm, then the κ saturates at 8.2 W/m-K, agreeing with the bulk value of SiGe. Although Fig. 8.6a alone seems to suggest $l = 8.3$ μm for SiGe nanowires, it is necessary to reexamine the data by plotting A/K_{total} vs. L in Fig. 8.6b. From Fig. 8.6b, A/K_{total} decreases linearly with L for $L > 8$ μm, indicating the expected diffusive transport behavior. Interestingly, the A/K_{total} vs. L changes the slope at $L = 8$ μm and it significantly deviates from the diffusive behavior for $L < 8$ μm. Applying the regression analysis gives $l = 8.05$ μm with standard deviation $= 0.31$ μm. Moreover, extrapolating the diffusive behavior to $L \rightarrow 0$ suggests that the (classical) contact thermal resistance is nearly zero, which indicates that the observed ballistic thermal conduction shown in Fig. 8.6a cannot be attributed to effects of finite contact thermal resistance. We emphasize that the x-axes of Fig. 8.6a, b are "length" rather than "effective length", which means that we have strictly followed the operational definition of ballistic thermal conduction

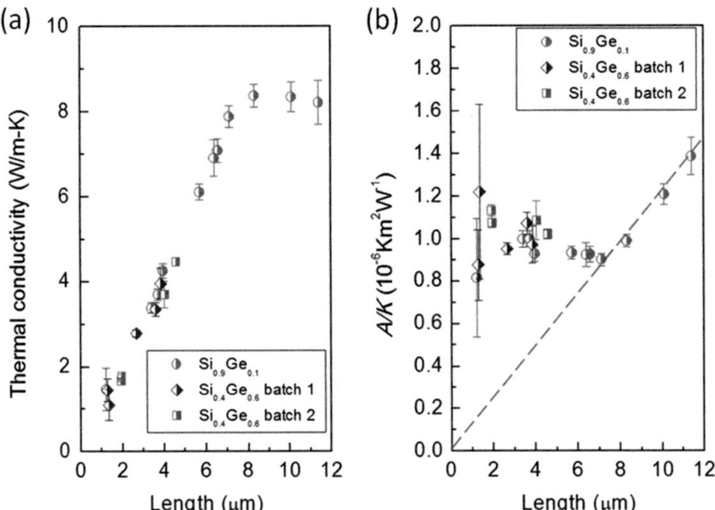

Fig. 8.6 (**a**) κ vs. L and (**b**) A/K_{total} vs. L relations for homogeneously-alloyed SiGe nanowires. Both κ vs. L and A/K_{total} vs. L relations deviate from the diffusive behavior for $L < 8$ μm, indicating the ballistic thermal conduction. The *gray dashed line* is a fitted line extrapolating from data of $L > 8$ μm [19, 23]

and made the variation of the length to be parallel to the temperature gradient. Thus our experimental investigations provide the direct determination of l.

The observed ballistic thermal conduction persisting over 8 μm is not only an unprecedented long distance ever found in thermal conductors at room temperature but also more than 9 times longer than those of nanotubes, graphene, or diamond reported so far [28, 29]. Remarkably, the room temperature averaged phonon mean free path observed in our experiment is even longer than the electronic counterparts ($l_{electron} < 1$ μm) of the highest-mobility (>200,000 $cm^2\ V^{-1}\ s^{-1}$) graphene devices for observing fractional quantum Hall effect at low temperatures [30, 31].

Theoretical calculations have pointed out the low-frequency acoustic phonons (<1 THz) to be the dominant carriers for transmitting heat in SiGe [26]. In fact, we observed a much weaker diameter-dependence of κ in SiGe nanowires ($(\Delta\kappa/(\kappa_{bulk}d) < 5.1 \times 10^{-4}$/nm for SiGe nanowires, whereas $\Delta\kappa/(\kappa_{bulk}d) = 2.3 \times 10^{-3}$/nm for Si nanowires and $\Delta\kappa/(\kappa_{bulk}d) = 2.2 \times 10^{-3}$/nm for Ge nanowires, here d is the diameter of nanowires) [10, 32–34]. Apparently, the criterion of the Casimir limit does not apply to SiGe nanowires and specular scatterings of the low-frequency phonons can happen at the surface of the SiGe nanowires.

We note that the contribution of the low frequency phonons can be estimated from the slope of κ vs. L relation in Fig. 8.6a. Because the low-frequency acoustic phonons are nearly dispersiveless, from the slope ($\kappa/L = 9.5 \times 10^5$ W/K-m^2) of Fig. 8.6a for $L < 8.3$ μm, we can have the relation from dispersiveless kinetic theory $\kappa/L = C_a v_a/3$ (where C_a and $v_a = 4108$ m/s are respectively the averaged specific heat and the averaged sound velocity of the low-frequency phonons). We obtain $C_a = 680$ J/K-m^3 and $C_a/C_{bulk} = 0.04$ % (where $C_{bulk} = 1.7 \times 10^6$ J/K-m^3 is the experimentally measured specific heat) [35]. The result indicates that the low frequency phonons carrying out the heat conduction in the SiGe nanowires only occupy 0.04 % of the excited phonons. Incorporating the density of states of SiGe [36], we further estimate that these phonons exhibit frequencies less than 0.3 THz. On the other hand, applying similar analyses to Si and Ge reveals 20–30 % of the excited phonon modes are responsible for the ballistic heat conductions [8, 20]. The result is consistent with theoretical calculations that phonons below 1 THz dominate the heat transfer in SiGe [26]. In contrast, phonons up to 6 THz contribute equally to the heat transfer in Si [26].

Although ballistic phonon transport could imply infinite thermal conductance in a classical physics picture, the maximum heat flow per channel is limited by fundamental constants in quantum mechanics [37]. In principle, ballistic thermal conduction should obey Landauer's formulation for quantum thermal conduction (K_Q) [37]:

$$K_Q = \frac{k_B^2}{h} \sum_m \int_{x_m}^{\infty} dx \frac{x^2 e^x}{(e^x - 1)^2} T^m (x k_B T/\hbar) \approx \frac{\pi^2 k_B^2 T}{3h} \qquad (8.6)$$

where is k_B Boltzmann constant, T_m is the transmission coefficient of the mth phonon mode. At low temperatures, Landauer's formula can be reduced to a simple form expressed by fundamental constants, as shown in Eq. (8.6). Because the total thermal conductance of a SiGe nanowire is proportional to its cross-sectional area and the number of channels (N) of quantum conductance ($K_Q \sim 10^{-10}$ W/K at 300 K), from $K/A = \kappa/L = 9.5 \times 10^5$ W/K-m$^2 = NK_Q/A$, we obtain the averaged area occupied by each quantum channel $A/N \sim 100$ nm^2. Furthermore, N can be obtained via integrating the density of states:

$$N = \int_0^{k_{\max}} D(k)d^2k = \frac{Ak_{\max}^2}{4\pi^2} \tag{8.7}$$

where $k_{\max} = 2\pi/\lambda_{\min}$ denotes the maximum wave vector (or equivalently, the shortest wavelength $\lambda_{\min}$) of phonons that carry out the heat conduction in a SiGe nanowire. From Eq. (8.7), we obtain $\lambda_{\min} \sim 10$ nm and maximum frequency of the phonons $f \sim v_a/\lambda_{\min} = 0.4$ THz, consistent with earlier estimates based on $C_a/C_{bulk} = 0.04$ %. Thus the independent analysis based on quantum thermal conductance suggests that the dominant acoustic phonon wavelength is larger than 10 nm (or the frequency is less than 0.4 THz), which is again consistent with previous estimates.

Now we discuss the issue of contact thermal resistance. From operational definition, contact thermal resistance the measured residue thermal resistance as $L \to 0$. In the diffusive (classical) regime, contact thermal resistance results from backscattering of phonons. In the ballistic (quantum) regime, contact thermal resistance occurs even if the phonons are scattering-free but the available quantum channels are geometrically constricted at the contacts. Remarkably, extrapolating the data in the diffusive regime ($L > 8.3$ μm) to $L \to 0$ in Fig. 8.6b gives negligible classical contact resistance (<2 % of the nanowires' thermal resistance), indicating nearly scattering-free transmission of phonons at the contacts. On the other hand, extrapolating the data in the ballistic regime ($L < 8.3$ μm) to $L \to 0$ shows that the contact thermal resistance ($1/K = L/A\kappa$) is inversely proportional to cross sectional area ($A = \pi d^2/4$) of the nanowire rather than the real contact area ($\sim \pi dL_c$, $L_c \sim 300$ nm is the physical contact length of the nanowire and the heater/sensor pads, d is the diameter of the nanowire). The counter-intuitive result is in fact due to the occurrence of ballistic thermal conduction that constrains the phonons to enter into limited heat conduction channels of a nanowire [38]. Because the number of phonon modes of a waveguide is constrained by the smallest cross-sectional area from a heat source to a heat sink, considering $\pi d^2/4 \ll \pi dL_{con}$ in our experiment, the total number of available channels for transmitting heat is thus limited by $\pi d^2/4$ rather than πdL_{con}. The effect also explains why so small deviations to the dashed line in Fig. 8.2 are observed for SiGe nanowires with different contact geometries. Furthermore, independent analyses suggest that the classical contact thermal resistance must be less than 10 % otherwise the κ of SiGe nanowires would be larger than that of bulk SiGe [14, 39].

So far we have provided two kinds of evidence (i.e. the length dependent thermal transport and the unconventional contact thermal resistance) supporting the room temperature ballistic thermal conduction over 8 μm in SiGe nanowires. Now we provide the third evidence that the ballistic conduction enables non-additivity of resistances in series, a phenomenon that was known to exclusively occur in low-dimensional ballistic electronic systems at ultralow temperatures in the past [40]. Similar to the electrical counterparts, connecting two diffusive or two ballistic thermal conductors in series is known to yield distinct results on the total thermal conductance (K_{total}):

$$K_{total} \leq \begin{cases} (K_1,\ K_2)_{\min} & \text{for ballistic conductors} \\ K_1K_2/\ (K_1 + K_2) & \text{for diffusive conductors} \end{cases} \tag{8.8}$$

where the inequality holds when the classical contact thermal resistance at the junction is nonzero. $(K_1,\ K_2)_{min}$ denotes the minimum value of the set $(K_1,\ K_2)$, a result originating from the geometrical constriction in ballistic thermal conductors mentioned above. Note that Eq. (8.8) has taken into account the classical contact resistance so the K_{total} should never exceed $K_1K_2/(K_1 + K_2)$ for two diffusive conductors connected in series. Therefore, any violations of the inequality of the diffusive thermal conductors in Eq. (8.8) will be the evidence of ballistic thermal conduction.

The upper graphs of Fig. 8.7 shows a series of SEM images when we rubbed, pressed, or rotated the two mechanically-touched SiGe nanowires against each other using a manipulator. The process allows us to survey many possible thermal contact configurations between two SiGe nanowires. Here a constant power (P) was supplied to the heater and the temperature rise of the heater before ($\varDelta T_b$) and after ($\varDelta T_a$) connecting the nanowires was measured. Because the thermal conductance of the tungsten tip is $\sim 10^4$ higher than the nanowire, it functions like a heat sink for the system. The thermal conductance (K) of the nanowire can be obtained via

$$K = P\left(\frac{1}{\Delta T_a} - \frac{1}{\Delta T_b}\right) \tag{8.9}$$

We then measured the K of the system for all possible thermal contact configurations, as shown in Fig. 8.7.

The lengths, diameters, and thermal conductances of the two SiGe nanowires are respectively $L_1 \sim L_2 = 3.65$ μm, $d_1 \sim d_2 = 183$ nm, $K_1 \sim K_2 = 2.81 \times 10^{-8}$ W/K. The corresponding measured K_{total} are displayed in Fig. 8.7. At sequence #3, 4, and 5, the data clearly exceed the limit predicted by diffusive conductors. It is because the total length ($L_1 + L_2 = 7.3$ μm < 8.3 μm) is within the ballistic transport regime, the whole system still behaves like a ballistic thermal conductor (with added phonon scatterings at the junction). Remarkably, perfect ballistic phonon transmission (i.e. $K_{total} = (K_1,\ K_2)_{min} = 2.81 \times 10^{-8}$ W/K) is observed at sequence #5, indicating that the phonon scatterings at the junction can be reduced to zero.

Figure 8.8a displays the measured K_{total} of another two SiGe nanowires of dissimilar diameters connected in series (the lengths, diameters, and thermal conductances are respectively $L_1 = L_2 = 3$ μm, $d_1 = 158$ nm, $d_2 = 140$ nm,

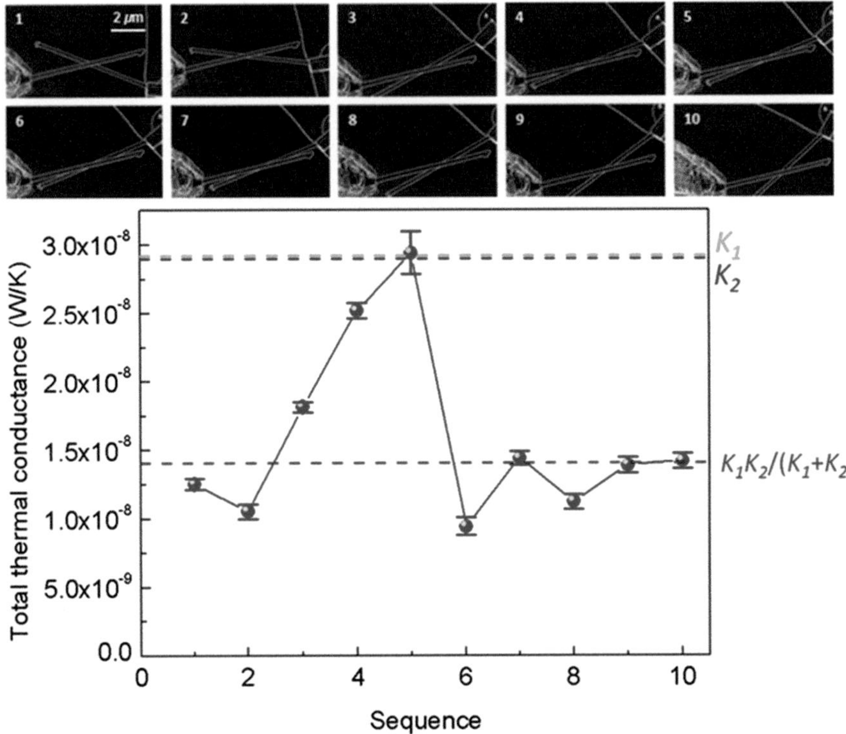

Fig. 8.7 (*Upper*) A series of SEM images when we rubbed, pressed, or rotated the two mechanically-touched SiGe nanowires against each other using a manipulator. The process allows us to survey many possible thermal contact configurations between two SiGe nanowires. (*Lower*) measured total thermal conductance (K_{total}) during the sequence. The *green*, *blue*, and *red dashed lines* denote the values of K_1, K_2, and $K_1K_2/(K_1 + K_2)$, respectively. The *blue/red dashed line* also denotes the maximum K_{total} allowed from Eq. (8.8) for ballistic/diffusive thermal conductors [23]

$K_1 = 1.96 \times 10^{-8}$ W/K, $K_2 = 1.53 \times 10^{-8}$ W/K). Remarkably, all the data shown in Fig. 8.8a disobey the inequality of Eq. (8.2) for diffusive conductors. Instead, the K_{total}'s follow the inequality for ballistic thermal conductors and the largest values (sequence #8, 9 in Fig. 8.8a) equals to $(K_1, K_2)_{min}$. Controlled experiments on connecting two Si nanowires in series indeed obey the expected relation for diffusive conductors, as shown in Fig. 8.8b.

We have provided three kinds of experimental evidence for the unusual room temperature ballistic thermal conduction persisting over 8.3 μm in SiGe nanowires. Curiously, can the observed phenomena be attributed to anomalous thermal conduction in 1D systems? We note that the above analyses are all based on bulk properties of SiGe. Even though the low-frequency phonons dominate the heat conduction processes in SiGe, their wavelengths are within the orders of 10 nm, which are smaller than the diameters (>40 nm) of the nanowires. Thus effects of quantum confinements do not seem to play a significant role here. Most importantly,

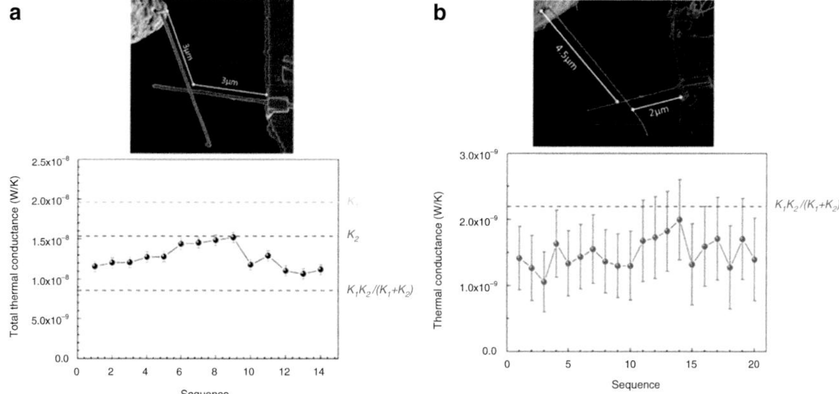

Fig. 8.8 (**a**) (*Upper*) A representative SEM image of two SiGe nanowires of dissimilar diameters mechanically connected and repeated the similar sequence described in Fig. 8.7. (*Lower*) The measured K_{total} during the sequence. The values of K_1, K_2, and $K_1K_2/(K_1+K_2)$ are denoted respectively at the figure. The K_2 and $K_1K_2/(K_1+K_2)$ lines also denote the maximum K_{total} allowed from Eq. (8.8) for ballistic/diffusive thermal conductors. (**b**) (*Upper*) A representative SEM image for a controlled experiment on two Si nanowires. (*Lower*) The measured K_{total} during the sequence, which displayed the expected diffusive behavior [23]

the κ stops increasing when L is larger than 8.3 μm, as shown in Fig. 8.6a. The result indicates that the theoretically proposed anomalous thermal conduction in 1D systems is not observed experimentally in SiGe nanowires. The observed non-Fourier thermal conduction phenomena in SiGe nanowires can be attributed to conventional ballistic thermal transport and they should be observed in bulk SiGe as well. Therefore, the $l \gg d$ proposed in the early section is not the criterion for defining a 1D system and the SiGe nanowires are not 1D materials, either.

8.4 Non-Fourier (Ballistic) Thermal Conduction in Si-Ge interfaces

Si-Ge core-shell nanowires consist of Si-core and Ge-shell of nanometer scales. They are heterogeneously-alloyed materials with tubular-structured interfaces. Recent theoretical works have suggested interface engineering could be more effective than surface roughness in reducing the thermal conductivity [41–46]. In a core-shell nanowire, atoms at the heterogeneous interfaces are stretched, inducing a strong coupling between the transverse and the longitudinal modes. It is suggested that because the transverse modes are quantized and localized, the coupled longitudinal modes are also not-propagating, resulting in a reduction of κ [44–46]. Aside from the effect, the added interface scatterings is supposed to reduce κ further [41, 42]. For electrical transport, the heterogeneous interface forms a tubular electronic conduction channel similar to carbon nanotubes and it support a high-mobility quasi-1D hole gas with mean free paths exceeding 170 nm at room temperature [47]. Many interesting electronic phenomena have been demonstrated, including the potential for high performance transistors [48], programmable circuits [49], and spin qubit manipulations [50]. Combining the dramatic reduction of κ and the high hole-mobility will make Si-Ge core-shell nanowires promising materials for thermoelectric applications. For experimental thermal transport measurements, M. C. Wingert et al. reported unexpected suppression of κ in Ge-Si core-shell nanowires have made the topic particularly interesting [23, 34, 51–53]. Naturally, it will be intriguing to ask whether the nanoscale tubular hetero-interfaces could mimic 1D systems and display unusual non-Fourier thermal conduction phenomena.

Experimentally, Si-Ge core-shell nanowires were synthesized by a chemical vapor deposition method [54]. The structures of the synthesized nanowires were characterized by STEM and TEM, revealing imperfect structures with many twin boundaries, defects, rough surfaces, and dislocations, as shown in Fig. 8.9a, b. The spatially-resolved energy dispersive X-ray spectroscopy (EDS) spectra in Fig. 8.9c, d and STEM-EDS spectral imaging across the nanowire in Fig. 8.9e–g confirm the presence of Si in the core region and Ge in the shell region. Interlayer diffusion was found to be less than 5 %, determined by the EDS resolution limit. The nanowires generally have lengths over 10 μm. The diameter of the Si-core can be as small as 12 nm while the outer diameter of the nanowire can be as large as 225 nm.

Ideally, if the thermal transport of Si-Ge core-shell nanowires exhibit universal properties similar to what we have found in homogeneously-alloyed SiGe nanowires, we could repeat the experimental method in the last section and measure

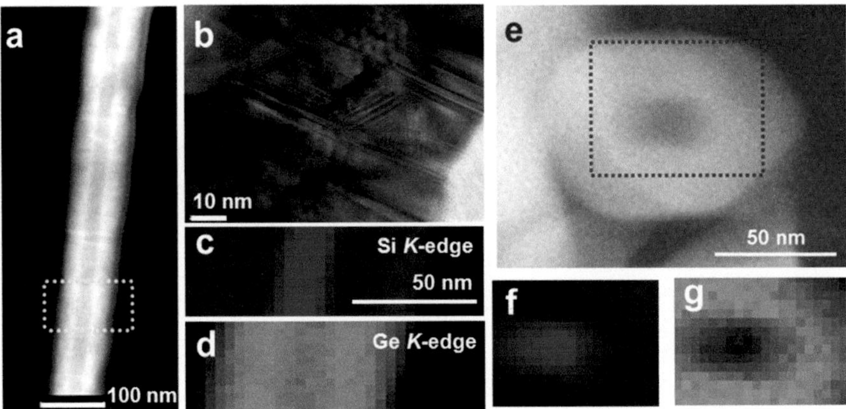

Fig. 8.9 (**a**) A representative high-angle annular dark field (HAADF) image of a Si-Ge core-shell nanowire. (**b**) A high resolution TEM image of the same nanowire. The corresponding (**c**) Si K-edge and (**d**) Ge K-edge elemental maps. (**e**) The cross-sectional HAADF image of the nanowire. The corresponding (**f**) Si K-edge and (**g**) Ge K-edge elemental maps confirm the Si-core and Ge-shell structure of the nanowire [23]

the $1/K$ vs. L and κ vs. L relations of different nanowires to unravel the ballistic thermal conduction phenomena. Unfortunately, due to the large variations of the Si-core diameters, Ge-shell thickness, concentrations of defects and impurities, or surface roughness between different samples, the investigated Si-Ge core-shell nanowires do not exhibit the desired universality. Thus new experimental methods must be invented to overcome the challenges.

We have employed two different methods to study the thermal transport behaviors of the Si-Ge core-shell nanowires [55]. Method I is similar to the one expressed in Eq. (8.9) that a constant power (P) was supplied to the heater and the temperature rise of the heater before (ΔT_b) and after (ΔT_a) connecting the nanowire was measured. But now the $1/K$ vs. L and κ vs. L relations are deliberately investigated on the same nanowire by repeatedly connecting/disconnecting a tungsten tip to different positions of a nanowire so as to vary the investigated L, as shown in Fig. 8.10a. The method will reduce the data variations inherent in different samples.

For Method II, we employed an electron-beam-heating technique similar to the method developed by D. Liu and her coworkers [56]. As shown in Fig. 8.10, a nanowire was suspended across two integrated Pt film resistors supported by suspended SiN_x pads. Under steady state, the total thermal conductance (K_{total}) of the system (including contributions from the thermal resistance of the nanowire (R_{NW}) and the (classical) contact thermal resistances at the left/right pads ($R_{c,left}$ and $R_{c,right}$)), can be obtained using the relation:

$$K_{total} = \frac{1}{R_{NW} + R_{c,left} + R_{c,right}} = \frac{P}{\Delta T_h - \Delta T_s}\left(\frac{\Delta T_h}{\Delta T_h + \Delta T_s}\right) \tag{8.10}$$

By switching the role of heater and sensor, we can respectively determine the thermal conductance for heat flux flowing from right to left ($K_{R\to L}$) and left to right ($K_{L\to R}$).

We then employed an electron beam (5 kV) of SEM and focused at position x of the nanowire ($x = 0$ is defined to be located at the edge of the left pad). Due to electron-phonon inelastic scatterings, the electron beam acted as a local heater generating total power $P_{total}(x)$ and raised the local temperature at position x to $T(x)$. Under steady state, we have the following relation:

$$\begin{cases} P_{total}(x) = P_{left}(x) + P_{right}(L-x) \\ P_{left}(x) = K_{R\to L}(x)\left(T(x) - T_{LeftPad}(x)\right) = K_{LeftPad}\left(T_{LeftPad}(x) - T_{bath}\right) \\ P_{right}(L-x) = K_{L\to R}(L-x)\left(T(x) - T_{RightPad}(L-x)\right) \\ \qquad\qquad = K_{RightPad}\left(T_{RightPad}(L-x) - T_{bath}\right) \end{cases} \tag{8.11}$$

where $P_{left}(x)$ and $P_{right}(L{-}x)$ are respectively the heat flux flowing to the left and to the right pads when the electron beam is positioned at x. $K_{R\to L}(x)$ and $K_{L\to R}(L{-}x)$ are the thermal conductance of the nanowire measured at distances x and $L{-}x$, for heat flux flowing from right to left and left to right, respectively. $T_{LeftPad}(x)$,

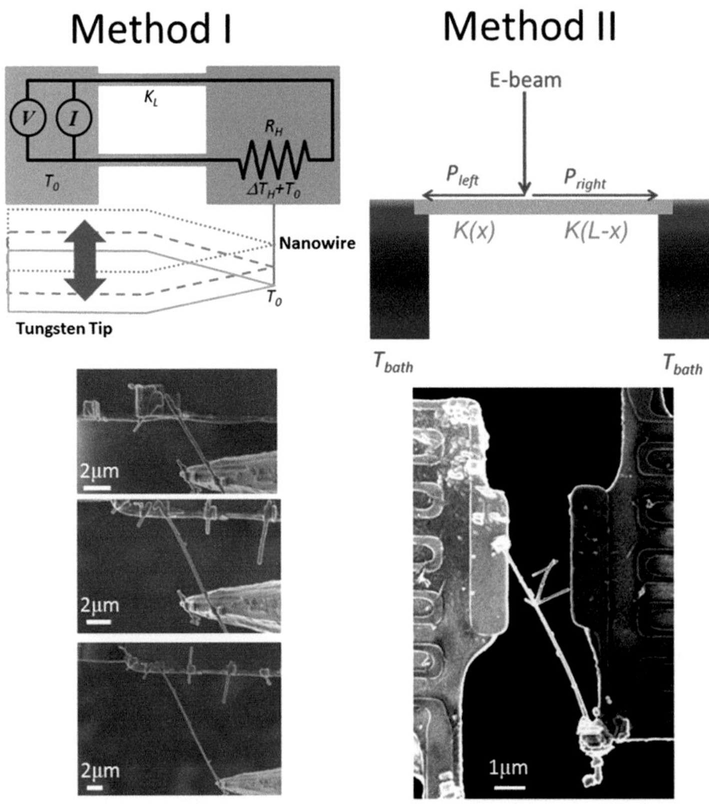

Fig. 8.10 Schematics (*upper*) and the corresponding SEM images (*lower*) of two different experimental methods for measuring length dependent thermal transport of nanowires. (Method I) One end of a nanowire is anchored to a microfabricated heater and the other end is attached to a tungsten tip serving as a heat sink. Here the heat sink is maintained at temperature T_0 and the Joule heating raises the temperature of the heater to $\Delta T_H + T_0$. The difference of ΔT_H before and after connecting the nanowire is measured so as to determine the thermal conductance (K) of the sample. The tungsten tip can repeatedly connect/disconnect to the nanowire so as to vary the investigated L. (Method II) A nanowire is anchored between two suspended sensors and an electron beam is focused on the nanowire serving as a local heater. The electron beam can be used for injecting heat at any positions along the nanowire [55]

$T_{RightPad}(x)$, and T_{bath} denote the temperature measured at the left pad, the right pad, and the heat bath, respectively.

Experimentally, we can position the electron beam at $x = -\varepsilon, L + \varepsilon$, ($\varepsilon < 1\ \mu$m), which is equivalent to heating the left pad and the right pad, respectively. Correspondingly, $K_{L\to R}(L+\varepsilon)$ and $K_{R\to L}(L+\varepsilon)$ can be independently determined by the thermal conductivity test fixture and thus we can obtain $T(-\varepsilon)$ and $T(L+\varepsilon)$ (note that $T(-\varepsilon)$ and $T(L+\varepsilon)$ are local temperatures at the ends of the nanowire anchored on the pads, but not the temperature of the pads). Therefore, unlike the previous

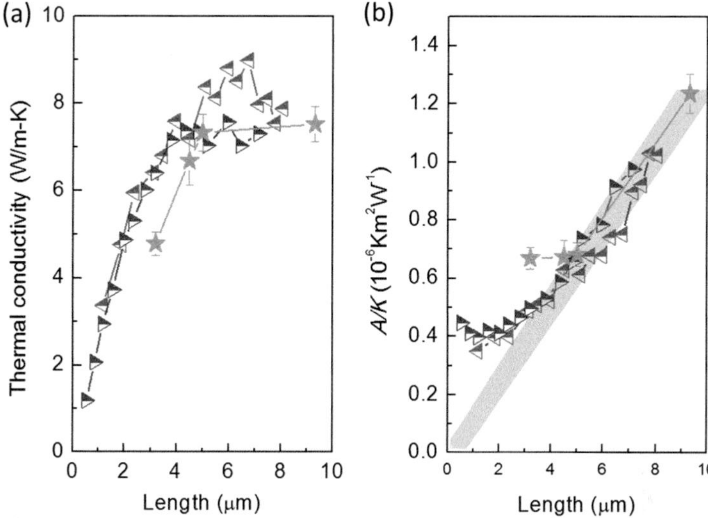

Fig. 8.11 (**a**) κ vs. L and (**b**) A/K_{total} vs. L relations for Si-Ge core-shell nanowires investigated by two different experimental methods. It can be seen that both κ vs. L and A/K_{total} vs. L relations deviate from the diffusive behavior for $L < 2.5$ μm, indicating the ballistic thermal conduction. The *gray* belt is a *fitted line* extrapolating from data of $L > 5$ μm [19, 55]

report where diffusive thermal conduction was presumed in the analyses [56], our method is capable of probing ballistic thermal conduction of a nanowire.

Putting the collected data together in Fig. 8.11, we can see the data start to deviate from Fourier's law for $L < 5$ μm, indicating micron-scale ballistic thermal conduction for the Si-Ge core-shell nanowires. Applying regression analysis gives $l = 5$ μm and 2.61 μm for the samples measured using Method I and Method II, respectively. The l of Si-Ge core-shell nanowires not only largely deviate from the indirect estimates that $l < 100$ nm for SiGe alloys [39, 45, 57–59], but is also much longer than $l \sim 1$ μm proposed for the best thermal conductors like diamond and graphene [28]. The finding also highlights the importance of our direct, rigorous, thermal transport measurements on l.

Similar to those of homogeneously-alloyed SiGe nanowires, we also find that κ displays much weaker dependence on the diameter ($\Delta\kappa/(\kappa_{bulk}d) < 3 \times 10^{-4}$/nm for Si-Ge core-shell nanowires for $d = 100 \sim 220$ nm). Because surface roughness are known to scatter high-frequency phonons [60], the much weaker d-dependence of κ in Si-Ge core-shell nanowires indicates the absence of high-frequency phonons in carrying out the heat conduction.

Note that the data for $L > 5$ μm shown in Fig. 8.11b can be extrapolated to zero when $L \to 0$, suggesting negligible classical contact thermal resistance. In contrast, finite quantum contact thermal resistance occurs for $L < 5$ μm and remains constant even when $L \to 0$, obeying Landauer's formulation for ballistic phonons. The small classical contact thermal resistance determined in our experiments is also consistent with the low-frequency phonon picture that shows insensitivities to nanoscale perturbations.

The hypothesis of low-frequency phonons mentioned above suggests that the phonon spectrum in carrying out the heat conduction is nearly monochromatic and dispersiveless. Similar to the analysis in the previous section, we can apply the dispersionless kinetic theory of phonons $\kappa = C_a v_a l/3$ to analyze our results. Surprisingly, we find the slope of the data in Fig. 8.3 ($\kappa/L = C_a v_a/3 \sim 2 \times 10^6$ W/K-m^2) is only 0.1 % of the bulk's value ($C_{bulk} v_{bulk}/3 = 1.7 \times 10^9$ W/K-m^2) [35]. The result suggests that only 0.1 % of the total excited phonons carry out the heat conduction process in the Si-Ge core-shell nanowires while the majorities of the high-frequency phonons are localized. The localized modes block most heat conduction channels, resulting in a reduced κ, similar to what has been observed in homogeneously-alloyed nanowires.

Comparing with $l > 8.3$ μm of for homogeneously-alloyed bulk SiGe [23], the reduced $l \sim 5$ μm observed in Si-Ge core-shell nanowires could be due to the phase segregation of Si and Ge elements that makes the phonon localization less effective. Because neither Si nor Ge channels exhibit l longer than 200 nm [8, 20], the pronounced elongation of l in Si-Ge core-shell nanowires must be attributed to phonons propagating along the tubular Si-Ge interfaces, with added small contributions from Si or Ge channels. Even if interlayer diffusion may cause the Si-Ge interface to be effectively thicker than theoretically anticipated, a 5 % alloy concentration set by the STEM detection limit is sufficient to induce the micron-scale ballistic thermal conduction.

In this section, we provide experimental evidence that non-Fourier thermal conduction can be found in Si-Ge core-shell nanowires in which the heterogeneous interface can induced high-frequency phonon localization and low-frequency ballistic phonons propagating over 5 μm. Although the discovery is remarkable, we find that κ converges to a constant value for large L and no anomalous thermal conduction is observed. It suggests that even though the Si-Ge interface is a few nanometers thick and its diameter is only tens of nanometers, the observed phenomena still belong to the conventional ballistic thermal conduction. We speculate that interlayer diffusion is the possible cause that makes the Si-Ge interface more similar to a 3D system rather than an ideal 1D system.

8.5 Non-Fourier (Ballistic) Thermal Conduction in Multiwall Nanotubes

Multiwall carbon or boron-nitride (BN) nanotubes comprise concentric cylindrical shells of strongly sp^2-bonded carbon or BN atoms, whereas the adjacent shells interact predominantly by weak van der Waals forces. The large disparity of the bonding strength also makes the thermal conduction in the multiwall nanotubes highly anisotropic. Because of the negligible interlayer diffusion and the atomic thin shells, they should be suited for experimental investigations of non-Fourier thermal conduction in 1D systems.

We synthesized multiwall carbon nanotubes using arc discharged method and multiwall BN nanotubes using chemical vapor deposition method. The samples exhibit outer diameter of 30–40 nm and lengths ~10 μm. The highly disparity of thermal conductivity along the nanotube axis and across the shells makes the heat current mostly concentrated on the outermost shell where the contacts are connected and the external heat current is injected. As shown in Fig. 8.12, if we assume the cross sectional area of a multiwall nanotube carrying the heat current to be the entire shell area ($A = \pi d^2/4$, where d is the outer diameter of the multiwall nanotube), κ would display an inverse, linear dependence on d [61]. The result suggest that if we assume that the outermost shell carries the heat current only (i.e. $A = \pi d\delta$, where $=$ 0.34 nm is the thickness of the shell), the equivalent κ would be independent of d. In addition, the equivalent κ (~5800 W/m-K at room temperature) would be close to the theoretical limit of a single-wall carbon nanotube [62]. The result suggests that heat currents are likely to flow at the outermost shell of a multiwall nanotube.

We note that the conventional two-probe method is inadequate to unravel non-Fourier thermal conduction phenomena from unwanted effects of contact thermal resistance. To overcome the difficulty, we invent a sequential multiprobe method that can establish a sample's deviation from Fourier's law behavior even with the presence of finite contact thermal resistances. Figure 8.13 display the experimental procedure for a BN nanotube attached to the test fixture via two Pt/C composites [trimethyl)methylcyclopentadienyl platinum, $(CH_3)_3(CH_3C_5H_4)Pt$] deposition. The right contact attaches the nanotube to the top of a preformed vertical "rib" on the right pad. The vertical ribs ensure that the nanotube is fully suspended along its entire length between the contacts. Then the K of the nanotube is measured. Following the measurement, an additional Pt/C composite is deposited, gradually reducing the suspended length, as shown in Fig. 8.13. Although it is known that the

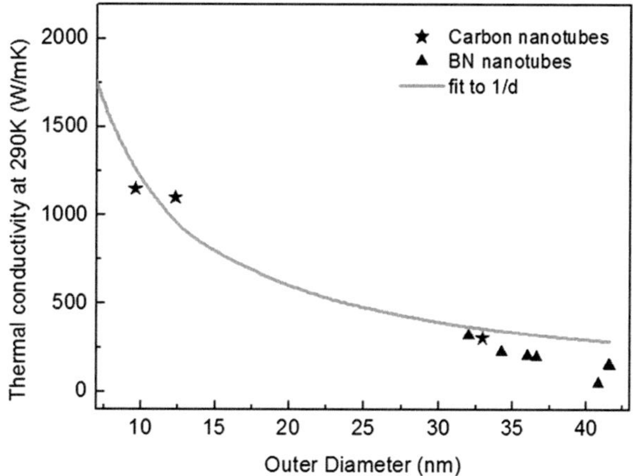

Fig. 8.12 Measured κ vs. d for various multiwall carbon nanotubes and BN nanotubes. The *solid curve* is a fit to $\kappa \sim 1/d$ relation [61]

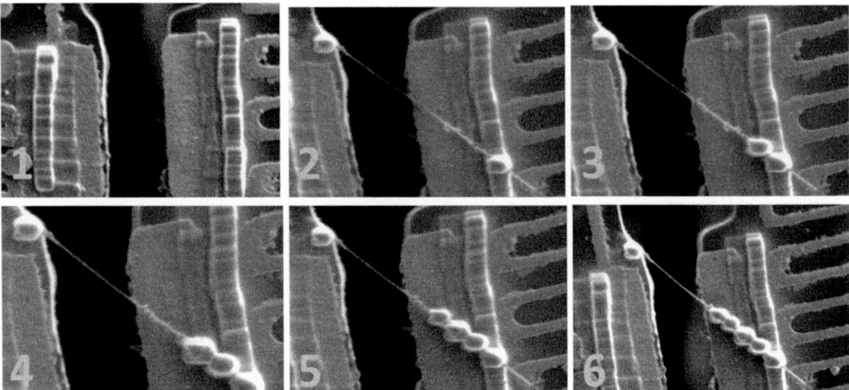

Fig. 8.13 A series of SEM images showing a sequence of Pt/C composite deposition that gradually reduces the suspended length of a BN nanotube whose original length is 7.01 μm [63]

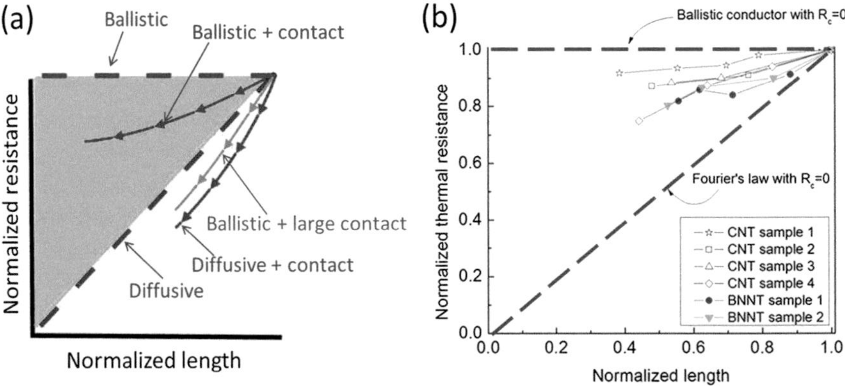

Fig. 8.14 (**a**) A schematic normalized resistance vs. normalized length relation showing the possible results of the multiprobe method; including ballistic conduction with zero contact resistance (*blue dashed line*), ballistic conductor with small contact resistance (*solid curve with symbols*), diffusive conductor with zero contact resistance (*dashed red line*), ballistic conductor with large contact resistance (*dark blue curve with symbols*), and diffusive conductor with finite contact resistance (*red solid line with symbols*). Data of non-Fourier thermal conductor (with small contact resistance) will fall into the cyan region. (**b**) Experimental data for four carbon nanotubes (*open symbols*) and two BN nanotubes (*solid red and green symbols*) [63]

Pt/C composite could diffuse along the nanotube, its contribution to the total thermal conductance is less than 2 %, determined by a controlled experiment.

The advantage of the multiprobe method is to discern the diffusive thermal conduction with ballistic transport phenomena even with finite contact thermal resistances. As illustrated in Fig. 8.14a, when plotting normalized thermal resistance vs. normalized L and assuming the contact thermal resistance is zero, ballistic transport will yield the horizontal blue dashed line while diffusive conduction will be at the red dashed line. When the contact thermal resistance is small, the results of ballistic transport will fall within the blue region shown in Fig. 8.14a while the results of diffusive conduction will be located below the red dashed line. Of course, when the contact thermal resistance is large, the results of ballistic transport may fall below the red dashed line as well, making it indistinguishable from diffusive conduction. Nevertheless, the multiprobe method can help us to experimentally discern the differences between ballistic transport and diffusive conduction if contact thermal resistances are not dominant.

Figure 8.14b shows the measured data for four carbon nanotubes and two BN nanotubes. It is clear that all of the data consistently lie above the red dashed line. The results demonstrate that the multiwall nanotubes are non-Fourier thermal conductors at room temperature. Furthermore, we note that electrical conduction of a multiwall carbon nanotube should be nearly diffusive. Thus simultaneous investigating the electrical/thermal resistance vs. L on the same sample will provide the best controlled experiment to check the validity of our findings. Figure 8.15 shows the normalized electrical/thermal resistance vs. normalized L, it can be seen that the electrical transport behavior is very close to the expected diffusive result. On the other hand, the thermal transport significantly deviates from the diffusive result. The experiment strongly supports that multiwall nanotubes are non-Fourier thermal conductors.

A violation of Fourier's law can be quantified by parametrizing $\kappa \sim L^{\alpha}$. To determine α, we will employ circuit models to analyze our measured data. Mathematically, because there are uncertainties on whether each Pt/C composite deposition may connect to the adjacent contacts, we utilize two thermal circuit models to consider the two extreme situations, as shown in Fig. 8.16. In model A, each contact is separated from others. The thermal resistance of the sample between each contact post is equivalent to a series of resistors $R_s(n)$ ($n =$ sequence number) connected to each other, with contact resistance $R_c(n)$ representing each deposition connected in

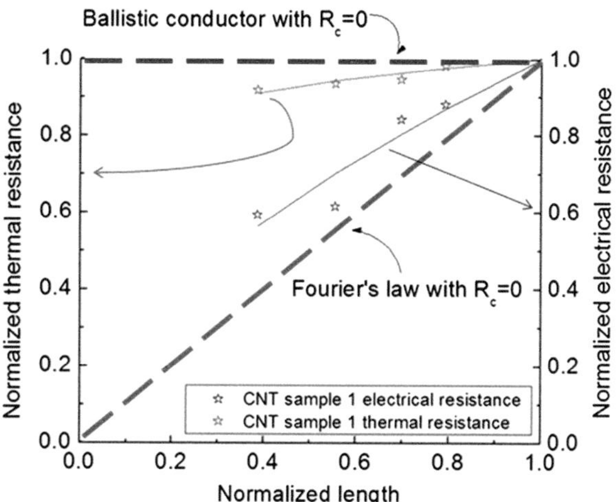

Fig. 8.15 Normalized electrical/thermal resistance vs. normalized length for a multiwall carbon nanotube that simultaneous length dependent electrical/thermal transport measurements were conducted [63]

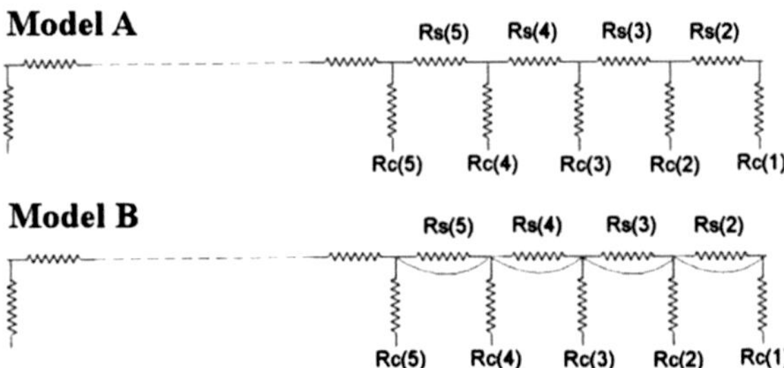

Fig. 8.16 Two circuit models for analyzing the experimental data, where $R_s(n)$ and $R_c(n)$ respectively denote the sample resistance and the contact resistance at each Pt/C composite deposition [63]

parallel to the $R_s(n)$'s. Each $R_s(n)$ and $R_c(n)$ has been adjusted for the corresponding geometrical factors, i.e., $R_s(n) = R_s l(n)^{1-\alpha}$ and $R_c(n) = R_c/l(n)$, where $l(n)$ is the length of the nth contact. In model B, we consider the contacts are shorted to each other. The two models qualitatively yield the expected results shown in Fig. 8.14a. However, we find that model B generally yields a higher α (i.e. a large deviation from Fourier's law) than does model A. In the following, we will take conservative estimation on α and focus on the results based on model A.

Figure 8.17a, b show the fitting results using circuit model A for a carbon nanotube and a BN nanotube. It is clear that Fourier's law gives poor fits to the experimental data. The best fit is found to occur at $\alpha = 0.6$ for the carbon nanotube and $\alpha = 0.5$ for the BN nanotube, respectively. The contact resistance of the first deposition contributes less than 28 % and 25 % to the original thermal resistance, respectively.

We have repeated the analyses for different nanotubes. Generally, we find $\alpha = 0.6$–0.8 for carbon nanotubes and $\alpha = 0.4$–0.6 for BN nanotubes. The difference may originate from the differences of isotopic disorder between BN nanotubes and carbon nanotubes (boron has 19.9 % ^{10}B and 80.1 % ^{11}B isotopic disorder, whereas carbon has 99 % ^{12}C).

The phonon mean free paths (l) of the investigated multiwall nanotubes were estimated to be $l = 20$–50 nm using a simplified relation $\kappa = Cvl$, where C and v are, respectively, the specific heat and averaged sound velocity ~15 km/s. But the values are now known to be underestimated due to the ignorance of phonon dispersion relation. In fact, experiments on inducing breakdown of multiwall carbon nanotubes by biasing high electrical currents have suggested that the phonon mean free path to be in the range of 800 nm at high temperature. If we use the effective $\kappa \sim 5800$ W/m-K of the outermost shell for the estimation, the l could be larger than 1 μm. Therefore, although we have provided solid evidence for non-Fourier thermal conduction in multiwall carbon nanotubes and BN nanotubes, it is possible that the above results are due to mixed effects from conventional ballistic thermal conduction. In fact, the sample length variation shorter than 6 μm and the large α's (0.6–0.8, much larger than theoretical predictions) have suggest that the condition for probing anomalous thermal conduction in 1D systems is not rigorously satisfied in the investigated multiwall nanotubes. The non-Fourier

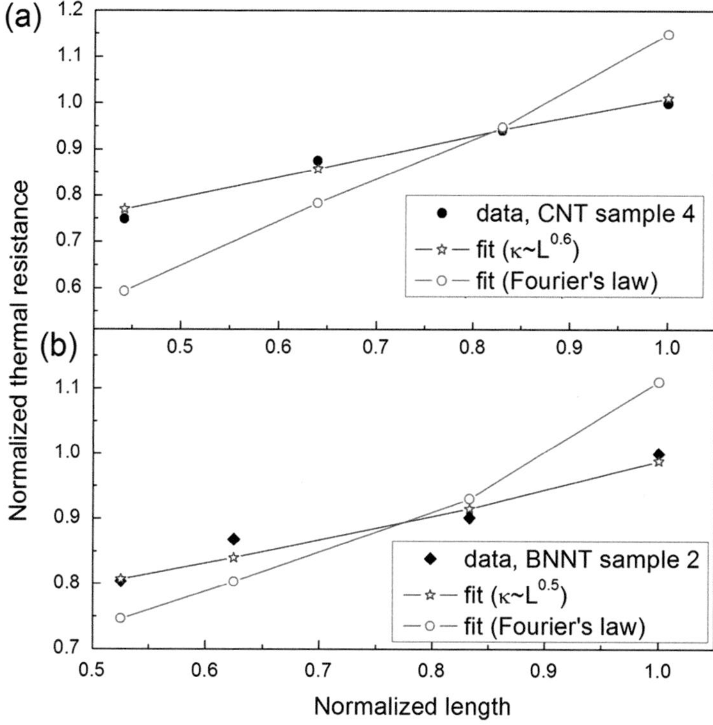

Fig. 8.17 Normalized thermal resistance vs. normalized sample length for (**a**) CNT sample 4 (*solid circles*), best fit using $\alpha = 0.6$ (*open stars*), and best fit using Fourier's law (*open circles*). (**b**) BN nanotube sample 2 (*solid diamonds*), best fit using $\alpha = 0.4$ (*open stars*), and best fit using Fourier's law (*open circles*) [63]

thermal conduction observed in multiwall nanotubes is possible to be mixed with conventional ballistic thermal transport phenomena.

8.6 Non-Fourier (Anomalous) Thermal Conduction in Ultralong Carbon Nanotubes

As we have explained in previous sections, although thermal transport measurements on SiGe nanowires, Si-Ge interfaces, and multiwall nanotube reveals interesting ballistic thermal conduction at room temperatures, the observed phenomena are still within the conventional knowledge of ballistic conduction. To experimentally probe anomalous thermal conduction phenomena in 1D systems, nanoscale samples with lengths much larger than l must be used for our investigations. However, it is not a trivial task. We not only need to evade contributions from ballistic phonons by studying sufficiently long nanomaterials, but also need to minimize unwanted effects such as impurity distributions or defect concentrations that may complicate the experimental results.

On the other hand, theoretical disputes on many issues of anomalous thermal conduction have not completely settled yet. For example, while the power law divergence of κ's in 1D models have been known to be insensitive to disorders, it is not clear whether there would exist a universal α for all 1D systems [64]. Even though the renormalization group analyses have suggested a universal $\alpha = 1/3$ [65], various numerical results have indicated otherwise [66, 67]. Moreover, mode-coupling theory and molecular dynamics simulations have suggested a wide range of α's depending on the temperatures, diameters, and chiralities of a quasi-1D system like a single-wall carbon nanotube (SWCNT) [68, 69]. Curiously, while the divergence of κ have been rigorously proved in 1D systems obeying momentum conservation [70], calculations based on solving the Boltzmann transport equation instead suggest that the κ will saturate at a constant value for a defect-free, sufficient long (10 μm ~ 1 mm) SWCNT [71, 72]. Contrary to 1D systems, the calculations also suggest that disorders can facilitate the saturation of κ at much shorter lengths (<100 nm) [71]. However, it is not clear why a quasi-1D system like a SWCNT could show normal heat conduction behavior even when both 1D and 2D systems are known to violate Fourier's law. The fundamental disagreements between different theories urgently await experimental inputs to resolve the controversy.

To rigorously study the fundamental heat transfer phenomena, we have considerably improved the material syntheses and the measurement sensitivities over previous setups. Ultralong SWCNTs with lengths exceeding 2 cm were grown using chemical vapor deposition methods [73, 74]. Individual SWCNTs were picked up by a custom-designed manipulator and placed on a thermal conductivity test fixture consisting of parallel suspended SiN_x beams, as shown in Fig. 8.18a. The suspended SiN_x beams with deposited Pt films were utilized as independent resistive thermometers (RT_i's) for generating heat or sensing temperature variations. From the thermal circuits shown in Fig. 8.18b, the thermal conductance (K) of the SWCNT anchored between RT_1 and RT_2 can be determined as follows:

$$K_{12} = \frac{2P(\overline{\Delta T}_2 + \overline{\Delta T}_3 + \ldots)}{3\overline{\Delta T}_1\left(\frac{3}{2}\overline{\Delta T}_1 - \overline{\Delta T}_2\right)} \tag{8.12}$$

where P is power, $\overline{\Delta T}_i$ ($i = 1, 2$) is the average temperature rise of each RT_i and can be determined by measuring the resistance (R) variations of the Pt films based on their linear R vs T behavior. During the experiment, an alternating current with frequency f ($f < 7$ Hz) was supplied to the heater (RT_1) and the corresponding changes of the temperature on the sensor (RT_2) was detected at frequency $2f$ using a lock-in amplifier. The background contribution due to radiation heat transfer had been carefully removed and the thermal conductance of the SWCNT anchored between any two neighboring RT_i's was obtained using a similar method (Fig. 8.18c). Due to the much reduced heat loss via SiN_x beams, we demonstrated that the test fixture was capable of measuring thermal conductance as

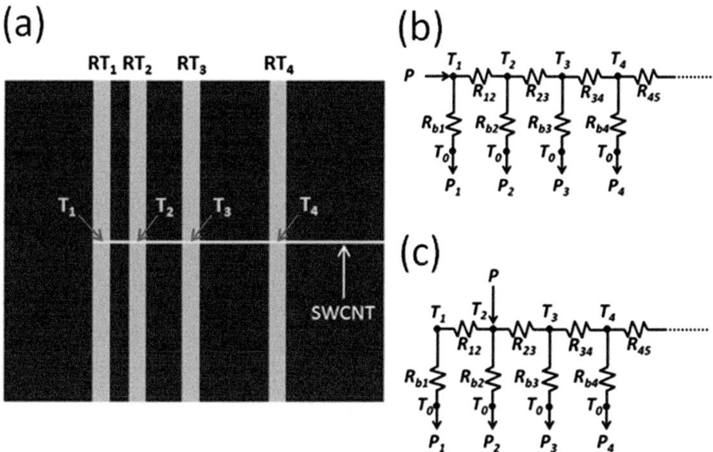

Fig. 8.18 (**a**) Schematic diagram of the thermal conductivity test fixture consisting of parallel resistive thermometers (RT_i's) made by Pt films on SiN_x beams. (**b** and **c**) The corresponding thermal circuits showing the respective SWCNT thermal resistance (R_{ij}), SiN_x beam thermal resistance (R_{bi}), heat currents at each SiN_x beam (P_i), and temperature rise ($T_i - T_0$) if a given power (P) is dissipated in (b) RT_1 or (c) RT_2, respectively

low as 1.6×10^{-12} W/K at room temperature. We assumed the diameters $d = 2$ nm of the investigated SWCNT and the corresponding $\kappa = KL/\pi d\delta$ can be determined (SWCNT thickness $\delta = 0.34$ nm). The L's of the investigated SWCNTs span from 2 μm (Fig. 8.19a) to 126 μm (Fig. 8.19b). Note that the latter scale is much longer than the theoretical l of a SWCNT [75, 76]. Therefore, any deviations from Fourier's law observed in our experimental data should not be confused with conventional ballistic thermal conduction.

Figure 8.20 shows κ vs. L relations for the investigated SWCNT. The L's of the investigated SWCNTs span two orders of magnitude, varying from 2 to 126 μm. Remarkably, the κ's of SWCNTs display divergent behavior with increasing L. As shown in Fig. 8.20, it is found that $\kappa = 2082$ W/m-K for $L = 4$ μm and it increases to 2757 W/m-K for $L = 7.7$ μm, 2879 W/m-K for $L = 15.5$ μm, 3738 W/m-K for $L = 96.2$ μm, and 4332 W/m-K for $L = 126$ μm. Note that the effect of contact thermal resistance has not been taken into account yet. Thus $\kappa = 4332$ W/m-K is simply a lower bound.

The violation of Fourier's law in SWCNTs can be quantified by expressing $\kappa \sim L^{\alpha}$. Various models have suggested $\alpha = 1/2$, 1/3, or 2/5 for different 1D systems [3]. Because it is known that the diameter and the chirality do not change in an ultralong SWCNT [73], the former two uncertainties can be minimized by analyzing the length dependence of κ of the same sample. From Fig. 8.20, the α is found to be 0.19. But since effects of contact resistance and radiation loss have not been considered here, the α could be a lower bound.

Fig. 8.19 SEM images of an individual SWCNT suspended on the test fixture. The suspended length varies from (**a**) 4 µm to (**b**) 126 µm

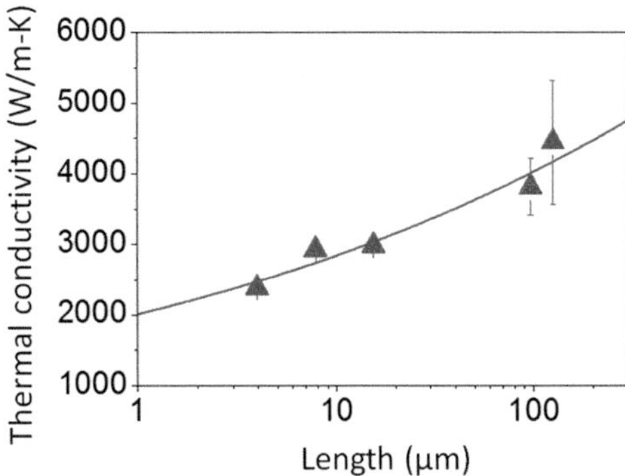

Fig. 8.20 The κ vs. L relation for the investigated SWCNT. The *solid curve* is a fit using $\alpha = 0.19$

We now analyze the effect of contact thermal resistance. Because the contact areas ($\sim dw$, where d is the diameter of the SWCNT and $w = 2$ µm is the width of a SiN_x beam) between the SWCNT and each RT_i's are identical, the contact thermal resistance ($1/K_c$) should be approximately a constant for each sample and its effect can be analyzed in terms of K_s/K_c, where K_s is the intrinsic thermal conductance of a 1 μm-long SWCNT. So SWCNT's measured thermal conductance (K) follows: $1/K = (L/L_0)^{1-\alpha}/K_s + 1/K_c$, and the measured κ is expressed as:

$$\kappa = \frac{K_s L}{\pi d \delta}\left(\frac{1}{(L/L_0)^{1-\alpha} + K_s/K_c}\right) \tag{8.13}$$

here $L_0 = 1\,\mu\text{m}$. We have incorporated the effect of contact resistance using Eq. (8.13) to analyze our data. The result suggests that $0.17 < \alpha < 0.23$ and $K_s/K_c < 0.3$ yield good fits to the experimental data. Therefore, the observed divergent behavior of κ originates from the intrinsic properties of the ultralong SWCNTs but not from artifacts of contact thermal resistance.

Now we discuss the effect of thermal radiation from the surface of a SWCNT to our measurements. Because of thermal radiation from the surface of a SWCNT, the heat received at the sensor will be always lower than that transmitted from the heater. Thus Eq. (8.12) underestimates the actual P delivered in the SWCNT and, correspondingly, the measured K_m simply sets the lower bound of the actual sample thermal conductance K_s.

The thermal radiation from the SWCNT surface gives an equivalent thermal resistance (R_r) connecting from a SWCNT to the environment, which can be expressed by:

$$R_r = \frac{T - T_\infty}{\varepsilon \sigma A \left(T^4 - T_\infty^4\right)} \tag{8.14}$$

where $\varepsilon = 0.98$ is the emissivity of SWCNTs [77], $\sigma = 5.67 \times 10^{-8}$ Wm^{-2} K^{-4} is Stefan-Boltzmann constant, $A = \pi dl$ is surface area of each segment, T is the temperature of the SWCNT segment, and $T_\infty \sim 300$ K is the temperature of the environment.

In our experiment, $T - T_\infty < 20$ K, applying the first-order approximation we then have:

$$R_r \approx \frac{1}{4\varepsilon\sigma A T_\infty^3} \tag{8.15}$$

The higher-order terms contribute less than 10 % of Eq. (8.15). Now we estimate the radio R_r/R (where R is the thermal resistance of the SWCNT) Assuming the thermal conductivity of a 1 μm-long SWCNT is ~3000 W/m-K and $\varepsilon = 0.98$, we have

$$\frac{R_r}{R} \approx 2.45 \times 10^5 \tag{8.16}$$

The result suggests that thermal radiation is negligibly small for short SWCNTs. In fact, we estimate that it would give less than 5 % increase to the measured thermal conductivity for a 100 μm-long SWCNT.

Because naturally abundant ethanol vapor was used as the synthetic source, isotopic impurities (98.9 % ^{12}C and 1.1 % ^{13}C) are expected in the SWCNTs. In addition, impurities and defects are unavoidable for the investigated SWCNTs. Furthermore, TEM images also reveal a thin layer (~2 nm) of amorphous carbon covering some parts of the SWCNTs. Yet the non-Fourier thermal conduction emerges regardless of these external perturbations, which is consistent with various 1D disordered models [64]. On the other hand, our data disagree with the theoretical result based on solving Boltzmann transport equation that the divergent behavior of κ would disappear when disorders are introduced in SWCNTs [71].

We emphasize that the data simply reflects our preliminary results for shading light on the non-Fourier (anomalous) thermal conduction in 1D systems. Although the current result is not sufficient to resolve the theoretical controversies on the universality of α, it nevertheless provides evidence that unconventional ballistic thermal conduction can persist over 100 μm in a SWCNT. Because impurities and defects are unavoidable in the investigated samples, the experimental result also rules out some theoretical predictions that the non-Fourier thermal conduction would disappear when defects are present. The observation of the anomalous thermal conduction with $\alpha < 1$ in ultralong SWCNTs also indicates that the proposed $\lambda \gg d$ (see the introduction section) is the correct criterion for defining 1D systems.

8.7 Conclusion

I have presented my personal journey of our experimentally searching for non-Fourier thermal conduction phenomena in nanoscale materials. The room temperature ballistic thermal conduction persisting over 8 μm in SiGe nanowires is found to be the phenomena of conventional ballistic conduction. Furthermore, due to the interlayer diffusion or the induced phonon localization at the heterogeneous interface, conventional, micron-scale ballistic thermal conduction is found in Si-Ge core-shell nanowires as well. These phenomena are very likely to be 3D effects and they could be found in bulk samples as well. On the other hand, multiwall carbon and BN nanotubes, due to their highly anisotropic thermal conductivities and the absence of interlayer diffusion, are very likely to display 1D thermal transport phenomena. Unfortunately, the micrometer-long samples have impeded clear signals unique to 1D systems to emerge from the conventional ballistic phonons. Thus the results of multiwall nanotubes are, rigorously speaking, falling in the category of conventional ballistic thermal conduction. Finally, the quest for unambiguous experimental evidence for the anomalous thermal conduction in 1D systems prompts us to overcome various experimental challenges and measure the thermal transport properties of ultralong single-wall carbon nanotubes. Our results show that a disordered, quasi-1D system like a single-wall carbon nanotube can display divergent κ for length up to 100 μm, disagreeing with some theoretical predictions. I believe that many unexpected phenomena can be discovered once experimental investigations are continuously improved to unravel the non-Fourier thermal conduction.

References

1. Pomeranchuk, I.: On the thermal conductivity of dielectrics. Phys. Rev. **60**, 820–821 (1941)
2. Lepri, S., Livi, R., Politi, A.: Heat conduction in chains of nonlinear oscillators. Phys. Rev. Lett. **78**, 1896–1899 (1997)
3. Dhar, A.: Heat transport in low-dimensional systems. Adv. Phys. **57**, 457–537 (2008)
4. Li, Y.Y., Liu, S., Li, N.B., Hanggi, P., Li, B.W.: 1D momentum-conserving systems: the conundrum of anomalous versus normal heat transport. New J. Phys. **17**, 043064 (2015)
5. Saito, K., Dhar, A.: Heat conduction in a three dimensional anharmonic crystal. Phys. Rev. Lett. **104**, 040601 (2010)
6. Wang, L., He, D.H., Hu, B.B.: Heat conduction in a three-dimensional momentum-conserving anharmonic lattice. Phys. Rev. Lett. **105**, 160601 (2010)
7. Asheghi, M., Leung, Y.K., Wong, S.S., Goodson, K.E.: Phonon-boundary scattering in thin silicon layers. Appl. Phys. Lett. **71**, 1798–1800 (1997)
8. Ju, Y.S., Goodson, K.E.: Phonon scattering in silicon films with thickness of order 100 nm. Appl. Phys. Lett. **74**, 3005–3007 (1999)
9. Liu, W.J., Asheghi, M.: Thermal conduction in ultrathin pure and doped single-crystal silicon layers at high temperatures. J. Appl. Phys. **98**, 123523 (2005)
10. Li, D.Y., Wu, Y.Y., Kim, P., Shi, L., Yang, P.D., Majumdar, A.: Thermal conductivity of individual silicon nanowires. Appl. Phys. Lett. **83**, 2934–2936 (2003)
11. Casimir, H.B.G.: Note on the conduction of heat in crystals. Physica **5**, 495–500 (1938)
12. Schwab, K., Henriksen, E.A., Worlock, J.M., Roukes, M.L.: Measurement of the quantum of thermal conductance. Nature **404**, 974–977 (2000)
13. Lee, J., Lim, J., Yang, P.D.: Ballistic phonon transport in holey silicon. Nano Lett. **15**, 3273–3279 (2015)
14. Koh, Y.K., Cahill, D.G.: Frequency dependence of the thermal conductivity of semiconductor alloys. Phys. Rev. B **76**, 075207 (2007)
15. Minnich, A.J., Johnson, J.A., Schmidt, A.J., Esfarjani, K., Dresselhaus, M.S., Nelson, K.A., Chen, G.: Thermal conductivity spectroscopy technique to measure phonon mean free paths. Phys. Rev. Lett. **107**, 095901 (2011)
16. Regner, K.T., Sellan, D.P., Su, Z., Amon, C.H., McGaughey, A.J., Malen, J.A.: Broadband phonon mean free path contributions to thermal conductivity measured using frequency domain thermoreflectance. Nat. Commun. **4**, 1640 (2013)
17. Wilson, R.B., Cahill, D.G.: Anisotropic failure of Fourier theory in time-domain thermoreflectance experiments. Nat. Commun. **5**, 5075 (2014)
18. da Cruz, C.A., Li, W., Katcho, N.A.: Role of phonon anharmonicity in time-domain thermoreflectance measurements. Appl. Phys. Lett. **101** (2012)
19. Huang, B.W., Hsiao, T.K., Lin, K.H., Chiou, D.W., Chang, C.W.: Length-dependent thermal transport and ballistic thermal conduction. AIP Adv. **5**, 053202 (2015)
20. Alvarez-Quintana, J., Rodriguez-Viejo, J., Alvarez, F.X., Jou, D.: Thermal conductivity of thin single-crystalline germanium-on-insulator structures. Int. J. Heat Mass Tran. **54**, 1959–1962 (2011)
21. Johnson, J.A., Maznev, A.A., Cuffe, J., Eliason, J.K., Minnich, A.J., Kehoe, T., Torres, C.M.S., Chen, G., Nelson, K.A.: Direct measurement of room-temperature nondiffusive thermal transport over micron distances in a silicon membrane. Phys. Rev. Lett. **110**, 025901 (2013)
22. Hu, Y.J., Zeng, L.P., Minnich, A.J., Dresselhaus, M.S., Chen, G.: Spectral mapping of thermal conductivity through nanoscale ballistic transport. Nature Nanotechnol. **10**, 701–706 (2015)

23. Hsiao, T.K., Chang, H.K., Liou, S.C., Chu, M.W., Lee, S.C., Chang, C.W.: Observation of room temperature ballistic thermal conduction persisting over 8.3 micrometers in SiGe nanowires. Nature Nanotechnol. **8**, 534–538 (2013)
24. Zhu, G.H., Lee, H., Lan, Y.C., Wang, X.W., Joshi, G., Wang, D.Z., Yang, J., Vashaee, D., Guilbert, H., Pillitteri, A., Dresselhaus, M.S., Chen, G., Ren, Z.F.: Increased phonon scattering by nanograins and point defects in nanostructured silicon with a low concentration of germanium. Phys. Rev. Lett. **102**, 196803 (2009)
25. Bera, C., Mingo, N., Volz, S.: Marked effects of alloying on the thermal conductivity of nanoporous materials. Phys. Rev. Lett. **104**, 115502 (2010)
26. Garg, J., Bonini, N., Kozinsky, B., Marzari, N.: Role of disorder and anharmonicity in the thermal conductivity of silicon-germanium alloys: a first-principles study. Phys. Rev. Lett. **106**, 045901 (2011)
27. Maldovan, M.: Narrow low-frequency spectrum and heat management by thermocrystals. Phys. Rev. Lett. **110**, 025902 (2013)
28. Balandin, A.A.: Thermal properties of graphene and nanostructured carbon materials. Nature Mater. **10**, 569–581 (2011)
29. Chiu, H.Y., Deshpande, V.V., Postma, H.W.C., Lau, C.N., Miko, C., Forro, L., Bockrath, M.: Ballistic phonon thermal transport in multiwalled carbon nanotubes. Phys. Rev. Lett. **95**, 226101 (2005)
30. Bolotin, K.I., Ghahari, F., Shulman, M.D., Stormer, H.L., Kim, P.: Observation of the fractional quantum Hall effect in graphene. Nature **462**, 196–199 (2009)
31. Du, X., Skachko, I., Duerr, F., Luican, A., Andrei, E.Y.: Fractional quantum Hall effect and insulating phase of Dirac electrons in graphene. Nature **462**, 192–195 (2009)
32. Chen, R.K., Hochbaum, A.I., Murphy, P., Moore, J., Yang, P.D., Majumdar, A.: Thermal conductance of thin silicon nanowires. Phys. Rev. Lett. **101**, 105501 (2008)
33. Hochbaum, A.I., Chen, R.K., Delgado, R.D., Liang, W.J., Garnett, E.C., Najarian, M., Majumdar, A., Yang, P.D.: Enhanced thermoelectric performance of rough silicon nanowires. Nature **451**, 163–165 (2008)
34. Wingert, M.C., Chen, Z.C.Y., Dechaumphai, E., Moon, J., Kim, J.-H., Xiang, J., Chen, R.: Thermal conductivity of Ge and Ge-Si core-shell nanowires in the phonon confinement regime. Nano Lett. **11**, 5507–5513 (2011)
35. Meddins, H.R., Parrott, J.E.: Thermal and thermoelectric properties of sintered germanium-silicon alloys. J. Phys. C Solid State **9**, 1263–1276 (1976)
36. Lu, Q., Adu, K.W., Gutierrez, H.R., Chen, G., Lew, K.-K., Nimmatoori, P., Zhang, X., Dickey, E.C., Redwing, J.M., Eklund, P.C.: Raman scattering from Si1-xGex alloy nanowires. J. Phys. Chem. C **112**, 3209–3215 (2008)
37. Rego, L.G.C., Kirczenow, G.: Quantized thermal conductance of dielectric quantum wires. Phys. Rev. Lett. **81**, 232–235 (1998)
38. Prasher, R.: Predicting the thermal resistance of nanosized constrictions. Nano Lett. **5**, 2155–2159 (2005)
39. Steele, M.C., Rosi, F.D.: Thermal conductivity and thermoelectric power of germanium-silicon alloys. J. Appl. Phys. **29**, 1517–1520 (1958)
40. Smith, C.G.: Low-dimensional quantum devices. Rep. Prog. Phys. **59**, 235–282 (1996)
41. Yang, R.G., Chen, G., Dresselhaus, M.S.: Thermal conductivity modeling of core-shell and tubular nanowires. Nano Lett. **5**, 1111–1115 (2005)
42. Prasher, R.: Thermal conductivity of tubular and core/shell nanowires. Appl. Phys. Lett. **89**, 063121 (2006)
43. Markussen, T., Jauho, A.P., Brandbyge, M.: Surface-decorated silicon nanowires: a route to high-ZT thermoelectrics. Phys. Rev. Lett. **103** (2009)
44. Hu, M., Giapis, K.P., Goicochea, J.V., Zhang, X.L., Poulikakos, D.: Significant reduction of thermal conductivity in Si/Ge core-shell nanowires. Nano Lett. **11**, 618–623 (2011)
45. Hu, M., Zhang, X.L., Giapis, K.P., Poulikakos, D.: Thermal conductivity reduction in core-shell nanowires. Phys. Rev. B **84**, 085442 (2011)

46. Chen, J., Zhang, G., Li, B.W.: Impacts of atomistic coating on thermal conductivity of germanium nanowires. Nano Lett. **12**, 2826–2832 (2012)
47. Lu, W., Xiang, J., Timko, B.P., Wu, Y., Lieber, C.M.: One-dimensional hole gas in germanium/silicon nanowire heterostructures. Proc. Natl. Acad. Sci. USA **102**, 10046–10051 (2005)
48. Xiang, J., Lu, W., Hu, Y.J., Wu, Y., Yan, H., Lieber, C.M.: Ge/Si nanowire heterostructures as high-performance field-effect transistors. Nature **441**, 489–493 (2006)
49. Yan, H., Choe, H.S., Nam, S.W., Hu, Y.J., Das, S., Klemic, J.F., Ellenbogen, J.C., Lieber, C.M.: Programmable nanowire circuits for nanoprocessors. Nature **470**, 240–244 (2011)
50. Hu, Y.J., Kuemmeth, F., Lieber, C.M., Marcus, C.M.: Hole spin relaxation in Ge-Si core-shell nanowire qubits. Nature Nanotechnol. **7**, 47–50 (2012)
51. Yin, L., Lee, E.K., Lee, J.W., Whang, D., Choi, B.L., Yu, C.: The influence of phonon scatterings on the thermal conductivity of SiGe nanowires. Appl. Phys. Lett. **101**, 043114 (2012)
52. Lee, E.K., Yin, L., Lee, Y., Lee, J.W., Lee, S.J., Lee, J., Cha, S.N., Whang, D., Hwang, G.S., Hippalgaonkar, K., Majumdar, A., Yu, C., Choi, B.L., Kim J.M., Kim, K.: Large thermoelectric figure-of-merits from SiGe nanowires by simultaneously measuring electrical and thermal transport properties. Nano Lett. **12**, 2918–2923 (2012)
53. Kim, H., Kim, I., Choi, H.J., Kim, W.: Thermal conductivities of Si1-xGex nanowires with different germanium concentrations and diameters. Appl. Phys. Lett. **96**, 233106 (2010)
54. Chang, H.-K., Lee, S.-C.: The growth and radial analysis of Si/Ge core-shell nanowires. Appl. Phys. Lett. **97**, 251912 (2010)
55. Hsiao, T.K., Huang, B.W., Chang, H.K., Liou, S.C., Chu, M.W., Lee, S.C., Chang, C.W.: Micron-scale ballistic thermal conduction and suppressed thermal conductivity in heterogeneously-interfaced nanowires. Phys. Rev. B **91**, 035406 (2015)
56. Lu, D., Xie, R.G., Yang, N., Li, B.W., Thong, J.T.L.: Profiling nanowire thermal resistance with a spatial resolution of nanometers. Nano Lett. **14**, 806–812 (2014)
57. Slack, G.A., Hussain, M.A.: The maximum possible conversion efficiency of silicon-germanium thermoelectric generators. J. Appl. Phys. **70**, 2694–2718 (1991)
58. Liu, C.-K., Yu, C.-K., Chien, H.-C., Kuo, S.-L., Hsu, C.-Y., Dai, M.-J., Luo, G.-L., Huang, S.-C., Huang, M.-J.: Thermal conductivity of Si/SiGe superlattice films. J. Appl. Phys. **104**, 114301 (2008)
59. Minnich, A.J., Lee, H., Wang, X.W., Joshi, G., Dresselhaus, M.S., Ren, Z.F., Chen, G., Vashaee, D.: Modeling study of thermoelectric SiGe nanocomposites. Phys. Rev. B **80**, 155327 (2009)
60. Martin, P., Aksamija, Z., Pop, E., Ravaioli, U.: Impact of phonon-surface roughness scattering on thermal conductivity of thin Si nanowires. Phys. Rev. Lett. **102**, 125503 (2009)
61. Chang, C.W., Fennimore, A.M., Afanasiev, A., Okawa, D., Ikuno, T., Garcia, H., Li, D.Y., Majumdar, A., Zettl, A.: Isotope effect on the thermal conductivity of boron nitride nanotubes. Phys. Rev. Lett. **97**, 085901 (2006)
62. Mingo, N., Broido, D.A.: Carbon nanotube ballistic thermal conductance and its limits. Phys. Rev. Lett. **95**, 096105 (2005)
63. Chang, C.W., Okawa, D., Garcia, H., Majumdar, A., Zettl, A.: Breakdown of Fourier's law in nanotube thermal conductors. Phys. Rev. Lett. **101**, 075903 (2008)
64. Dhar, A., Saito, K.: Heat conduction in the disordered Fermi-Pasta-Ulam chain. Phys. Rev. E **78**, 061136 (2008)
65. Narayan, O., Ramaswamy, S.: Anomalous heat conduction in one-dimensional momentum-conserving systems. Phys. Rev. Lett. **89**, 200601 (2002)
66. Garrido, P.L., Hurtado, P.I., Nadrowski, B.: Simple one-dimensional model of heat conduction which obeys Fourier's law. Phys. Rev. Lett. **86**, 5486–5489 (2001)
67. Hurtado, P.I.: Breakdown of hydrodynamics in a simple one-dimensional fluid. Phys. Rev. Lett. **96**, 010601 (2006)
68. Zhang, G., Li, B.W.: Thermal conductivity of nanotubes revisited: effects of chirality, isotope impurity, tube length, and temperature. J. Chem. Phys. **123**, 114714 (2005)

69. Wang, J.S., Li, B.W.: Intriguing heat conduction of a chain with transverse motions. Phys. Rev. Lett. **92**, 074302 (2004)
70. Prosen, T., Campbell, D.K.: Momentum conservation implies anomalous energy transport in 1D classical lattices. Phys. Rev. Lett. **84**, 2857–2860 (2000)
71. Mingo, N., Broido, D.A.: Length dependence of carbon nanotube thermal conductivity and the "problem of long waves". Nano Lett. **5**, 1221–1225 (2005)
72. Lindsay, L., Broido, D.A., Mingo, N.: Lattice thermal conductivity of single-walled carbon nanotubes: beyond the relaxation time approximation and phonon-phonon scattering selection rules. Phys. Rev. B **80**, 125407 (2009)
73. Zheng, L.X., O'Connell, M.J., Doorn, S.K., Liao, X.Z., Zhao, Y.H., Akhadov, E.A., Hoffbauer, M.A., Roop, B.J., Jia, Q.X., Dye, R.C., Peterson, D.E., Huang, S.M., Liu, J., Zhu, Y.T.: Ultralong single-wall carbon nanotubes. Nature Mater. **3**, 673–676 (2004)
74. Hofmann, M., Nezich, D., Reina, A., Kong, J.: In-situ sample rotation as a tool to understand chemical vapor deposition growth of long aligned carbon nanotubes. Nano Lett. **8**, 4122–4127 (2008)
75. Yu, C.H., Shi, L., Yao, Z., Li, D.Y., Majumdar, A.: Thermal conductance and thermopower of an individual single-wall carbon nanotube. Nano Lett. **5**, 1842–1846 (2005)
76. Donadio, D., Galli, G.: Thermal conductivity of isolated and interacting carbon nanotubes: comparing results from molecular dynamics and the Boltzmann transport equation. Phys. Rev. Lett. **99**, 255502 (2007)
77. Fainchtein, R., Brown, D.M., Siegrist, K.M., Monica, A.H., Hwang, E.R., Milner, S.D., Davis, C.C.: Time-dependent near-blackbody thermal emission from pulsed laser irradiated vertically aligned carbon nanotube arrays. Phys. Rev. B **85**, 125432 (2012)

Chapter 9
Thermal Transport in Graphene, Few-Layer Graphene and Graphene Nanoribbons

Denis L. Nika and Alexander A. Balandin

Abstract The discovery of unusual heat conduction properties of graphene has led to a surge of theoretical and experimental studies of phonon transport in two-dimensional material systems. The rapidly developing graphene thermal field spans from theoretical physics to practical engineering applications. In this invited review we outline different theoretical approaches developed for describing phonon transport in graphene and provide comparison with available experimental thermal conductivity data. A special attention is given to analysis of the recent theoretical results for the phonon thermal conductivity of graphene and few-layer graphene, the effects of the strain, defects, isotopes and edge scattering on the acoustic phonon transport in these material systems.

9.1 Introduction

Thermal management has become a crucial issue for continuing progress in electronic industry owing to increased levels of dissipated power density and speed of electronic circuits [1]. Self-heating is a major problem in electronics, optoelectronics and photonics [2]. These facts stimulated practical interest in thermal properties of materials. Acoustic phonons are the main heat carriers in a variety of material systems. The phonon and thermal properties of nanostructures are substantially different from those of bulk crystals [3–15]. Semiconductor nanostructures do not conduct heat as well as bulk crystals due to increased phonon—boundary scattering [4, 5] as well as changes in the phonon dispersion and density of states (DOS) [3–8]. From the other side, theoretical studies suggested that phonon transport in strictly two-dimensional (2D) and one-dimensional (1D) systems can reveal exotic

D.L. Nika
Department of Physics and Engineering, Moldova State University, Chisinau MD-2009, Republic of Moldova
e-mail: dlnika@yahoo.com

A.A. Balandin (✉)
Department of Electrical and Computer Engineering and Materials Science and Engineering Program, University of California, Riverside, CA 92521, USA
e-mail: balandin@ee.ucr.edu

S. Lepri (ed.), *Thermal Transport in Low Dimensions*, Lecture Notes in Physics 921, DOI 10.1007/978-3-319-29261-8_9

behavior, leading to infinitely large *intrinsic* thermal conductivity [9–12]. These theoretical results have led to discussions of the validity of Fourier's law in low-dimensional systems [16, 17] and further stimulated interest in the acoustic phonon transport in 2D systems.

In this Chapter, we focus on the specifics of the acoustic phonon transport in graphene. The Chapter is mostly based on our original and review papers dedicated to various aspects of heat conduction in graphene [18–28]. After a summary of the basics of thermal physics in nanostructures and experimental data for graphene's thermal conductivity, we discuss, in more detail, various theoretical approaches to calculation of the phonon thermal conductivity in graphene. Special attention is given to the analysis of the most recent theoretical results on the relative contributions of different phonon polarization branches to the thermal conductivity of graphene. The readers interested in the experimental thermal conductivity values of graphene and related materials are referred to a complementary review [18].

9.2 Basics of Phonon Transport and Thermal Conductivity

The main experimental technique for investigation of the acoustic phonon transport in a given material system is the measurement of its lattice thermal conductivity [29, 30]. The thermal conductivity is introduced through Fourier's law [31, 32]:

$$\vec{\phi} = -K\nabla T, \tag{9.1}$$

where $\vec{\phi}$ is the heat flux, ∇T is the temperature gradient and $K = (K_{\alpha\beta})$ is the thermal conductivity tensor. In the isotropic medium, thermal conductivity does not depend on the direction of the heat flow and K is treated as a constant. The latter is valid for the small temperature variations only. In a wide temperature range, thermal conductivity is a function of temperature, i.e. $K \equiv K(T)$. In general, in solid materials heat is carried by phonons and electrons so that $K = K_p + K_e$, where K_p and K_e are the phonon and electron contributions, respectively. In metals or degenerately-doped semiconductors, K_e is dominant due to the large density of free carriers. The value of K_e can be determined from the measurement of the electrical conductivity σ via the Wiedemann-Franz law [33]:

$$\frac{K_e}{\sigma T} = \frac{\pi^2 k_B^2}{3e^2}, \tag{9.2}$$

where k_B is the Boltzmann's constant and e is the charge of an electron. Phonons are usually the main heat carriers in carbon materials. Even in graphite, which has metal-like properties [34], the heat conduction is dominated by acoustic phonons [35]. This fact is explained by the strong covalent sp^2 bonding, resulting in high in-plane phonon group velocities and low crystal lattice anharmonicity for in-plane vibrations.

The phonon thermal conductivity can be written as

$$K_p = \Sigma_j \int C_j(\omega)\, \upsilon_{x,j}(\omega)\, \upsilon_{x,j}(\omega)\, \tau_j(\omega)\, d\omega, \tag{9.3}$$

where summation is performed over the phonon polarization branches j, which include two transverse acoustic branches and one longitudinal acoustic branch, $\upsilon_{x,j}$ is the projection of the phonon group velocity $\vec{\upsilon}_j = d\omega_j / d\vec{q}$ on the X-axis for the jth branch, which, in many solids, can be approximated by the sound velocity, τ_j is the phonon relaxation time, $C_j = \hbar\omega_j \partial N_0 \left(\hbar\omega_j / k_B T\right) / \partial T$ is the contribution to heat capacity from the jth branch, and $N_0\left(\frac{\hbar\omega_j}{k_B T}\right) = \left[\exp\left(\frac{\hbar\omega_j}{k_B T}\right) - 1\right]^{-1}$ is the Bose-Einstein phonon equilibrium distribution function. The phonon mean-free path (MFP) Λ is related to the relaxation time through the expression $\Lambda = \tau\upsilon$. In the relaxation-time approximation (RTA), various scattering mechanisms, which limit the MFP, usually considered as additive, i.e. $\tau_j^{-1} = \sum_i \tau_{i,j}^{-1}$, where i denotes scattering mechanisms. In typical solids, acoustic phonons, which carry the bulk of heat, are scattered by other phonons, lattice defects, impurities, conduction electrons, and interfaces [36–39].

In ideal crystals, i.e. crystals without defects or rough boundaries, Λ is limited by the phonon–phonon scattering due to the crystal lattice anharmonicity. In this case, thermal conductivity is referred to as intrinsic. The anharmonic phonon interactions, which lead to the finite thermal conductivity in three dimensions, can be described by the Umklapp processes [36]. The Umklapp scattering rates depend on the Gruneisen parameter γ, which determines the degree of the lattice anharmonicity [36, 37]. Thermal conductivity is extrinsic when it is mostly limited by the extrinsic effects such phonon—boundary or phonon—defect scattering.

In nanostructures, the phonon energy spectra are quantized due to the spatial confinement of the acoustic phonons. The quantization of the phonon energy spectra, typically, leads to decreasing phonon group velocity. The modification of the phonon energies, group velocities and density of states, together with phonon scattering from boundaries affect the thermal conductivity of nanostructures. In most of cases, the spatial confinement of acoustic phonons results in a reduction of the phonon thermal conductivity [40, 41]. However, in some cases, the thermal conductivity of nanostructures embedded within the acoustically hard barrier layers can be increased via spatial confinement of acoustic phonons [6, 7, 10, 42].

The phonon boundary scattering can be evaluated as [39]

$$\frac{1}{\tau_{B,j}} = \frac{\upsilon_{x,j}}{D} \frac{1-p}{1+p}, \tag{9.4}$$

where D is the nanostructure or grain size and p is the specularity parameter defined as a probability of specular scattering at the boundary. The momentum-conserving specular scattering ($p = 1$) does not add to thermal resistance. Only diffuse phonon scattering from rough interfaces ($p \rightarrow 0$), which changes the phonon momentum, limits the phonon MFP. The commonly used expression for the phonon specularity is given by [39, 43, 44]

$$p(\lambda) = \exp\left(-\frac{16\pi^2\eta^2}{\lambda^2}\right), \tag{9.5}$$

where η is the root mean square deviation of the height of the surface from the reference plane and λ is the phonon wavelength.

When the phonon—boundary scattering is dominant, the thermal conductivity scales with the nanostructure or grain size D as $K_p \sim C_p \upsilon \Lambda \sim C_p \upsilon^2 \tau_B \sim C_p \upsilon D$. In nanostructures with $D \ll \Lambda$, the thermal conductivity dependence on the physical size of the structure becomes more complicated due to the strong quantization of the phonon energy spectra [6, 40, 42]. The specific heat C_p depends on the phonon density of states, which leads to different $C_p(T)$ dependences in three-dimensional (3D), two-dimensional and one-dimensional systems, and reflected in $K(T)$ dependence at low T [36, 39]. In bulk at low T, $K(T) \sim T^3$ while it is $K(T) \sim T^2$ in 2D systems.

The thermal conductivity K defines how well a given material conducts heat. The thermal diffusivity, α , defines how fast the material conducts heat. It is given by the expression

$$\alpha = \frac{K}{C_p \rho_m}, \tag{9.6}$$

where ρ_m is the mass density. Many experimental techniques directly measure thermal diffusivity rather than thermal conductivity.

9.3 Experimental Data for Thermal Conductivity of Graphene and Few-Layer Graphene

The first measurements of heat conduction in graphene [19–22, 45, 46] were carried out at the University of California—Riverside (see Fig. 9.1). The experimental study was made possible by the development of the optothermal technique. The experiments were performed with the large-area suspended graphene layers exfoliated from the high-quality Kish and highly ordered pyrolytic graphite. It was found that the thermal conductivity varies in a wide range and can exceed that of the bulk graphite, which is ~2000 W/mK at room temperature (RT). It was also determined that the electronic contribution to heat conduction in the un-gated graphene near RT

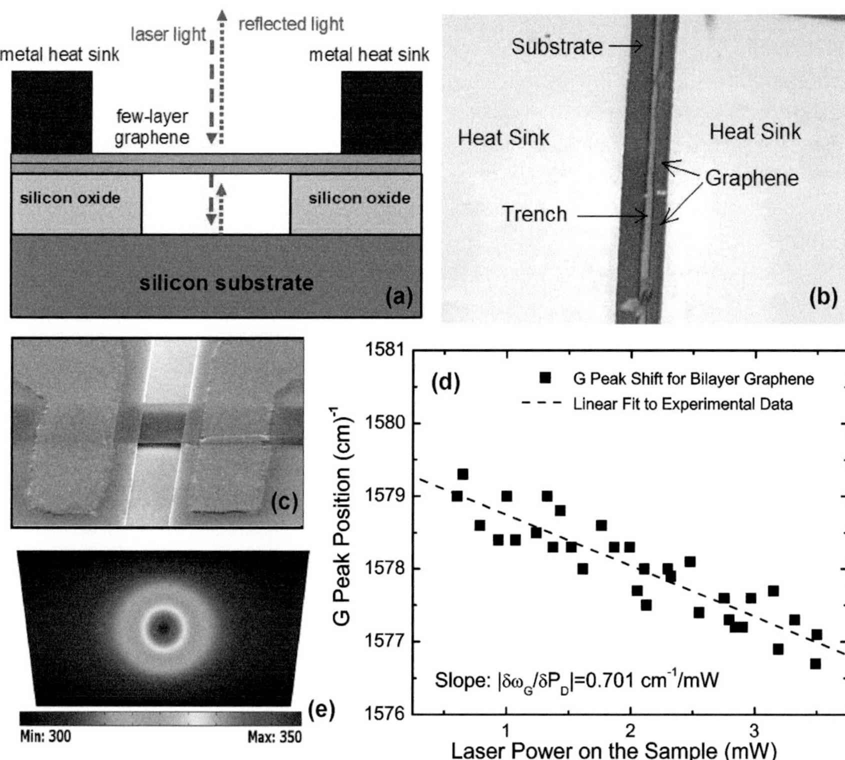

Fig. 9.1 Illustration of optothermal micro-Raman measurement technique developed for investigation of phonon transport in graphene. (**a**) Schematic of the thermal conductivity measurement showing suspended FLG flakes and excitation laser light. (**b**) Optical microscopy images of FLG attached to metal heat sinks. (**c**) Colored scanning electron microscopy image of the suspended graphene flake to clarify typical structure geometry. (**d**) Experimental data for Raman G-peak position as a function of laser power, which determines the local temperature rise in response to the dissipated power. (**e**) Finite-element simulation of temperature distribution in the flake with the given geometry used to extract the thermal conductivity. Figure is after Ref. [22] reproduced with permission from the Nature Publishing Group

is much smaller than that of phonons, i.e. $K_e \ll K_p$. The phonon MFP in graphene was estimated to be on the order of 800 nm near RT [20].

Several independent studies, which followed, also utilized the Raman optothermal technique but modified it via addition of a power meter under the suspended portion of graphene. It was found that the thermal conductivity of suspended high-quality chemical vapour deposited (CVD) graphene exceeded $\sim$2500 W/mK at 350 K, and it was as high as $K \approx 1400$ W/mK at 500 K [47]. The reported value was also larger than the thermal conductivity of bulk graphite at RT. Another Raman optothermal study with the suspended graphene found the thermal conductivity in the range from $\sim$1500 to $\sim$5000 W/mK [48]. Another group that repeated the Raman-based measurements found $K \approx 630$ W/mK for a suspended graphene membrane [49]. The differences in the actual temperature of graphene under laser heating, strain distribution in the suspended graphene of various sizes and geometries can explain the data variation.

Another experimental study reported the thermal conductivity of graphene to be ~1800 W/mK at 325 K and ~710 W/mK at 500 K [50]. These values are lower than that of bulk graphite. However, instead of measuring the light absorption in graphene under conditions of their experiment, the authors of Ref. [50] assumed that the optical absorption coefficient should be 2.3 %. It is known that due to many-body effects, the absorption in graphene is the function of wavelength λ, when $\lambda > 1$ eV [51–53]. The absorption of 2.3 % is observed only in the near-infrared at ~1 eV. The absorption steadily increases with decreasing λ (increasing energy). The 514.5-nm and 488-nm Raman laser lines correspond to 2.41 and 2.54 eV, respectively. At 2.41 eV the absorption is about $1.5 \times 2.3\ \% \approx 3.45\ \%$ [51]. The value of 3.45 % is in agreement with the one reported in another independent study [54]. Replacing the assumed 2.3 % with 3.45 % in the study reported in [50] gives ~2700 W/mK at 325 K and 1065 W/mK near 500 K. These values are higher than those for the bulk graphite and consistent with the data reported by other groups [47, 54], where the measurements were conducted by the same Raman optothermal technique but with the measured light absorption.

The data for suspended or partially suspended graphene is closer to the intrinsic thermal conductivity because suspension reduces thermal coupling to the substrate and scattering on the substrate defects and impurities. The thermal conductivity of fully supported graphene is smaller. The measurements for exfoliated graphene on SiO_2/Si revealed in-plane $K \approx 600$ W/mK near RT [55]. Solving the Boltzmann transport equation (BTE) and comparing with their experiments, the authors determined that the thermal conductivity of free graphene should be ~3000 W/mK near RT.

Despite the noted data scatter in the reported experimental values of the thermal conductivity of graphene, one can conclude that it is very large compared to that for bulk silicon ($K = 145$ W/mK at RT) or bulk copper ($K = 400$ W/mK at RT)—important materials for electronic applications. The differences in K of graphene can be attributed to variations in the graphene sample lateral sizes (length and width), thickness non-uniformity due to the mixing between single-layer and few-layer graphene, material quality (e.g. defect concentration and surface contaminations), grain size and orientation, as well as strain distributions. Often the reported thermal conductivity values of graphene corresponded to different sample temperatures T, despite the fact that the measurements were conducted at ambient temperature. The strong heating of the samples was required due to the limited spectral resolution of the Raman spectrometers used for temperature measurements. Naturally, the thermal conductivity values determined at ambient but for the samples heated to $T \sim 350$ K and $T \sim 600$ K over a substantial portion of their area would be different and cannot be directly compared. One should also note that the data scatter for thermal conductivity of carbon nanotubes (CNTs) is much larger than that for graphene. For a more detail analysis of the experimental uncertainties the readers are referred to a comprehensive review [18].

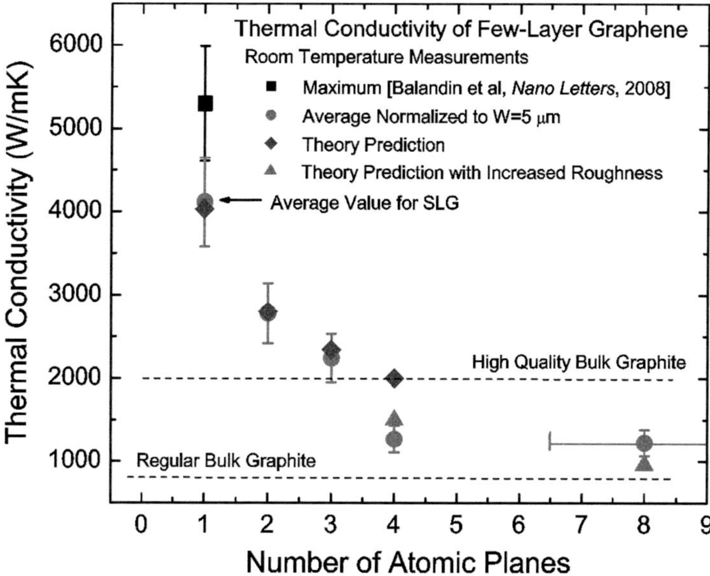

Fig. 9.2 Measured thermal conductivity as a function of the number of atomic planes in FLG. The *dashed straight lines* indicate the range of bulk graphite thermal conductivities. The *diamonds* were obtained from the first-principles theory of thermal conduction in FLG based on the actual phonon dispersion and accounting for all allowed three-phonon Umklapp scattering channels. The *triangles* are Callaway–Klemens model calculations, which include extrinsic effects characteristic for thicker films. Figure is after Ref. [22] reproduced with permission from the Nature Publishing Group

The phonon thermal conductivity undergoes an interesting evolution when the system dimensionality changes from 2D to 3D. This evolution can be studied with the help of suspended few-layer graphene (FLG) with increasing thickness H—number of atomic planes n. It was reported in [22] that thermal conductivity of suspended uncapped FLG decreases with increasing n approaching the bulk graphite limit (see Fig. 9.2). This trend was explained by considering the intrinsic quasi-2D crystal properties described by the phonon Umklapp scattering [22]. As n in FLG increases—the phonon dispersion changes and more phase-space states become available for phonon scattering leading to thermal conductivity decrease. The phonon scattering from the top and bottom boundaries in suspended FLG is limited if constant n is maintained over the layer length. The small thickness of FLG ($n < 4$) also means that phonons do not have transverse cross-plane component in their group velocity leading to even weaker boundary scattering term for the phonons. In thicker FLG films the boundary scattering can increase due to the non-zero cross-plane phonon velocity component. It is also harder to maintain the constant thickness through the whole area of FLG flake. These factors can lead to a thermal conductivity below the graphite limit. The graphite value is recovered for thicker films.

The experimentally observed evolution of the thermal conductivity in FLG with n varying from 1 to $n \sim 4$ [22] is in agreement with the theory for the crystal lattices described by the Fermi-Pasta-Ulam Hamiltonians [56]. The molecular-dynamics (MD) calculations for graphene nanoribbons with the number of planes n from 1 to 8 [57] also gave the thickness dependence of the thermal conductivity in agreement with the UC Riverside experiments [22]. The strong reduction of the thermal conductivity as n changes from 1 to 2 is in line with the earlier theoretical predictions [58]. In another reported study, the Boltzmann transport equation was solved under the assumptions that in-plane interactions are described by Tersoff potential while the Lennard-Jones potential models interactions between atoms belonging to different layers [59, 60]. The obtained results suggested a strong thermal conductivity decrease as n changed from 1 to 2 and slower decrease for $n > 2$.

The thermal conductivity dependence on the FLG is entirely different for the encased FLG where thermal transport is limited by the acoustic phonon scattering from the top and bottom boundaries and disorder. The latter is common when FLG is embedded between two layers of dielectrics.

An experimental study [61] found $K \approx 160$ W/mK for encased single-layer graphene (SLG) at $T = 310$ K. It increases to ~1000 W/mK for graphite films with the thickness of 8 nm. It was also found that the suppression of thermal conductivity in encased graphene, as compared to bulk graphite, was stronger at low temperatures where K was proportional to T^{β} with $1.5 < \beta < 2$ [61]. Thermal conduction in encased FLG was limited by the rough boundary scattering and disorder penetration through graphene.

Recently the measurements of K in twisted bilayer graphene (T-BLG) were performed using an optothermal Raman technique [62]. The obtained values of $K = 1400$–700 W/mK in a temperature range 320–750 K are almost by a factor of 2 smaller than in SLG and by a factor of 1.4 smaller than in Bernal-stacked bilayer graphene (BLG). The twisting affects phonon energy spectra, changes selection rules for phonon transitions and opens up new paths for phonon relaxation [62–64] (see Fig. 9.3). The experimental data on thermal conductivity in graphene and FLG is presented in Table 9.1.

9.4 Theories of Phonon Thermal Conductivity in Graphene, Few-Layer Graphene and Graphene Nanoribbons

The first experimental investigations of the thermal properties in graphene materials [19, 20, 22, 47, 48, 55] stimulated numerous theoretical and computational works in the field. Here, we briefly review the state-of-the-art in theory of thermal transport in graphene and GNRs. Many different theoretical models have been proposed for the prediction of the phonon and thermal properties in graphite, graphene and GNRs during the last few years. The phonon energy spectra have

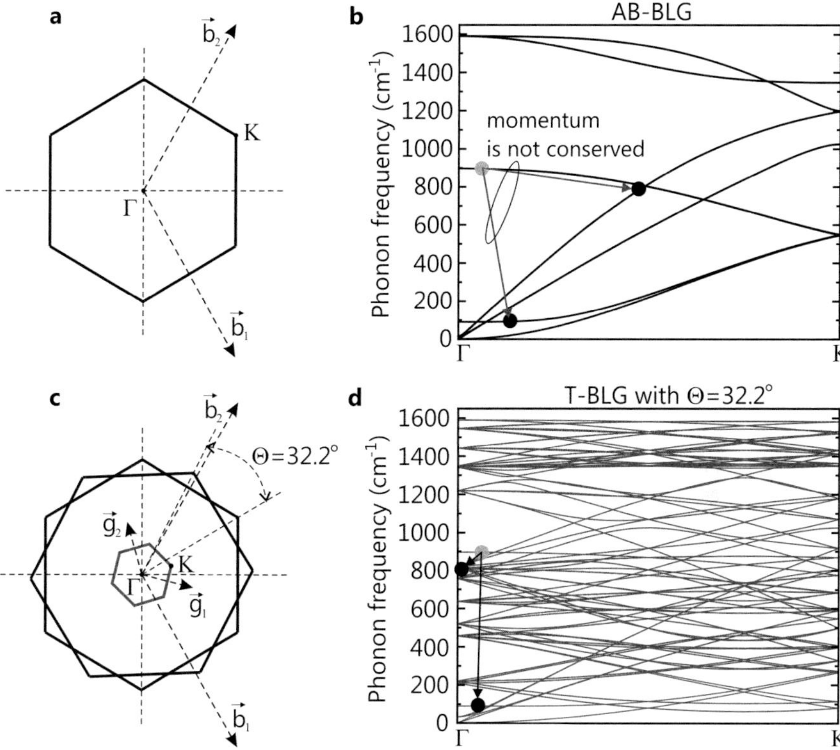

Fig. 9.3 (**a–b**) Brillouin zone and calculated phonon dispersions for Bernal-stacked bilayer graphene and (**c–d**) Brillouin zone and calculated phonon dispersion for twisted bilayer graphene. The twist angle in the calculation was assumed to be 32.2°. Note that new channels of phonon relaxation appear in twisted graphene: the normal decay of the phonon with $\omega = 900$ cm^{-1} (*gray point*) into two phonons with $\omega' = 800$ cm^{-1} and $\omega'' = 100$ cm^{-1} (*black points*) is allowed by the momentum conversation law in TBLG and is not allowed in AB-BLG (*panel* (**d**)). Figure is after Ref. [62] reproduced with permission from the Royal Society of Chemistry

been theoretically investigated using Perdew-Burke-Ernzerhof generalized gradient approximation (GGA) [71–73], valence-force-field (VFF) and Born-von Karman models of lattice vibrations [23, 24, 26, 63, 74, 75], continuum approach [76–78], first-order local density approximation [72, 79, 80], fifth- and fourth-nearest neighbor force constant approaches [73, 81] or utilized the Tersoff, Brenner or Lennard-Jones potentials [59, 60, 82]. The thermal conductivity investigations have been performed within molecular dynamics simulations [57, 83–98], density functional theory [99, 100], Green's function method [101, 102] and Boltzmann-transport-equation (BTE) approach [21–26, 35, 45, 46, 59, 60, 82, 103–106]. It has been shown that phonon energies strongly depend on the interatomic force constants (IFCs)—fitting parameters of interatomic interactions, used in the majority of the models. Therefore a proper choice of interatomic force constants is crucial for the

Table 9.1 Thermal conductivity of graphene and graphene nanoribbons: experimental data

Sample	K (W/mK)	Method	Description	Refs.
SLG	~2000–5000	Raman optothermal	Suspended; exfoliated	[19, 20]
	~2500	Raman optothermal	Suspended; chemical vapor deposition (CVD) grown	[47]
	~1500–5000	Raman optothermal	Suspended; CVD grown	[48]
	600	Raman optothermal	Suspended; exfoliated; $T \sim 660$ K	[49]
	600	Electrical	Supported; exfoliated;	[55]
	310–530	Electrical self-heating	Exfoliated and chemical vapor deposition grown; $T \sim 1000$ K	[65]
	2778 ± 569	Raman optothermal	Suspended, CVD-grown	[62]
	~1700	Electrical self heating	Suspended; CVD-grown; flake length ~ 9 μm; strong length dependence	[66]
Bilayer graphene	~1900	Raman optothermal	Suspended; $T \sim 320$ K	[62]
	560–620	Electrical self-heating	Suspended; polymeric residues on the surface.	[67]
Twisted bilayer	~1400	Raman optothermal	Suspended; $T \sim 320$ K	[62]
FLG	1300–2800	Raman optothermal	Suspended; exfoliated; $n = 2$–4	[22]
	50–970	Heat-spreader method	FLG, encased within SiO_2; $n = 2, \ldots, 21$	[61]
	150–1200	Electrical self-heating	Suspended and supported FLG; polymeric residues on the surface.	[68]
FLG nanoribbons	1100	Electrical self-heating	Supported; exfoliated; $n < 5$	[69]
	80–150	Electrical self-heating	Supported	[70]

accurate description of phonon energy spectra and thermal conductivity in graphene, twisted graphene and graphene nanoribbons [18, 27, 28, 63].

Although various models predicted different values of thermal conductivity, they demonstrated consistent results on the strong dependence of graphene lattice thermal conductivity on extrinsic parameters of flakes: edge quality, FLG thickness, lateral size and shape, lattice strain, isotope, impurity and grain concentration. The molecular dynamic (MD) simulations give usually smaller values of thermal conductivity in comparison with BTE model and experimental data due to exclusion of long wavelength phonons from the model by a finite size of the simulation domain [27]. The effect of the edge roughness on the thermal conductivity in graphene and GNRs has been investigated in [21, 23–26, 45, 46, 78, 83, 103, 106–108]. The rough edge can suppress the thermal conductivity by an order of magnitude as compared to that in graphene or GNRs with perfect edges due to the boundary scattering of phonons. Impurities, single vacancies, double vacancies and Stone-Wales defects decrease the thermal conductivity of graphene and GNRs by more than 50–80 % in dependence of the defect concentration [21, 23, 24, 26, 89–93].

A study of thermal conductivity of graphene and GNRs under strain was performed in [87, 99–102, 109]. An enhancement of the thermal conductivity of up to 36 % for the strained 5-nm armchair or zigzag GNRs was found in the ballistic transport regime [102]. In the diffusive transport regime, the applied strain enhanced the Umklapp scattering and thermal conductivity diminishes by $\sim$1.4 orders of magnitude at RT in comparison with the unstrained graphene [100]. The authors of Ref. [87] have found that when the strain is applied in both directions—parallel and perpendicular to the heat transfer path—the graphene sheets undergo complex reconstructions. As a result, some of the strained graphene structures can have higher thermal conductivity than that of SLG without strain [87]. The discrepancy between theoretical findings and experiments requires additional investigations of thermal transport in strained graphene and GNRs. The isotope composition is another key parameter for thermal conductivity engineering in these materials [18, 27, 28, 110–114]. Naturally occurring carbon materials are made up of two stable isotopes of 12C ($\sim$99 %) and 13C ($\sim$1 %). The change in the isotope composition significantly influences the crystal lattice properties. Increasing the "isotope doping" leads to a suppression of the thermal conductivity in graphene and GNRs of up to two orders of magnitude at RT due to the enhanced phonon-point defect (mass-difference) scattering [27, 105, 110–114].

Graphene and GNRs also demonstrated an intriguing dependence of the thermal conductivity on their geometrical parameters: lateral sizes and shapes [23, 24, 26, 78, 94–97, 106]. Using BTE approach, Nika et al. [26] have demonstrated that RT thermal conductivity of a rectangular graphene flake with 5 μm width increases with length L up to $L \sim$ 40–200 μm and converges for $L >$ 50–1000 μm in dependence on the phonon boundary scattering parameter p (see Fig. 9.4a). The dependence of the thermal conductivity on L is non-monotonic, which is explained by the interplay between contribution to the thermal conductivity from two groups of phonons: participating and non-participating in the edge scattering [26]. The exceptionally large mean free path (MFP) of the acoustic phonons in graphene is essential for this

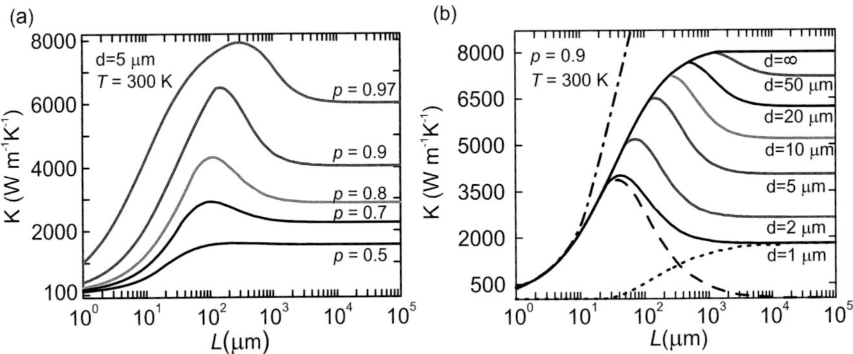

Fig. 9.4 (**a**) Dependence of the thermal conductivity of the rectangular graphene ribbon on the ribbon length L shown for different specular parameters p. The ribbon width is fixed at $d = 5$ μm. (**b**) Dependence of the thermal conductivity of the rectangular graphene ribbon on the ribbon length L shown for different ribbon width d. The specular parameter is fixed at $p = 0.9$. Note in both panels an unusual non-monotonic length dependence of the thermal conductivity, which results from the exceptionally long phonon mean free path of the low-energy phonons and their angle-dependent scattering from the ribbon edge. Figure is after Ref. [26] reproduced with permission from the American Chemical Society

effect. The increase in the flake width or phonon edge scattering (see Fig. 9.4a, b) attenuates the non-monotonic behavior. It disappears in circular flakes or flakes with very rough edges (with specular parameter $p < 0.5$).

A number of studies [94–96] employed the MD simulations to investigate the length dependence of the thermal conductivity in graphene and GNRs. The converged thermal conductivity in graphene was found for $L > 16$ μm in [94]. In [95, 96] the thermal conductivity increases monotonically with an increase of the length up to 2.8 μm in graphene [96] and 800 nm in GNRs [95]. The obvious length dependence in graphene and GNRs can be attributed to the extremely large phonon mean free path $\Lambda \sim 775$ nm [20], which provides noticeable length dependence even for flakes with micrometer lengths.

Keblinsky and co-workers [83] found from the MD study that the thermal conductivity of graphene is $K \approx 8000$–10,000 W/mK at RT for the square graphene sheet. The K value was size independent for $L > 5$ nm [83]. For the ribbons with fixed $L = 10$ nm and width W varying from 1 to 10 nm, K increased from ~1000 to 7000 W/mK. The thermal conductivity in GNR with rough edges can be suppressed by orders of magnitude as compared to that in GNR with perfect edges [83, 107]. The isotopic superlattice modulation of GNR or defects of crystal lattices also significantly decreases the thermal conductivity [110, 115]. The uniaxial stretching applied in the longitudinal direction enhances the low-temperature thermal conductance for the 5 nm arm-chair or zigzag GNR up to 36 % due to the stretching-induced convergence of phonon spectra to the low-frequency region [102].

Aksamija and Knezevic [103] calculated the dependence of the thermal conductivity of GNR with the width 5 nm and RMS (root mean square) edge roughness $\Delta = 1$ nm on temperature. The thermal conductivity was calculated taking into account the three-phonon Umklapp, mass-defect and rough edge scatterings [103]. The authors obtained RT thermal conductivity $K \sim 5500$ W/mK for the graphene nanoribbon. The study of the nonlinear thermal transport in rectangular and triangular GNRs under the large temperature biases was reported in [116]. The authors found that in short (~6 nm) rectangular GNRs, the negative differential thermal conductance exists in a certain range of the applied temperature difference. As the length of the rectangular GNR increases the effect weakens. A computational study reported in [117] predicted that the combined effects of the edge roughness and local defects play a dominant role in determining the thermal transport properties of zigzag GNRs. The experimental data on thermal transport in GNRs is very limited. In [69] the authors used an electrical self-heating methods and extracted the thermal conductivity of sub 20-nm GNRs to be more than 1000 W/mK at 700–800 K. A similar experimental method but with more accurate account of GNRs thermal coupling to the substrate has been used in [70]. Pop and co-workers [70] found substantially lower values of thermal conductivity of ~80–150 W/mK at RT.

Ong and Pop [86] examined thermal transport in graphene supported on SiO_2 using MD simulations. The approach employed by the authors utilized the reactive empirical bond order (REBO) potential to model the atomic interaction between the C atoms, Munetoh potential to model the atomic interactions between the Si and O atoms and Lennard-Jones potential to model the van der Waals type C-Si and C-O couplings. Authors suggested that thermal conductivity in supported graphene is by an order of magnitude smaller than that in suspended graphene due to damping of the out-of-plane ZA phonons.

The strong dependence of the thermal conductivity of graphene on the defect concentration was established in the computational studies reported in [84, 88]. Both studies used MD simulations. According to Hao et al. [88] 2 % of the vacancies or other defects can reduce the thermal conductivity of graphene by as much as a factor of five to ten. Zhang et al. [84] determined from their MD simulations that the thermal conductivity of pristine graphene should be ~2903 W/mK at RT. According to their calculations the thermal conductivity of graphene can be reduced by a factor of 1000 at the vacancy defect concentration of ~9 %. The numeric results of Refs. [84, 88] suggest another possible explanation of the experimental data scatter, which is different defect density in the examined graphene samples. For example, if the measurements of the thermal conductivity of graphene by the thermal bridge technique give smaller values than those by the Raman optothermal technique, one should take into account that the thermal bridge technique requires substantial number of fabrication steps, which result in residual defects.

The available theoretical values of phonon thermal conductivity in SLG, few-layer graphene and GNRs are presented in Tables 9.2 and 9.3 at RT (if not indicated otherwise). Readers interested in a more detailed description of theoretical models for the heat conduction in graphene materials are referred to review papers [27, 28, 118].

Table 9.2 Thermal conductivity of graphene and few-layer graphene: theoretical data

Sample	K (W/mK)	Method	Description	Refs.
SLG	1000–8000	BTE, γ_{LA}, γ_{TA}	Strong size dependence	[24]
	2000–8000	BTE, $\gamma_s(q)$	Strong edge, width and grunaisen parameter dependence	[23]
	~2430	BTE, 3rd-order interatomic force constants (IFCs)	K(graphene) $\geq K$ (carbon nanotube)	[119]
	1500–3500	BTE, 3rd-order IFCs	Strong size dependence	[59]
	100–8000	BTE	Strong length, size, shape and edge dependence	[26]
	2000–4000	Continuum approach + BTE	Strong isotope, point-Defects and strain influence	[78, 120]
	~4000	Ballistic	Strong width dependence	[121]
	~2900	MD simulation	Strong dependence on the vacancy concentration	[84]
	~20,000	VFF + MD simulation	Ballistic regime; flake length ~ 5 μm; strong width and length dependence	[122]
	100–550	MD simulation	Flake length $L < 200$ nm; strong length and defect dependence	[91]
	~3000	MD simulation	Sheet length ~ 15 μm; strong size dependence	[94]
	2360	MD simulation	$L \sim 5$ μm; strong length dependence	[96]
	4000–6000	MD simulation	Strong strain dependence	[100]
	~3600	Boltzmann-Peierls equation + density functional perturbation theory	$L = 10$ μm; insensitivity to small isotropic strain	[123]
	~1250	MD simulation	$L = 100$ μm; strong length dependence for $L < 100$ μm	[124]
	1800	MD simulation	6 nm × 6 nm sheet; isolated	[98]
	1000–1300	MD simulation	6 nm × 6 nm sheet; Cu-supported; strong dependence on the interaction strength between graphene and substrate	
FLG	1000–4000	BTE, $\gamma_s(q)$	n = 8 − 1, strong size dependence	[22]
	1000–3500	BTE, 3rd-order IFCs	n = 5 − 1, strong size dependence	[59]
	2000–3300	BTE, 3rd-order IFCs	n = 4 − 1	[60]
	580–880	MD simulation	n = 5 − 1, strong dependence on the Van-der Vaals bond strength	[85]

Table 9.3 Thermal conductivity of GNRs: theoretical data

K (W/mK)	Method	Description	Refs.
1000–7000	Theory: molecular dynamics, Tersoff	Strong ribbon width and edge dependence	[83]
~5500	BTE	GNR with width of 5 μm; strong dependence on the edge roughness	[103]
~2000	MD simulation	$T = 400$ K; 1.5 nm × 5.7 nm zigzag GNR; strong edge chirality influence	[108]
30–80	AIREBO potential + MD simulation	10—zigzag and 19 -arm-chair nanoribbons; strong defect dependence	[90, 92]
3200–5200	MD simulation	Strong GNRs width (W) and length dependence; 9 nm $\leq L \leq$ 27 nm and 4 nm $\leq W \leq$ 18 nm	[93]
400–600	MD simulation	$K \sim L^{0.24}$; 100 nm $\leq L \leq$ 650 nm	[95]
100–1000	BTE	GNRs supported on SiO_2; strong edge and width dependence	[106]
500–300	MD simulation	Few-layer GNRs; 10-ZGNR, $n = 1, \ldots, 5$	[97]

9.4.1 *Specifics of Two-Dimensional Phonon Transport*

We now address in more detail some specifics of the acoustic phonon transport in 2D systems. Investigation of the heat conduction in graphene [19, 20] and CNTs [125] raised the issue of ambiguity in the definition of the intrinsic thermal conductivity for 2D and 1D crystal lattices. It was theoretically shown that the intrinsic thermal conductivity limited by the crystal anharmonicity has a finite value in 3D bulk crystals [12, 56]. However, many theoretical models predict that the intrinsic thermal conductivity reveals a logarithmic divergence in strictly 2D systems, $K \sim ln(N)$, and the power-law divergence in 1D systems, $K \sim N^{\alpha}$, with the number of atoms N $(0 < \alpha < 1)$ [12, 16, 56, 125–129]. The logarithmic divergence can be removed by introduction of the *extrinsic* scattering mechanisms such as scattering from defects or coupling to the substrate [56]. Alternatively, one can define the *intrinsic* thermal conductivity of a 2D crystal for a given size of the crystal.

Graphene is not an ideal 2D crystal, considered in most of the theoretical works, since graphene atoms vibrate in three directions. Nevertheless, the intrinsic graphene thermal conductivity strongly depends on the graphene sheet size due to weak scattering of the low-energy phonons by other phonons in the system. Therefore, the phonon boundary scattering is an important mechanism for phonon relaxation in graphene. Different studies [26, 130, 131] also suggested that an accurate accounting of the higher-order anharmonic processes, i.e. above three-phonon Umklapp scattering, and inclusion of the normal phonon processes into consideration allow one to limit the low-energy phonon MFP. The normal phonon processes do not contribute directly to thermal resistance but affect the phonon mode distribution [59, 119]. However, even these studies found that the graphene sample has to be very large (>10 μm) to obtain the size-independent thermal conductivity.

In BTE approach within relaxation time approximation the thermal conductivity in quasi-2D system are given by [23, 26]:

$$K = \frac{1}{4\pi k_B T^2 h} \times \sum_s \int_0^{q_{\max}} \left\{ \left[\hbar\omega_s(q) \frac{d\omega_s(q)}{dq} \right]^2 \tau_{tot}(s,q) \frac{\exp[\hbar\omega_s(q)/kT]}{[\exp[\hbar\omega_s(q)/kT]-1]^2} q \right\} dq. \tag{9.7}$$

Here $\hbar\omega_s(q)$ is the phonon energy, $h = 0.335$ nm is the graphene layer thickness, τ_{tot} is the total phonon relaxation time, q is the phonon wavenumber, T is the temperature and k_B is the Boltzmann constant.

The specific phonon transport in the quasi—2D system such as graphene can be illustrated with a simple expression for Umklapp—limited thermal conductivity derived by us in [24]:

$$K_U = \frac{M}{4\pi T h} \sum_{s=TA,LA} \frac{\omega_{s,\max}\overline{\upsilon}_s^2}{\gamma_s^2} F(\omega_{s,\min}, \omega_{s,\max}), \tag{9.8}$$

where

$$\begin{aligned} F(\omega_{s,\min}, \omega_{s,\max}) &= \int_{\hbar\omega_{s,\min}/k_BT}^{\hbar\omega_{s,\max}/k_BT} \xi \frac{exp(\xi)}{[exp(\xi)-1]^2} d\xi \\ &= \left[ln\{exp(\xi)-1\} + \frac{\xi}{1-exp(\xi)} - \xi \right] \Bigg|_{\hbar\omega_{s,\min}/k_BT}^{\hbar\omega_{s,\max}/k_BT}. \end{aligned} \tag{9.9}$$

In the above equation, $\xi = \hbar\omega/k_BT$, and the upper cut-off frequencies $\omega_{s,\max}$ are defined from the actual phonon dispersion in graphene [23]: $\omega_{LA,\max} = 2\pi f_{LA,\max}(\Gamma K) = 241$ rad/ps, $\omega_{TA,\max} = 2\pi f_{TA,\max}(\Gamma K) = 180$ rad/ps. The integrand in Eq. (9.9) can be further simplified near RT when $\hbar\omega_{s,\max} > k_BT$, and it can be expressed as

$$F(\omega_{s,\min}) \approx -ln\{|exp(\hbar\omega_{s,\min}/k_BT)-1|\} + \frac{\hbar\omega_{s,\min}}{k_BT} \frac{exp(\hbar\omega_{s,\min}/k_BT)}{exp(\hbar\omega_{s,\min}/k_BT)-1} \tag{9.10}$$

In Eqs. (9.7, 9.8, 9.9, and 9.10) the contribution of ZA phonons to thermal transport has been neglected [24, 35, 132] because of their low group velocity and large Gruneisen parameter γ_{ZA} [24, 72].

There is a clear difference between the heat transport in basal planes of bulk graphite and in single layer graphene [35, 132]. In the former, the heat transport is approximately two-dimensional only up to some lower-bound cut-off frequency $\omega_{\min}$. Below $\omega_{\min}$ there appears to be strong coupling with the cross-plane phonon modes and heat starts to propagate in all directions, which reduces the contributions of these low-energy modes to heat transport along basal planes to negligible values. In bulk graphite, there is a physically reasonable reference point for the on-set of the cross-plane coupling, which is the ZO′ phonon branch near ~4 THz observed in the spectrum of bulk graphite [35, 133]. The presence of the ZO′ branch and corresponding $\omega_{\min} = \omega_{ZO'}$ $(q = 0)$ allows one to avoid the logarithmic divergence in the Umklapp-limited thermal conductivity integral [see Eqs. (9.7, 9.8, 9.9, and 9.10)] and calculate it without considering other scattering mechanisms.

The physics of heat conduction is principally different in graphene where the phonon transport is 2D all the way to zero phonon frequency. Therefore the lower-bound cut-off frequencies $\omega_{s,\min}$ for each s are determined from the condition that the phonon MFP cannot exceed the physical size L of the flake, i.e.

$$\omega_{s,\min} = \frac{\overline{\upsilon}_s}{\gamma_s}\sqrt{\frac{M\,\overline{\upsilon}_s}{k_B T}\frac{\omega_{s,\max}}{L}}. \tag{9.11}$$

We would like to emphasize here that using size-independent graphite $\omega_{\min}$ for SLG or FLG (as has been proposed in [134]) is without scientific merit and leads to an erroneous calculation of thermal conductivity, as described in detail in [25]. Equations (9.8, 9.9, 9.10, and 9.11) constitute a simple analytical model for the calculation of the thermal conductivity of the graphene layer, which retains such important features of graphene phonon spectra as different $\overline{\upsilon}_s$ and γ_s for *LA* and *TA* branches. The model also reflects the two-dimensional nature of heat transport in graphene all the way down to zero phonon frequency.

In Fig. 9.5a, we present the dependence of thermal conductivity of graphene on the dimension of the flake L. The data is presented for the averaged values of the Gruneisen parameters $\gamma_{LA} = 1.8$ and $\gamma_{TA} = 0.75$ obtained from *ab initio* calculations, as well as for several other close sets of $\gamma_{LA,TA}$ to illustrate the sensitivity of the result to the Gruneisen parameters. For small graphene flakes, the K dependence on L is rather strong. It weakens for flakes with $L \geq 10$ μm. The calculated values are in good agreement with experimental data for suspended exfoliated [19, 20] and CVD graphene [47, 48]. The horizontal dashed line indicates the experimental thermal conductivity for bulk graphite, which is exceeded by graphene's thermal conductivity at smaller L. Thermal conductivity, presented in Fig. 9.5, is an *intrinsic* quantity limited by the three-phonon Umklapp scattering only. But it is determined for a specific graphene flake size since L defines the lower-bound (long-wavelength) cut-off frequency in Umklapp scattering through Eq. (9.11). In experiments, thermal conductivity is also limited by defect scattering. When the size of the flake becomes very large with many polycrystalline grains, the scattering on their boundaries will also lead to phonon relaxation. The latter can

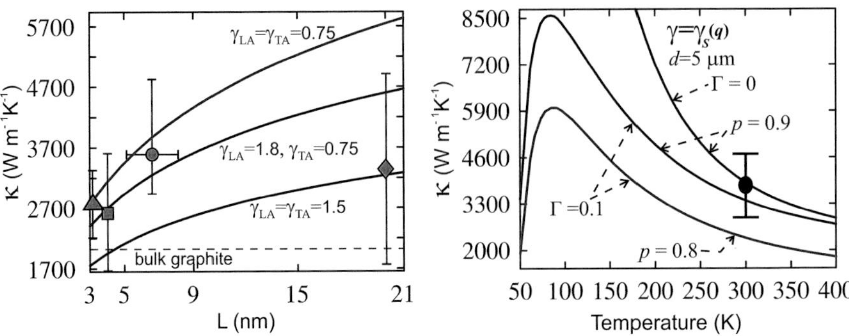

Fig. 9.5 (**a**) Calculated room temperature thermal conductivity of graphene as a function of the lateral size for several values of the Gruneisen parameter.(**b**) Calculated thermal conductivity of suspended graphene as a function of the temperature. Note a strong dependence on the size of the graphene flakes. Experimental data points from [19, 20] (*circle*), [47] (*square*), [48] (*rhomb*) and [62] (*triangle*) are shown for comparison. Figure 9.5b is after Ref. [23] reproduced with permission from the American Physical Society

be included in this model through adjustment of L. The extrinsic phonon scattering mechanisms or high-order phonon–phonon scatterings prevent indefinite growth of thermal conductivity of graphene with L [26].

The simple model described above is based on the Klemens-like expressions for the relaxation time (see [35, 36]). Therefore it does not take into account all peculiarities of the 2D three-phonon Umklapp processes in SLG or FLG, which are important for the accurate description of thermal transport. There are two types of the three-phonon Umklapp scattering processes [23, 36]. The first type is the scattering when a phonon with the wave vector $\vec{q}\,(\omega)$ absorbs another phonon from the heat flux with the wave vector $\vec{q}\,'(\omega')$, i.e. the phonon leaves the state $\vec{q}$. For this type of scattering processes the momentum and energy conservation laws are written as:

$$\begin{aligned} &\vec{q}\,(\omega) + \vec{q}\,'(\omega') = \vec{b}_i + \vec{q}\,''(\omega''), \quad i = 1,2,3 \\ &\omega + \omega' = \omega''. \end{aligned} \tag{9.12}$$

The processes of the second type are those when the phonons $\vec{q}\,(\omega)$ of the heat flux decay into two phonons with the wave vectors $\vec{q}\,'(\omega')$ and $\vec{q}\,''(\omega'')$, i.e. leaves the state $\vec{q}\,(\omega)$, or, alternatively, two phonons $\vec{q}\,'(\omega')$ and $\vec{q}\,''(\omega'')$ merge together forming a phonon with the wave vector $\vec{q}\,(\omega)$, which correspond to the phonon coming to the state $\vec{q}\,(\omega)$. The conservation laws for this type are given by:

$$\begin{aligned} &\vec{q}\,(\omega) + \vec{b}_i = \vec{q}\,'(\omega') + \vec{q}\,''(\omega''), \quad i = 4,5,6 \\ &\omega = \omega' + \omega'', \end{aligned} \tag{9.13}$$

In Eqs. (9.12 and 9.13) $\overrightarrow{b}_i = \overrightarrow{\Gamma\Gamma}_i$, $i = 1, 2, \ldots, 6$ is one of the vectors of the reciprocal lattice. Calculations of the thermal conductivity in graphene taking into account all possible three-phonon Umklapp processes allowed by the Eqs. (9.12 and 9.13) and actual phonon dispersions were carried out in [23]. For each phonon mode (q_i, s), were found all pairs of the phonon modes $(\overrightarrow{q}', s')$ and $(\overrightarrow{q}'', s'')$ such that the conditions of Eqs. (9.12 and 9.13) are met. As a result, in $(\overrightarrow{q}')$-space were constructed the *phase diagrams* for all allowed three-phonon transitions [23]. Using the long-wave approximation for a matrix element of the three-phonon interaction one can obtain for the Umklapp scattering rates:

$$\begin{aligned}\frac{1}{\tau_U^{(I),(II)}\left(s,\overrightarrow{q}\right)} = {} & \frac{\hbar\gamma_s^2\left(\overrightarrow{q}\right)}{3\pi\rho\upsilon_s^2\left(\overrightarrow{q}\right)} \\ & \times \sum_{s's'';\overrightarrow{b}_i} \iint \omega_s\left(\overrightarrow{q}\right)\omega'_{s'}\left(\overrightarrow{q}'\right)\omega''_{s''}\left(\overrightarrow{q}''\right) \times \left\{N_0\left[\omega'_{s'}\left(\overrightarrow{q}'\right)\right]\right. \\ & \left.\mp N_0\left[\omega''_{s''}\left(\overrightarrow{q}''\right)\right] + \frac{1}{2} \mp \frac{1}{2}\right\} \times \\ & \times\delta\left[\omega_s\left(\overrightarrow{q}\right) \pm \omega'_{s'}\left(\overrightarrow{q}'\right) - \omega''_{s''}\left(\overrightarrow{q}''\right)\right] dq'_l dq'_\perp. \end{aligned} \quad (9.14)$$

Here q'_l and $q'_\perp$ are the components of the vector $\overrightarrow{q}'$ parallel or perpendicular to the lines defined by Eqs. (9.12 and 9.13), correspondingly, $\gamma_s\left(\overrightarrow{q}\right)$ is the mode-dependent Gruneisen parameter, which is determined for each phonon wave vector and polarization branch and ρ is the surface mass density. In Eq. (9.14) the upper signs correspond to the processes of the first type while the lower signs correspond to those of the second type. The integrals for $q_l, q_\perp$ are taken along and perpendicular to the curve segments, correspondingly, where the conditions of Eqs. (9.12 and 9.13) are met.

The main mechanisms of phonon scattering in graphene are phonon–phonon Umklapp (U) scattering, rough edge scattering (boundary (B)) and point-defect (PD) scattering:

$$\frac{1}{\tau_{tot}(s,q)} = \frac{1}{\tau_U(s,q)} + \frac{1}{\tau_B(s,q)} + \frac{1}{\tau_{PD}(s,q)}, \quad (9.15)$$

where $1/\tau_U = 1/\tau_U^I + 1/\tau_U^{II}$, $1/\tau_B(s,q) = (\upsilon_s/L)\left((1-p)/(1+p)\right)$ and $1/\tau_{PD}(s,q) = S_0\Gamma q_s\omega_s^2/(4\upsilon_s)$. Here $\upsilon_s = d\omega_s/dq$ is the phonon group velocity, p is the specularity parameter of rough edge scattering, S is the surface per atom and Γ is the measure of the strength of the point defect scattering.

The sensitivity of the thermal conductivity, calculated using Eqs. (9.7, 9.12, 9.13, 9.14, and 9.15), to the value of p and Γ is illustrated in Fig. 9.5b. The data is presented for different sizes (widths) of the graphene flakes.

9.5 Conclusions

We reviewed theoretical and experimental results pertinent to 2D phonon transport in graphene. Phonons are the dominant heat carriers in the ungated graphene samples near room temperature. The unique nature of 2D phonons translates to unusual heat conduction in graphene and related materials. Recent computational studies suggest that the thermal conductivity of graphene depends strongly on the concentration of defects and strain distribution. Investigation of the physics of 2D phonons in graphene can shed light on the thermal energy transfer in low-dimensional systems. The results presented in this review are important for the proposed practical applications of graphene in heat removal and thermal management of advanced electronics.

Acknowledgments The work at UC Riverside was supported, in part, by the National Science Foundation (NSF) project CMMI-1404967 Collaborative Research Genetic Algorithm Driven Hybrid Computational Experimental Engineering of Defects in Designer Materials; NSF project ECCS-1307671 Two-Dimensional Performance with Three-Dimensional Capacity: Engineering the Thermal Properties of Graphene, and by the STARnet Center for Function Accelerated nano-Material Engineering (FAME)—Semiconductor Research Corporation (SRC) program sponsored by The Microelectronics Advanced Research Corporation (MARCO) and the Defense Advanced Research Project Agency (DARPA). The work at Moldova State University was supported, in part, by the Moldova State Project 15.817.02.29 F and ASM-STCU project #5937.

References

1. Balandin, A.A.: Better computing through CPU cooling. IEEE Spectrum. 29–33 (2009)
2. Pop, E.: Energy dissipation and transport in nanoscale devices. Nano Res. **3**, 147 (2010)
3. Balandin, A., Wang, K.L.: Effect of phonon confinement on the thermoelectric figure of merit of quantum wells. J. Appl. Phys. **84**, 6149 (1998)
4. Li, D., Wu, Y., Kim, P., Shi, L., Yang, P., Majumdar, A.: Thermal conductivity of individual silicon nanowires. Appl. Phys. Lett. **83**, 2934 (2003)
5. Balandin, A.A., Pokatilov, E.P., Nika, D.L.: Phonon engineering in hetero- and nanostructures. J. Nanoelectron. Optoelectron. **2**, 140 (2007)
6. Pokatilov, E.P., Nika, D.L., Balandin, A.A.: Acoustic-phonon propagation in rectangular semiconductor nanowires with elastically dissimilar barriers. Phys. Rev. B **72**, 113311 (2005)
7. Pokatilov, E.P., Nika, D.L., Balandin, A.A.: Acoustic phonon engineering in coated cylindrical nanowires. Superlattice Microst. **38**, 168 (2005)
8. Liu, W., Asheghi, M.: Thermal conductivity measurements of ultra-thin single crystal silicon layers. J. Heat Transfer **128**, 75 (2006)
9. Piazza, F., Lepri, S.: Heat wave propagation in a nonlinear chain. Phys. Rev. B **79**, 094306 (2009)
10. Lepri, S., Livi, R., Politi, A.: Studies of thermal conductivity in Fermi-Pasta-Ulam-like lattices. Chaos **15**, 015118 (2005)
11. Lepri, S., Livi, R., Politi, A.: Thermal conduction in classical low-dimensional lattices. Phys. Rep. **377**, 1 (2003)
12. Basile, G., Bernardin, C., Olla, S.: Momentum conversion model with anomalous thermal conductivity in low dimensional system. Phys. Rev. Lett. **96**, 204303 (2006)

13. Pokatilov, E.P., Nika, D.L., Balandin, A.A.: Phonon spectrum and group velocities in AlN/GaN/AlN and related heterostructures. Superlattice Microst. **33**, 155 (2003)
14. Pernot, G., Stoffel, M., Savic, I., Pezzoli, F., Chen, P., Savelli, G., Jacquot, A., Schumann, J., Denker, U., Mönch, I., Deneke, G., Schmidt, O.G., Rampnoux, J.M., Wang, S., Plissonnier, M., Rastelli, A., Dilhaire, S., Mingo, N.: Precise control of thermal conductivity at the nanoscale through individual phonon-scattering barriers. Nature Mater. **9**, 491 (2010)
15. Nika, D.L., Pokatilov, E.P., Balandin, A.A., Fomin, V.M., Rastelli, A., Schmidt, O.G.: Reduction of lattice thermal conductivity in one-dimensional quantum-dot superlattices due to phonon filtering. Phys. Rev. B **84**, 165415 (2011)
16. Chang, C.W., Okawa, D., Garcia, H., Majumdar, A., Zettl, A.: Breakdown of Fourier's law in nanotube thermal conductors. Phys. Rev. Lett. **101**, 075903 (2008)
17. Narayan, O., Ramaswamy, S.: Anomalous heat conduction in one dimensional momentum-conserving systems. Phys. Rev. Lett. **89**, 200601 (2002)
18. Balandin, A.A.: Thermal properties of graphene and nanostructured carbon materials. Nature Mater. **10**, 569 (2011)
19. Balandin, A.A., Ghosh, S., Bao, W., Calizo, I., Teweldebrhan, D., Miao, F., Lau, C.N.: Superior thermal conductivity of single layer graphene. Nano Lett. **8**, 902 (2008)
20. Ghosh, S., Calizo, I., Teweldebrhan, D., Pokatilov, E.P., Nika, D.L., Balandin, A.A., Bao, W., Miao, F., Lau, C.N.: Extremely high thermal conductivity in graphene: Prospects for thermal management application in nanoelectronic circuits. Appl. Phys. Lett. **92**, 151911 (2008)
21. Ghosh, S., Nika, D.L., Pokatilov, E.P., Balandin, A.A.: Heat conduction in graphene: experimental study and theoretical interpretation. New J. Phys. **11**, 095012 (2009)
22. Ghosh, S., Bao, W., Nika, D.L., Subrina, S., Pokatilov, E.P., Lau, C.N., Balandin, A.A.: Dimensional crossover of thermal transport in few-layer graphene. Nature Mater. **9**, 555 (2010)
23. Nika, D.L., Pokatilov, E.P., Askerov, A.S., Balandin, A.A.: Phonon thermal conduction in graphene: role of Umklapp and edge roughness scattering. Phys. Rev. B **79**, 155413 (2009)
24. Nika, D.L., Ghosh, S., Pokatilov, E.P., Balandin, A.A.: Lattice thermal conductivity of graphene flakes: comparison with bulk graphite. Appl. Phys. Lett. **94**, 203103 (2009)
25. Nika, D.L., Pokatilov, E.P., Balandin, A.A.: Theoretical description of thermal transport in graphene: the issues of phonon cut-off frequencies and polarization branches. Phys. Stat. Sol. B **248**, 2609 (2011)
26. Nika, D.L., Askerov, A.S., Balandin, A.A.: Anomalous size dependence of the thermal conductivity of graphene ribbons. Nano Lett. **12**, 3238 (2012)
27. Nika, D.L., Balandin, A.A.: Two-dimensional phonon transport in graphene. J. Phys. Cond. Matter. **24**, 233203 (2012)
28. Balandin, A.A., Nika, D.L.: Phonons in low-dimensions: engineering phonons in nanostructures and graphene. Mater Today **15**, 266 (2012)
29. Cahill, D.G., Ford, W.K., Goodson, K.E., Mahan, G.D., Majumdar, A., Maris, H.J., Merlin, R., Sr, P.: Nanoscale thermal transport. J. Appl. Phys. **93**, 793 (2003)
30. Bhandari, C.M., Rowe, D.M.: Thermal Conduction in Semiconductors. Wiley (1988)
31. Srivastava, G.P.: The Physics of Phonons, p. 127. Taylor & Francis Group, LLC (1990)
32. Mills, A.F.: Heat and Mass Transfer, p. 9. Richard D Irwin Inc. 9 (1995)
33. Ziman, J.M.: Principles of the Theory of Solids, 2nd edn. Cambridge University Press, New York (2001)
34. Pierson, H.O.: Handbook of Carbon, Graphite, Diamonds and Fullerenes: Processing, Properties and Applications. Noyes Publications (2010)
35. Klemens, P.G.: Theory of the a-plane thermal conductivity of graphite. J. Wide Bandgap Mater. **7**, 332 (2000)
36. Klemens, P.G.: Thermal conductivity and lattice vibrational modes. In: Seitz, F., Turnbull, D (eds.) Solid State Physics,vol. 7, p. 1. Academic, New York (1958)
37. Callaway, J.: Model for lattice thermal conductivity at low temperatures. Phys. Rev. **113**, 1046 (1959)
38. Parrott, J.E., Stuckes, A.D.: Thermal Conductivity of Solids. Methuen, New York (1975)

39. Ziman, J.M.: Electrons and Phonons: The Theory of Transport Phenomena in Solids. Oxford University Press, New York (2001)
40. Balandin, A., Wang, K.L.: Significant decrease of the lattice thermal conductivity due to phonon confinement in a free-standing semiconductor quantum well. Phys. Rev. B **58**, 1544 (1998)
41. Zou, J., Balandin, A.: Phonon heat conduction in a semiconductor nanowires. J. Appl. Phys. **89**, 2932 (2001)
42. Balandin, A.A.: Nanophononics: phonon engineering in nanostructures and nanodevices. J. Nanosci. Nanotech. **5**, 1015 (2005)
43. Aksamija, Z., Knezevic, I.: Anisotropy and boundary scattering in the lattice thermal conductivity of silicon nanomembranes. Phys. Rev. B **82**, 045319 (2010)
44. Soffer, S.B.: Statistical model for the size effect in electrical conduction. J. Appl. Phys. **38**, 1710 (1967)
45. Balandin, A.A., Ghosh, S., Nika, D.L., Pokatilov, E.P.: Extraordinary thermal conductivity of graphene: possible applications in thermal management. ECS Trans. **28**, 63 (2010)
46. Balandin, A.A., Ghosh, S., Nika, D.L., Pokatilov, E.P.: Thermal conduction in suspended graphene layers. Fuller. Nanotub. Car. N. **18**, 1 (2010)
47. Cai, W., Moore, A.L., Zhu, Y., Li, X., Chen, S., Shi, L., Ruoff, R.S.: Thermal transport in suspended and supported monolayer graphene grown by chemical vapor deposition. Nano Lett. **10**, 1645 (2010)
48. Jauregui, L.A., Yue, Y., Sidorov, A.N., Hu, J., Yu, Q., Lopez, G., Jalilian, R., Benjamin, D.K., Delk, D.A., Wu, W., Liu, Z., Wang, X., Jiang, Z., Ruan, X., Bao, J., Pei, S.S., Chen, Y.P.: Thermal transport in graphene nanostructures: experiments and simulations. ECS Trans. **28**, 73 (2010)
49. Faugeras, C., Faugeras, B., Orlita, M., Potemski, M., Nair, R.R., Geim, A.K.: Thermal conductivity of graphene in Corbino membrane geometry. ACS Nano **4**, 1889 (2010)
50. Lee, J.U., Yoon, D., Kim, H., Lee, S.W., Cheong, H.: Thermal conductivity of suspended pristine graphene measured by Raman spectroscopy. Phys. Rev. B **83**, 081419 (2011)
51. Mak, K.F., Shan, J., Heinz, T.F.: Seeing many-body effects in single and few layer graphene: observation of two-dimensional saddle point excitons. Phys. Rev. Lett. **106**, 046401 (2011)
52. Kim, K.S., Zhao, Y., Jang, H., Lee, S.Y., Kim, J.M., Kim, K.S., Ahn, J., Kim, P., Choi, J., Hong, B.H.: Large-scale pattern growth of graphene films for stretchable transparent electrodes. Nature **457**, 706 (2009)
53. Kravets, V.G., Grigorenko, A.N., Nair, R.R., Blake, P., Anissimova, S., Novoselov, K.S., Geim, A.K.: Spectroscopic ellipsometry of graphene and an exciton-shifted van Hove peak in absorption. Phys. Rev. B **81**, 155413 (2010)
54. Chen, S., Moore, A.L., Cai, W., Suk, J.W., An, J., Mishra, C., Amos, C., Magnuson, C.W., Kang, J., Shi, L., Ruoff, R.S.: Raman measurement of thermal transport in suspended monolayer graphene of variable sizes in vacuum and gaseous environments. ACS Nano **5**, 321 (2011)
55. Seol, J.H., Jo, I., Moore, A.L., Lindsay, L., Aitken, Z.H., Pettes, M.T., Li, X., Yao, Z., Huang, R., Broido, D., Mingo, N., Ruoff, R.S., Shi, L.: Two-dimensional phonon transport in supported graphene. Science **328**, 213 (2010)
56. Saito, K., Dhar, A.: Heat conduction in a three dimensional anharmonic crystal. Phys. Rev. Lett. **104**, 040601 (2010)
57. Zhong, W.R., Zhang, M.P., Ai, B.Q., Zheng, D.Q.: Chirality and thickness-dependent thermal conductivity of few-layer graphene: a molecular dynamics study. Appl. Phys. Lett. **98**, 113107 (2011)
58. Berber, S., Kwon, Y.-K., Tomanek, D.: Unusually high thermal conductivity in carbon nanotubes. Phys. Rev. Lett. **84**, 4613 (2000)
59. Lindsay, L., Broido, D.A., Mingo, N.: Flexural phonons and thermal transport in multilayer graphene and graphite. Phys. Rev. B **83**, 235428 (2011)
60. Singh, D., Murthy, J.Y., Fisher, T.S.: Mechanism of thermal conductivity reduction in few-layer graphene. J. Appl. Phys. **110**, 044317 (2011)

61. Jang, W., Chen, Z., Bao, W., Lau, C.N., Dames, C.: Thickness-dependent thermal conductivity of encased graphene and ultrathin graphite. Nano Lett. **10**, 3909 (2010)
62. Li, H., Ying, H., Chen, X., Nika, D.L., Cocemasov, I.V., Cai, W., Balandin, A.A., Chen, S.: Thermal conductivity of twisted bilayer graphene. Nanoscale **6**, 13402 (2014)
63. Cocemasov, A.I., Nika, D.L., Balandin, A.A.: Phonons in twisted bilayer graphene. Phys. Rev. B **88**, 035428 (2013)
64. Nika, D.L., Cocemasov, A.I., Balandin, A.A.: Specific heat of twisted bilayer graphene: engineering phonons by plane rotations. Appl. Phys. Lett. **105**, 031904 (2014)
65. Dorgan, V.E., Behnam, A., Conley, H.J., Bolotin, K.I., Pop, E.: High-filed electric and thermal transport in suspended graphene. Nano Lett. **13**, 4581 (2013)
66. Fu, X., Pereira, F.L.C., Yu, W., Zhang, K., Zhao, X., Bae, S., Bui, H.H., Loh, K.P., Donadio, D., Li, B., Ozyilmaz, B.: Length-dependent thermal conductivity in suspended single-layer graphene. Nature Commun. **5**, 3689 (2015)
67. Pettes, M.T., Jo, I., Yao, Z., Shi, L.: Influence of polymeric residue on the thermal conductivity of suspended bilayer graphene. Nano Lett. **11**, 1195 (2011)
68. Wang, Z., Xie, R., Bui, C.T., Liu, D., Ni, X., Li, B., Thong, J.T.L.: Thermal transport in suspended and supported few-layer graphene. Nano Lett. **11**, 113 (2011)
69. Murali, R., Yang, Y., Brenner, K., Beck, T., Meindl, J.D.: Breakdown current density of graphene nanoribbons. Appl. Phys. Lett. **94**, 243114 (2009)
70. Liao, A.D., Wu, J.Z., Wang, X., Tahy, K., Jena, D., Dai, H., Pop, E.: Thermally limited current carrying ability of graphene nanoribbons. Phys. Rev. Lett. **106**, 256801 (2011)
71. Maultzsch, J., Reich, S., Thomsen, C., Requardt, H., Ordejon, P.: Phonon dispersion in graphite. Phys. Rev. Lett. **92**, 075501 (2004)
72. Mounet, N., Marzari, N.: First-principles determination of the structural, vibrational and thermodynamic properties of diamond, graphite, and derivatives. Phys. Rev. B **71**, 205214 (2005)
73. Wirtz, L., Rubio, A.: The phonon dispersion of graphite revisited. Solid State Commun. **131**, 141 (2004)
74. Falkovsky, L.A.: Symmetry constraints on phonon dispersion in graphene. Phys. Lett. A **372**, 5189 (2008)
75. Perebeinos, V., Tersoff, J.: Valence force model for phonons in graphene and carbon nanotubes. Phys. Rev. B **79**, 241409(R) (2009)
76. Droth, M., Burkard, G.: Acoustic phonon and spin relaxation in graphene nanoribbons. Phys. Rev. B **84**, 155404 (2011)
77. Qian, J., Allen, M.J., Yang, Y., Dutta, M., Stroscio, M.A.: Quantized long-wavelength optical phonon modes in graphene nanoribbon in the elastic continuum model. Superlattice Microst. **46**, 881 (2009)
78. Alofi, A., Srivastva, G.P.: Phonon conductivity in graphene. J. Appl. Phys. **112**, 013517 (2012)
79. Yan, J.-A., Ruan, W.Y., Chou, M.Y.: Phonon dispersions and vibrational properties of monolayer, bilayer, and trilayer graphene: density-functional perturbation theory. Phys. Rev. B **77**, 125401 (2008)
80. Dubay, O., Kresse, G.: Accurate density functional calculations for the phonon dispersion relation of graphite layer and carbon nanotubes. Phys. Rev. B **67**, 035401 (2003)
81. Wang, H., Wang, Y., Cao, X., Feng, M., Lan, G.: Vibrational properties of graphene and graphene layers. J. Raman Spectrosc. **40**, 1791 (2009)
82. Lindsay, L., Broido, D.: Optimized Tersoff and Brenner empirical potential parameters for lattice dynamics and phonon thermal transport in carbon nanotubes and graphene. Phys. Rev. B **81**, 205441 (2010)
83. Evans, W.J., Hu, L., Keblinsky, P.: Thermal conductivity of graphene ribbons from equilibrium molecular dynamics: effect of ribbon width, edge roughness, and hydrogen termination. Appl. Phys. Lett. **96**, 203112 (2010)
84. Zhang, H., Lee, G., Cho, K.: Thermal transport in graphene and effects of vacancies. Phys. Rev. B **84**, 115460 (2011)

85. Wei, Z., Ni, Z., Bi, K., Chen, M., Chen, Y.: In-plane lattice thermal conductivities of multilayer graphene films. Carbon **49**, 2653 (2011)
86. Ong, Z.-Y., Pop, E.: Effect of substrate modes on thermal transport in supported graphene. Phys. Rev. B **84**, 075471 (2011)
87. Wei, N., Xu, L., Wang, H.-Q., Zheng, J.-C.: Strain engineering of thermal conductivity in graphene sheets and nanoribbons: a demonstration of magic flexibility. Nanotechnology **22**, 105705 (2011)
88. Hao, F., Fang, D., Xu, Z.: Mechanical and thermal transport properties of graphene with defects. Appl. Phys. Lett. **99**, 041901 (2011)
89. Mortazavi, B., Ahzi, S.: Thermal conductivity and tensile response of defective graphene: A molecular dynamics study. Carbon **63**, 460 (2013)
90. Ng, T., Yeo, J.J., Liu, Z.S.: A molecular dynamics study of the thermal conductivity of graphene nanoribbons containing dispersed Stone–Thrower–Wales defects. Carbon **50**, 4887 (2012)
91. Jang, Y.Y., Cheng, Y., Pei, Q.X., Wang, C.W., Xiang, Y.: Thermal conductivity of defected graphene. Phys. Lett. A **376**, 3668 (2012)
92. Yeo, J.J., Liu, Z., Ng, T.Y.: Comparing the effects of dispersed Stone–Thrower–Wales defects and double vacancies on the thermal conductivity of graphene nanoribbons. Nanotechnology **23**, 385702 (2012)
93. Yang, D., Ma, F., Sun, Y., Hu, T., Xu, K.: Influence of typical defects on thermal conductivity of graphene nanoribbons: An equilibrium molecular dynamics simulation. Appl. Surf. Sci. **258**, 9926 (2012)
94. Park, M., Lee, S.C., Kim, Y.S.: Length-dependent thermal conductivity of graphene and its macroscopic limit. J. Appl. Phys. **114**, 053506 (2013)
95. Yu, C., Zhang, G.: Impacts of length and geometry deformation on thermal conductivity of graphene nanoribbons. J. Appl. Phys. **113**, 044306 (2013)
96. Cao, A.: Molecular dynamics simulation study on heat transport in monolayer graphene sheet with various geometries. J. Appl. Phys. **111**, 083528 (2012)
97. Cao, H.-Y., Guo, Z.-X., Xiang, H., Gong, Z.-G.: Layer and size dependence of thermal conductivity in multilayer graphene. Phys. Lett. A **373**, 525 (2012)
98. Cheng, L., Kumar, S.: Thermal transport in graphene supported copper. J. Appl. Phys. **112**, 043502 (2012)
99. Yeo, P.S.E., Loh, K.P., Gan, C.K.: Strain dependence of the heat transport properties of graphene nanoribbons. Nanotechnology **23**, 495702 (2012)
100. Ma, F., Zheng, H.B., Sun, Y.J., Yang, D., Xu, K.W., Chu, P.K.: Strain effect on lattice vibration, heat capacity, and thermal conductivity of graphene. Appl. Phys. Lett. **101**, 111904 (2012)
101. Huang, Z., Fisher, T.S., Murthy, J.Y.: Simulation of phonon transmission through graphene and graphene nanoribbons with a Green's function method. J. Appl. Phys. **108**, 094319 (2010)
102. Zhai, X., Jin, G.: Stretching-enhanced ballistic thermal conductance in graphene nanoribbons. EPL **96**, 16002 (2011)
103. Aksamija, Z., Knezevic, I.: Lattice thermal conductivity of graphene nanoribbons: anisotropy and edge roughness scattering. Appl. Phys. Lett. **98**, 141919 (2011)
104. Klemens, P.G., Pedraza, D.F.: Thermal conductivity of graphite in basal plane. Carbon **32**, 735 (1994)
105. Adamyan, V., Zavalniuk, V.: Lattice thermal conductivity of graphene with conventionally isotopic defects. J. Phys. Cond. Matter. **24**, 415406 (2012)
106. Aksamija, Z., Knezevic, I.: Thermal transport in graphene nanoribbons supported on SiO_2. Phys. Rev. B **86**, 165426 (2012)
107. Savin, A.V., Kivshar, Y.S., Hu, B.: Suppression of thermal conductivity in graphene nanoribbons with rough edges. Phys. Rev. B **82**, 195422 (2010)
108. Hu, J., Ruan, X., Chen, Y.P.: Thermal conductivity and thermal rectification in graphene nanoribbons: a molecular dynamic study. Nano Lett. **9**, 2730 (2009)

109. Guo, Z., Zhang, D., Gong, X.-G.: Thermal conductivity of graphene nanoribbons. Appl. Phys. Lett. **95**, 163103 (2009)
110. Ouyang, T., Chen, Y.P., Yang, K.K., Zhong, J.X.: Thermal transport of isotopic-superlattice graphene nanoribbons with zigzag edge. EPL **88**, 28002 (2009)
111. Chen, S., Wu, Q., Mishra, C., Kang, J., Zhang, H., Cho, K., Cai, W., Balandin, A.A., Ruoff, R.S.: Thermal conductivity of isotopically modified graphene. Nature Mater. **11**, 203 (2012)
112. Jiang, J.W., Lan, J.H., Wang, J.S., Li, B.W.: Isotopic effects on the thermal conductivity of graphene nanoribbons: localization mechanism. J. Appl. Phys. **107**, 054314 (2010)
113. Zhang, H., Lee, G., Fonseca, A.F., Borders, T.L., Cho, K.: Isotope effect on the thermal conductivity of graphene. J. Nanomater. **53**, 7657 (2010)
114. Hu, J., Schiffli, S., Vallabhaneni, A., Ruan, X., Chen, Y.P.: Tuning the thermal conductivity of graphene nanoribbons by edge passivation and isotope engineering: a molecular dynamics study. Appl. Phys. Lett. **97**, 133107 (2010)
115. Jinag, J.-W., Wang, B.-S., Wang, J.-S.: First principle study of the thermal conductance in graphene nanoribbon with vacancy and substitutional silicon defects. Appl. Phys. Lett. **98**, 113114 (2011)
116. Hu, J., Wang, Y., Vallabhaneni, A., Ruan, X., Chen, Y.: Nonlinear thermal transport and negative differential thermal conductance in graphene nanoribbons. Appl. Phys. Lett. **99**, 113101 (2011)
117. Xie, Z.-X., Chen, K.-Q., Duan, W.: Thermal transport by phonons in zigzag graphene nanoribbons with structural defects. J. Phys. Cond. Matter **23**, 315302 (2011)
118. Wemhoff, A.P.: A review of theoretical techniques for graphene and graphene nanoribbon thermal conductivity prediction Int. J. Transp. Phenom. **13**, 121 (2012)
119. Lindsay, L., Broido, D.A., Mingo, N.: Diameter dependence of carbon nanotube thermal conductivity and extension to the graphene limit. Phys. Rev. B **82**, 161402 (2010)
120. Alofi, A., Srivastava, G.P.: Thermal conductivity of graphene and graphite. Phys. Rev. B **87**, 115421 (2013)
121. Munoz, E., Lu, J., Yakobson, B.I.: Ballistic thermal conductance of graphene ribbons. Nano Lett. **10**, 1652 (2010)
122. Jiang, J.-W., Wang, J.-S., Li, B.: Thermal conductance of graphite and dimerite. Phys. Rev. B **79**, 205418 (2009)
123. Lindsay, L., Li, W., Carrete, J., Mingo, N., Broido, D.A., Reinecke, T.L.: Phonon thermal transport in strained and unstrained graphene from first principles. Phys. Rev. B **89**, 155426 (2015)
124. Barbarino, G., Melis, C., Colombo, L.: Intrinsic thermal conductivity in monolayer graphene is ultimately upper limit. Phys. Rev. B **91**, 035416 (2015)
125. Kim, P., Shi, L., Majumdar, A., Mc Euen, P.L.: Thermal transport measurement of individual multiwalled nanotubes. Phys. Rev. Lett. **87**, 215502 (2001)
126. Lippi, A., Livi, R.: Heat conduction in two-dimensional nonlinear lattices. J. Stat. Phys. **100**, 1147 (2000)
127. Yang, L.: Finite heat conductance in a 2D disorder lattice. Phys. Rev. Lett. **88**, 094301 (2002)
128. Dhar, A.: Heat conduction in the disordered harmonic chain revisited. Phys. Rev. Lett. **86**, 5882 (2001)
129. Casher, A., Lebowitz, J.L.: Heat flow in regular and disordered harmonic chains. J. Math. Phys. **12**, 1701 (1971)
130. Ecsedy, D.J., Klemens, P.G.: Thermal resistivity of dielectric crystals due to 4-phonon processes and optical modes. Phys. Rev. B **15**, 5957 (1977)
131. Mingo, N., Broido, D.: Length dependence of carbon nanotube thermal conductivity and "the problem of long waves". Nano Lett. (5) 1221 (2005)
132. Klemens, P.G.: Theory of thermal conduction in the ceramic films. Int. J. Thermophys. **22**, 265 (2001)
133. Kelly, B.T.: Physics of Graphite. Applied Science Publishers, London (1981)
134. Kong, B.D., Paul, S., Nardelli, M.B., Kim, K.W.: First-principles analysis of lattice thermal conductivity in monolayer and bilayer graphene. Phys. Rev. B **80**, 033406 (2009)

Chapter 10
From Thermal Rectifiers to Thermoelectric Devices

Giuliano Benenti, Giulio Casati, Carlos Mejía-Monasterio, and Michel Peyrard

Abstract We discuss thermal rectification and thermoelectric energy conversion from the perspective of nonequilibrium statistical mechanics and dynamical systems theory. After preliminary considerations on the dynamical foundations of the phenomenological Fourier law in classical and quantum mechanics, we illustrate ways to control the phononic heat flow and design thermal diodes. Finally, we consider the coupled transport of heat and charge and discuss several general mechanisms for optimizing the figure of merit of thermoelectric efficiency.

10.1 Dynamical Foundations of Fourier Law

The possibility to manipulate the heat current represents a fascinating challenge for the future, especially in view of the need of future society of providing a sustainable supply of energy and due to the strong concerns about the environmental

G. Benenti (✉)
Dipartimento di Scienza e Alta Tecnologia, Center for Nonlinear and Complex Systems, Università degli Studi dell'Insubria, via Valleggio 11, 22100 Como, Italy

Istituto Nazionale di Fisica Nucleare, Sezione di Milano, via Celoria 16, 20133 Milano, Italy
e-mail: giuliano.benenti@uninsubria.it

G. Casati
Dipartimento di Scienza e Alta Tecnologia, Center for Nonlinear and Complex Systems, Università degli Studi dell'Insubria, via Valleggio 11, 22100 Como, Italy

International Institute of Physics, Federal University of Rio Grande do Norte, Natal, Brazil
e-mail: giulio.casati@uninsubria.it

C. Mejía-Monasterio
Laboratory of Physical Properties, School of Agricultural, Food and Biosystems Engineering, Technical University of Madrid, Av. Complutense s/n, 28040 Madrid, Spain
e-mail: carlos.mejia@upm.es

M. Peyrard
Laboratoire de Physique CNRS UMR 5672, Ecole Normale Supérieure de Lyon, 46 allée d'Italie, 69364 Lyon Cedex 7, France
e-mail: michel.peyrard@ens-lyon.fr

S. Lepri (ed.), *Thermal Transport in Low Dimensions*, Lecture Notes in Physics 921, DOI 10.1007/978-3-319-29261-8_10

impact of the combustion of fossil fuels. However along these lines there are severe difficulties both of theoretical and experimental nature. In particular it turns out that manipulation of the heat current is much more difficult than the manipulation of the electric current.

It is therefore necessary to start from first principles in order to get a deep and systematic understanding of the properties of heat transport. Namely we would like to understand these properties starting from the microscopic dynamical equations of motion.

Along these lines a necessary step is the derivation of the Fourier heat law from dynamical equations of motion. In particular we would like to understand under what conditions Fourier law is valid. What are the dynamical properties needed to have normal transport in a given system? This is a nontrivial question and for many years it has been addressed according to different perspectives. It concerns, on one hand, the foundations of nonequilibrium statistical mechanics and, on the other hand, the practical issue of constructing microscopic models which agree with the macroscopic equations which describe transport. For example, for a class of hyperbolic systems (transitive Anosov) a guiding principle was proposed (the so-called chaotic hypothesis [60]) as a prescription for extending equilibrium methods to nonequilibrium situations. We remark that in these works the randomness needed to obtain a consistent description of the irreversible macroscopic phenomena comes from the exponential instability of the microscopic chaotic dynamics.

First numerical evidence of the validity of Fourier heat conduction law in an exponentially unstable system was reported in [36] where the so-called "ding-a-ling" model was considered (Fig. 10.1). The model consists of harmonic oscillators which exchange their energy via elastic, hard-core collisions, with intermediate hard spheres. The even-numbered particles in Fig. 10.1 form a set of equally spaced lattice oscillators with each oscillator being harmonically bound to its individual lattice site and with all oscillators vibrating at the same frequency ω. The odd-numbered particles are free particles constrained only by the two adjacent even-numbered oscillators. It can be shown that the dynamics of this model is uniquely determined by the parameter ω^2/E (where E is the energy per particle) and that the dynamics becomes exponentially unstable when this parameter is $\gg 1$. The validity of Fourier law was established in the standard way by putting the two end particles in contact with thermal reservoirs, taken as Maxwellian gases, at different temperatures. The system was then numerically integrated until the stationary state was reached and the energy exchange at the left and right reservoir became equal. This gives the average heat flux j. After defining the particle temperature to be twice its average kinetic energy, the value of the steady-state internal temperature gradient ∇T was computed. Then the thermal conductivity κ was computed via the heat

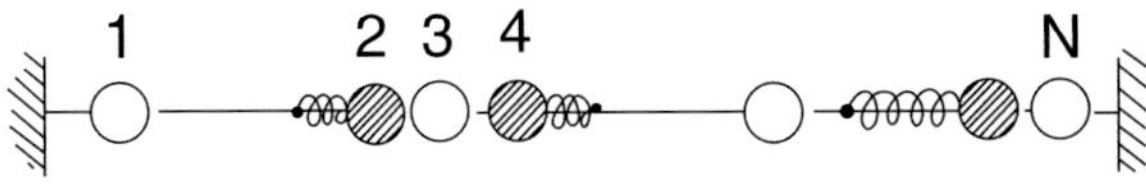

Fig. 10.1 The ding-a-ling model. Here the springs merely symbolize the harmonic restoring force

Fourier law $j = -\kappa \nabla T$. A normal conductivity independent on the system length was found.

It is important to stress however that hard chaos with exponential instability is not a necessary condition to induce normal transport properties. Moreover rigorous results are still lacking and in spite of several efforts, the connection between Lyapunov exponents, correlations decay and diffusive properties is still not completely clear. As a matter of facts it turns out that mixing property is sufficient to ensure normal heat transport [85]. This might be an important step in the general attempt to derive macroscopic statistical laws from the underlying deterministic dynamics. Indeed, systems with zero Lyapunov exponent have zero algorithmic complexity and, at least in principle, are analytically solvable.

A particular case is given by total momentum conserving systems which typically exhibit anomalous conductivity. This type of systems is largely discussed in other contributions of this volume and therefore will not be considered here any longer. Here we would like to add only a word of caution, and to suggest that anomalous behavior in such systems is perhaps more general than so far believed. The point is that our present understanding of the heat conduction problem is mainly based on numerical empirical evidence while rigorous analytical results are difficult to obtain. Numerical analysis consists of steady state, nonequilibrium simulations or of equilibrium simulations based on linear response theory and Green-Kubo formula. Typically, if both methods give reasonable evidence for Fourier law and if, moreover, they lead to the same numerical value of the coefficient of thermal conductivity κ, then this is generally considered as an almost conclusive evidence that Fourier law is indeed valid.

This conclusion, however, might be not correct as shown in [42], where the heat conductivity of the one-dimensional diatomic hard-point gas model was studied. As shown in Fig. 10.2, the Fourier-like behavior, seen in both equilibrium and nonequilibrium simulations, turns out to be a finite-size effect and Fourier law appear to hold up to some size N after which anomalous behavior sets in. This

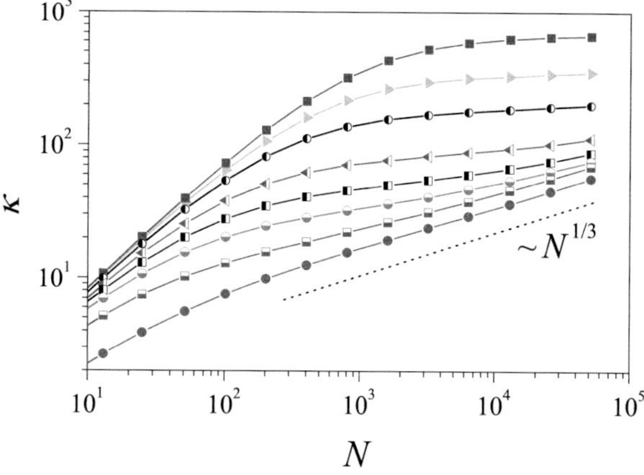

Fig. 10.2 The heat conductivity κ versus the system size N for the one-dimensional diatomic gas model, with alternative mass M and m. From *top* to *bottom*, the mass ratio M/m is respectively 1.07, 1.10, 1.14, 1.22, 1.30, 1.40, the golden mean (≈ 1.618), and 3

behavior requires a better understanding. Indeed, while it is natural to expect an initial ballistic behavior for larger and larger N as one approaches the integrable limit, it is absolutely not clear why the value of κ appears to saturate to a constant value before entering the anomalous regime $\kappa \sim N^{\alpha}$ (with $\alpha \approx 1/3$) at an even larger system size N.

To summarize, while establishing a complete connection between ergodic properties and macroscopic transport features is still beyond the reach of present understanding, we may conclude that apart some particular notable exceptions, dynamical mixing property induces deterministic diffusion and hence Fourier law.

10.2 Fourier Law in Quantum Mechanics

The next step is to discuss whether or not Fourier heat law can be derived from quantum dynamics without a priori statistical assumptions. This calls directly in question the issue of "Quantum chaos". The first attempts to provide a microscopic description of heat transport in quantum systems dates back to the beginning of the twentieth century with the work of Debye in 1912 [50] and subsequently of Peierls in 1929 [111]. Based on modifications of the kinetic Boltzmann equation these theories are classical in essence by considering classical-like quasi-particles, and fail to describe systems out of equilibrium with dimensions comparable to the electron and phonon mean free paths.

The recent achievements in the miniaturisation of devices have boosted the interest in understanding the conditions under which heat is transported diffusively in quantum systems. In spite of the many efforts a rigorous derivation of a quantum Fourier law for general Hamiltonians remains an unsolved problem. A main difficulty to the study of heat conduction in quantum mechanics is the lack of appropriate definitions of local quantities such as the temperature and the heat current [54], and calls in question the problem of thermalization, namely the relaxation to a state in local equilibrium, in isolated [1] and open quantum systems [30]. It has been found that the conditions for thermalization are essentially related to the systems' integrability and localization properties (e.g. due to disorder). Non-ergodic systems, undergo relaxation to a generalized Gibbs state [14], so that the application of standard statistical mechanics methods is possible.

Quantum systems in contact with external heat baths can be treated by using the Lindblad-Gorini-Kossakowski-Sudarshan equation [63, 87] in a convenient setup in which only boundary degrees of freedom are coupled with the environment. Within the Markovian approximation, the system's many-body density matrix evolves according to

$$\frac{d}{dt}\rho(t) = \hat{\mathscr{L}}\rho(t), \tag{10.1}$$

where the Liouvillian superoperator is defined as

$$\hat{\mathscr{L}}\rho := -\frac{i}{\hbar}[\mathscr{H},\rho] + \sum_{\mu}\left(L_\mu \rho L_\mu^\dagger - \frac{1}{2}\{L_\mu^\dagger L_\mu, \rho\}\right). \tag{10.2}$$

We assume here that the Hamiltonian $\mathscr{H}$ can be written as a sum of locally interacting terms, $\mathscr{H} = \sum_n H_n$ and L_μ are the Lindblad (or so-called quantum jump) operators, which are assumed to act only at the boundary sites of the system. This setup provides a fully coherent bulk dynamics and incoherent boundary conditions, which is particularly suited for studying nonequilibrium heat transport in a setup similar to the classical case [151].

The Quantum Master Equation (QME) approach can be used to study not only heat transport, but also nonequilibrium processes in general (particle transport, spin transport, etc.). Depending on the process in question the Lindblad operators L_μ target specific canonical states, creating a local equilibrium state near the boundaries of the system. The conductivities are then obtained by measuring expectation values of the current observables in the steady states of the Lindblad equation, in the thermodynamic limit $N \to \infty$ (see e.g., [115]). This approach has been extensively used in recent years to study heat transport in one-dimensional models of quantum spin chains coupled at their ends with Lindblad heat baths [2, 54, 93, 103, 107, 138, 153], as well as in chains of quantum oscillators [62, 160] (for a recent review see e.g., [20, 106]).

The Lindblad equation (10.1) allows efficient numerical simulation of the steady state of locally interacting systems, in terms of the time-dependent-density-matrix-renormalization-group method (tDMRG) [46, 130, 149] in the Liouville space of linear operators acting on wave functions [115]. In cases where the tDMRG method cannot be applied, like when the interaction is long-range (e.g., Coulomb), the QME can be solved using the method of quantum trajectories, (see, for example, [103]). In the latter case the idea is to represent the density operator as an expectation of $|\Psi\rangle\langle\Psi|$ where the many-body wave function Ψ is a solution of a stochastic Schrödinger equation $d\Psi(t) = -(i/\hbar)H\Psi dt + d\xi$, with $d\xi$ being an appropriate stochastic process simulating the action of the baths. In addition, this method has the advantage that non-Markovian effects can be treated easily and intuitively. In more general settings, the QME can always be solved exactly through numerical integration where the quantum canonical heat baths are often modeled in terms of the Redfield equation [117]. Such approach has found a broad applicability in many-body systems and has been used to investigate heat transport [119, 121, 131, 153].

One alternative of using the QME approach is the Keldysh formalism of nonequilibrium Green's functions, where one essentially discusses the scattering of elementary quasi-particle excitations between two or more infinite non-interacting Hamiltonian reservoirs. The Keldysh formalism considers an initial product state density matrix describing the finite system and two infinite baths in thermal equilibrium at e.g., different temperatures. The system and the reservoirs are then coupled and the density matrix is evolved according to the full Hamiltonian. In the steady state, currents and local densities can be obtained in terms of the so-called Keldysh Green's functions. This approach has been used, among other things, to study heat transport in driven nanoscale engines [6, 8] and spin heterostructures [7]. Another commonly used approach to study heat transport in quantum systems is based on the Green-Kubo formula, originally developed to study electric transport [81].

Within linear response theory, the current is taken as the system's response to an external perturbative potential which can be included within the Hamiltonian of the system. First order perturbation theory yields the Green-Kubo formula relating the nonequilibrium conductivity with the equilibrium current-current correlation. This formula is naturally extended to study heat transport, where the heat current appears as the response to an external temperature gradient. This ad-hoc generalization remains conceptually troublesome since there is no potential term in the system's Hamiltonian representing a temperature gradient situation [82, 88]. In spite of this, the Green-Kubo approach has become a widely employed [61, 65, 158, 159].

In spite of all efforts, a microscopic derivation of Fourier law in quantum mechanics is still lacking, and only partial understanding concerning the conditions under which this is expected to hold has been gained. Particularly, in analogy with the studies at the classical level the relation between the validity of Fourier law and the onset of quantum chaos has been investigated in recent years [102, 104, 105, 115, 116, 121, 137].

As it has been shown in the previous section for classical systems, diffusive heat transport is directly related to the chaoticity of the dynamics. While such relation is not strict, classical deterministic chaos is yet expected to yield diffusive behavior. It is nowadays well established that quantum systems for which their classical analogues are chaotic, exhibit characteristic signatures in the spectra and the eigenfunctions that are different from those observed in systems that are classically integrable [21, 35]. The global manifestation of the onset of chaos in quantum systems consists of a very complex structure of the quantum states as well as in spectral fluctuations that are statistically described by Random Matrix Theory [64]. In the following we discuss the relation among the validity of quantum Fourier law and the onset of quantum chaos.

10.2.1 Fourier Law and the Onset of Quantum Chaos

The relation between the validity of the quantum Fourier law and the onset of quantum chaos in a genuinely nonequilibrium situation was studied in [102]. There a quantum Ising chain of N spins $1/2$ subjected to a uniform magnetic field $\mathbf{h} = (h_x, 0, h_z)$ and coupled at its extremes with quantum heat baths, was considered. The Hamiltonian of this system is

$$\mathscr{H} = \sum_{n=0}^{N-2} H_n + \frac{h}{2}(\sigma_{\mathrm{L}} + \sigma_{\mathrm{R}}) \,, \qquad (10.3)$$

Fig. 10.3 The finite quantum spin chain model coupled to external heat baths at different temperatures. The *dotted lines* represent the nearest neighbour interaction. The *double dashed lines* represent the coupling Ξ with the baths. The angle θ of the magnetic field is measured with respect to the direction x of the chain

where H_n are local energy density operators appropriately defined as

$$H_n = -Q\sigma_n^z\sigma_{n+1}^z + \frac{\mathbf{h}}{2}\cdot(\boldsymbol{\sigma}_n + \boldsymbol{\sigma}_{n+1}) \ , \tag{10.4}$$

and $\sigma_{\mathrm{L}} = \mathbf{h}\cdot\boldsymbol{\sigma}_0/h$, $\sigma_{\mathrm{R}} = \mathbf{h}\cdot\boldsymbol{\sigma}_{N-1}/h$ are the spin operators along the direction of the magnetic field of s_0 and s_{N-1} respectively. The operators $\boldsymbol{\sigma}_n = (\sigma_n^x, \sigma_n^y, \sigma_n^z)$ are the Pauli matrices for the nth spin, $n = 0, 1, \ldots N-1$. A schematic representation of this model is shown in Fig. 10.3.

In this model, the angle $\theta = \arctan(h_z/h_x)$ of the magnetic field makes with the chain affects the dynamics of the system. If $\theta = 0$, the Hamiltonian (10.3) corresponds to the Ising chain in a transversal magnetic field, which is integrable as (10.3) can be mapped into a model of free fermions through standard Wigner-Jordan transformations. For $\theta > 0$, the system is no longer integrable and for $\theta \approx \pi/4$, quantum chaos sets in. The system becomes again (nearly) integrable when $\theta \approx \pi/2$. Therefore, by tuning θ one can explore different regimes of quantum dynamics and study the relation between the integrability of the system and the validity of Fourier's law.

The integrability of a quantum system can be characterised by the Nearest Neighbour Level Spacing (NNLS) distribution $P(s)$, which is the probability density to find two adjacent levels at a distance s. For an integrable system the distribution $P(s)$ has typically a Poisson distribution:

$$P_{\mathrm{P}}(s) = \exp(-s) \ . \tag{10.5}$$

In contrast, in the quantum chaos regime, Hamiltonians obeying time-reversal invariance exhibit a NNLS distribution that corresponds to the Gaussian Orthogonal Ensemble of random matrices (GOE). This distribution is well-approximated by the Wigner surmise, which reads

$$P_{\mathrm{WD}}(s) = \frac{\pi s}{2}\exp\left(-\frac{\pi s^2}{4}\right) , \tag{10.6}$$

exhibiting "level repulsion".

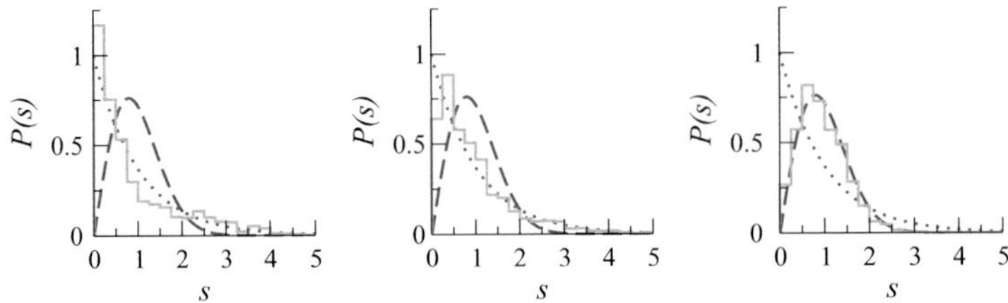

Fig. 10.4 NNLS distribution $P(s)$ for the integrable (*left panel*), intermediate (*middle panel*) and chaotic (*right panel*) spin chains. The histogram was numerically obtained for a chain of $N = 12$ spins by diagonalizing Hamiltonian (10.3) and averaging over the spectra of even and odd parity. The *dotted curve* corresponds to P_{P} and the *dashed curve* to P_{WD}

Figure 10.4 shows the NNLS distribution $P(s)$ for three different directions of the magnetic field: (a) *integrable case* $\mathbf{h} = (3.375, 0, 0)$, at which $P(s)$ is well described by $P_{\mathrm{P}}(s)$, (b) *intermediate case* $\mathbf{h} = (7.875, 0, 2)$ at which the distribution $P(s)$ shows a combination of (weak) level repulsion and exponential decay, and (c) *chaotic case* $\mathbf{h} = (3.375, 0, 2)$ at which the distribution $P(s)$ agrees with $P_{\mathrm{WD}}(s)$ and thus corresponds to the regime of quantum chaos.

In Mejía-Monasterio et al. [102] a numerical method to solve the dynamics of open quantum spin chains was introduced. This method consists in periodically and stochastically collapsing the state of the spins at the boundaries of the chain to a state that is consistent with local equilibrium states at different temperatures. These stochastic quantum heat baths are analogous to the stochastic baths used in classical simulations and even when this method does not yield a stochastic unravelling of QME, it is numerically simple to implement and analyse (for more details see [102, 103]).

Using this method, local thermal equilibrium was first checked by computing time averages of the density matrix of the system:

$$\overline{\rho} = \lim_{t\to\infty} \int_0^t |\psi(s)\rangle\langle\psi(s)|\mathrm{d}s \, , \tag{10.7}$$

where $\psi(s)$ is the state of the system at time s. Setting both heat baths to the same temperature $T_{\mathrm{L}} = T_{\mathrm{R}} = T$, it was found that $\overline{\rho}$ is diagonal within numerical accuracy and consistent with

$$\langle\phi_n|\rho|\phi_m\rangle = \frac{e^{-\beta E_n}}{\mathscr{Z}}\delta_{m,n} \, , \tag{10.8}$$

inside each symmetry band. Here $|\phi_n\rangle$ are the eigenfunctions in the energy basis, $\mathscr{H}|\phi_n\rangle = E_n|\phi_n\rangle$, and $\mathscr{Z} = \sum_n e^{-\beta E_n}$ is the canonical partition function. From a best fit to exponential of Eq. (10.8) a value of the local temperature in the bulk of the system can be extracted. The results are shown in Fig. 10.5 as a function of the temperature of the heat baths. It can be seen that for large enough temperatures

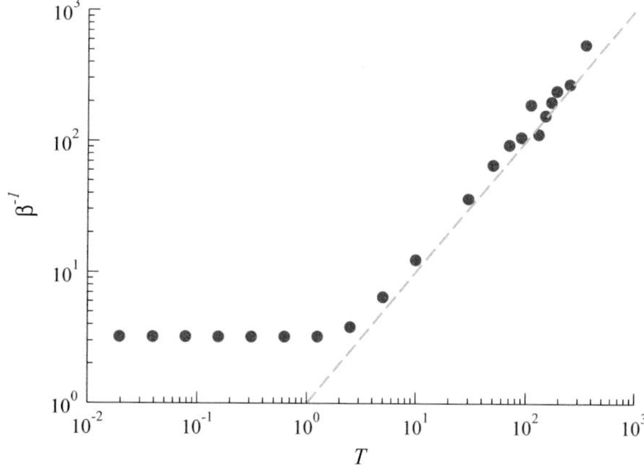

Fig. 10.5 Local temperature in the center of the chain β^{-1} as a function of the temperature of the baths T, obtained from equilibrium simulations in the chaotic chain of seven spins, as the best fit to exponential of the local density matrix $\rho_n(E_n)$ in the central symmetry band. The *dashed line* stands for the identity

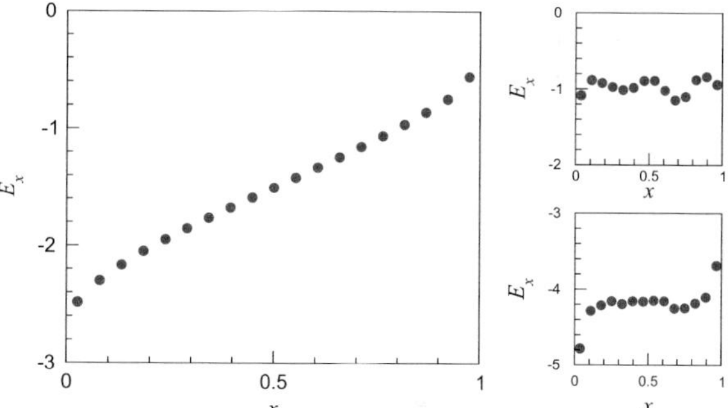

Fig. 10.6 Energy profile obtained from a time average of the expectation value of the energy density operator $E_x = \langle H_x \rangle$, with $x = n/N$ for a chain of 20 spins. The temperatures of the baths are $T_L = 5$ and $T_R = 50$. The different panels are for the chaotic (*left*), intermediate (*bottom right*) and integrable (*upper right*) chains

of the heat baths ($T \gtrsim 5$) the system thermalizes to exactly the same temperature [103].

Out of equilibrium expectation values in the nonequilibrium steady state were obtained as follows: for each realization, the initial wave function $|\psi(0)\rangle$ of the system is chosen at random. The system is then evolved for some relaxation time $\tau_{\rm rel}$ after which it is assumed to fluctuate around a unique steady state. Measurements are then performed as time averages of the expectation value of the observables, that are further averaged over different random realizations.

Figure 10.6 shows the energy profile obtained from the time average of the local energy density operator $E_x = \langle H_x \rangle$ (with $x = n/N$), for the above three different spin chains. Interestingly, for the chaotic chain, a linear energy profile in the bulk of the chain was found. This indicates that the chaotic chain is able to sustain a heat current which depends on the nonequilibrium imposed by the external heat baths. In contrast, the integrable chain shows a flat constant energy profile. The intermediate chain which is neither chaotic nor integrable is not able to sustain a diffusive heat current and shows and energy profile which is flat except near the boundaries.

To directly check the validity of Fourier's law it is possible to define local heat current operators using the continuity equation for the local energy density operators

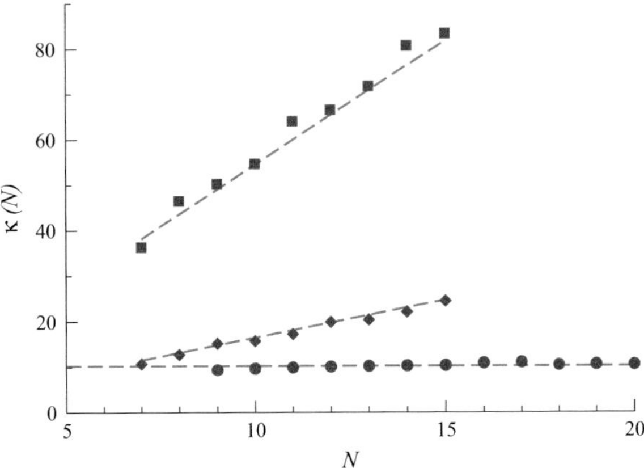

Fig. 10.7 Heat conductivity $\kappa = -j/\nabla T$ as a function of the size of the spin chain N, for the chaotic (*circles*), intermediate (*diamonds*), and integrable (*squares*) chains. The temperatures of the baths are $T_L = 5$ and $T_R = 50$. The *dashed curves* correspond to the best linear fit for each of the data sets

$\partial_t H_n = i[\mathscr{H}, H_n] = -(j_{n+1} - j_n)$, requiring that $j_n = [H_n, H_{n-1}]$. From Eqs. (10.3) and (10.4) the local current operators are explicitly given by

$$j_n = h_x Q \left(\sigma_{n-1}^z - \sigma_{n+1}^z\right) \sigma_n^y, \quad 1 \le n \le N-2. \tag{10.9}$$

Figure 10.7 shows the heat conductivity as a function of the system size N, calculated as $\kappa = -j/\nabla T$. The mean current j was calculated as an average of $\langle j_n \rangle$ over time and over the $N - 8$ central spins. For the particular choice of the energy density operator (10.4), its averaged expectation value is related to the local temperature as $\langle H_n \rangle \propto -1/T$ [102]. The temperature difference was thus obtained as $\Delta T = -1/\langle H_{N-5} \rangle + 1/\langle H_3 \rangle$. For large N the heat conductivity of the chaotic chain was found to converge to a constant value, thus confirming the validity of the Fourier's law. On the contrary, for the integrable and an intermediate chains, κ diverges linearly with N, which is a signature of ballistic transport.

These results represent a solid suggestion that, in analogy to what is observed in classical systems, in the quantum realm Fourier law holds once quantum chaos has set in. Arguably, quantum chaos yields diffusive heat transport as it leads to exponential decay of the "dynamic" correlations, in particular the energy current-current correlation that defines the heat conductivity through a Green-Kubo formula.

The crucial relation between diffusive transport and quantum chaos was later investigated in [137] for models of isolated quantum chains made of interacting subunits, each containing a finite number of energy levels. These models representing single-particle multi-channel quantum wires, exhibit a transition to quantum chaos as the strength of the interaction between the subunits increases. By solving the corresponding Schrödinger equation, it was found that the evolution of the local energy density operators is in agreement with the corresponding diffusion equation only when the system level statistics is chaotic. There, a Heisenberg spin chain in an external magnetic field was also studied, yielding the same result.

Later, in [115] heat and spin transport in several open quantum spin chains was considered and numerically solved by means of the tDMRG method. The same model considered in [102] was studied for much larger system sizes and the relation between quantum Fourier law and quantum chaos put forward there, was reconfirmed with high accuracy.

10.3 Controlling the Heat Flow: Thermal Rectifiers

Contrary to the case of electronic transport, where the concept of diode is well known, when one thinks of heat flow and Fourier law, the idea of directed transport does not come to mind at all. It is even counter intuitive. However, as shown in Sect. 10.3.1, the concept of a thermal diode is perfectly compatible with the usual Fourier law, provided one builds a device with materials having a temperature-dependent thermal conductivity. With a simple one-dimensional model system, Sect. 10.3.2 shows how such materials could be obtained. As shown in Sect. 10.3.3 the same results can be extended in higher dimensions. The actual realization of thermal rectifiers is briefly discussed in Sect. 10.3.4.

10.3.1 The Fourier Law and the Design of a Thermal Rectifier

Thermal rectification is everyday's experience: due to thermal convection a fluid heated from below efficiently transfers heat upwards, while the same fluid, heated from the top surface shows a much weaker transfer rate downwards. In this case this is because the heat flow is due to a transfer of matter. The idea that one could build a solid-state device that lets heat flow more easily in one way than in the other is less intuitive, and may even appear in contradiction with thermodynamics at first examination. However this is not so, and the design of a thermal rectifier is perfectly compatible with the Fourier law [114].

Let us consider the heat flow along the x-direction, in a material in contact with two different heat baths at temperature T_1 for $x = 0$ and T_2 for $x = L$. A rectification can only be expected if the device has some spatial dependence which allows us to distinguish its two ends, i.e. if the local thermal conductivity depends on x. This can either come from an inhomogeneity of the material or from its geometry. Moreover the thermal conductivity $\kappa(x, T)$ may also depend on temperature so that the Fourier law relates the heat flux j_f to the local temperature $T(x)$ by

$$T(x) = T_1 + \int_0^x \frac{j_f}{\kappa[\xi, T(\xi)]}\, d\xi \ . \tag{10.10}$$

Solving this equation with the boundary condition $T(x = L) = T_2$ determines the value of j_f. If the boundary conditions are reversed, imposing temperature T_2 for $x = 0$ and temperature T_1 for $x = L$, solving the same equation leads to another temperature distribution and another distribution of the local thermal conductivity

$\kappa(x, T)$. Therefore the reverse flux j_r is not equal to the forward flux j_f. The rectifying coefficient can be defined as

$$R = \left|\frac{j_r}{j_f}\right| . \tag{10.11}$$

In general, for arbitrary $\kappa(x, T)$, there is no condition that imposes that R should be unity.

Figure 10.8 shows a simple example where the spatial dependence is obtained by juxtaposing two different homogenous materials, each one having a thermal conductivity that strongly depends on temperature. In this case $\kappa(x, T)$ is a sigmoidal function in both cases, but on one side κ is large at low temperature while, on the other side it is large at high temperature. An even simpler device can be obtained by combining one material with a temperature dependent thermal conductivity with another one which has a constant thermal conductivity [114]. Such a device has a lower rectifying coefficient but nevertheless behaves as a thermal diode.

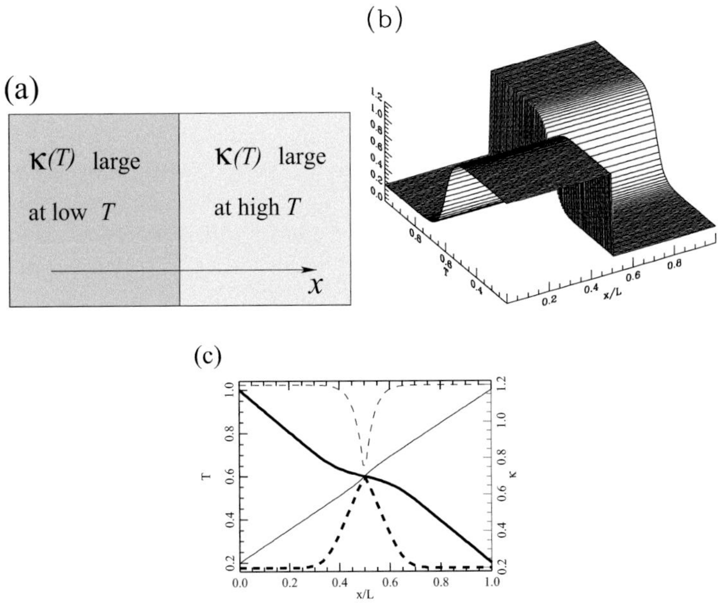

Fig. 10.8 Thermal rectifier made by the juxtaposition of two different homogeneous materials which have a thermal conductivity that highly depends on temperature. The boundary temperatures are $T_1 = 1.0$ and $T_2 = 0.2$ in arbitrary scale. (**a**) Schematic view of the device. (**b**) Variation of $\kappa(x, T)$. (**c**) The temperature distributions [solution of Eq. (10.10)] (*full lines*) and the variation versus space of the local conductivity $\kappa[x, T(x)]$ (*dashed lines*) are shown for the forward boundary condition ($T(x = 0) = T_1$, $T(x = L) = T_2$) (*thick lines*) and reverse boundary condition (*thin lines*). For this choice of $\kappa(x, T)$, the rectifying coefficient is $R = |j_r/j_f| = 4.75$

10.3.2 A One-Dimensional Model for a Thermal Rectifier

As shown in Sect. 10.3.1, in order to obtain a thermal rectifier, we need two basic ingredients, a temperature dependent thermal conductivity and the breaking of the inversion symmetry of the device in the direction of the flow. In this section we show how this can be obtained in a simple model system.

In a solid the heat transfer by conduction is a transfer of energy without a transport of matter. Heat can be carried either by the propagation of atomic vibrations, i.e. phonons, or by the diffusion of the random fluctuations of mobile particles, which are generally charged so that electrical and heat conductivity are closely related as stated by the Wiedemann-Franz law for metals [10]. Here we consider the case of electrical insulators in which heat is only carried by lattice vibrations. The simplest model of a thermal diode can be designed with a one dimensional lattice of interacting particles having a single degree of freedom. However, in the search of simplicity, one should make sure that the model does not lead to unphysical properties. In particular we want to select a model system that obeys the macroscopic Fourier law, with a well defined thermal conductivity κ, which may not be the case for a one-dimensional lattice [84]. However if the translational invariance is broken by a substrate potential, so that momentum is not a constant of the motion, a simple one-dimensional lattice of harmonically coupled particles subjected to an external potential, known as a Klein-Gordon model, can show a well defined thermal conductivity while allowing an easy analysis of the properties of the system. Such a lattice can form the basis for a thermal rectifier [141].

As an example, let us consider the model schematized in Fig. 10.9, i.e. a chain of N particles with harmonic coupling constant K and a Morse on-site potential $V_n = D_n\big[\exp(-\alpha_n y_n) - 1\big]^2$. The variable y_n designates the displacement of the particles with respect to their equilibrium positions, p_n their momentum, and H_n is the local energy density. This model was introduced as a simple one-dimensional model of

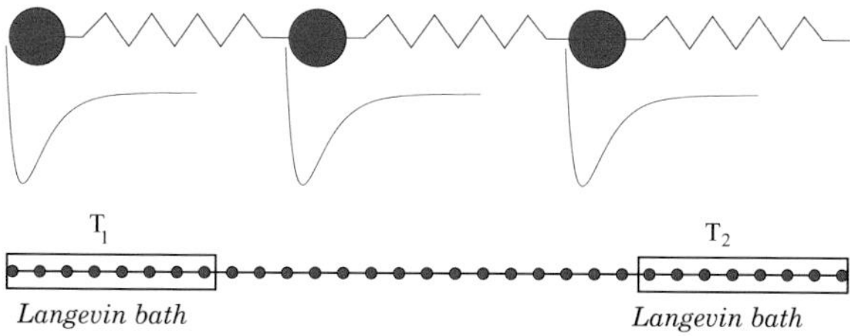

Fig. 10.9 Schematic picture of the one-dimensional model used as the basis for a simple thermal rectifier. *Upper part*: the harmonically coupled particles are subjected to an on-site potential, here a Morse potential. *Lower part*: the model used in numerical simulations to measure the heat flow. The two end-segments (*boxes*) are in contact with a numerical Langevin thermostat, while the central part of the lattice is evolving according to the equations of motions that derive from the Hamiltonian (10.12)

DNA [49]. In this case the on-site potential describes the interaction between the two strands of DNA.

In the present context this model can simply be viewed as a simple example to study heat transfer in a one-dimensional lattice, with Hamiltonian

$$\mathscr{H} = \sum_{n=1}^{N} H_n = \sum_{n=1}^{N} \left[\frac{p_n^2}{2m} + \frac{1}{2} K (y_n - y_{n-1})^2 + D_n (e^{-\alpha_n y_n} - 1)^2 \right]. \tag{10.12}$$

In such a system we can define a local temperature by $T_n = \langle p_n^2/m \rangle$ where the brackets designate a statistical average. Expressing dH_n/dt with the Hamilton equations and using the continuity equation for the energy flux,

$$\frac{dH(x,t)}{dt} + \frac{\partial j(x,t)}{\partial x} = 0, \tag{10.13}$$

in a finite difference form leads to a discrete expression for the local heat flux:

$$j_n = K \left\langle \dot{y}_n (y_{n+1} - y_n) \right\rangle . \tag{10.14}$$

The thermal properties of the model subjected to a temperature difference, in a steady state, can be studied by molecular dynamics simulations by imposing fixed temperatures T_1 and T_2 at the two ends with Langevin thermostats. The simulations have to be carried long enough to reach a steady state in which the heat flux j is constant along the lattice.

If the system is homogeneous ($D_n = D$ and $\alpha_n = \alpha$ for all n) such calculations show that, as expected, a well defined uniform thermal gradient is observed along the lattice, except in the immediate vicinity of the thermostats where a sharp temperature change is observed due to a Kapiza resistance between the thermostats and the bulk lattice (Fig. 10.10, circles). For large N the effect of the contact resistance becomes negligible. The calculation shows that, with a fixed temperature difference the flux decreases as $1/N$, where N is the number of lattice sites, which indicates that the model has a well defined thermal conductivity per unit length [114].

If the system is inhomogeneous, by including a central region in which the parameters are different from those in the two side domains, as shown in Fig. 10.10 (top), the heat flow is determined by the overlap of the phonon bands in the different regions. For the example shown in Fig. 10.10, with $T_1 = 0.16$ and $T_2 = 0.15$, the flux is equal to $j = 0.35 \times 10^{-3}$ for $D_1 = 0.5$ (corresponding to an homogeneous lattice) and decreases to $j = 0.18 \times 10^{-3}$ for $D_1 = 0.8$ for which the phonon bands partly overlap and $j = 0.48 \times 10^{-5}$ when there is no overlap between the phonon bands.

This provides a clue on a possible way to get the temperature dependent thermal conductivity needed to build a thermal rectifier as shown in Sect. 10.3.1: a nonlinearity of the on-site potential amounts to having temperature dependent

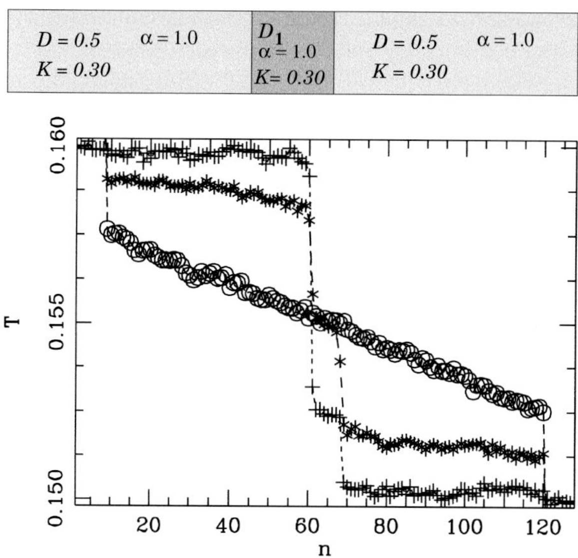

Fig. 10.10 *Top*: Model parameters in the case of an inhomogeneous lattice. *Bottom*: Variation of the local temperature along a lattice of 128 particles described by Hamiltonian (10.12) in contact with thermostats at temperatures $T_1 = 0.16$, $T_2 = 0.15$, in energy units, applied to the first and last eight particles. Results for different values of the parameter D_1 : $D_1 = 0.5$ (case of a homogeneous lattice: *circles*), $D_1 = 0.8$ (*stars*), and $D_1 = 1.2$ (*crosses*)

phonon bands. In the case of the model with Hamiltonian (10.12), this can easily be checked by a self-consistent phonon approximation [49]. The idea is to expand the free energy by separating the mean value of y_n, $\eta = \langle y_n \rangle$, and the deviations u_n around this value $y_n = \eta + u_n$. Then the Hamiltonian is approximated by $\mathscr{H} = \mathscr{H}_0 + \mathscr{H}_1$, where

$$\mathscr{H}_0 = \sum_n \left[\frac{1}{2} m \dot{u}_n^2 + \frac{1}{2} \phi (u_n - u_{n-1})^2 + \frac{1}{2} \Omega_2 u_n^2 \right] \tag{10.15}$$

describes an effective harmonic lattice. The free energy can be expanded as $\mathscr{F} = \mathscr{F}_0 + \mathscr{F}_1$, where $\mathscr{F}_1 = \langle \mathscr{H}_1 \rangle_0$. Then, by minimizing $\mathscr{F}_1$ with the variational parameters $\eta, \phi = K, \Omega_2$, one gets the lower bound of the effective phonon band of the lattice as $\Omega_2 = 2\alpha^2 D \exp[-2\alpha\eta/3]$. As T grows so does η, so that the band shifts downwards. Therefore, if the central region has a value $D_1 > D$ so that the phonon bands do not overlap at low temperature, as T increases the decay of the effective lower bound of the phonon band leads to an increased overlap, and therefore an increased thermal conductivity.

Expanding on these ideas one can build a thermal rectifier by introducing the necessary asymmetry pointed out in Sect. 10.3.1. Using left and right side regions with a weak nonlinearity ($\alpha = 0.5$) and different values of the parameter D

($D_{\text{left}} = 4.5$ and $D_{\text{right}} = 2.8$) and a harmonic coupling constant $K = 0.18$, one gets two domains with phonon bands that do not overlap. Nevertheless a good thermal conductivity can be restored with a central region with a high nonlinearity ($\alpha = 1.1$) and $D_{\text{center}} = 1.1338$ which is such that, when the highest temperature is on the right side of the device the variation versus space of the effective phonon band in the central region provides a match between the left and right phonon bands, while, if the highest temperature is on the left side of the device, the variation versus space of the effective phonon band in the central region leads to a large phonon-band mismatch, as shown in Fig. 10.11. Figure 10.11 shows that such a system does indeed lead to thermal rectification because, when the hot side is on the left, the mismatch of the phonon bands leads to temperature jumps at the junctions between the different parts. This is due to a large contact thermal resistance. When the hot side is on the right the temperature evolves continuously along the device. The contact resistances are then low, and the energy flux is 2.4 times larger in this configuration. The calculation of the theoretical phonon bands, based on the self-consistent phonon approximation, is only approximately correct, first because the method itself is only approximate but also because the calculation is made by assuming a linear temperature variation inside the device, which is a crude approximation. However this method provides a first step to design a rectifier, which has to be improved with the results of the numerical simulations.

The results shown in Fig. 10.11 only provide a simple illustration of what can be done with the idea of phonon-band matching, combined to nonlinearity to allow the local phonon frequency spectra to vary with temperature. One can imagine many possible improvements, for instance by stacking devices, or increasing the number of interfaces, to increase the rectifying coefficient. Another approach is to design a system with a continuous variation of the vibrational properties versus space, which amounts to stacking an infinity of interfaces which have temperature dependent properties and therefore have different transmissivities when the direction of the temperature gradient is reversed. This allows a better control of the rectifying effect. Figure 10.12 shows such an example, which has a rectifying coefficient $R = 4.95$ and exhibits an effective phonon band which is almost flat when the thermal gradient is in the favorable direction.

In spite of these achievements there are still several problems which are difficult to overcome. In particular in the model thermal rectifiers described above, the rectification power is small and rapidly decays to zero as the system size increases. A possible way to overcome this difficulty has been discussed in [43] where, by considering one dimensional anharmonic chains of oscillators, empirical evidence is provided that *graded mass distribution* and *long range interparticle interactions*, lead to a substantial improvement of the thermal rectification phenomenon which moreover does not decay to zero with increasing system size.

The system is a one-dimensional chain of N oscillators described by the Hamiltonian

$$\mathcal{H} = \sum_{j=1}^{N} \left(\frac{p_j^2}{2m_j} + \frac{q_j^4}{4} \right) + \sum_{i,j} \frac{(q_j - q_i)^2}{2 + 2|i-j|^\lambda}, \tag{10.16}$$

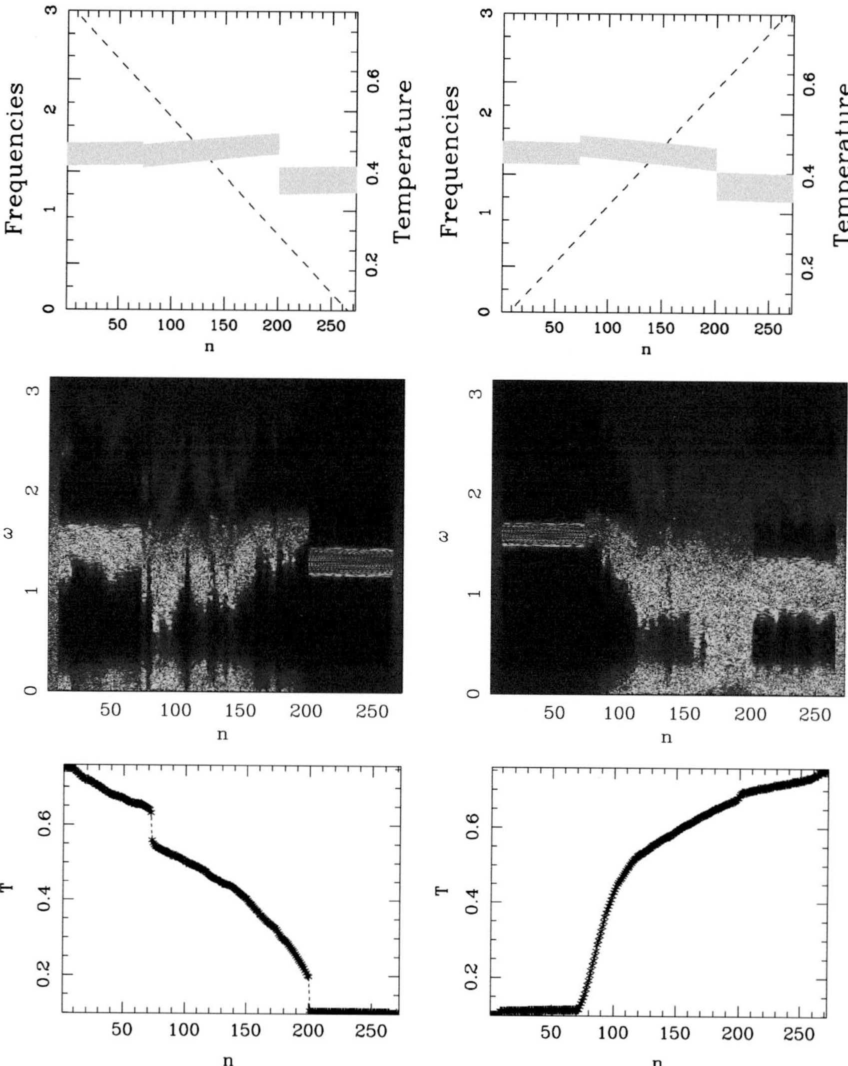

Fig. 10.11 Properties of a model for a thermal rectifier with two different boundary conditions: *left figures*: energy flow from left to right $T_{\text{left}} = 0.7$, $T_{\text{right}} = 0.1$, *right figures*: energy flow from right to left $T_{\text{right}} = 0.7$, $T_{\text{left}} = 0.1$. The *upper figures* show the theoretical phonon bands along the device obtained from the self-consistent phonon approximation with the assumed temperature distribution shown by the *dash line*. The *middle figures* show the actual distribution of the phonon frequencies deduced from numerical simulations, and the *lower figures* show the variation of the local temperature along the system, determined from the numerical results. The ratio of the flux in the two directions is $|j_{\text{right}\to\text{left}}|/j_{\text{left}\to\text{right}} = 2.4$

where q_j is the displacement of the jth particle with mass m_j and momentum p_j from its equilibrium position. A graded mass distribution is used. The exponent λ controls the decay of the interparticle interactions with distance.

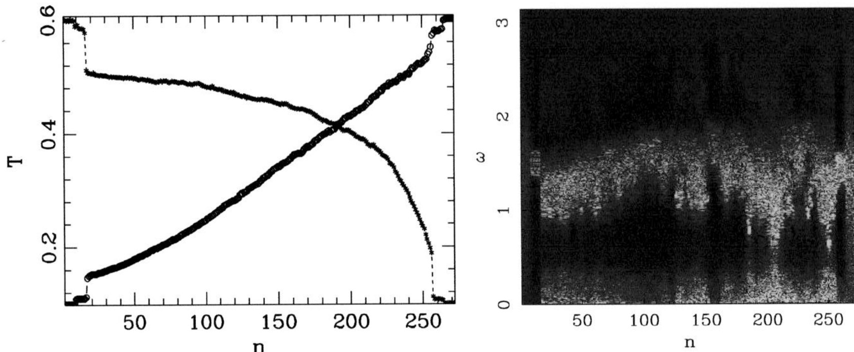

Fig. 10.12 Thermal rectifier using a continuous variation of the local properties. *Left*: Temperature variation versus space inside the device in the two possible orientations of the thermal gradient. *Right*: Observed variation of the local phonon spectra along the device when the hot side is on the right

In view of previous results [147, 154], it is expected that in a system with graded mass distribution, e.g. $m_1 < m_2 < \ldots < m_N$, thermal rectification will be present, even for the simple case of nearest neighbor interaction (NN). Long range interactions (LRI) introduce new channels for the heat transport through the new links (interactions) between the different sites. Moreover in a graded system, the new channels connect distant particles with very different masses. Therefore new, asymmetric channels, are created which in turn favors the asymmetric flow, i.e., rectification. Hence, by introducing long range interactions in a graded system, an increase of the thermal rectification is expected. Moreover, as we increase the system size, new particles are introduced that, in the case of long range interactions, create new channels for the heat current. This may avoid the usual decay of rectification with increasing system length.

In Fig. 10.13 we plot the rectification factor as a function of the system size. Here the mass gradient is fixed. It is seen that the presence of LRI leads to a very large rectification and prevents the decay of the rectification factor with the system size. Strictly speaking we cannot make any claim for larger system sizes. However it is clear from Fig. 10.13 that the N-dependence for the LRI case is qualitatively different from the NN case where the decay of the rectification factor with N is observed.

10.3.3 Model in Higher Dimension

Extending the same concept to higher dimension is of course important for actual applications. For instance two dimensional models could describe smart conducting layers to carry heat out of some nano-devices. The same idea of playing with the phonon bands is indeed also valid in two dimensions.

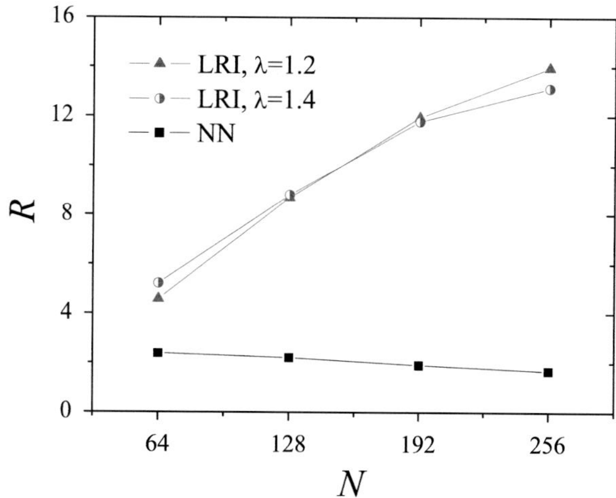

Fig. 10.13 Dependence of rectification factor on the system size N. Here $T_1 = 9.5$, $T_2 = 0.5$, $m_1 = 1$. *Triangles* are for LRI with $\lambda = 1.2$, *circles* are for LRI with $\lambda = 1.4$, *squares* are for the NN case. The mass gradient is fixed ($m_N = 2$ for N $= 64$; $m_N = 3$ for N $= 128$; $m_N = 4$ for $N = 192$; $m_N = 5$ for $N = 256$)

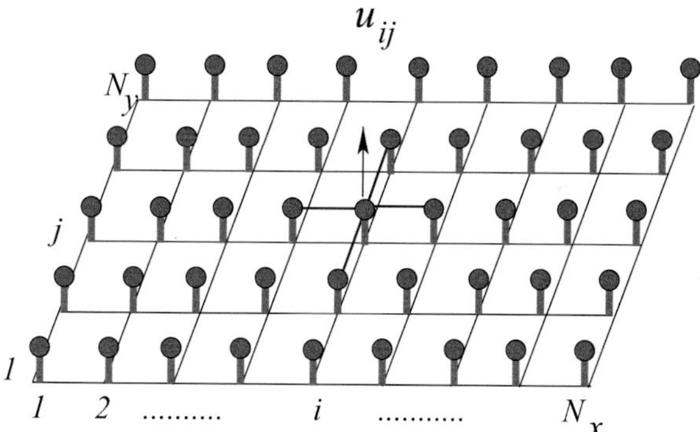

Fig. 10.14 Two dimensional lattice described by Hamiltonian (10.17)

Figure 10.14 shows a two-dimensional lattice of local oscillators which is described by the Hamiltonian

$$\mathscr{H} = \sum_{i=1,N_x j=1,N_y} \left[\frac{p_{ij}^2}{2m} + \frac{1}{2}C_x(i,j)\big(u_{i+1,j} - u_{i,j}\big)^2 + \frac{1}{2}C_x(i-1,j)\big(u_{i,j} - u_{i-1,j}\big)^2 \right.$$
$$+ \frac{1}{2}C_y(i,j)\big(u_{i,j+1} - u_{i,j}\big)^2 + \frac{1}{2}C_y(i,j-1)\big(u_{i,j} - u_{i,j-1}\big)^2$$
$$\left. +D(i,j)\left(\exp[-\alpha(i,j)u_{i,j}] - 1\right)^2 \right]. \tag{10.17}$$

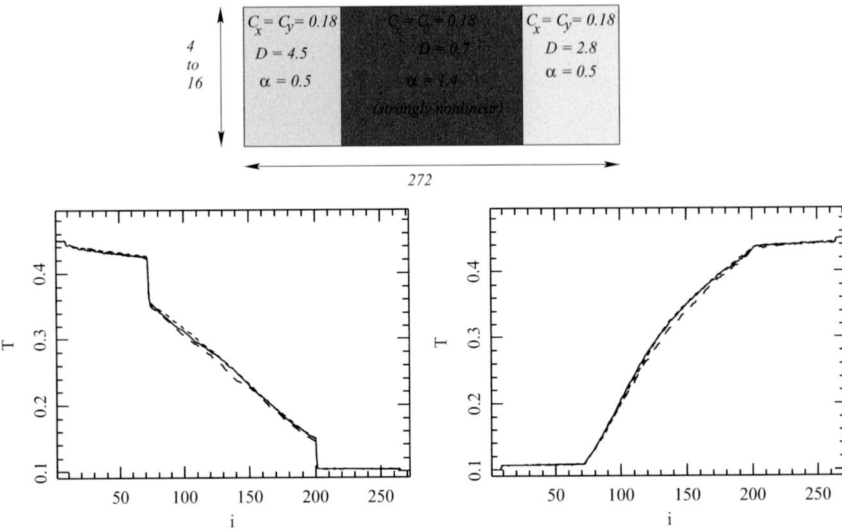

Fig. 10.15 *Top*: Schematic plot of a two-dimensional model for a thermal rectifier indicating the values of the parameters of Hamiltonian (10.17) in the different regions. *Bottom*: Temperature profiles along the x axis of the device for two opposite thermal gradients. The various lines (*continuous* and *dash lines*) correspond to different lattice sizes in the y direction (from 4 to 16)

With appropriate parameters, as indicated in Fig. 10.15 (top), this system can operate as a rectifier with a rectifying ratio $R = 1.69$ because, while the interfaces show a large thermal resistance causing a sharp temperature drop if the left edge of the lattice is connected to the hot bath, when the gradient is reversed, the temperature varies smoothly along the device because there are no interfacial thermal resistances [see Fig. 10.15 (bottom left and right)].

10.3.4 Building an Actual Thermal Rectifier

Actually the experimental observation of heat flow rectification has a long history [95]. The early observations made in 1975 with a *GaAs* crystal found a small rectification effect ($R \approx 1$), which was strongly dependent on the location of the contacts on the sample, but an asymmetry of the heat flow was nevertheless clear. A first analysis made by assuming that the thermal conductivity was the sum of a space-dependent term and a temperature dependent term showed that the observation was compatible with the Fourier law [96], and provided the first view of the ideas presented in Sect. 10.3.1. These results were followed by some debates over the actual origin of the observed rectification [12], and a rectification coefficient $R \approx 1.35$ could then be obtained with a two-component sample made of tin in contact with α-brass. Heat flow rectification could also be observed with a carbon nanotube [40], loaded with $C_9H_{16}Pt$ molecules on one part of its length, but the origin of the rectification was still not clearly established.

More recently the ideas presented in Sect. 10.3.1 were systematically exploited to build rectifiers [78], using two cobalt oxides with different thermal conductivities. The vicinity of a structural phase transition could be used to enhance the temperature dependence of the thermal conductivity [79] and the asymmetry of the shape has been exploited to vary the spatial dependence of $\kappa(x, T)$ [128]. The measurements show a good quantitative agreement with the results of Sect. 10.3.1 and [114], if one takes into account the experimental data for the thermal conductivity of the materials used in the device. A quantitative microscopic calculation of $\kappa(T)$ is however a harder challenge. The control of the temperature dependence of κ, trough a control of the phonon bands, discussed in Sect. 10.3.2 is only one possibility but other mechanisms can be considered whether they use a structural change through a phase transition, or variations of the density of mobile carriers in materials which are also electrical conductors. And of course, in such materials the use of an electric field to manipulate the spatial distribution of the carriers in a solid state device can also open other possibilities to control the heat flow. It is also worth mentioning recent experimental implementations of thermal rectifiers, exploiting phononic [142], electronic [94], or photonic [41] thermal currents. Possibilities to manipulate phonons and devise heat diodes, transistors, thermal logic gates and thermal memories are reviewed in [86].

10.4 Thermoelectric Efficiency

Thermoelectricity concerns the conversion of temperature differences into electric potential or vice-versa. It can be used to perform useful electrical work or to pump heat from a cold to a hot place, thus performing refrigeration. Although thermoelectricity was discovered about 200 years ago, a strong interest of the scientific community arose only in the 1950s when Abram Ioffe discovered that doped semiconductors exhibit relatively large thermoelectric effect. This initiated an intense research activity in semiconductors physics which was not motivated by microelectronics but by Ioffe's suggestion that home refrigerators could be built with semiconductors [90, 91]. As a result of these efforts the thermoelectric material Bi_2Te_3 was developed for commercial purposes. However this activity lasted only few years until the mid 1960s since, in spite of all efforts and consideration of all type of semiconductors, it turned out that thermoelectric refrigerators have still poor efficiency as compared to compressor based refrigerators. Nowadays Peltier refrigerators are mainly used in situations in which reliability and quiet operation, and not the cost and conversion efficiency, is the main concern, like equipments in medical applications, space probes, etc.

In the last two decades thermoelectricity has experienced a renewed interest [16, 26, 32, 53, 55, 80, 132, 134] due to the perspectives of using tailored thermoelectric nanomaterials, where a dramatic enhancement of the energy harvesting performances can be envisaged [66]. Indeed layering in low-dimensional systems may reduce the phonon thermal conductivity as phonons can be scattered by the interfaces between layers. Moreover, sharp features in the electronic density of states, favorable for thermoelectric conversion (see the discussion below) are in principle possible due to quantum confinement. Recent efforts have focused on one hand, on the study of nanostructured materials and on the other hand, in understanding the fundamental dynamical mechanisms which control the coupled transport of heat and particles [20].

10.4.1 The Thermoelectric Figure of Merit ZT

For a material subject to a temperature gradient ∇T and a external uniform electric field $\mathcal{E}$, within linear response the equations describing thermoelectric transport are

$$\begin{aligned} j_q &= -\kappa' \nabla T + \sigma \Pi \mathcal{E}, \\ j_e &= -\sigma S \nabla T + \sigma \mathcal{E}, \end{aligned} \tag{10.18}$$

where j_q and j_e denote the heat and electric local currents appearing in the material due to the external forcing, σ is the coefficient of electrical conductivity, S is the thermopower (or Seebeck coefficient), Π is the Peltier coefficient, and κ' is the heat conductivity measured at zero electric field and is related to the usual heat conductivity κ measured at zero electric current as $\kappa' = \kappa + \sigma S \Pi$. From (10.18) the usual phenomenological relations follow: if the temperature gradient vanishes, $\nabla T = 0$, then $j_e = \sigma \mathcal{E}$ is Ohm's law and the Peltier coefficient $\Pi = j_q / j_e$. If the electric current vanishes, $j_e = 0$, then $j_q = -\kappa \nabla T$ is Fourier's law, and $\mathcal{E} = S \nabla T$, which is the definition of the thermopower. We start by considering systems with time-reversal symmetry, for which the Onsager reciprocity relations imply $\Pi = TS$ (see Sect. 10.4.2).

The suitability of a thermoelectric material for energy conversion or electronic refrigeration is evaluated by the dimensionless thermoelectric figure of merit ZT [72]

$$ZT = \frac{\sigma S^2}{\kappa} T, \tag{10.19}$$

as follows. Consider a material maintained on one end at temperature T_{H} and on the other at temperature T_{C}, and subject to an external electric field $\mathcal{E}$. Then ZT is related to the efficiency $\eta = P / j_q$ of converting the heat current j_q (flowing between the thermal baths) into electric power $P \equiv \mathcal{E} j_e$, generated by attaching the thermoelectric element to an Ohmic impedance. If we optimize the efficiency over $\mathcal{E}$ we obtain the *maximum efficiency*

$$\eta_{\max} = \eta_{\mathrm{C}} \frac{\sqrt{ZT+1}-1}{\sqrt{ZT+1}+1}, \tag{10.20}$$

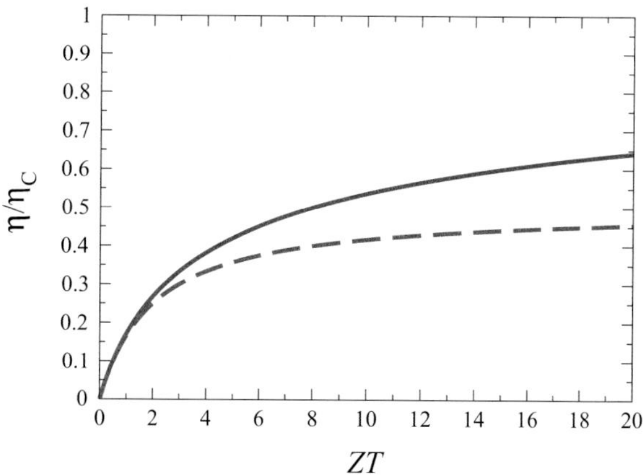

Fig. 10.16 Linear response efficiency for heat to work conversion, in units of the Carnot efficiency η_C, as a function of the figure of merit ZT. The *top* and the *bottom* curve correspond to the maximum efficiency η_{max} and to the efficiency at the maximum power $\eta(P_{max})$, respectively

where $\eta_C = 1 - T_C/T_H$ is the Carnot efficiency and $T = (T_H + T_C)/2$ is the average temperature. Thermodynamics only imposes (see Sect. 10.4.2) $ZT \geq 0$ and η_{max} is a monotonous growing function of ZT (see Fig. 10.16), with $\eta_{max} = 0$ when $ZT = 0$ and $\eta_{max} \to \eta_C$ when $ZT \to \infty$.

The Carnot efficiency is obtained for reversible quasi-static transformations, which require infinite time and consequently the extracted power is zero. An important question is how much the efficiency deteriorates when transformations are operated in a finite time. This is a central question in the field of *finite-time thermodynamics* [4]. Hence, the notion of *efficiency at maximum power* $\eta(P_{max})$ was introduced: it is obtained by optimizing over $\mathscr{E}$ the power P rather than the efficiency η. Within linear response we obtain [143]

$$\eta(P_{max}) = \frac{\eta_C}{2} \frac{ZT}{ZT + 2} \,. \qquad (10.21)$$

Note that also $\eta(P_{max})$ is a growing function of ZT (see Fig. 10.16). In the limit $ZT \to \infty$, $\eta(P_{max})$ takes its maximum value of $\eta_C/2$. Such value also corresponds to the linear response expansion of the so-called Curzon-Ahlborn upper bound [5, 39, 45, 59, 108, 129, 143, 155]. Therefore, high values of ZT are favorable for thermoelectric conversion.

Nowadays, most efficient thermoelectric devices operate at around $ZT \approx 1$, whilst it is generally accepted that $ZT > 3$ is the target value for efficient, commercially competing, thermoelectric technology [91]. The great challenge to increase thermoelectric efficiency relies on understanding the microscopic mechanisms that may allow to control individually S, σ and κ. However, the different transport coefficients are generally interdependent making optimisation extremely difficult and so far, no clear paths exist which may lead to reach that target. A particular example of this interdependence is the Wiedemann-Franz law [10] which states that for metallic materials, σ and κ are, as a matter of fact, proportional, thus making metals poor thermoelectric materials in general.

Note that ZT is related to the heat conductivities defined above as $\frac{\kappa'}{\kappa} = 1 + ZT$, which has been used in [145] to make an analogy between a classical heat engine and a thermoelectric material. The used correspondence is $N \to V$ and $\mu_e \to -p$, with N number of charge carriers, μ_e the electrochemical potential, and V, p volume and pressure of the gas in the engine. As a consequence, $\frac{\kappa'}{\kappa} \to \frac{c_P}{c_V}$, with c_p and c_V specific heat at constant pressure and volume, respectively. The ratio $\frac{c_P}{c_V}$ is bounded for ideal (non interacting) gases, but diverges at the gas-liquid critical conditions. These considerations suggest that large values of ZT could be expected near electronic phase transitions, for systems with strong interactions between the charge carriers [110].

10.4.2 The Onsager Matrix

Let us consider a system of particles enclosed in a chamber, coupled to two particle reservoirs. Calling the energy balance for the thermoelectric process, the energy current can be written in terms of the heat and electric currents as $j_u = j_q + \frac{\mu_e}{e} j_e$, where μ_e is the electrochemical potential. For particles having electric charge e the electrochemical potential is simply $\mu_e = e\phi$, where ϕ is the ordinary electrostatic potential ($\mathscr{E} = -\nabla\phi$). Assuming that the particles are the only carriers of heat, one may interchange the electrochemical potential with the chemical potential μ corresponding to the work generated by the exchange of particles between the system and the reservoirs. Within the linear response regime, the energy current and the particle current $j_\rho = \frac{1}{e} j_e$ are related to the conjugated thermodynamic forces (gradient of chemical potential μ and gradient of temperature T) as [33, 51]

$$\mathbf{j} = \mathbb{L}\mathbf{F}\,, \tag{10.22}$$

where $\mathbf{j} \equiv (j_\rho, j_u)^t$, $\mathbf{F} \equiv (\nabla(-\mu/T), \nabla(1/T))^t$, and

$$\mathbb{L} \equiv \begin{pmatrix} L_{\rho\rho} & L_{\rho u} \\ L_{u\rho} & L_{uu} \end{pmatrix} \tag{10.23}$$

is the Onsager matrix of kinetic transport coefficients. In the absence of magnetic fields (or other effects breaking time reversibility), the Onsager reciprocity relations state that the crossed kinetic coefficients are equal: $L_{\rho u} = L_{u\rho}$. Moreover, the second law of thermodynamics imposes that the entropy production rate $\dot{s} = \mathbf{j} \cdot \mathbf{F} = j_\rho \nabla(-\mu/T) + j_u \nabla(1/T) \geq 0$. Therefore $\mathbb{L}$ has to be nonnegative, i.e. $L_{\rho\rho}, L_{uu} \geq 0$ and $\det \mathbb{L} \geq 0$.

The kinetic coefficients L_{ij} are related to the thermoelectric transport coefficients as

$$\sigma = \frac{e^2}{T} L_{\rho\rho}\,, \quad \kappa = \frac{1}{T^2} \frac{\det \mathbb{L}}{L_{\rho\rho}}, \quad S = \frac{1}{eT}\left(\frac{L_{\rho u}}{L_{\rho\rho}} - \mu\right) = \frac{\Pi}{T}\,, \tag{10.24}$$

where the temperature T and chemical potential μ are taken as mean values in the bulk. Moreover, using Eqs. (10.19) and (10.24), the thermoelectric figure of merit reads

$$ZT = \frac{(L_{u\rho} - \mu L_{\rho\rho})^2}{\det \mathbb{L}} . \tag{10.25}$$

Note that the limit $ZT \to \infty$ can be reached only if the Onsager matrix $\mathbb{L}$ is ill-conditioned, namely when the ratio

$$\frac{[\mathrm{tr}\,(\mathbb{L})]^2}{\det \mathbb{L}} \to \infty \tag{10.26}$$

and therefore the linear system (10.22) becomes singular. That is, the Carnot efficiency is obtained when the energy current and the particle current become proportional: $j_u = c j_\rho$, with the proportionality factor c independent of the values of the applied thermodynamic forces. Such condition is refereed to as *tight coupling* condition.

10.4.3 Non-interacting Systems

We consider a system whose ends are in contact with left/right baths (reservoirs), which are able to exchange energy and particles with the system, at fixed temperature T_α and chemical potential μ_α, where $\alpha = L, R$ denotes the left/right bath. The reservoirs are modeled as infinite ideal gases, and therefore particle velocities are described by the Maxwell-Boltzmann distribution. We use a stochastic model of the thermochemical baths [83, 101]: Whenever a particle of the system crosses the boundary which separates the system from the left or right reservoir, it is removed. On the other hand, particles are injected into the system from the boundaries, with rates γ_α computed by counting how many particles from reservoir α cross the reservoir-system boundary per unit time. For one-dimensional reservoirs we obtain $\gamma_\alpha = \frac{1}{h\beta_\alpha} e^{\beta_\alpha \mu_\alpha}$, where $\beta_\alpha = 1/(k_B T_\alpha)$ (k_B is the Boltzmann constant and h is the Planck's constant). Assuming that both energy and charge are carried only by non-interacting particles, like in a dilute gas, we arrive at simple expressions for the particle and heat currents [122]:

$$j_\rho = \frac{1}{h} \int_0^\infty d\varepsilon \left(e^{-\beta_L(\varepsilon-\mu_L)} - e^{-\beta_R(\varepsilon-\mu_R)} \right) \tau(\varepsilon) , \tag{10.27}$$

$$j_{q,\alpha} = \frac{1}{h} \int_0^\infty d\varepsilon (\varepsilon - \mu_\alpha) \left(e^{-\beta_L(\varepsilon-\mu_L)} - e^{-\beta_R(\varepsilon-\mu_R)} \right) \tau(\varepsilon) , \tag{10.28}$$

where $j_{q,\alpha}$ is the heat current from reservoir α and $\tau(\varepsilon)$ denotes the transmission probability for a particle with energy ε to transit from one end to the other end of the system ($0 \leq \tau(\varepsilon) \leq 1$). The thermoelectric efficiency is then given by (we assume $T_L > T_R$, $\mu_R > \mu_L$, and $j_\rho, j_{q,L} \geq 0$)

$$\eta = \frac{j_{q,L} - j_{q,R}}{j_{q,L}} = \frac{(\mu_R - \mu_L) \int_0^\infty d\varepsilon \left(e^{-\beta_L(\varepsilon-\mu_L)} - e^{-\beta_R(\varepsilon-\mu_R)}\right) \tau(\varepsilon)}{\int_0^\infty d\varepsilon(\varepsilon - \mu_L) \left(e^{-\beta_L(\varepsilon-\mu_L)} - e^{-\beta_R(\varepsilon-\mu_R)}\right) \tau(\varepsilon)} . \quad (10.29)$$

When the transmission is possible only within a tiny energy window around $\varepsilon = \varepsilon_\star$, the efficiency reads

$$\eta = \frac{\mu_R - \mu_L}{\varepsilon_\star - \mu_L} . \quad (10.30)$$

In the limit $j_\rho \to 0$, corresponding to reversible transport [68, 69], we get $\varepsilon_\star$ from Eq. (10.27):

$$\varepsilon_\star = \frac{\beta_L \mu_L - \beta_R \mu_R}{\beta_L - \beta_R} . \quad (10.31)$$

Substituting such $\varepsilon_\star$ in Eq. (10.30), we obtain the Carnot efficiency $\eta = \eta_C = 1 - T_R/T_L$. Such delta-like energy-filtering mechanism for increasing thermoelectric efficiency has been pointed out in [68, 69, 89]. As remarked above, Carnot efficiency is obtained in the limit of zero particle current, corresponding to zero entropy production and zero output power. However, high values of *ZT* can still be achieved with sharply-peaked transmission functions without greatly reducing the output power [109, 144].

In the linear response regime, using a delta-like energy filtering, i.e. $\tau(\varepsilon) = 1$ in a tiny interval of width $\delta\varepsilon$ around some energy $\bar{\varepsilon}$ and 0 otherwise, we obtain

$$L_{\rho\rho} = \frac{\Lambda(\delta\varepsilon)}{hk_B} e^{-\beta(\bar{\varepsilon}-\mu)}, \; L_{u\rho} = L_{\rho u} = \frac{\Lambda\bar{\varepsilon}(\delta\varepsilon)}{hk_B} e^{-\beta(\bar{\varepsilon}-\mu)}, \; L_{uu} = \frac{\Lambda\bar{\varepsilon}^2(\delta\varepsilon)}{hk_B} e^{-\beta(\bar{\varepsilon}-\mu)} , \quad (10.32)$$

where Λ is the length of system. From these relations we immediately derive that the Onsager matrix is ill-conditioned and therefore $ZT = \infty$ and $\eta = \eta_C$. We point out that the parameters $\bar{\varepsilon}$ and $\delta\varepsilon$ characterizing the transmission window, appear in the Onsager matrix elements (10.32) and therefore are assumed to be independent of the applied temperature and chemical potential gradients. On the other hand, the energy $\varepsilon_\star$ in Eqs. (10.30) and (10.31) depends on the applied gradients. There is of course no contradiction since (10.30) and (10.31) have general validity beyond the linear response regime.

A dynamical realization of the energy-filtering mechanism was discussed in [37]. We start by writing for a gas of non-interacting particles the microscopic instantaneous charge and energy currents per particle at position $\mathbf{r}^*$ and time t:

$$\iota_\rho(\mathbf{r}^*, t) = v_x \delta(\mathbf{r}^* - \mathbf{r}(t)) , \quad (10.33)$$

$$\iota_u(\mathbf{r}^*, t) = \varepsilon(t) v_x(\mathbf{r}(t), t) \delta(\mathbf{r}^* - \mathbf{r}(t)) , \quad (10.34)$$

where ε is the energy of the particle, $\mathbf{r}$ its position and v_x its velocity along the direction of the currents. The thermodynamic averages of the two currents become proportional precisely when the variables ε and v_x are uncorrelated:

$$j_u = \langle \iota_u \rangle = \langle \varepsilon v_x \rangle = \langle \varepsilon \rangle \langle v_x \rangle = \langle \varepsilon \rangle \langle \iota_\rho \rangle = \langle \varepsilon \rangle j_\rho \, . \tag{10.35}$$

Therefore, $ZT = \infty$ follows from the fact that the average particle's energy $\langle \varepsilon \rangle$ does not depend on the thermodynamic forces. In the context of classical physics this happens for instance in the limit of large number of internal degrees of freedom, provided the dynamics is ergodic.

This observation was used in [37] where an ergodic gas of non-interacting particles with d_{int} internal degrees of freedom in a *d-dimensional* chamber connected to reservoirs was studied. It was shown that for such system the thermoelectric figure of merit becomes

$$ZT = \frac{1}{c_V} \left(c_V - \frac{\mu}{T} \right)^2 , \tag{10.36}$$

where $c_V = c_V^* + 1/2$ and $c_V^* = D/2$ ($D = d + d_{\text{int}}$) is the dimensionless heat capacity at constant volume of the gas. Figure 10.17 shows the figure of merit *ZT* numerically computed for a gas of noninteracting point-like particles as a function of the specific heat (internal degrees of freedom are modeled as free rotating modes). The particles evolve inside a Lorentz gas channel with finite horizon, so that the particles motion is diffusive (see the inset of Fig. 10.17). The channel is connected at its boundaries to stochastic reservoirs at different temperatures and chemical potentials. The numerical results confirm the analytical expression of Eq. (10.36). The simple mechanism for the growth of *ZT* also implies that the equilibrium distribution of the particle energy per degree of freedom becomes more sharply peaked as *D* increases.

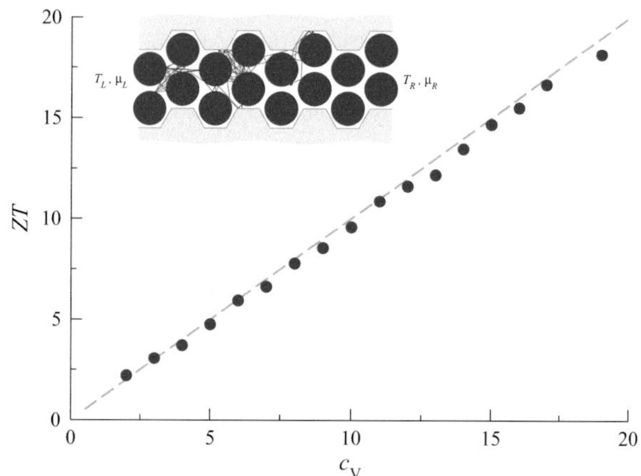

Fig. 10.17 Figure of merit *ZT* as a function of the heat capacity c_V, at $\mu = 0$. Numerical results are obtained from nonequilibrium simulations (for the details of the simulations see [37]). The *dashed line* corresponds to the analytical expression of Eq. (10.36). *Inset*: schematic drawing of the model used in the numerical simulations

We point out that, while the discussion in this section was focused on classical systems, noninteracting systems can be easily treated in quantum mechanics too by means of the Landauer-Büttiker formalism, [48, 71]. In this approach, the particle and heat currents are given, similarly to Eqs. (10.27) and (10.28), in terms of integrals over the energy distribution of the particles injected from the reservoirs and the scattering transmission probability of the system (for the use of this formalism in thermoelectricity see [20]). Implementations of the energy filtering mechanism may be possible in, e.g., nanowires or nanostructured materials for which the shape of the transmission function can be controlled more easily than in bulk materials. Finally, we note that the results of this section are obtained in the absence of phonon heat leaks.

10.4.4 Interacting Systems

The thermoelectric properties of strongly interacting systems are of fundamental interest since their efficiency is not bounded by inherent limitations of non-interacting systems, such as the Wiedemann-Franz law. Experiments on some strongly correlated materials such as sodium cobalt oxides revealed unusually large thermopower values [140, 146], due in part to the strong electron-electron interactions [113]. Very little is known about the thermoelectric properties of interacting systems: analytical results are rare and numerical simulations are challenging. The linear response Kubo formalism has been used to investigate the thermoelectric properties of one-dimensional integrable and nonintegrable strongly correlated quantum lattice models [9, 113, 133, 156]. With regard to the simulation of classical dynamical models, an extension of the model discussed in Sect. (10.4.3), with inter-particle interactions added by substituting the Lorentz lattice with the rotating Lorentz gas model [83, 101] was studied in [34]. It was shown that while ZT is bounded from above by its value obtained at zero interaction, it still increases with c_V. On the other hand, for a one-dimensional di-atomic disordered hard-point gas coupled to particle reservoirs (see the upper panel in Fig. 10.18 for a schematic representation of the model), it was numerically found [38] that ZT diverges in the thermodynamic limit as a power-law, $ZT \sim \langle N \rangle^{\alpha}$, where $\langle N \rangle$ is the average number of particles in the system and $\alpha \approx 0.79$. Note that if the masses of all particles are the same, the dynamics is integrable and one can find analytically that ZT is independent of $\langle N \rangle$ (in particular, $ZT = 1$ when the chemical potential $\mu = 0$). Later [122] showed that the numerically observed large values of ZT could not be explained in terms of the energy filtering mechanism. Indeed, the particle current at the position $x \in [0, \Lambda]$ (Λ is the system size) can be expressed as $j_\rho = \int_0^\infty d\varepsilon D(\varepsilon)$, where $D(\varepsilon) \equiv D_L(\varepsilon) - D_R(\varepsilon)$ plays the role of "transmission function": $D_L(\varepsilon)$ is the density of particles with energy ε crossing x and coming from the left side, while $D_R(\varepsilon)$ is the density of particles with energy ε from the right side. If the divergence of ZT with Λ was due to energy filtering, then $D(\varepsilon)$ would sharpen with increasing the system size. Conversely, no sign of narrowing of $D(\varepsilon)$ was found in [122], As

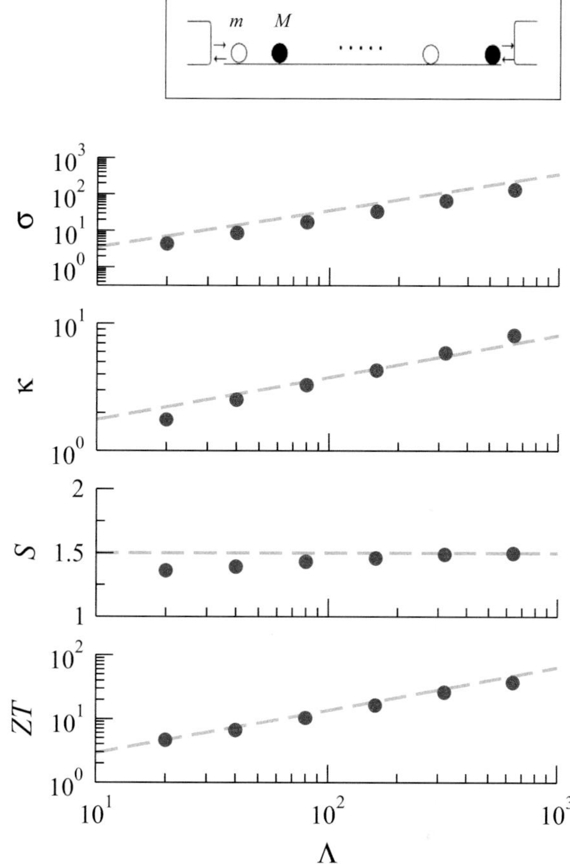

Fig. 10.18 Thermoelectric transport coefficients for the one-dimensional di-atomic disordered hard-point gas model, as a function of the system size Λ. The *dashed curves* correspond from *top* to *bottom* to $\sigma \sim \Lambda$, $\kappa \sim \Lambda^{0.33}$, $S = 1.5$, and $ZT \sim \Lambda^{0.67}$. In the *upper panel* a schematic representation of the model is shown

discussed below in Sect. 10.4.4.2, the divergence of ZT can be explained on the basis of a theoretical argument [18] applicable to non-integrable systems with momentum conservation.

10.4.4.1 Green-Kubo Formula

While the Landauer-Büttiker approach cannot be applied to interacting systems, the linear response regime can be numerically investigated in equilibrium simulations by using the Green-Kubo formula. This formula expresses the Onsager kinetic coefficients in terms of equilibrium dynamic correlation functions of the corresponding current at finite temperature β^{-1} [81, 88] as

$$L_{ij} = \lim_{\omega \to 0} \mathrm{Re} L_{ij}(\omega) , \tag{10.37}$$

where

$$L_{ij}(\omega) \equiv \lim_{\varepsilon\to 0}\int_0^\infty dt e^{-i(\omega-i\varepsilon)t} \lim_{\Omega\to\infty}\frac{1}{\Omega}\int_0^\beta d\tau \langle J_i(0)J_j(t+i\tau)\rangle, \qquad (10.38)$$

where $\langle\ \cdot\ \rangle = \{\mathrm{tr}\,[(\,\cdot\,)\exp(-\beta\mathscr{H})]\}/\mathrm{tr}\,[\exp(-\beta\mathscr{H})]$ denotes the equilibrium expectation value at temperature T, $\mathscr{H}$ is the system's Hamiltonian, Ω is the system's volume, and $J_i(t) = \int_\Omega d\mathbf{r} j_i(\mathbf{r},t)$ is the total current ($i = \rho, u$).

Within the framework of Kubo linear response approach, the real part of $L_{ij}(\omega)$ can be decomposed into a singular contribution at zero frequency and a regular part $L_{ij}^{\mathrm{reg}}(\omega)$ as

$$\mathrm{Re}L_{ij}(\omega) = 2\pi D_{ij}\delta(\omega) + L_{ij}^{\mathrm{reg}}(\omega)\,. \qquad (10.39)$$

The coefficient of the singular part defines the generalized Drude weights D_{ij} (for $i = j = \rho$, we have the conventional Drude weight $D_{\rho\rho}$). Importantly, it has been shown that non-zero Drude weights, $D_{ij} \neq 0$, are a signature of ballistic transport [61, 65, 158, 159], namely in the thermodynamic limit the kinetic coefficients L_{ij} diverge linearly with the system size. Moreover, it has been conjectured that at finite temperature, an integrable system is an ideal conductor characterised by a finite Drude weight if at zero temperature the Drude weight is positive, while the system remains an insulator if the zero temperature Drude weight is zero. On the other hand nonintegrable systems are believed to have a vanishing Drude weight and thus, to exhibit normal transport.

10.4.4.2 Conservation Laws and Thermoelectric Efficiency

The way in which the dynamic correlation functions in Eq. (10.38) decay, determines the ballistic, anomalous or diffusive character of the energy and particle transport, and it has been understood that this decay is directly related to the existence of conserved dynamical quantities [158, 159]. For quantum spin chains and under suitable conditions, it has been proved that systems possessing conservation laws exhibit ballistic transport at finite temperature [70].

However, the role that the existence of conserved quantities plays on the thermoelectric efficiency has been considered only recently [18, 19, 38, 44, 122].

The decay of time correlations for the currents can be related to the existence of conserved quantities by using Suzuki's formula [139], which generalizes an inequality proposed by Mazur [98]. Consider a system of size Λ and Hamiltonian $\mathscr{H}$, with a set of M *relevant conserved quantities* Q_m, $m = 1,\ldots,M$, namely the commutators $[\mathscr{H}, Q_m] = 0$. A constant of motion Q_m is by definition relevant if it is not orthogonal to the currents under consideration, in our case $\langle J_\rho Q_m\rangle \neq 0$ and $\langle J_u Q_m\rangle \neq 0$. It is assumed that the M constants of motion are orthogonal, i.e., $\langle Q_m Q_n\rangle = \langle Q_n^2\rangle\delta_{m,n}$ (this is always possible via a Gram-Schmid procedure). Furthermore, we assume that the set $\{Q_m\}$ exhausts all relevant conserved quantities.

Then using Suzuki's formula [139], we can express the finite-size Drude weights

$$d_{ij}(\Lambda) \equiv \frac{1}{2\Lambda} \lim_{t\to\infty} \frac{1}{t} \int_0^t dt' \langle J_i(t') J_j(0) \rangle \tag{10.40}$$

in terms of the relevant conserved quantities:

$$d_{ij}(\Lambda) = \frac{1}{2\Lambda} \sum_{m=1}^{M} \frac{\langle J_i Q_m \rangle \langle J_j Q_m \rangle}{\langle Q_m^2 \rangle} \,. \tag{10.41}$$

On the other hand, the thermodynamic Drude weights can also be expressed in terms of time-averaged current-current correlations as

$$D_{ij} = \lim_{t\to\infty} \lim_{\Lambda\to\infty} \frac{1}{2\Lambda t} \int_0^t dt' \langle J_i(t') J_j(0) \rangle \,. \tag{10.42}$$

If the thermodynamic limit $\Lambda \to \infty$ commutes with the long-time limit $t \to \infty$, then the thermodynamic Drude weights D_{ij} can be obtained as

$$D_{ij} = \lim_{\Lambda\to\infty} d_{ij}(\Lambda) \,. \tag{10.43}$$

Moreover, if the limit does not vanish we can conclude that the presence of relevant conservation laws yields non-zero generalized Drude weights, which in turn imply that transport is ballistic, $L_{ij} \sim \Lambda$. As a consequence, the electrical conductivity is ballistic, $\sigma \sim L_{\rho\rho} \sim \Lambda$, while the thermopower is asymptotically size-independent, $S \sim L_{u\rho}/L_{\rho\rho} \sim \Lambda^0$.

We can see from Suzuki's formula that for systems with a single relevant constant of motion ($M = 1$), the ballistic contribution to $\det \mathbb{L}$ vanishes, since it is proportional to $D_{\rho\rho} D_{uu} - D_{\rho u}^2$, which is zero from Eqs. (10.41) and (10.43). Hence, $\det \mathbb{L}$ grows slower than Λ^2, and therefore the thermal conductivity $\kappa \sim \det \mathbb{L}/L_{\rho\rho}$ grows sub-ballistically, $\kappa \sim \Lambda^\alpha$, with $\alpha < 1$. Since $\sigma \sim \Lambda$ and $S \sim \Lambda^0$, we can conclude that $ZT \sim \Lambda^{1-\alpha}$ [18]. Hence ZT diverges in the thermodynamic limit $\Lambda \to \infty$. This general theoretical argument applies for instance to systems where momentum is the only relevant conserved quantity.

It has been recently shown that these expectations fully describe the results obtained for the one-dimensional disordered hard-point gas, see Fig. 10.18 and [18]. This enhancement of ZT has also been verified for more realistic models in [19], where the nonequilibrium steady state properties of a two-dimensional gas of particles interacting through elastic collisions and enclosed in a box connected to reservoirs at both ends were studied numerically. The inter-particle collisions were modeled by the method of Multiparticle Collision Dynamics (MPC) [92]. Similarly to [18], it was found that the generalized Drude weights are finite, leading to non decaying current-current time correlations. As a consequence, the transport coefficients exhibit an anomalous scaling yielding a figure of merit that for this

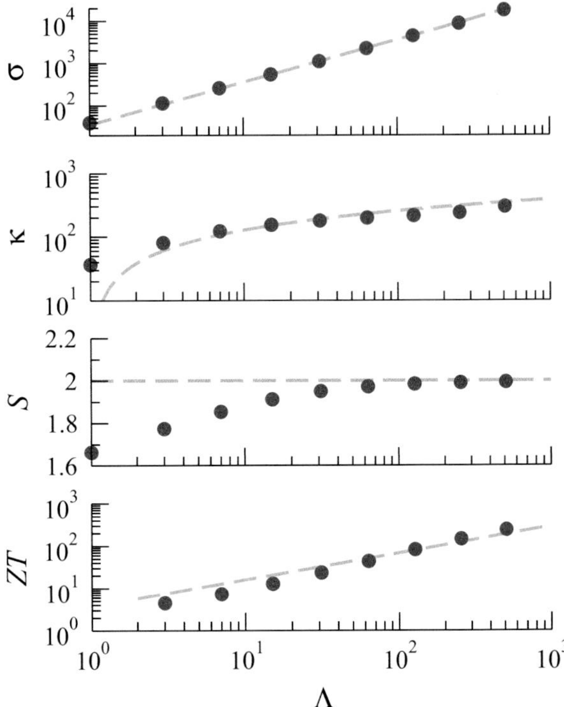

Fig. 10.19 Thermoelectric transport coefficients for the two-dimensional MPC gas of interacting particles as a function of the system size Λ (for details see [19]). The *dashed curves* correspond from *top* to *bottom* to $\sigma = (\pi \langle N \rangle / 2m) \Lambda$ with $\langle N \rangle$ the mean number of particles, $\kappa \sim \log \Lambda$, $S = 2$, and $ZT \sim \Lambda / \log \Lambda$

model diverges as $ZT \sim \Lambda / \log \Lambda$. The logarithmic term appears as a consequence of the existence of long time tails in the decay of the energy current-current time correlation, typically observed in two-dimensional hydrodynamic systems [3]. The dependence of the thermoelectric transport coefficients as a function of the system size is shown in Fig. 10.19. Finally, results consistent with the above model have been obtained not only for instantaneous collision models but also for a system with finite range of the interaction, more precisely for a one-dimensional gas of particles with nearest-neighbor Coulomb interaction, modeling a screened Coulomb interaction between electrons [44]. This latter model takes advantage of the recently reported Fourier-like behavior of thermal conductivity [42, 47, 127, 148, 157], namely, of the appearance of a very broad range of system size where the thermal conductivity behaves normally according to the Fourier law, i.e., κ is size-independent, see Fig. 10.2. As a consequence, ZT exhibits a rapid, liner growth with the system size. While the Fourier-like regime might be an intermediate (in the system size) regime, followed by an asymptotic regime of anomalous thermal conductivity $\kappa \sim \Lambda^{1/3}$ [52, 84], the range of validity of such regime may expand rapidly as an integrable limit is approached [42]. We point out that it is a priori not excluded that there exist models where the long-time limit $t \to \infty$ and the thermodynamical limit $\Lambda \to \infty$ do not commute when computing the Drude weights. However, numerical evidence shows that for the models so far considered these two limits commute [18, 19, 44].

It is interesting to note the contrasting behavior obtained when more than one conserved quantities exist. For $M > 1$, in general $D_{\rho\rho}D_{uu} - D_{u\rho}^2 \neq 0$. As a consequence, $\det \mathbb{L} \sim \Lambda^2$, and therefore the heat conductivity becomes ballistic and ZT asymptotically independent of the system size. This situation is commonly found in integrable systems, for which infinite constants of motion exist at the thermodynamic limit. For instance, in noninteracting systems, momentum conservation implies that all transport coefficients are ballistic, thus leading to a constant ZT. The enhancement in the efficiency due to the existence of conserved quantities is limited to systems of interacting particles.

10.4.5 Breaking Time-Reversibility

When time-reversal symmetry is broken, typically by a magnetic field $\boldsymbol{B}$, Onsager-Casimir reciprocity relations no longer imply $L_{ji} = L_{ij}$, but rather $L_{ji}(\boldsymbol{B}) = L_{ij}(-\boldsymbol{B})$. While these relations imply $\sigma(\boldsymbol{B}) = \sigma(-\boldsymbol{B})$ and $\kappa(\boldsymbol{B}) = \kappa(-\boldsymbol{B})$, the thermopower is not bounded to be a symmetric function under the exchange $\boldsymbol{B} \to -\boldsymbol{B}$. This simple remark has deep consequences on thermoelectric efficiency.

The maximum efficiency and the efficiency at maximum power are now determined by two parameters [17]: the asymmetry parameter

$$x = \frac{S(\boldsymbol{B})}{S(-\boldsymbol{B})} = \frac{S(\boldsymbol{B})}{\Pi(\boldsymbol{B})}T \tag{10.44}$$

and the "figure of merit"

$$y = \frac{\sigma(\boldsymbol{B})S(\boldsymbol{B})S(-\boldsymbol{B})}{\kappa(\boldsymbol{B})}T = \frac{\sigma(\boldsymbol{B})S(\boldsymbol{B})\Pi(\boldsymbol{B})}{\kappa(\boldsymbol{B})}. \tag{10.45}$$

In terms of these variables, the maximum efficiency reads

$$\eta_{\max} = \eta_C\, x\, \frac{\sqrt{y+1}-1}{\sqrt{y+1}+1}, \tag{10.46}$$

while the efficiency at maximum power is

$$\eta(P_{\max}) = \frac{\eta_C}{2}\frac{xy}{2+y}. \tag{10.47}$$

In the particular case $x = 1$, y reduces to the ZT figure of merit of the time-symmetric case, Eq. (10.46) reduces to Eq. (10.20), and Eq. (10.47) to Eq. (10.21). While thermodynamics does not impose any restriction on the attainable values of the asymmetry parameter x, the positivity of entropy production implies $h(x) \leq y \leq 0$ if $x \leq 0$ and $0 \leq y \leq h(x)$ if $x \geq 0$, where the function $h(x) = 4x/(x-1)^2$. Note that $\lim_{x\to 1} h(x) = \infty$ and therefore there is no upper bound on $y(x=1) = ZT$. For a given value of the asymmetry x, the maximum (over y) $\bar{\eta}(P_{\text{max}})$ of $\eta(P_{\text{max}})$ and the maximum $\bar{\eta}_{\text{max}}$ of η_{max} are obtained for $y = h(x)$:

$$\bar{\eta}(P_{\text{max}}) = \eta_C \frac{x^2}{x^2+1} , \tag{10.48}$$

$$\bar{\eta}_{\text{max}} = \begin{cases} \eta_C x^2 & \text{if } |x| \leq 1 , \\ \eta_C & \text{if } |x| \geq 1 . \end{cases} \tag{10.49}$$

The functions $\bar{\eta}(P_{\text{max}})(x)$ and $\bar{\eta}_{\text{max}}(x)$ are drawn in Fig. 10.20. In the case $|x| > 1$, it is in principle possible to overcome the Curzon-Ahlborn limit within linear response (that is, to have $\eta(P_{\text{max}}) > \eta_C/2$) and to reach the Carnot efficiency, for increasingly smaller and smaller figure of merit y as the asymmetry parameter x increases. The Carnot efficiency is obtained for $\det \mathbb{L} = (L_{\rho u} - L_{u\rho})^2/4 > 0$ when $|x| > 1$, that is, the tight coupling condition is not fulfilled.

The output power at maximum efficiency reads

$$P(\bar{\eta}_{\text{max}}) = \frac{\bar{\eta}_{\text{max}}}{4} \frac{|L_{\rho u}^2 - L_{u\rho}^2|}{L_{\rho\rho}} \frac{T_{\text{H}} - T_{\text{C}}}{T^2} . \tag{10.50}$$

Therefore, always within linear response, it is allowed from thermodynamics to have Carnot efficiency and nonzero power simultaneously when $|x| > 1$. Such a possibility can be understood on the basis of the following argument [27, 29]. We first split the particle and energy currents into a reversible part (which changes sign by reversing $\boldsymbol{B} \to -\boldsymbol{B}$) and an irreversible part (invariant with respect to the

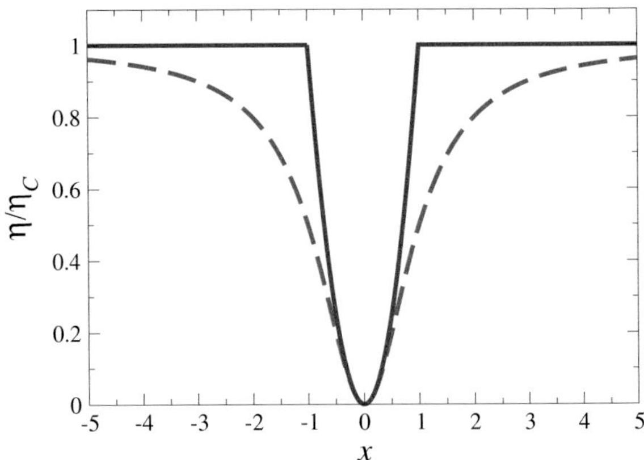

Fig. 10.20 Efficiency η in units of the Carnot efficiency η_C as a function of the asymmetry parameter x, with $\eta = \bar{\eta}(P_{\text{max}})$ (*dashed curve*) and $\eta = \bar{\eta}_{\text{max}}$ (*full curve*). For $x = 1$, $\bar{\eta}(P_{\text{max}}) = \eta_C/2$ and $\bar{\eta}_{\text{max}} = \eta_C$ are obtained for $y(x=1) = ZT = \infty$

inversion $\boldsymbol{B} \to -\boldsymbol{B}$), defined by

$$\mathbf{j}^{\mathrm{rev}}(\boldsymbol{B}) = \frac{\mathbb{L}(\boldsymbol{B}) - \mathbb{L}^t(\boldsymbol{B})}{2}\,\mathbf{F}, \quad \mathbf{j}^{\mathrm{irr}}(\boldsymbol{B}) = \frac{\mathbb{L}(\boldsymbol{B}) + \mathbb{L}^t(\boldsymbol{B})}{2}\,\mathbf{F}\,. \tag{10.51}$$

Only the irreversible part of the currents contributes to the entropy production: $\dot{s} = \mathbf{j}^{\mathrm{irr}} \cdot \mathbf{F} = j_\rho^{\mathrm{irr}} \nabla(-\mu/T) + j_u^{\mathrm{irr}} \nabla(1/T)$. The reversible currents vanish for $\boldsymbol{B} = 0$. On the other hand, for broken time-reversal symmetry the reversible currents can in principle become arbitrarily large, giving rise to the possibility of dissipationless transport.

It is interesting to compare the performances of a system as a thermal machine or as a refrigerator. For a refrigerator, the most important benchmark is the *coefficient of performance* $\eta^{(r)} = j_q/P$ ($j_q < 0$, $P < 0$), given by the ratio of the heat current extracted from the cold system over the absorbed power. The efficiency of an ideal, dissipationless refrigerator is given by $\eta_C^{(r)} = T_{\mathrm{C}}/(T_{\mathrm{H}} - T_{\mathrm{C}})$. While in the time-reversal case the linear response normalized maximum efficiency $\eta_{\max}/\eta_C$ and coefficient of performance $\eta_{\max}^{(r)}/\eta_C^{(r)}$ for power generation and refrigeration coincide, this is no longer the case with broken time-reversal symmetry. For refrigeration the maximum value of the coefficient of performance reads

$$\eta_{\max}^{(r)} = \eta_C^{(r)}\,\frac{1}{x}\,\frac{\sqrt{y+1}-1}{\sqrt{y+1}+1}. \tag{10.52}$$

For small fields, x is in general a linear function of the magnetic field, while y is by construction an even function of the field. As a consequence, a small external magnetic field either improves power generation and worsens refrigeration or vice-versa, while the average efficiency

$$\frac{1}{2}\left[\frac{\eta_{\max}(\boldsymbol{B})}{\eta_C} + \frac{\eta_{\max}^{(r)}(\boldsymbol{B})}{\eta_C^{(r)}}\right] = \frac{\eta_{\max}(\mathbf{0})}{\eta_C} = \frac{\eta_{\max}^{(r)}(\mathbf{0})}{\eta_C^{(r)}}, \tag{10.53}$$

up to second order corrections. Due to the Onsager-Casimir relations, $x(-\boldsymbol{B}) = 1/x(\boldsymbol{B})$ and therefore by inverting the direction of the magnetic field one can improve either power generation or refrigeration.

With regard to the practical relevance of the results presented in this section, we should note that, as a consequence of the symmetry properties of the scattering matrix [48] (see Sect. 10.4.6), in the non-interacting case the thermopower is a symmetric function of the magnetic field, thus implying $x = 1$. On the other hand, as we shall discuss in Sect. 10.4.6, this symmetry may be violated when electron-phonon or electron-electron interactions are taken into account. Non-symmetric thermopowers have been reported in measurements for certain orientations of a bismuth crystal [152] and in Andreev interferometer experiments [58] (for a theoretical analysis of these latter experiments see [75]).

10.4.6 Inelastic Scattering and Probe Terminals

Inelastic scattering events like electron-phonon interactions, can be conveniently modeled by means of a third terminal (or conceptual probe), whose parameters (temperature and chemical potential) are chosen self-consistently so that there is no net *average* flux of particles and heat between this terminal and the system (see Fig. 10.21, left panel). In mesoscopic physics, probe reservoirs are commonly used to simulate phase-breaking processes in partially coherent quantum transport, since they introduce phase-relaxation without energy damping [31]. The advantage of such approach lies in its simplicity and independence from microscopic details of inelastic processes. Probe terminals have been widely used in the literature and proved to be useful to unveil nontrivial aspects of phase-breaking processes [48], heat transport and rectification [13, 22–24, 52, 112, 118, 120], and thermoelectric transport [11, 15, 25, 29, 56, 57, 67, 73, 76, 77, 123, 124, 126, 135, 136].

The approach can be generalized to any number n_p of probe reservoirs. We call $\mathbf{j}_k \equiv (j_{k,\rho}, j_{k,u})^t$ the particle and energy currents from the kth terminal (at temperature T_k and chemical potential μ_k), with $k = 3, \ldots, n$ denoting the $n_p = n - 2$ probes. Due to the steady-state constraints of charge and energy conservation, $\sum_k j_{k,\rho} = \sum_k j_{k,u} = 0$, we can express, for instance, the currents from the second reservoir as a function of the remaining $2(n-1)$ currents. The corresponding generalized forces are given by $\mathbf{X}_k \equiv (\Delta(\mu_k/T), \Delta T_k/T^2)^t$, with $\Delta\mu_k = \mu_k - \mu$, $\Delta T_k = T_k - T$, $\mu = \mu_2$, and $T = T_2$. The linear response relations between currents and thermodynamic forces read as follows:

$$\mathbf{j}_i = \sum_{\substack{j=1 \\ (j \neq 2)}}^{n} \mathbb{L}_{ij} \mathbf{X}_j, \tag{10.54}$$

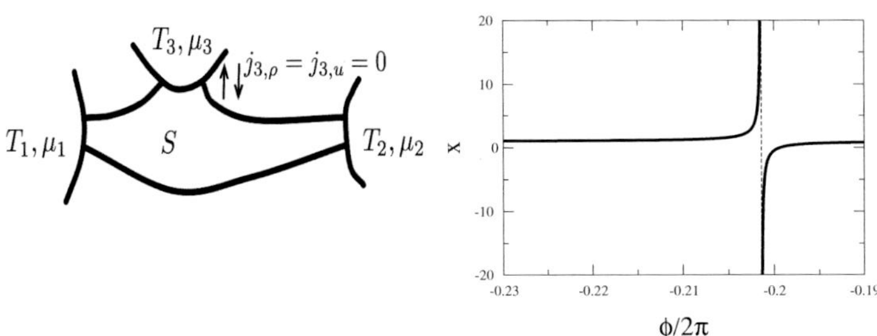

Fig. 10.21 *Left panel*: schematic drawing of thermoelectric transport, with a third terminal acting as a probe reservoir mimicking inelastic scattering. The temperature T_3 and the chemical potential μ_3 of the third reservoir are such that the net average electric and energy currents through this reservoir vanish: $j_{3,\rho} = J_{3,u} = 0$. This setup can be generalized to any number of probe reservoirs, $k = 3, \ldots, n$, by setting $j_{k,\rho} = J_{k,u} = 0$ for all probes. *Right panel*: asymmetry parameter x for a three-terminal Aharonov-Bohm interferometer, with one of the terminals acting as a probe, see [123] for details

where $\mathbb{L}_{ij}$ are 2×2 matrices, so that the overall Onsager matrix $\mathbb{L}$ has size $2(n-1)$. We then impose the condition of zero average currents through the probes, $j_{k,\rho} = j_{k,u} = 0$ for $k = 3, \ldots, n$ to reduce the Onsager matrix to a 2×2 matrix $\mathbb{L}'$ connecting the fluxes $\mathbf{j}_1$ through the first reservoir and the conjugated forces $\mathbf{X}_1$ as $\mathbf{j}_1 = \mathbb{L}'\mathbf{X}_1$. The reduced matrix $\mathbb{L}'$ fulfills the Onsager-Casimir relations and represents the Onsager matrix for two-terminal inelastic transport modeled by means of probe reservoirs. The transport coefficients and the thermodynamic efficiencies can then be computed in the usual way from the reduced matrix $\mathbb{L}'$.

The particle and energy currents can be conveniently computed, for any number of probes, by means of the multi-terminal Landauer-Büttiker formula [48]:

$$j_{k,\rho} = \frac{1}{h}\int_{-\infty}^{\infty} d\varepsilon \sum_l [\tau_{l\leftarrow k}(\varepsilon) f_k(\varepsilon) - \tau_{k\leftarrow l}(\varepsilon) f_l(\varepsilon)], \tag{10.55}$$

$$j_{k,u} = \frac{1}{h}\int_{-\infty}^{\infty} d\varepsilon\, \varepsilon \sum_l [\tau_{l\leftarrow k}(\varepsilon) f_k(\varepsilon) - \tau_{k\leftarrow l}(\varepsilon) f_l(\varepsilon)], \tag{10.56}$$

where $\tau_{l\leftarrow k}(\varepsilon)$ is the transmission probability from terminal k to terminal l at the energy ε. Charge conservation and the requirement of zero current at zero bias impose

$$\sum_k \tau_{k\leftarrow l} = \sum_k \tau_{l\leftarrow k} = M_l, \tag{10.57}$$

with M_l being the number of modes in the lead l. Moreover, in the presence of a magnetic field $\boldsymbol{B}$ we have

$$\tau_{k\leftarrow l}(\boldsymbol{B}) = \tau_{l\leftarrow k}(-\boldsymbol{B}). \tag{10.58}$$

The last relation is a consequence of the unitarity of the scattering matrix $\mathbb{S}(\boldsymbol{B})$ that relates the outgoing wave amplitudes to the incoming wave amplitudes at the different leads. The time-reversal invariance of unitary dynamics leads to $\mathbb{S}(\boldsymbol{B}) = \mathbb{S}^t(-\boldsymbol{B})$, which in turn implies (10.58) [48]. In the two-terminal case, Eq. (10.57) means $\tau_{1\leftarrow 2} = \tau_{2\leftarrow 1}$. Hence, we can conclude from this relation and Eq. (10.58) that $\tau_{2\leftarrow 1}(\boldsymbol{B}) = \tau_{2\leftarrow 1}(-\boldsymbol{B})$, thus implying that the Seebeck coefficient is a symmetric function of the magnetic field.

Probe terminals can break the symmetry of the Seebeck coefficient. We can have $S(-\boldsymbol{B}) \neq S(\boldsymbol{B})$, that is, $L'_{12} \neq L'_{21}$ in the reduced Onsager matrix $\mathbb{L}$. Arbitrarily large values of the asymmetry parameter $x = S(\boldsymbol{B})/S(-\boldsymbol{B})$ were obtained in [123] (see Fig. 10.21, right panel) by means of a three-dot Aharonov-Bohm interferometer model. The asymmetry was found also for chaotic cavities, ballistic microjunctions [126], and random Hamiltonians drawn from the Gaussian unitary ensemble [11], and also in the framework of classical physics, for a three-terminal deterministic railway switch transport model [67]. In the latter model, only the values zero and one are allowed for the transmission functions $\tau_{j\leftarrow i}(\varepsilon)$, i.e., $\tau_{j\leftarrow i}(\varepsilon) = 1$ if particles injected from terminal i with energy ε go to terminal j and $\tau_{j \leftarrow i}(\varepsilon) = 0$ is such particles go to a terminal other than j. The transmissions $\tau_{j \leftarrow i}(\varepsilon)$ are piecewise constant in the intervals $[\varepsilon_i, \varepsilon_{i+1}]$, $(i = 1, 2, \ldots)$, with switching $\tau_{j \leftarrow i} = 1 \to 0$ or viceversa possible at the threshold energies ε_i, with the constraints (10.57) always fulfilled.

In all the above instances, it was not possible to find at the same time large values of asymmetry parameter (10.44) and high thermoelectric efficiency. Such failure was explained by [29] and is generic for non-interacting three-terminal systems. In that case, when the magnetic field $\boldsymbol{B} \neq 0$, current conservation, which is mathematically expressed by unitarity of the scattering matrix $\mathbb{S}$, imposes bounds on the Onsager matrix stronger than those derived from the positivity of entropy production. As a consequence, Carnot efficiency can be achieved in the three-terminal setup only in the symmetric case $x = 1$. On the other hand, the Curzon-Ahlborn linear response bound, $\eta_C/2$, for the efficiency at maximum power can be overcome for moderate asymmetries, $1 < x < 2$, with a maximum of $4\eta_C/7$ at $x = 4/3$. The bounds obtained by Brandner et al. [29] are in practice saturated in a quantum transmission model reminiscent of the above described railway switch model [11] (see Fig. 10.22). The generic multi-terminal case was also discussed for noninteracting electronic transport [27]. By increasing the number n_p of probe terminals, the constraint from current conservation on the maximum efficiency and the efficiency at maximum power becomes weaker. However, the bounds (10.48) and (10.49) from the second law of thermodynamics are saturated only in the limit $n_p \to \infty$. Moreover, numerical evidence suggests that the power vanishes when the maximum efficiency is approached [28]. It is an interesting open question whether similar bounds on efficiency, tighter that those imposed by the positivity of entropy production, exist in more general transport models for interacting systems.

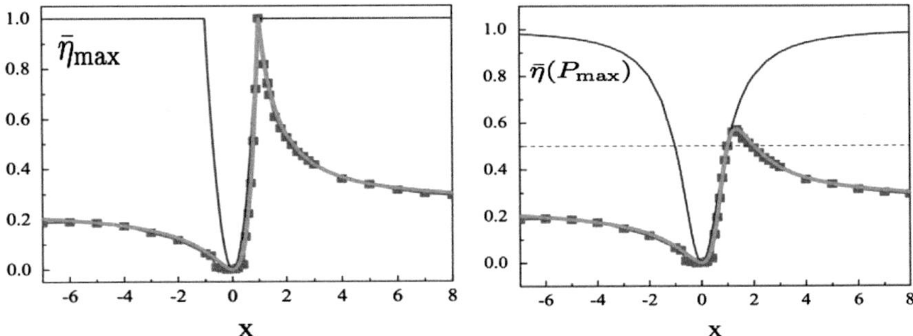

Fig. 10.22 Maximum efficiency $\bar{\eta}_{\max}$ (*left panel*) and efficiency at maximum power $\bar{\eta}(P_{\max})$ (*right panel*), both in units of η_C. *Upper curves* correspond to the thermodynamics bounds [17], *lower curves* to the more restrictive bounds [29] from the unitarity of the scattering matrix for three terminals, *squares* are obtained from a transmission model whose details are described in [11]. *Dotted-dashed line* corresponds to the Curzon-Ahlborn linear-response limit $\eta_C/2$. Note that such limit is exceeded in the interval [1,2] with the transmission model

Finally, we point out that in a genuine multi-terminal device all terminals should be treated on equal footing, without necessarily declaring some of them as probes. First investigations for a generic three-terminal system have shown that in some instances the coupling to a third terminal can improve both the extracted power and the efficiency of a thermoelectric device [99]. Moreover, with three terminals one can separate the currents, with charge and heat flowing to different reservoirs. As a result, it is possible to violate in a controlled fashion the Wiedemann-Franz law, greatly enhancing thermoelectric performances [100].

10.5 Concluding Remarks

In this chapter we have discussed several microscopic mechanisms for the design of a thermal rectifier and the increase of the efficiency of thermoelectric energy conversion. Although not intuitive, solid-state thermal rectifiers do exist and there have already been the first experimental implementations. With regard to thermoelectricity, basic concepts to improve the efficiency have been identified: energy filtering for non-interacting systems and momentum conservation in non-integrable interacting systems.

Several questions remain open. An important point for thermal rectification is the need to have a strongly temperature dependent thermal conductivity. Some ideas have already been explored, but the microscopic theory is still incomplete. It appears promising in this connection to work in the vicinity of a structural phase transition. Moreover, the above discussed rectifiers are based on insulating materials. It would be interesting, in order to combine thermal rectification with thermoelectric power generation or cooling, to include and understand the role of mobile charge carriers. Recent experimental investigations are moving forward in this direction [94].

In spite of the long history of thermoelectricity, from the viewpoint of statistical physics the theory of the coupled transport of heat and charge is still in its infancy. With regard to the challenging problem of improving the efficiency of heat to work conversion, for non-interacting systems we have a quite complete theoretical picture and understand the limitations imposed by nature (notably, the Wiedemann-Franz law). On the other hand, the understanding of general mechanisms connected to strongly interacting systems, for which the Wiedemann-Franz law does not apply, are only beginning to emerge. In particular, regimes near electronic phase transitions might be favorable for thermoelectric conversion [110, 145]. A deeper understanding of the nonlinear regime is also needed [74, 125, 150], since, as observed experimentally in mesoscopic devices [97], the Onsager-Casimir reciprocity relations break down and this fact could in principle allow for improved thermoelectric efficiencies. Furthermore, in the nonlinear regime rectification effects occur and their impact on thermoelectricity is still not well understood.

Acknowledgements G.B. and G.C. acknowledge the support by MIUR-PRIN. C.M.-M. acknowledges the support by the Spanish MICINN grant MTM2012-39101-C02-01 and MTM2015-63914-P.

References

1. Aberg, S.: Phys. Rev. Lett. **64**, 3119 (1990)
2. Ajisaka, S., Barra, F., Mejía-Monasterio, C., Prosen, T.: Phys. Rev. B **86**, 125111 (2012)
3. Alder, J., Wainwright, T.E.: Phys. Rev. Lett. **18**, 988 (1967)
4. Andresen, B.: Angew. Chem. Int. Ed. **50**, 2690 (2011)
5. Apertet, Y., Ouerdane, H., Goupil, C., Lecoeur, Ph.: Phys. Rev. E **85**, 031116 (2012)
6. Arrachea, L., Moskalets, M., Martin-Moreno, L.: Phys. Rev. B **75**, 245420 (2007)
7. Arrachea, L., Lozano, G.S., Aligia, A.A.: Phys. Rev. B **80**, 014425 (2009)
8. Arrachea, L., Mucciolo, E.R., Chamon, C., Capaz, R.B.: Phys. Rev. B **86**, 125424 (2012)
9. Arsenault, L.-F., Shastry, B.S., Sémon, P., Tremblay, A.- M.S.: Phys. Rev. B **87**, 035126 (2013)
10. Aschcroft, N.W., Mermin, N.D.: Solid State Physics. Saunders College, Philadelphia (1976)
11. Balachandran, V., Benenti, G., Casati, G.: Phys. Rev. B, **87**, 165419 (2013)
12. Balcerek, K., Tyc, T.: Phys. Status Solidi (a) **47**, K125 (1978)
13. Bandyopadhyay, M., Segal, D.: Phys. Rev. E **84**, 011151 (2011)
14. Barthel, T., Schollwöck, U.: Phys. Rev. Lett. **100**, 100601 (2008)
15. Bedkihal, S., Bandyopadhyay, M., Segal, D.: Eur. Phys. J. B **86**, 506 (2013)
16. Bell, L.E.: Science **321**, 1457 (2008)
17. Benenti, G., Saito, K., Casati, G.: Phys. Rev. Lett. **106**, 230602 (2011)
18. Benenti, G., Casati, G., Wang, J.: Phys. Rev. Lett. **110**, 070604 (2013)
19. Benenti, G., Casati, G., Mejía-Monasterio, C.: New J. Phys. **16**, 015014 (2014)
20. Benenti, G., Casati, G., Prosen, T., Saito, K.: `arXiv:1311.4430`
21. Bohigas, O., Giannoni, M.-J., Schmit, C.: Phys. Rev. Lett. **52**, 1 (1984)
22. Bolsterli, M., Rich, M., Visscher, W.M.: Phys. Rev. A **1**, 1086 (1970)
23. Bonetto, F., Lebowitz, J.L., Lukkarinen, J.: J. Stat. Phys. **116**, 783 (2004)
24. Bonetto, F., Lebowitz, J.L., Lukkarinen, J., Olla, S.: J. Stat. Phys. **134**, 1097 (2009)
25. Bosisio, R., Gorini, C., Fleury, G., Pichard, J.-L.: New J. Phys. **16**, 095005 (2014)
26. Boukai, A.I., et al.: Nature **451**, 168 (2008)
27. Brandner, K., Seifert, U.: New J. Phys. **15**, 105003 (2013)
28. Brandner, K., Seifert, U.: Phys. Rev. E **91**, 012121 (2015)
29. Brandner, K., Saito, K., Seifert, U.: Phys. Rev. Lett. **110**, 070603 (2013)
30. Breuer, H.-P., Petruccione, F.: The Theory of Open Quantum Systems. Oxford University Press, New York (2007)
31. Büttiker, M.: IBM J. Res. Dev. **32**, 63 (1988)
32. Cahill, D.G., et al.: J. Appl. Phys. **93**, 793 (2003)
33. Callen, H.B.: Thermodynamics and an Introduction to Thermostatics, 2nd edn. Wiley, New York (1985)
34. Casati, G., Mejía-Monasterio, C.: AIP Conf. Proc. **1076**, 18 (2008)
35. Casati, G., Valz-Gris, F., Guarneri, I.: Lett. Nuovo Cimento **28**, 279 (1980)
36. Casati, G., Ford, J., Vivaldi, F., Visscher, W.M.: Phys. Rev. Lett. **52**, 1861 (1984)
37. Casati, G., Mejía-Monasterio, C., Prosen, T.: Phys. Rev. Lett. **101**, 016601 (2008)
38. Casati, G., Wang, L., Prosen, T.: J. Stat. Mech. L03004 (2009)
39. Chambadal, P.: Les Centrales Nucléaires. Armand Colin, Paris (1957)
40. Chang, C.W., Okawa, D., Majumdar, A., Zettl, A.: Science **314**, 1121 (2006)
41. Chen, Z., et al.: Nat. Commun. **5**, 5446 (2014)
42. Chen, S., Wang, J., Casati, G., Benenti, G.: Phys. Rev. E **90**, 032134 (2014)

43. Chen, S., Pereira, E., Casati, G.: Europhys. Lett. **111**, 30004 (2015)
44. Chen, S., Wang, J., Casati, G., Benenti, G.: Phys. Rev. E **92**, 032139 (2015)
45. Curzon, F., Ahlborn, B.: Am. J. Phys. **43**, 22 (1975)
46. Daley, A., Kollath, C., Schollwöck, U.: J. Stat. Mech. P04005 (2004)
47. Das, S.G., Dhar, A., Narayan, O.: J. Stat. Phys. **154**, 204 (2013)
48. Datta, S.: Electronic Transport in Mesoscopic Systems. Cambridge University Press, Cambridge (1995)
49. Dauxois, T., Peyrard, M., Bishop, A.R.: Phys. Rev. E **47**, 684 (1993)
50. Debye, P.: Ann. Phys. **39**, 789 (1912)
51. de Groot, S.R., Mazur, P.: Nonequilibrium Thermodynamics. North-Holland, Amsterdam (1962)
52. Dhar, A.: Adv. Phys. **57**, 457 (2008)
53. Dresselhaus, M.S., et al.: Adv. Mater. **19**, 1043 (2007)
54. Dubi, Y., Di Ventra, M.: Phys. Rev. B **79**, 115415 (2009)
55. Dubi, Y., Di Ventra, M.: Rev. Mod. Phys. **83**, 131 (2011)
56. Entin-Wohlman, O., Aharony, A.: Phys. Rev. B **85**, 085401 (2012)
57. Entin-Wohlman, O., Imry, Y., Aharony, A.: Phys. Rev. B **82**, 115314 (2010)
58. Eom, J., Chien, C.-J., Chandrasekhar, V.: Phys. Rev. Lett. **81**, 437 (1998)
59. Esposito, M., Kawai, R., Lindenberg, K., Van den Broeck, C.: Phys. Rev. Lett. **105**, 150603 (2010)
60. Gallavotti, G., Cohen, E.G.D.: Phys. Rev. Lett. **74**, 2694 (1995); J. Stat. Phys. **80**, 931 (1995); Gallavotti, G.: J. Stat. Phys. **84**, 8996 (1996); Phys. Rev. Lett. **77**, 4334 (1996)
61. Garst, M., Rosch, A.: Europhys. Lett. **55**, 66 (2001)
62. Gaul, C., Büttner, H.: Phys. Rev. E **76**, 011111 (2007)
63. Gorini, V., Kossakowski, A., Sudarshan, E.C.G.: J. Math. Phys. **17**, 821 (1976)
64. Guhr, T., Müller-Groeling, A., Weidenmüller, H.A.: Phys. Rep. **299**, 189 (1998)
65. Heidrich-Meisner, F., Honecker, A., Brenig, W.: Phys. Rev. B **71**, 184415 (2005)
66. Hicks, L.D., Dresselhaus, M.S.: Phys. Rev. B **47**, 12727 (1993); 16631(R) (1993)
67. Horvat, M., Prosen, T., Benenti, G., Casati, G.: Phys. Rev. E **86**, 052102 (2012)
68. Humphrey, T.E., Linke, H.: Phys. Rev. Lett. **94**, 096601 (2005)
69. Humphrey, T.E., Newbury, R., Taylor, R.P., Linke, H.: Phys. Rev. Lett. **89**, 116801 (2002)
70. Ilievski, E., Prosen, T.: Commun. Math. Phys. **318**, 809 (2013)
71. Imry, Y.: Introduction to Mesoscopic Physics. Oxford University Press, New York (1997)
72. Ioffe, A.F.: Semiconductor Thermoelements, and Thermoelectric Cooling. Infosearch, London (1957)
73. Jacquet, P.A.: J. Stat. Phys. **134**, 709 (2009)
74. Jacquod, Ph., Meair, J.: J. Phys. Condens. Matter **25**, 082201 (2013)
75. Jacquod, Ph., Whitney, R.S.: Europhys. Lett. **91**, 67009 (2010)
76. Jiang, J.H., Entin-Wohlman, O., Imry, Y.: Phys. Rev. B **85**, 075412 (2012)
77. Jiang, J.H., Entin-Wohlman, O., Imry, Y.: Phys. Rev. B **87**, 205420 (2013)
78. Kobayashi, W., Teraoka, Y., Terasaki, I.: Appl. Phys. Lett. **95**, 171905 (2009)
79. Kobayashi, W., et al.: Appl. Phys. Express **5**, 027302 (2012)
80. Koumoto, K., Mori, T. (eds.): Thermoelectric Nanomaterials: Materials Design and Applications. Springer Series in Material Science, vol. 182. Springer, Berlin (2013)
81. Kubo, R.: J. Phys. Soc. Jpn. **12**, 570 (1957)
82. Kubo, R., Toda, M., Hashitsume, N.: Statistical Physics II: Nonequilibrium Statistical Mechanics. Springer, Berlin (1985)
83. Larralde, H., Leyvraz, F., Mejía-Monasterio, C.: J. Stat. Phys. **113**, 197 (2003)
84. Lepri, S., Livi, R., Politi, A.: Phys. Rep. **377**, 1 (2003)
85. Li, B., Casati, G., Wang, J.: Phys. Rev. E **67**, 021204 (2003)
86. Li, N., et al.: Rev. Mod. Phys. **84**, 1045 (2012)
87. Lindblad, G.: Commun. Math. Phys. **48**, 119 (1976)
88. Mahan, G.D.: Many-Particle Physics. Plenum Press, New York (1990)
89. Mahan, G.D., Sofo, J.O.: Proc. Natl. Acad. Sci. USA **93**, 7436 (1996)

90. Mahan, G.D., Sales, B., Sharp, J.: Phys. Today **50**, 42 (1997)
91. Majumdar, A.: Science **303**, 778 (2004)
92. Malevanets, A., Kapral, R.: J. Chem. Phys. **110**, 8605 (1999)
93. Manzano, D., Tiersch, M., Asadian, A., Briegel, H.J.: Phys. Rev. E **86**, 061118 (2012)
94. Martínez-Pérez, M.J., Fornieri, A., Giazotto, F.: Nat. Nanotechnol. **10**, 303 (2015)
95. Marucha, Cz., Mucha, J., Rafalowicz, J.: Phys. Status Solidi (a) **31**, 269 (1975)
96. Marucha, Cz., Mucha, J., Rafalowicz, J.: Phys. Status Solidi (a) **37**, K5 (1976)
97. Matthews, J., et al.: Phys. Rev. B **90**, 165428 (2014)
98. Mazur, P.: Physica **43**, 533 (1969)
99. Mazza, F., et al.: New J. Phys. **16**, 085001 (2014)
100. Mazza, F., et al.: Phys. Rev. B **91**, 245435 (2015)
101. Mejía-Monasterio, C., Larralde, H., Leyvraz, F.: Phys. Rev. Lett. **86**, 5417 (2001)
102. Mejía-Monasterio, C., Prosen, T., Casati, G.: Europhys. Lett. **72**, 520 (2005)
103. Mejía-Monasterio, C., Wichterich, H.: Eur. Phys. J. Spec. Top. **151**, 113 (2007)
104. Michel, M., Hartmann, M., Gemmer, J., Mahler, G.: Eur. Phys. J. B **34**, 325 (2003)
105. Michel, M., Mahler, G., Gemmer, J.: Phys. Rev. Lett. **95**, 180602 (2005)
106. Michel, M., Gemmer, J., Mahler, G.: Int. J. Mod. Phys. B **20**, 4855 (2006)
107. Michel, M., Hess, O., Wichterich, H., Gemmer, J.: Phys. Rev. B **77**, 104303 (2008)
108. Novikov, I.I.: J. Nucl. Energy **7**, 125 (1958)
109. O'Dwyer, M.F., Lewis, R.A., Zhang, C., Humphrey, T.E.: Phys. Rev. B **72**, 205330 (2005)
110. Ouerdane, H., et al.: Phys. Rev. B **91**, 100501(R) (2015)
111. Peierls, R.: Ann. Phys. **3**, 1055 (1929)
112. Pereira, E.: Phys. Lett. A **374**, 1933 (2010)
113. Peterson, M.R., Mukerjee, S., Shastry, B.S., Haerter, J.O.: Phys. Rev. B **76**, 125110 (2007)
114. Peyrard, M.: Europhys. Lett. **76**, 49 (2006)
115. Prosen, T., Žnidarič, M.: J. Stat. Mech. Theory Exp. **02**, P02035 (2009)
116. Prosen, T., Žnidarič, M.: New J. Phys. **12**, 025016 (2010)
117. Redfield, A.G.: IBM J. Res. Dev. **1**, 19 (1957)
118. Roy, D., Dhar, A.: Phys. Rev. B **75**, 195110 (2007)
119. Saito, K.: Europhys. Lett. **61**, 34 (2003)
120. Saito, K.: J. Phys. Soc. Jpn. **75**, 034603 (2006)
121. Saito, K., Takesue, S., Miyashita, S.: Phys. Rev. E **61**, 2397 (2000)
122. Saito, K., Benenti, G., Casati, G.: Chem. Phys. **375**, 508 (2010)
123. Saito, K., Benenti, G., Casati, G., Prosen, T.: Phys. Rev. B **84**, 201306(R) (2011)
124. Sánchez, R., Büttiker, M.: Phys. Rev. B **83**, 085428 (2011)
125. Sánchez, D., López, R.: Phys. Rev. Lett. **110**, 026804 (2013)
126. Sánchez, D., Serra, L.: Phys. Rev. B **84**, 201307(R) (2011)
127. Savin, A.V., Kosevich, Y.A.: Phys. Rev. E **89**, 032102 (2014)
128. Sawaki, D., Kobayashi, W., Morimoto, Y., Terasaki, I.: Appl. Phys. Lett. **98**, 081915 (2011)
129. Schmiedl, T., Seifert, U.: Europhys. Lett. **81**, 20003 (2008)
130. Schollwöck, U.: Ann. Phys. **326**, 96 (2011)
131. Segal, D., Nitzan, A.: Phys. Rev. Lett. **94**, 034301 (2005)
132. Shakouri, A.: Annu. Rev. Mater. Res. **41**, 399 (2011)
133. Shastry, B.S.: Rep. Prog. Phys. **72**, 016501 (2009)
134. Snyder, G.J., Toberer, E.S.: Nat. Mater. **7**, 105 (2008)
135. Sothmann, B., Büttiker, M.: Europhys. Lett. **99**, 27001 (2012)
136. Sothmann, B., Sánchez, R., Jordan, A.N., Büttiker, M.: Phys. Rev. B **85**, 205301 (2012)
137. Steinigeweg, R., Gemmer, J., Michel, M.: Europhys. Lett. **75**, 406 (2006)
138. Sun, K.-W., Wang, C., Chen, Q.-H.: Europhys. Lett. **92**, 24002 (2010)
139. Suzuki, M.: Physica **51**, 277 (1971)
140. Terasaki, I., Sasago, Y., Uchinokura, K.: Phys. Rev. B **56**, R12685 (1997)
141. Terraneo, M., Peyrard, M., Casati, G.: Phys. Rev. Lett. **88**, 094302 (2002)
142. Tian, H., et al.: Sci. Rep. **2**, 523 (2012)
143. Van den Broeck, C.: Phys. Rev. Lett. **95**, 190602 (2005)

144. Vashaee, D., Shakouri, A.: Phys. Rev. Lett. **92**, 106103 (2004)
145. Vining, C.B.: Mater. Res. Soc. Symp. **478**, 3 (1997)
146. Wang, Y., Rogado, N.S., Cava, R.J., Ong, N.P.: Nature **423**, 425 (2003)
147. Wang, J., Pereira, E., Casati, G.: Phys. Rev. E **86**, 010101(R) (2012)
148. Wang, L., Hu, B., Li, B.: Phys. Rev. E **88**, 052112 (2013)
149. White, S., Feiguin, A.E.: Phys. Rev. Lett. **93**, 076401 (2004)
150. Whitney, R.S.: Phys. Rev. B **87**, 115404 (2013)
151. Wichterich, H., et al.: Phys. Rev. E **76**, 031115 (2007)
152. Wolfe, R., Smith, G.E., Haszko, S.E.: Appl. Phys. Lett. **2**, 157 (1963)
153. Wu, L.-A., Segal, D.: Phys. Rev. E **77**, 060101(R) (2008)
154. Yang, N., Li, N., Wang, L., Li, B.: Phys. Rev. B **76**, 020301(R) (2007)
155. Yvon, J.: In: Proceedings of the International Conference on Peaceful Uses of Atomic Energy, p. 387. United Nations, Geneva (1955)
156. Zemljič, M.M., Prelovšek, P.: Phys. Rev. B **71**, 085110 (2007)
157. Zhong, Y., Zhang, Y., Wang, J., Zhao, H.: Phys. Rev. E **85**, 060102(R) (2012)
158. Zotos, X., Prelovšek, P.: In: Baeriswyl, D., Degiorgi, L. (eds.) Strong Interactions in Low Dimensions. Kluwer Academic, Dordrecht (2004)
159. Zotos, X., Naef, F., Prelovšek, P.: Phys. Rev. B **55**, 11029 (1997)
160. Žunkovič, B., Prosen, T.: AIP Conf. Proc. **1468**, 350 (2012)

Index

S. Lepri (ed.), *Thermal Transport in Low Dimensions*, Lecture Notes in Physics 921, DOI 10.1007/978-3-319-29261-8